MPLS Network Management

MIBs, Tools, and Techniques

The Morgan Kaufmann Series in Networking

Series Editor, David Clark, M.I.T.

MPLS Network Management: MIBs, Tools, and Techniques
Thomas D. Nadeau

Developing IP-Based Services: Solutions for Service Providers and Vendors
Monique Morrow and Kateel Vijayananda

Telecommunications Law in the Internet Age
Sharon K. Black

Optical Networks: A Practical Perspective, 2e
Rajiv Ramaswami and Kumar N. Sivarajan

Internet QoS: Architectures and Mechanisms
Zheng Wang

TCP/IP Sockets in Java: Practical Guide for Programmers
Michael J. Donahoo and Kenneth L. Calvert

TCP/IP Sockets in C: Practical Guide for Programmers
Kenneth L. Calvert and Michael J. Donahoo

Multicast Communication: Protocols, Programming, and Applications
Ralph Wittmann and Martina Zitterbart

MPLS: Technology and Applications
Bruce Davie and Yakov Rekhter

High-Performance Communication Networks, 2e
Jean Walrand and Pravin Varaiya

Computer Networks: A Systems Approach, 2e
Larry L. Peterson and Bruce S. Davie

Internetworking Multimedia
Jon Crowcroft, Mark Handley, and Ian Wakeman

Understanding Networked Applications: A First Course
David G. Messerschmitt

Integrated Management of Networked Systems: Concepts, Architectures, and their Operational Application
Heinz-Gerd Hegering, Sebastian Abeck, and Bernhard Neumair

Virtual Private Networks: Making the Right Connection
Dennis Fowler

Networked Applications: A Guide to the New Computing Infrastructure
David G. Messerschmitt

Modern Cable Television Technology: Video, Voice, and Data Communications
Walter Ciciora, James Farmer, and David Large

Switching in IP Networks: IP Switching, Tag Switching, and Related Technologies
Bruce S. Davie, Paul Doolan, and Yakov Rekhter

Wide Area Network Design: Concepts and Tools for Optimization
Robert S. Cahn

Practical Computer Network Analysis and Design
James D. McCabe

Frame Relay Applications: Business and Technology Case Studies
James P. Cavanagh

For further information on these books and for a list of forthcoming titles, please visit our website at www.mkp.com.

MPLS Network Management

MIBs, Tools, and Techniques

Thomas D. Nadeau

Cisco Systems

MORGAN KAUFMANN PUBLISHERS

AN IMPRINT OF ELSEVIER SCIENCE

AMSTERDAM BOSTON LONDON NEW YORK
OXFORD PARIS SAN DIEGO SAN FRANCISCO
SINGAPORE SYDNEY TOKYO

Senior Editor	Rick Adams
Publishing Services Manager	Edward Wade
Production Editor	Howard Severson
Developmental Editor	Karyn Johnson
Cover Design	Yvo Riezebos
Cover Image	PhotoDisc
Text Design	Mark Ong
Composition and Illustration	Technologies 'N Typography
Copyeditor	Ken DellaPenta
Proofreader	Jennifer McClain
Indexer	Steve Rath
Printer	The Maple-Vail Book Manufacturing Group

Morgan Kaufmann Publishers
An imprint of Elsevier Science
340 Pine Street, Sixth Floor
San Francisco, CA 94104–3205
www.mkp.com

07 06 05 04 03   5 4 3 2 1

Library of Congress Control Number: 2002108920
ISBN: 1–55860–751-X

This book is printed on acid-free paper.

For my father Clem.
Writing a book was a childhood dream for both of us.

Foreword

Dr. Bruce Davie
Cisco Systems, Inc.

It is, it seems, a sad fact of networking that network management always comes last. Networking protocols are designed, implemented, and standardized, and the issue of how to manage networks that use these protocols is too often treated as an afterthought. The standardization of Management Information Bases (MIBs)—the fundamental building blocks of network management—is frequently the last task that a standardization-working group undertakes. Of course, there are some good reasons for that (it's hard to define a MIB for a protocol that is not yet fully specified), but nevertheless, network management often seems to receive less attention than it deserves.

It appears that the world of publishing mimics protocol design in this respect. As I write, there are at least a dozen books in print on Multiprotocol Label Switching (MPLS), and yet none, until this one, has addressed the management of MPLS networks. The publication of this book could hardly be more timely.

It is particularly ironic that so little has been written on MPLS network management when one considers that much of the motivation for the design of MPLS was to improve the manageability of service provider networks. One of the major applications of MPLS is traffic engineering, which is all about giving an operator of a large network more control over how traffic flows over links. MPLS gives a network manager tools by which traffic can be precisely routed from congested links to uncongested links. Network management is critical to this task. An operator needs to know which links are congested, and how much load needs to be carried between various points in the network; he must also be able to configure and monitor the MPLS paths that carry the traffic. So while the basic forwarding and control plane machinery of MPLS enables traffic engineering, traffic engineering only becomes practical in real networks when network management capabilities are provided.

Probably the most widely deployed application of MPLS is provider-provisioned virtual private networks. In a sense, the motivation for the invention of MPLS VPNs was to improve the manageability of large VPNs. Providers had found that large meshes of virtual circuits were difficult to manage, and MPLS combined with BGP-based routing mechanisms provided a natural way to build large VPNs with less management overhead. However, as soon as providers began to deploy MPLS/ BGP VPNs, it became apparent that sophisticated management tools would still be needed. MPLS VPNs are deployed in more than a hundred service provider networks today, but it is likely that the number would be much lower had the management tools not been provided in a timely way.

Tom Nadeau has thus written a much-needed book that should be welcomed by several audiences. As more service providers deploy the technology, they will clearly want to know what options exists for managing it. MPLS is now starting to show up in large enterprise networks as well; this book will be valuable to anyone who wants to deploy MPLS. The book will also help engineers who build MPLS equipment (especially those tasked with making the equipment manageable) as well as those who build network management tools.

One aspect of this book that I found particularly attractive is that it breaks away from being a dry description of MIBs. It's awfully tempting to think of network management as little more than MIBs, but there is much more to it than that in the real world. Tom clearly has experience in the real world of networking and writes about it well. All sorts of methods beyond SNMP are used to manage networks, such as command line interfaces and Netflow, and they are well covered here. He also explains about overall paradigms for network management and relates the tools to typical scenarios encountered by network managers.

Tom's background has prepared him well for the task of writing this book. As one of the leading authors of MPLS MIBs, he is a recognized authority in the field. He also knows enough about the real world networks in which MPLS equipment is deployed to recognize how much bigger and more complex the management problem is than simply defining MIBs. This book deserves to be read not just because it is the first book on MPLS network management, but because it focuses on solving the real problems network managers face.

Contents

List of Tables

Preface

Several years ago, there were only a handful of deployments of Multi-Protocol Label Switching (MPLS) technology, and those were restricted to researchers or nonproduction networks. Even in practice, MPLS was largely still theory. Over the past several years, MPLS has matured as a technology, and its acceptance and popularity in the marketplace has grown by leaps and bounds. Today a large percentage of the Internet's traffic traverses MPLS-enabled networks run by service providers, and this number seems to only continue to grow. Many of the largest service providers are now using MPLS technology within their networks not only to carry basic core network traffic, but also for the advanced applications and services that can be deployed by using MPLS. For example, there has been a large push by providers recently to deploy virtual private network (VPN) services, both so-called layer-2 and layer-3 VPN services. The demand for this has been driven by customers seeing the benefits and cost savings afforded by using this technology. A sampling of the providers that now deploy MPLS in much or all of their networks includes ATT, British Telecom, UUNet, NTT, France Telecom, and Global Crossing, among others.[1]

As with most new networking technologies, during the initial development and deployments of MPLS, network management functions were largely an afterthought and took mostly a backseat role in both vendor development and standards body efforts. The focus at that time was to deploy the technology and certify that MPLS was capable of doing what those who invented it advertised it could. As confidence in MPLS grew, more and more providers began to introduce it into their production networks. It was at this time that their operations staffs began to query

---

1 These service providers have identified the fact that they have deployed MPLS in a public forum. Furthermore, some have even declared their deployment of some MPLS applications such as MPLS VPN. These are but a few of the service providers who have done so publicly. The list of providers who have kept this information private is even larger.

their MPLS vendors for network management capabilities. By this time it was relatively late in the game, especially where the Internet Engineering Task Force—the standards body that standardized MPLS—was concerned. By that time, MPLS was mature enough that it was nearly ready to be adopted as an official standard. Unfortunately, a paltry two standards-based Management Information Bases (MIBs) existed to manage it. A similar number of proprietary MIBs were deployed by any one vendor, and in most cases, these were merely to manage the nonstandard versions of MPLS those vendors had deployed for some time. In essence, the proprietary command-line interface provided by device vendors was the sole management interface available for MPLS.

The tools and techniques used to manage MPLS networks are quite interesting and important given their history. Since standards-based network management interfaces and tools lagged behind the standardization and deployment of MPLS, and since invention is the mother of necessity, service providers were forced, in many cases, to develop their own tools to manage their MPLS-enabled networks. Therefore, many tools and techniques for best practices for managing MPLS-enabled networks existed well before any standards-based tools or interfaces. In many cases, these tools are quite simple, yet they get the job done in a manner that is satisfactory for the provider. For example, many of the homegrown tools that were developed were simply shell scripts of various types that accessed the device's command-line interface and interacted with it by "scraping" its screen, and then feeding it commands. The problem with this approach is that screen formats can change, and when they do, the scripts fail. Furthermore, there is no standard for command-line interface syntax, so scripts must be customized for each new piece of equipment being deployed and managed. What is really needed are standard network management interfaces.

It was at that time that the authors of the MPLS MIBs that exist today got into the game and started working on network management for MPLS. This work largely dealt with the IETF standard MIBs, as well as implementing them on our devices. Much of this work was a game of catch-up and fill in the blanks, and was impeded by the fact that each service provider other device vendors and I spoke with seemed to have a different idea about how to manage an MPLS network. Once a common idea of how to manage an MPLS network was obtained, the MIBs began to take shape. Not long after, the IETF received input from many vendors and service providers on these MIBs. This input resulted in the MIBs stabilizing to the point where today the core set of MIBs for managing MPLS are nearly standards themselves. Furthermore, all the while the core set was being developed, we and others kept uncovering new MIBs and tools that could be used to manage the larger picture of an MPLS-enabled network.

In addition to the numerous standards-based management interfaces, vendors are beginning to provide proprietary management interfaces to augment the standard interfaces. This is the next step in providing a comprehensive solution that network operators can use to successfully manage heterogeneous MPLS networks, as well as single-vendor networks. Device vendors are also modifying existing monitoring and measuring tools so that these tools can be better utilized within an MPLS-enabled network. For example, NetFlow has been modified extensively in the past few years to facilitate the monitoring of MPLS flows. These efforts can only lead to networks that are more manageable.

The tools and techniques of MPLS network management are allowing the theory and ideas of how to achieve effective, productive, and profitable MPLS-enabled networks to quickly become reality. A discussion of many of the MIBs, tools, and techniques will be the focus of this book. We will also discuss other management interfaces that can be used to manage MPLS, as well as other tools such as offline traffic engineering and NetFlow. We will shed some light (hopefully a lot!) on the current as well as evolving standards that these tools and interfaces are based upon, some proprietary extensions to these standards-based approaches, as well as the techniques of implementing and/or using all of these effectively and productively.

Objectives

There are several reasons behind our rationale for writing this text. First, we would like to provide existing or potential MPLS network operators, or device vendors deploying MPLS-enabled equipment, with a detailed examination of standards-based and proprietary tools and techniques currently available for monitoring, debugging, and optimizing MPLS-enabled networks. Those who are familiar with the existing standards-based tools will appreciate the detailed examples and tutorials of how to use and deploy these tools since the standards documents can sometimes be guilty of providing anything other than the dry rules for the standard. We hope that our examples and suggestions that extend your understanding of the tools and techniques can improve or enhance implementations or deployments. We will also touch upon how these standards are evolving, as well as those that might be on the bleeding edge today, but which might be inside of devices in the not too distant future.

Those using nonstandard tools such as offline traffic engineering tools will also benefit from our detailed discussions of these topics since only a few, if any, sources for this material exist today outside of vendor literature. Unfortunately, this literature is sometimes biased toward one approach over another. It is our hope to give an even-handed account of these tools or techniques that might not only assist in your purchase of these tools, but also in your understanding of them and how they might potentially improve your MPLS-enabled network or device. Lastly, it is our goal to combine in one place the relevant standard-based and proprietary tools and techniques. In doing so, it is our hope that we might not only provide a better context for all of these tools, but putting them together in a single place might simply make it easier to learn about, review, or reference these topics.

Multivolume Approach

We have taken a multivolume approach to the investigation of the area of MPLS network management. The reasons for our approach are twofold. First, from a practical perspective, we feel that it makes more sense to document the most often used and deployed tools and techniques in this text and continue with those that are just being deployed or standardized now in a subsequent edition when they are more mature. This will give operators time to deploy these tools so that we can later show how they can be best used given their feedback. Second, it makes more sense from a pedagogical perspective to separate out operational specifics from the how-to of the MIBs, tools, and techniques. This will allow you to first explore and understand the MIBs, tools, and techniques, and then you can follow this activity up in another volume that includes descriptions of the most common approaches or so-called best practices for applying them in a real operational environment.

Given these motivations, this book focuses on the presentation of *what* standard and sometimes proprietary MIBs, tools, and techniques are available for the management of MPLS-enabled networks. This book does not delve into how the tools and techniques are specifically deployed, but does give in-depth examples of how they can be used in general, as well as how they might be best implemented. Subsequent volumes will focus on the demonstration and investigation of best practice case studies and will show you *how* these tools are currently being used in real operational networks. The goal of these volumes is to enhance and improve existing implementations by sharing the knowledge of the best ways of managing MPLS-enabled networks. Subsequent volumes are also likely to further extend the tools and techniques described in this volume, since it seems their numbers grow almost daily.

It is not our intent for any one volume to provide an all-inclusive MPLS deployment cookbook or configuration guide. Instead, taken as a whole, the multiple volumes covering MPLS network management will provide you not only with a toolbox of MIBs, tools, and techniques for managing MPLS-enabled networks, but also will provide examples of how these tools are best applied to real operational

MPLS-enabled networks. It is our hope that vendors and operators alike can utilize these MIBs, tools, techniques, and ideas together to improve and enhance their products and/or the management of their MPLS-enabled networks.

Intended Audience

This book has been geared toward those interested in MPLS network management. Operators, vendors, their managers, and investors are all interested in the tools and techniques that can be used to manage MPLS-enabled networks because the bottom line is that they can improve profits by making deployments easier and more cost-effective. To this end, we have approached this material from several perspectives and geared it to the people who fall into those categories. First, this book has been crafted toward those network operators who are currently deploying MPLS within their networks and require the use of MPLS-related network management tools and techniques. In addition, this book will be a valuable aid to those operators who are considering deploying MPLS in their networks. Some operators may also wish to consider using this text as an aid in planning which types of management-related infrastructure will have to be replaced or upgraded given the ways in which this new technology can be managed once it has been deployed. This book is not intended to be a strict deployment or configuration manual for operators or service providers. That is a topic suitable for a book on its own.

The second group who should benefit from reading this book is third-party network management system (NMS) or operational system software (OSS) vendors. The management and engineers responsible for developing these products will find this book especially interesting. Third-party network management system vendors are quickly trying to produce and deploy network management applications that are used to monitor, configure, and provision MPLS systems for use by network operators. In most cases, it is much to the benefit of the third-party NMS vendor to provide the capability of managing the most diverse selection of MPLS hardware using the same management interface. It is for this reason that the majority of operators and NMS vendors are pursuing standards-based management interfaces. However, it is inevitable that hardware vendors, (e.g., Cisco, Juniper, etc.) will provide special, nonstandard features as part of their particular implementations to provide a competitive or strategic advantage over their competitors. These features are usually[2] not included in any standard, as some features remain proprietary. It is

---

2 Due to the evolving nature of MPLS, it is possible, and in fact very probable, for a feature that is currently deemed as proprietary to a specific vendor's implementation to become a standard feature of MPLS if enough vendors agree to adopt it. It is therefore our recommendation that you follow the evolving work of the standards bodies closely.

our intent to highlight some of these differences using examples from some of the more widely deployed MPLS devices and provide some strategies for managing them. It is, however, ultimately in the best interest of third-party software vendors to get device vendors to implement standards-based management interfaces.

The third audience we are targeting is engineers who are implementing MPLS for various network devices. Due to customer demands, these engineers will have to eventually implement the various management interfaces to expose the internal MPLS features of their device. We will provide details of our implementation experience, tips, and guidance as to the best ways of implementing and deploying those interfaces.

Finally, the last group of people who will find this book useful is in what we describe as the "interest" category. The people who fall into this category are the managers of the aforementioned engineers, technology analysts, investors of start-ups creating MPLS-enabled devices, investors of service providers offering MPLS-enabled network services, and students or researchers who are interested in understanding MPLS network management. These parties may be interested in MPLS network management either because they need to have some higher-level understanding of the technology in order to plan future product features or corporate strategy, answer a Request for Product (RFP), or are simply in need of an easy-to-understand introduction to the topic. We have structured the book such that these people may grasp a quick understanding of MPLS network management by reading the first few sections of each chapter without having to read much more. Others interested in more details of each topic may read on further. To this end, each chapter contains a high-level introduction to the topic, as well as where the tool or technique might fit into the larger picture of MPLS network management. Some chapters have been structured to contain a high-level or simplified example that may be helpful without going into too many details. Finally, each chapter ends with a summary of the contents of the chapter so that those skimming the material can quickly and easily determine whether or not to dive into a particular chapter without requiring much time.

Organization of This Book

This book is organized into chapters that fall into three basic parts: non-VPN or traffic-engineered MPLS-enabled networks, traffic engineering in MPLS-enabled networks, and finally VPN-enabled MPLS networks. Each chapter investigates mechanisms and techniques that can be used to manage that type of MPLS-enabled network. Some chapters focus on specific IETF MIBs, while others focus on a specific tool such as NetFlow that can be deployed within an MPLS-enabled network.

However, we intend this book to be viewed by the reader as a toolbox of sorts, containing the various measurement, construction, and optimization tools and techniques available today for managing MPLS-enabled networks.

Chapters 1–6 (the first part of the book) apply to any MPLS-enabled network and form the first partition of the book. Chapter 1 introduces you to Multi-Protocol Label Switching and ends with an explanation of why the management of this technology is so critical to its successful deployment. A model of how the MPLS-related MIBs fit together is also given and is highlighted at the beginning of each MIB-related chapter to refresh your view of the MIB interdependencies. Chapter 2 introduces management interfaces. This chapter first introduces the concept of a management interface, and then goes on to introduce several popular management interfaces including the command-line interface (CLI), CORBA, XML, bulk file transfer, and SNMP. Chapter 3 discusses the MPLS Label Switching Router Management Information Base (MPLS-LSR MIB). This MIB represents the basic Label Switching Router (LSR) label forwarding information base (LFIB). Chapter 4 introduces the MPLS Label Distribution Protocol MIB (MPLS-LDP MIB) and provides extensive examples as to how it can be both implemented and used. Chapter 5 presents the MPLS Forward Equivalency Class to Next Hop Label Forward Entry MIB (MPLS-FTN MIB). Chapter 6 introduces the reader to the IF-MIB. This chapter goes into many of the details of the IF-MIB that, from what we can tell, have not been covered in any other textbook. The chapter concludes by showing how the IF-MIB applies to MPLS-enabled networks.

Chapters 7–10 form the second part of the book and are concerned with MPLS-enabled networks that have traffic engineering (TE) enabled. Chapter 7 introduces you to traffic engineering in general as well as how it applies to MPLS. The remaining chapters introduce you to tools and techniques that can be used to gather data that can be input into a traffic engineering system or management station.

Chapter 11 introduces the PPVPN-MPLS-VPN MIB and shows how it can be used as an effective tool for managing MPLS-enabled networks. This part of the book may be viewed as being small compared to the other two parts, but you should keep in mind that a complete picture of management for VPN-enabled networks should also include those tools already introduced in the book.

Finally, Chapter 12 wraps up the book by first providing an extensive overview of what is on the horizon of MPLS network management. The chapter gives a brief overview of each topic, as well as pointers that you can use to investigate each topic in further detail.

The organization of the book is such that the chapters in the first part can, in general, be applied in networks employing any form of MPLS. This includes MPLS-enabled networks that only use LDP, but also covers ones that further add traffic engineering and/or VPN applications. However, the specifics of the TE and

VPN applications are introduced in the later chapters. The idea is that the MIBs, tools, and techniques presented in the beginning of the book form the foundation of management for the other applications of MPLS that will be built upon in the later chapters.

A summary of key terms and important points introduced in each chapter is provided at the end of the book, as well as an extensive listing of resources. A list of resources relevant to each chapter is listed at the end of each chapter after the chapter summary. Other resources, such as a resource guide for MPLS-enabled network management applications, an introduction to the IETF, and a complete glossary of terms and acronyms can be found at the end of the text.

A Note about Tables

Many of the chapters in the text focus on and describe relevant MPLS MIB modules in quite a bit of detail. In particular, we have tried to exhaustively enumerate all of the details of each MIB within each chapter. In many cases, large tables were required to capture this information completely either for its own sake, or for the purposes of illustrating an extensive example. Thus, the material in these chapters can be quite dense and potentially confusing for someone approaching the material for the first time. It is for this reason that we suggest that both novice and experienced readers approach these chapters using two passes. First, when initially reading a chapter, ignore the details presented in the tables. Instead, make yourself comfortable with the ideas and concepts behind the MIBs by focusing on the text *around* the tables. Only after you have a good idea of the mechanics and purposes of using the MIB, then go back and review the details presented in the tables. Taking this approach will help cement the ideas and concepts presented in each chapter and later the specific details.

Interviews

We have included short interviews with some notable figures involved in MPLS following each chapter. We have attempted to provide you with interviews from both operational and engineering areas as both areas of focus have much to offer in the way of learning about MPLS network management. It is our hope that these interviews will aid you in understanding the motivation and reasons for why MPLS network management is so important, as well as to augment your understanding of how some of the available components or practices surrounding MPLS and its corresponding network management came into being. Finally, some interviewers were instrumental in the creation and design of MPLS, so their opinions of how MPLS started and where it is going should provide some interesting insight into the state of the art and beyond.

The Web Site

A comprehensive Web site has been created to accompany this text and is available at *www.mkp.com/*. Additional information about the book, such as errata, updates, and Web links to useful resources, can be found at *www.lucidvision.com/mplsnmbook*.

Acknowledgments

Five hundred pots of green tea and 18 months of time have been consumed since I began this project, and it is finally finished. However, this accomplishment would not have been possible but for the help and support of many people. I would first like to thank my family and friends for their courage and support through the years, and especially during the time I worked on this book. I could not have done it without you. In particular, I would like to thank my parents Clement and Janina Nadeau: I would not be here at this point without your love and patience. Martha and Calvin Cole, without your assistance during my early days I might not be sitting here now writing this. I would also like to thank John Allan La Padula for being my best friend since we were about nine years old and for his help and unwitting suggestions during this process. I really do believe there is permanence in change. And the most thanks goes to my wonderful wife Katie and *number one son* Henry for putting up with the long evenings and weekends I spent sitting at my desk typing instead of spending time with you. In the end, nothing matters without you two and I am grateful for your patience and understanding.

This book would not have been possible without the technical input of many people. I would like to thank all of you for your assistance and efforts on this project and apologize if for some reason I miss you here. I would like to specifically thank my reviewers: Adrian Farrel from Movaz and Harmen Van Der Linde from AT&T, for their relentlessly helpful guidance and comments; Marco Caruggi from France Telecom; Kevin D'Souza, AT&T IP backbone, especially during the final weeks of writing; Kevin Santamaria, Global Crossing; and Bert Wijnen, from Lucent Technologies, for his thorough comments on my introduction to SNMP. I would also like to thank some of my colleagues at Cisco Systems who contributed input to the contents, reviewed parts of the manuscript, or gave me ideas for the material in the book: Mike Piecuch, Adrien Grise, George Swallow, Eric Osborne, Monique Morrow, and Anne-Marie Lambert. I would also like to thank all of the people who agreed to provide interviews for the book: Cheenu Srinivasan, Paramanet; Arun Vishwanathan, Force10 Networks; Joan Cucchiara, Crecent Networks; Kireeti Kompella, Juniper Networks; George Swallow, Cisco Systems; Andy Malis, Vivaci Networks; Harmen Van Der Linde, AT&T; Danny McPhereson; XiPeng Xiao, Photurus; and Ross Callon, Juniper Networks. A big thanks to Bruce Davie from Cisco Systems for his contribution to an interview as well as for a wonderful foreword.

Finally, I would like to thank all of the wonderful people at Morgan Kaufmann that helped me through this project when at times it could have been stopped in its tracks: Diane Cerra, Karyn Johnson, Rick Adams, and the rest of the publishing staff. You made it possible for me to realize one of my childhood dreams of writing a book.

Feedback and Comments

Thank you for taking the time to read this book. It is our hope that you find it to be a useful resource wherever you deploy, manage, or study MPLS.

Numerous questions and comments during the review phase of this project have allowed us to produce a better book than we had ever imagined. It is our opinion that the more comments about the text that we receive the better. If you have questions or comments about the book, please feel free to contact us via email at tnadeau@lucidvision.com. Alternatively, we can be reached via snail mail at Morgan Kaufmann Publishers, 340 Pine Street, 6th floor, San Francisco, CA, 94104.

1

Introduction

"It is a mistake to look too far ahead. Only one link in the chain of destiny can be handled at a time."

—Winston Churchill

Introduction

In this chapter, we look at the origins of Multi-Protocol Label Switching (MPLS) and introduce some of its basic concepts, including the separation of the control and forwarding planes of MPLS, the Forward Equivalence Class, and the MPLS label. After this introduction, we then introduce and discuss some of the new applications of MPLS networks such as traffic engineering and virtual private networks.

After an introduction to MPLS, we explain the basic premise behind why MPLS-enabled networks need to be managed to provide scalable, usable, and most importantly *profitable* MPLS networks. Given this motivation, we introduce how MPLS networks can be managed effectively using both standards-based and nonstandard tools, many of which are described in this book. This discussion serves as an introduction to the remainder of the book.

It is not our goal for this discussion to be an in-depth introduction to MPLS. We assume you have a good level of understanding of MPLS already and that the introduction given in this chapter can be used as a refresher. Advanced readers may skim the beginning of the chapter, but we recommend at least glancing at the latter half of the chapter. If you are in need of a more in-depth introduction to MPLS and SNMP, consult the references given in the Further Reading section at the end of the chapter as well as those related to MPLS and SNMP in the Bibliography at the end of the book.

1.1 A Brief Introduction to MPLS

In the past, routing devices were designed with the control and forwarding components commingled, which led to many shortcomings including low performance and scalability issues. In particular, routing lookups, especially those involving so-called longest-prefix match lookups, were quite complex and expensive in nature—in fact, quite a deal more complex than any layer-2 switching or bridging operation. Further complicating this process was the fact that many routers were required to forward packets from many different routing protocols. By accepting packets from different protocols, the positions of fields in packet headers could potentially be different for nearly every packet received, potentially further degrading forwarding performance. In contrast, nonrouting devices such as layer-2 bridges and switches were able to forward traffic at relatively high speeds because they based their forwarding decisions not on variable-length packet headers and network addresses of varying lengths, but on a short, fixed-length field. For example, all ATM cells have a fixed length and well-defined format. Devices switching ATM cells only need to examine a short identifier and can immediately forward the cell based on this simple piece of information. There is no question as to the position of the forwarding information in a cell. However, layer-2 devices suffered from the lack of routing information, which ultimately limited their scope and effectiveness. Let us now examine the control and forwarding planes in more detail, and then investigate how they can form the basis of an efficient and scalable MPLS label switching router (LSR).

The control component of a router is responsible for the exchange of routing information between other network nodes. It is this information that is used to form the router's routing database. This database paints a picture of the network from which a router can discern what it considers to be the most optimal path to any given destination in the network. Once stabilized, this database of best paths can be used to program the router's forwarding table. In contrast, the forwarding function of a router focuses exclusively on the actual decision of moving packets between ports on a network node. Each packet contains a header with source, destination, and other information. When a node receives a packet on a port, it needs to decide which port (or ports) it needs to forward that packet to. The forwarding process is quite mechanical by nature. When a node receives a packet, the forwarding component in that node will first examine the destination address contained in the incoming packet as well as perhaps other fields in the header. This information is then compared with entries in its forwarding database. It is this simple process that allows the forwarding component to make quick and simple decisions as to where the packet needs to be forwarded.

In some devices, the forwarding component is tightly coupled with the routing component. This approach sometimes results in limited portability of that technology to other types of forwarding planes. It also sometimes results in difficulties in extending the protocol with additional functions. MPLS is built on both the premise of a clean separation of the control and forwarding functions to take advantage of their individual advantages, as well as using them together in concert to provide additional advantages not possible with other technologies. The control and routing functions of MPLS are based on the Internet Protocol suite of protocols, which includes IP, RSVP, BGP, OSPF, and so on. The basic device in an MPLS-enabled network is the LSR. This device implements both the MPLS control and forwarding planes. The control function of an MPLS label switching router is responsible for distributing routing information to other LSRs, as well as the information required to convert this information into forwarding tables that can then be used by the forwarding function. The MPLS forwarding function is based on the use of a short, fixed-length label. This concept comes from the use of the same concept in layer-2 technologies such as ATM and Frame Relay, which base forwarding actions on a short, fixed-length identifier.

1.1.1 Forward Equivalency Classes

The forwarding function of a router is responsible for forwarding traffic toward its ultimate destination. The information in the forwarding table is programmed based on information from the control plane. If a packet is not delivered via a local interface directly to the destination, the router must forward the packet toward the ultimate destination using a port that will steer that traffic on a path considered most optimal by the routing function. For this reason, a router must forward traffic toward its destination via a next-hop router. This next-hop router may be the next-hop along the most optimal path for more than one destination subnetwork, so many packets with different network layer headers may be forwarded to the same next-hop router via the same output port. The packets traversing that router can then be organized into sets based on equivalent next-hop network nodes. We call such a set a *Forward Equivalency Class* (FEC). Thus, any packet that is forwarded to a particular next-hop is considered part of the FEC, and can thus be forwarded to the same next-hop. One important feature of the FEC is the granularity of the classification of traffic it can encompass. Since the FEC is based on a routing next-hop, it can include different classifications of packets. For example, since the routing information for a particular next-hop classification can be based on a destination prefix, it might include every packet traveling toward that destination. In this way, the granularity of packets classified by that FEC is quite coarse. However, if the routing database has programmed some next-hops for some traffic based on an application layer, for example, the traffic granularity might be much finer.

1.1.2 The MPLS Shim Header

MPLS packets are encapsulated using an MPLS shim header. The header has this name because it defines an additional header that is placed—or shimmed—between existing layer-2 and layer-3 headers. Curiously, the verb comes from the noun, not the other way around. Therefore, the shim header is so called because it is a small object that is inserted—shimmed—in between the existing layer-2 and layer-3 headers. Figure 1.1 shows the MPLS shim header format. The shim header comprises a sequence of one or more label stack entries. The entries in the sequence can be viewed together as a conceptual stack. A label stack entry comprises several components: label, EXP bits, the bottom of the stack bit, and TTL. The first element is the MPLS label. The label is a fixed-length 20-bit quantity that represents the label used to switch a packet. This label has local significance on a given interface between two neighboring LSRs only. That is, a label taken out of the context of a specific interface between two LSRs may or may not be found to be useful, or may be assigned to a different segment of an LSP. The second portion of the header is 3 bits called the experimental (EXP) bits. These bits are reserved for experimental use, such as for the purposes of classifying LSPs using Differentiated Services code points. The next element of the shim header is a single bit used to indicate the "bottom of the stack." This bit is set to 1 for the last entry in the label stack (i.e., for the bottom of the stack) and 0 for all other label stack entries. The fourth and final element in the stack is an 8-bit quantity called the time-to-live (TTL) field. The format of a label stack entry is detailed in Figure 1.2.

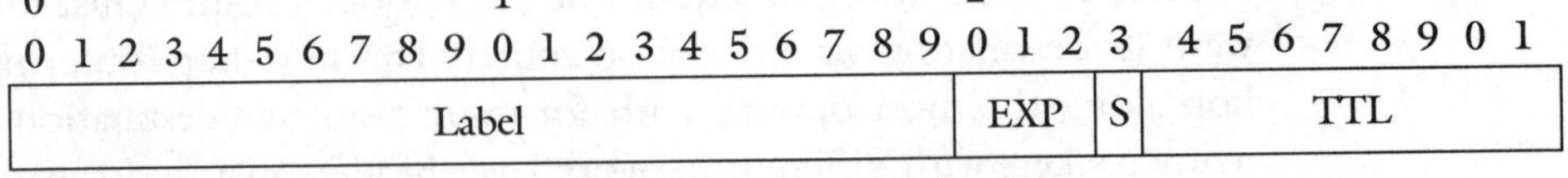

```
0                   1                   2                   3
0 1 2 3 4 5 6 7 8 9 0 1 2 3 4 5 6 7 8 9 0 1 2 3 4 5 6 7 8 9 0 1
+---------------------------------------+-----+-+---------------+
|                 Label                 | EXP |S|      TTL      |
+---------------------------------------+-----+-+---------------+
```

Figure 1.1 MPLS shim header format.

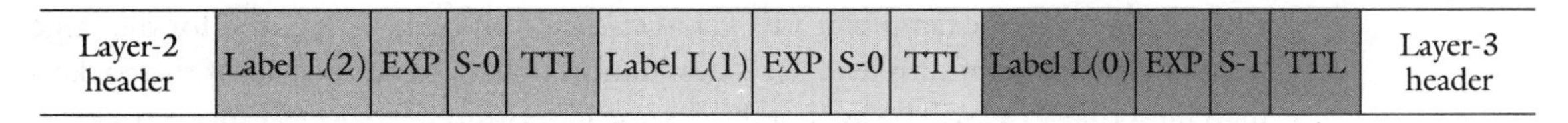

Figure 1.2 The MPLS label stack as it appears within the MPLS shim header.

The Label Stack

MPLS packets may contain more than one label. Depending on the application, it may be desirable to nest label switched paths (LSPs). For example, some TE and VPN operations find it useful to nest LSPs. When labels are nested, they are represented in the MPLS shim header as a stack structure, that is, a LIFO queue. The label stack is represented as a sequence of "label stack entries" in this stack. The topmost label appears closest to the layer-2 header, and the bottommost closest to the layer-3 header. Figure 1.2 demonstrates the label stack as a sequence of label stack entries. Each label stack entry is represented by 4 octets or 32 bits of data. Only the topmost label stack entry is used for any single lookup in the MPLS Label Forwarding Information Base (LFIB).

1.1.3 MPLS Label Switching

The MPLS forwarding plane is responsible for forwarding traffic based on an MPLS label. An MPLS label is a short, fixed-length 20-bit value (see Figure 1.1). The MPLS label has no structure. The MPLS label only has local significance between any two LSRs; therefore, the same label can be reused simultaneously within an MPLS-enabled network. In order for an MPLS LSR to be able to switch an MPLS packet, the label used in that packet's header must represent an entry in the MPLS LFIB of that LSR. The LFIB is essentially the label-to-label switching database used to program the LSR's forwarding plane. Once a packet is received, its label will be used by the forwarding plane to make a decision on where to forward the packet. At the edges of an MPLS-enabled network, label switching routers will map IP packets into FECs based on information provided by the MPLS control plane. Once classified into a FEC, the forwarding plane will be able to encapsulate any packet it receives that matches that FEC using the next-hop MPLS label assigned to that FEC.

Although assigned to a particular packet, the MPLS label does not necessarily encode the packet's network layer address, just its next-hop that will allow the packet to be forwarded to its destination, because many packets that are in the same FEC will be assigned the same label. This means that the next-hop choice may span multiple packets to many destinations. Thus, an MPLS label really encodes a

FEC identifier. For example, if a FEC has classified all packets destined for the same next-hop based on multiple layer-3 destination network prefixes, all of the packets matching that FEC will be assigned the same label (and next-hop). Once a packet is assigned a label, it will be switched based on this label until it reaches its ultimate destination. At that point, the MPLS header is removed and the packet forwarded using its original encapsulation.

When an MPLS packet is received, the LSR attempts to find a matching forwarding entry in its LFIB based on the packet's label and the interface on which the packet was received.

There are three operations that may be executed on the label stack when an MPLS packet is received and an entry matching this label is found in the LFIB: *push, pop,* or *swap*. All operations are executed on the top entry of the stack. When the topmost label is "popped" from the label stack, its label stack entry is completely removed from the MPLS shim header. When a label is "pushed" onto the stack, it moves all of the existing labels down by one relative index in the stack and inserts itself at the top of the stack. When a swap operation is executed, the topmost label entry is replaced with a different label, but the size of the stack remains the same. The "S" bit is set to indicate the last or bottommost entry in the label stack. All other entries in the label stack must set the "S" bit to 0.

The example shown in Table 1.1 demonstrates what an MPLS LFIB might look like. In the example, labels that are received on this LSR's MPLS interface "MPLSEth1/2" are switched to various other interfaces based on the incoming label. For example, when a packet containing label "1000" is received on interface "MPLSEth1/2," it is swapped for label "1050" and is forwarded on interface "MPLSEth1/3" to next-hop address 10.20.0.1. Note that this happens in all but the second-to-last row. The outgoing label in this case is noted as "pop." This refers to the removal of the MPLS shim header from the packet. Packets that have their MPLS headers stripped or "popped" are then forwarded on using their layer-3 encapsulation.

Table 1.1 Simplified Label Forwarding Information Base.

Incoming interface	Incoming label	Outgoing label	Next-hop	Outgoing interface
MPLSEth1/2	1000	1050	10.20.0.1	MPLSEth1/3
MPLSEth1/2	1002	1070	10.30.0.1	MPLSEth1/6
MPLSEth1/2	1006	"pop"	—	—
MPLSEth1/2	1005	1080	10.40.0.1	MPLSEth1/7

The MPLS Domain

An MPLS domain is composed of one or more MPLS label switching routers. A label switching router is any router or switch that supports the forwarding of MPLS-encapsulated packets based solely on the incoming interface and the information in the shim header. An LSR that sits at the edges of an MPLS domain and forwards traffic into and out of the MPLS domain is called a label edge router (LER). An LER maintains at least one interface into and out of the MPLS domain and acts as the point where the MPLS shim header is first *imposed* onto the incoming packet, and where the header is ultimately stripped and the packet forwarded using its original layer-3 encapsulation.[1] In effect, the LER must connect between the incoming technology and MPLS or vice versa. This process, in effect, tunnels the incoming technology through the MPLS network by encapsulating it within the MPLS packets.

Figure 1.3 shows a simple MPLS domain as well as the basic components of an MPLS-enabled network. The figure shows how MPLS LERs connect to external IP networks that may or may not contain customer sites. LERs are interconnected with other LERs within an MPLS-enabled domain. Other MPLS LSRs are interconnected in various ways within the MPLS domain.

The Label Switched Path

The path taken through the MPLS domain by a packet is referred to as a *label switched path* (LSP). The path taken may not be understood or completely stored by any one LSR within the MPLS domain, although in some cases it is. For example, traffic engineering allows the complete path to be stored at all LSRs along the path. This is because the labels swapped at each LSR have only local significance with regard to any two adjacently connected LSRs. Each LSR simply makes a local forwarding decision based on the incoming label of a packet, and switches the packet to a known outgoing label on a different interface. We should note that once the LFIBs have been established on all LSRs along the path of an LSP, the LSP is uniquely associated with the label and interface it is associated with, and therefore it is uniquely associated with a FEC.

An example of a label switched path is demonstrated in Figure 1.4. IP traffic to a destination reachable via the second LSR from the left is bound to a FEC at the leftmost LSR. All traffic entering the leftmost LSR will be classified using this FEC and will subsequently have MPLS shim headers imposed with a specific label

---

1 From this point on, we will assume that the layer-3 payload is always IP. Other protocols are equally supported and handled by routing, switching, or forwarding engines specific to their characteristics. However, to avoid confusion in the text we will limit our view to the most common payload, which is IP.

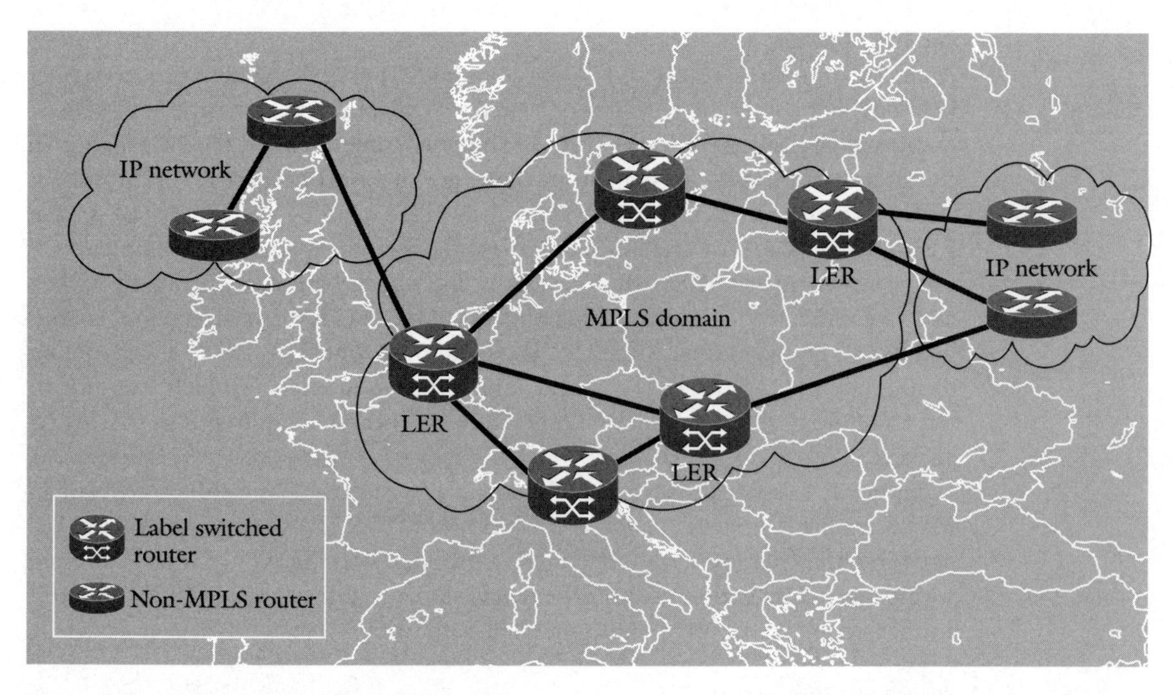

Figure 1.3 Components of an MPLS network.

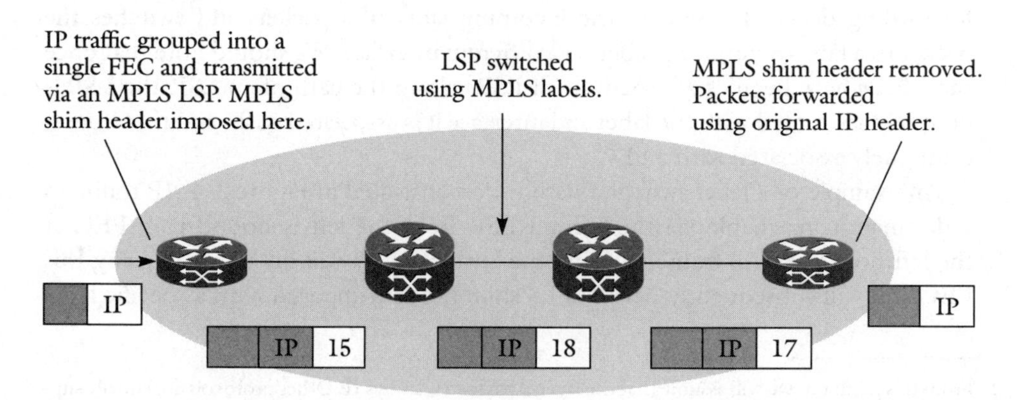

Figure 1.4 Example of an MPLS label switched path.

associated with this FEC—in this case 15. The MPLS-encapsulated packet will leave the leftmost LSR and will have its label swapped with the one indicated on the link as it traverses the LSP. When it arrives at the rightmost LSR, the shim header is removed and the packet forwarded out the rightmost interface using its original encapsulation.

1.2 Applications of MPLS

Currently, the two most important applications of MPLS are traffic engineering (TE) and virtual private networks (VPN). However, other new applications are taking shape such as Differentiated Services–aware traffic engineering that will enable voice-over-IP applications over MPLS, as well as virtual circuit emulation and virtual private LAN services over MPLS networks that will allow existing MPLS networks to be leveraged to offer additional emulated services. Although some of the applications of MPLS, such as traffic engineering and virtual private networks, technically can and are, in fact, currently being implemented and deployed using existing non-MPLS-based protocols, MPLS makes these applications simpler and more scalable. At the heart of the reasons for why MPLS is able to achieve these goals is because it takes advantage of the separation of the routing and forwarding functions, and because of its integrated signaling mechanisms. This has the advantage of reducing or eliminating many of the limitations of traditional routing and provisioning. For example, in the use of virtual private networks, MPLS simplifies the act of configuring a VPN by only requiring that the operator configure the edge devices connecting the customer edge networks into the VPN. MPLS signaling takes care of the actual connection to other pieces of the VPN. MPLS further improves the scalability by obviating the need for state information about the VPN to be stored anywhere within the core of the network. An example of an MPLS network that supports VPN is depicted in Figure 1.5. In this example two VPNs are supported: VPN A and VPN B. In order to support each VPN, the provider edge (PE) devices that connect the VPN sites must be configured. The core of the network is composed of provider core LSRs, or "P" routers. For example, P1 denotes a core "P" router in the figure. The core "P" routers do not have their configurations modified to support new sites of VPNs.

Another example of an important application of MPLS is in traffic engineering, where through the use of MPLS, it is possible to specify explicit routes during the process of setting up a path such that some specific data may be routed around network hot spots. Current technologies use routing protocols that tend to converge on a single, least-cost path to each possible (aggregate) destination. This occurs

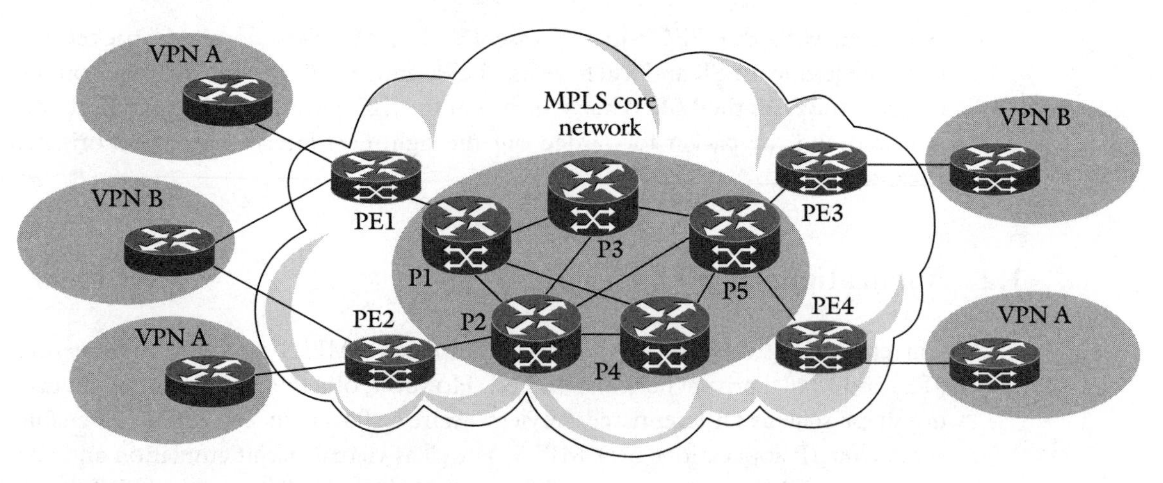

Figure 1.5 Example of an MPLS VPN.

even if there are multiple least-cost paths to the same destination. There are several problems with this approach. First, in many cases, parallel equal-cost paths exist to the same destinations, but all but one is preferred by the routing protocol. Second, since protocols generally prefer a single path that is considered most optimal, the routing protocols will direct all of the traffic destined to that destination onto that path. This often results in network hot spots at points in the network where many paths cross a single node.

It is possible to overcome these shortcomings with MPLS traffic engineering. MPLS TE allows an operator to specify an explicit route to direct some fraction of traffic through other parts of the network that are not selected by the routing process. These alternate paths may or may not be parallel least-cost paths. The important point is that the operator has the ability to override the routing protocol and choose which path certain flows of traffic take. Furthermore, TE allows an operator to create alternate backup paths, which bypass network trouble spots (i.e., disabled nodes or links). Given this mechanism, it is also straightforward to establish MPLS TE tunnels that transport packets that would not otherwise be correctly routed across a backbone network. For example, this is sometimes necessary in order to support virtual private networks across a backbone network between VPN end points, thereby making address translation and more costly tunneling approaches unnecessary.

An example of MPLS traffic engineering is depicted in Figure 1.6. Assume that each link carries an equal cost that is given to the routing protocol. Notice that given this assumption, two equal-cost paths that traverse the same number of network nodes exist. The thick dotted line represents the path through the network

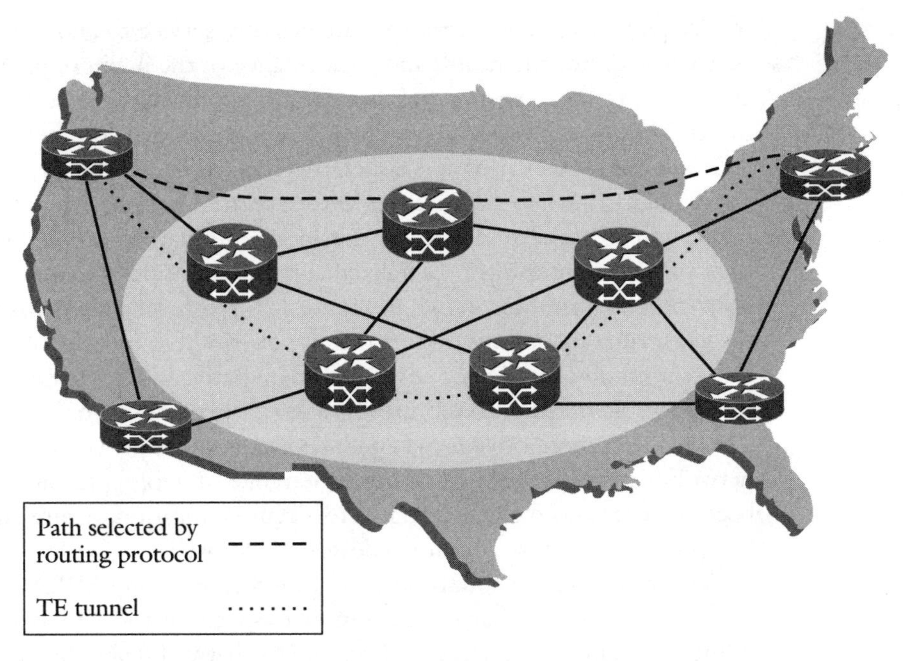

Figure 1.6 Example of a MPLS traffic-engineered tunnel.

that the routing protocol has chosen as most optimal. The thin dotted line represents an MPLS traffic-engineered tunnel that has been configured to override the path chosen by the routing protocol. This allows some of the traffic that would have taken the default path to be steered across the alternate path. In this example, the TE tunnel has been configured to use an alternative, unused path through the network in an effort to better utilize network resources.

1.3 Key Aspects of MPLS Network Management

Networks need to be managed for several reasons. First, from an entirely practical perspective, devices need to be monitored to ensure that they are functioning properly. Devices may also alert the operator to fault conditions, but if no corrective action is taken by the operator, then the device may continue to malfunction. For example, if a router's routing table has grown to a size that will soon exceed its available memory, it may be beneficial for the device to inform the operator of this condition. Services that are offered by a network also need to be managed, particularly when they are provisioned. In these cases, devices are contacted and configured. Managed services also require monitoring and maintenance. For example, if a service provider offers a virtual private network service to end users, it may be necessary to monitor the health and performance of the network paths that carry that customer's traffic to ensure that they are getting the network services that they paid for. In fact, this monitoring arrangement is sometimes a contractual necessity.

In all of these scenarios, it is either extremely difficult or nearly impossible for operators of medium to large networks to monitor every device in their networks by hand; instead, most prefer to do this in an automated manner. Some accomplish their management using a centralized approach, as is demonstrated by the sophisticated operations center shown in Figure 1.7. However, others may choose to have several smaller operations centers that are distributed. In either case, it is extremely time-consuming, and hence costly, for an operator to manually connect to each device's console in order to monitor its status, isolate faults, or configure the device. This becomes more obvious when you consider those provider networks where the network devices are located over a wide range of geographic areas. In this case, it becomes even more costly to travel to a remote location or hire additional staff to be on site where those additional devices are located.

Second, to make a sound business case for deploying MPLS, it must be made fully manageable so that the operational aspects of the network can scale up to numbers of devices, services, and customers that will make the network profitable. For example, the money spent debugging a problem by sending an operator into the field or by having the same operator go from router/switch to router/switch scratching their head might be better spent in building an automated system that can listen for alarms (see Section 2.6.6) that the router/switch can emit when in

Figure 1.7 A network management operations center.

distress. These alarms can then be used to pinpoint and isolate the scope of the problem. Once isolated, a management system can even take automated actions to correct the situation or simply alert an operator. An automated system can even be smart enough to not bother an operator if it deems a problem insignificant. Furthermore, management of the MPLS network becomes paramount when placed within the context of service level agreements and MPLS VPN services (see Chapter 11). When service level agreements are made between customers and providers, the service provider will not earn any money from that customer unless the services provided meet the agreement. The monitoring of the agreed-upon terms such as bandwidth, latency, delay, or service availability can be best accomplished using a network management system.

1.3.1 Origins of Network Management for MPLS

Once MPLS began to become mature and operational experience began to be gained by service providers deploying the technology, it was clear that MPLS was not very manageable given the lack of standard tools and management interfaces available at the time. In particular, the majority of MPLS vendors including Juniper and Cisco had only provided proprietary command-line interface extensions for the configuration and monitoring of MPLS features. When MPLS deployments were in early stages, it was acceptable for these and other vendors to provide minimal management capabilities for the MPLS features since operators were largely interested in simply having the protocol function up to specifications. However, as deployments became more mature and providers were more comfortable with the notion of using this protocol, it was clear that management of the protocol and its many features was now a priority. Furthermore, in heterogeneous networks where devices from multiple vendors had to coexist, an even larger problem existed. Since vendors had only deployed proprietary command-line interfaces, providers deploying devices from more than one vendor had to contend with more than one management interface for MPLS. This approach is expensive because it requires duplication of effort to manage the configuration and monitoring of the same features. The duplication of resources often ultimately translates into lost revenues for service providers. It was these requirements that began the push for standard interfaces for MPLS. In particular, the work on the IETF MIBs began in earnest during this time.

1.3.2 Configuration

One sore point for many operators is how to configure each one of the potentially hundreds of devices in their network. Further complicating the picture of configuration is the fact that many, if not most, provider networks are not comprised of devices made by a single vendor. This results in the service provider having to learn at least one different configuration language for each vendor from which it purchases equipment. Even further compounding this situation is that, through the magic of mergers and acquisitions, many vendors actually supply devices that have different configuration languages depending on which product line of theirs you choose to deploy.

It should be obvious from this description of the problems inherent in configuring a network of devices that it is a difficult situation at best. What would alleviate this situation would be the use of a common language and associated interfaces that can be used for the configuration of devices. There are many such languages available, yet no single one is used ubiquitously. Perhaps the closest contenders are SNMP—that is, SNMPv1 (RFCs 1155, 1157, and 1213), SNMPv2c (RFCs 1901–1906), and SNMPv3 (RFCs 2571–2575)—CORBA, and XML. Unfortunately, today the clear winner, at least for configuration, is the proprietary command-line interface (CLI), although SNMP is generally regarded as the best option for monitoring. The difficulty with a proprietary CLI is that it is generally accessible only via telnet or hardwired connections and generally has no standards-based schema. This results in every vendor implementation having a different management interface, which is clearly not something that excites a provider deploying a multivendor network. Although the CLI represents a majority of management interfaces, at least in the configuration area, the tide is turning toward standardized interfaces as networks grow ever more complex. These interfaces are commonly used for monitoring, and in many cases for provisioning as well. We will delve into the details of these various standard mechanisms for configuration in the pages to come.

1.3.3 Service Level Agreements

Typically, when a user signs up for access service (e.g., DSL, cable modem, dial-up), the service provider only agrees to provide that user with access to their network, and sometimes eventual access to the Internet. This agreement typically only specifies a minimum amount of bandwidth and provides no specifics about the average delay between access points and any other point in the network, or generally any other guarantees of service. Furthermore, there is typically no minimum response time during which outages in the network will be corrected by the service provider. This generally means that the user of a service is out of luck if their service does not function as advertised.

Some operators take their level of service a step further. These operators choose to monitor and maintain what some refer to as the "user experience." Although many operators strive to have networks simply function (i.e., route and switch a lot of traffic), others wish to ensure that their network is performing at levels acceptable to its customers. For example, this can mean that if user access to the Internet is unacceptably slow, the service provider will take some action to correct the situation—sometimes automatically. This approach is in direct contrast to other providers who would be content with end users just having access to the Internet at any speed.

The notion of service assurance and verification can be taken a step further beyond a provider assuring that they will monitor the health of user services. Frequently, end users and service providers will enter a formal contract called a *service level guarantee* or *agreement* (SLA). This agreement is an official agreement or contract between the service provider and a customer that specifies that the provider will sell a certain service to an end user for a certain price. If this service is provided as agreed upon, the end user must pay a certain fee for the service. However, if the service is not provided, typical recourses for the user are a reduction or refund of the fee they pay for the service during that period. Often the amount of additional work that a provider must perform to ensure that a service is functioning according to the service level agreement is significant. This elevated cost is precisely why SLAs are typically only signed between service providers and higher-paying customers such as large corporations or other service providers.

For example, in the United States, the service provider market is largely focused on selling bandwidth. This bandwidth is sometimes sold with guarantees of quality such as minimum delay and jitter. In other parts of the world, service providers concentrate instead on selling VPN services where site-to-site access quality is most important. All of these deployments typically contain SLA agreements with guarantees on the components of the service that the customers find most important, as well as the things a provider is willing to assure.

Given the motivation and elevated revenues from SLA agreements, providers are motivated to offer these premium services. However, these services do not come without additional effort on their part to verify the service quality and take corrective action when it does not meet the specified quality. In this regard, manual verification of SLAs is highly undesirable from a provider's perspective. This is simply because of its repetitive and frequent nature, especially when performed on a large scale. SLA agreements may also require that the operator take corrective action within some short period of time after a fault is detected. It is for these reasons that SLA monitoring and verification can be cumbersome or impossible if done manually, and therefore is a driver for the task to be performed by a fully or semi-automated network management system. In order to realize a management system that can verify SLAs in an automated fashion, network management functions must be integrated into devices that must be monitored. In particular, common management interfaces allow a provider to effectively monitor the data points of a service. This is especially important for heterogeneous networks and is also important in cases where customers insist on having independent third parties verify the SLA, since these companies often prefer not to build SLA verification software that is customized to a particular provider's network. Instead, they prefer to build software that is able to talk to a large set of devices in order to service many different service provider networks.

1.3.4 Service Level Agreement Verification

One often-overlooked aspect of service level agreement contracts is called service level agreement verification. The agreement of services between the end user and provider can be verified in several ways. The simplest form might be to issue Internet Protocol (IP) "pings" that emanate from the customer access points to other points in their networks or to locations within the Internet. This simulates user traffic traveling along the data path that all traffic takes through the network. If this traffic takes too long to traverse the network—or worse, is not getting to certain points within or external to the service provider's network—then the user experience suffers. Monitoring of the user experience might also be as sophisticated as monitoring the performance of many key network devices, collecting this information at a central location, and then making dynamic adjustments to the network using this information.

More sophisticated SLA verification is typically accomplished using network management tools that are specifically designed for the task. These tools include remote monitoring (RMON) or simply monitoring various counters on the network devices. Figure 1.8 illustrates how SLA monitoring and verification might be accomplished within an MPLS VPN deployment. A network management system (NMS) is positioned at key points, monitors certain traffic and quality of service (QoS) statistics, and reports them to the operator and customer. SLA verification can be done by the service provider, the customer, or by an unbiased third party. Use of standard network management interfaces to expose variables within the often-diverse population of network devices present in service provider networks is critical, especially when a third party is contracted to do the verification. The reason for this is simple: interoperability. SLA verification becomes quite cumbersome and costly if the party performing the verification is required to customize the verification suite for every device in a network. This is important if a third-party SLA verification company either sells software/hardware to service providers or performs the SLA verification service directly.

1.3.5 Fault Isolation

Fault isolation and detection are simply a means by which operators can detect, isolate, and report on defects discovered within their networks. The operator can use

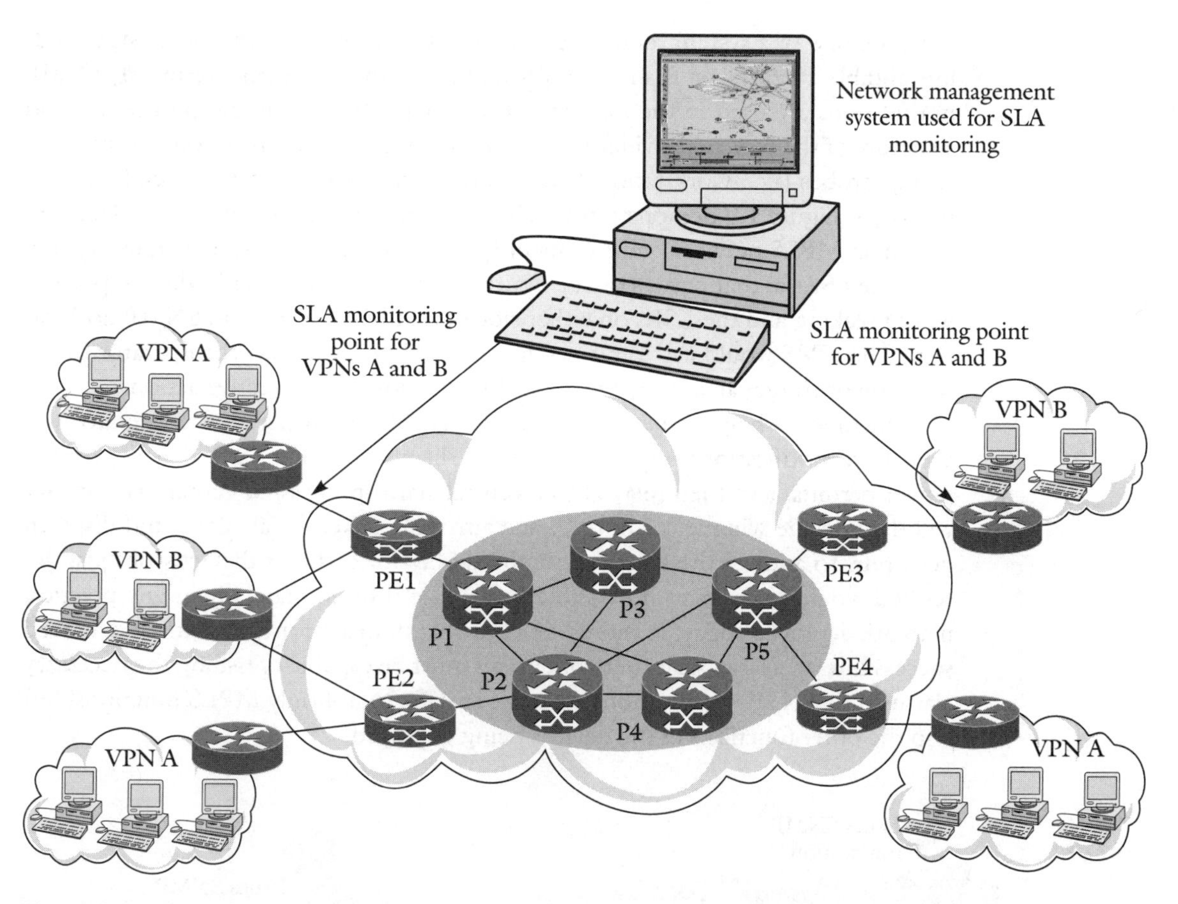

Figure 1.8 Example of a network of VPNs where SLAs are monitored by gathering information at certain key points within the network.

the information to repair the defect(s) found manually or automatically. When a device detects a problem, it will emit one or more messages as an alarm to alert the operator of the fault condition. These messages can be emitted under many conditions, including loss of service, device in distress (e.g., low on memory), or when the device has rebooted. Fault isolation is usually accomplished in modern networks in a three-part process that includes devices emitting asynchronous alarms, operators receiving those alarms, and then taking possible action because of those alarms. When a network device such as a router or switch discovers that an event of interest has occurred, it may issue an alarm. This alarm can be of the form of a system console message or an SNMP notification, which can be transmitted to the operators as an inform or notification. The reason for raising the alarms can include a configured threshold being exceeded, an internal fault condition such as low memory, or a system reboot. Although other forms of alarms do exist, including audible buzzers or flashing notifications on the command terminal, SNMP notifications are used in the majority of deployments. Depending on the size and structure of the service provider's network, the operator may place one or more listening probes (i.e., workstations) around their network to listen and collect these messages. Figure 1.9 demonstrates such a configuration where an NMS is deployed within an MPLS network. One of its purposes is to listen for notifications emitted from the LSRs in that network. The figure shows one of the links in the sample network breaking and the LSRs on either side of that link emitting an SNMP notification. The NMS would catch this notification and possibly alert the operator to the situation or trigger an automated procedure for possible corrective action.

Sometimes, when the networks are large and/or multitiered, the operator will even have notifications aggregated and perhaps even summarized if processing power permits, and then relayed to a central alarm-processing center. This center will then decide whether or not to issue a trouble ticket for an alarm and dispatch personnel to address the situation. It should be obvious that the activities just described would be next to impossible to achieve if done manually in any practical network deployed today. It should be clear that in order for MPLS to be deployed successfully on a large scale, network nodes must be capable of issuing the necessary alarms (i.e., SNMP notifications) that are specific to not only MPLS functions, but also the other functions in the devices being deployed.

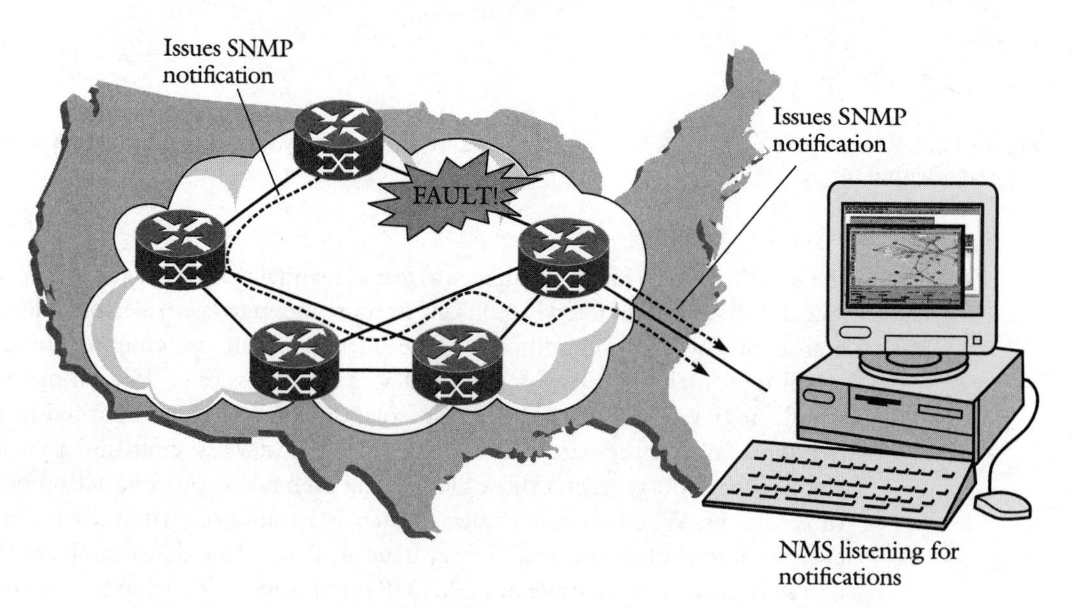

Figure 1.9 MPLS fault isolation using SNMP.

1.4 Management Information Base Modules for MPLS

The IETF, ITU, ATM Forum, and other standards bodies define documents called Management Information Base (MIB) modules that provide an external management interface for protocols and other features that are standardized within those organizations. Each MIB module can be thought of as a form of a data model used to manage the protocol or feature. The MIB module also defines the syntax, maximum access levels, and object interactions between those objects defined in that and other MIBs. The collection of MIB modules comprises the conceptual MIB that defines the entirety of MIB modules. We should also note that a MIB module is sometimes referred to as "a MIB" within certain contexts; thus care should be taken to discern when you mean a single MIB module or a collection of MIB modules that comprise a MIB.

The MPLS Traffic Engineering MIB module (MPLS-TE MIB) and the MPLS Label Distribution Protocol MIB (MPLS-LDP MIB) were the first MIBs proposed at the IETF in 1998. As standards-related work on these MIBs continued within the Working Group, and implementation and operational deployment by both device vendors and service providers continued, the MPLS-TE MIB had grown significantly in both size and scope. The primary reason for this was due to feedback and requirements from those deploying the MIB. In essence, the MIB has grown to encompass the functionality of both general LSR functions as well as traffic engineering functions. It was at this point that the MPLS Working Group decided that the MPLS-TE-MIB needed to be split into two MIBs: one to encompass general LSR switching functions, and one to encompass the general MPLS traffic engineering capabilities. It was at this point that the MPLS Label Switching Router MIB (MPLS-LSR MIB) was split from the MPLS-TE MIB and chartered as a separate Working Group item.

As time went on, the feedback process from service providers continued. It was at this time that the MPLS FEC-to-Next-Hop Label Forwarding Entry MIB (MPLS-FTN MIB) was proposed to expose the FEC-to-NHLFE mapping within LERs. In addition, MPLS BGP/VPNs were proposed, and implementation of this new application of MPLS had begun as well. Not long after this, the Provider-Provisioned MPLS Virtual Private Network (PPVPN-MPLS-VPN MIB) was proposed to the IETF and was adopted. It is likely that there will be many other standard MIBs provided by the IETF that cover all of the essential and common MPLS functionality, thereby making the manageability of MPLS networks far easier and straightforward for those who choose to utilize this technology.

Table 1.2 MPLS MIB module drafts.

MIB title	Date started	Current status	Description
draft-ietf-mpls-ldp-mib-*.txt	August 1998	Currently under IESG review	LDP protocol
draft-ietf-mpls-te-mib-*.txt	November 1998	Currently under IESG review	Traffic engineering
draft-ietf-mpls-lsr-mib-*.txt	June 1999	Currently under IESG review	Active TFIB of an LSR
draft-ietf-mpls-ftn-mib-*.txt	November 2000	MPLS WG last call.	FEC-to-NHLFE mapping
draft-ietf-ppvpn-mpls-vpn-mib-*.txt	August 2001	Adopted as Working Group draft by PPVPN WG	MPLS VPN
draft-nadeau-mpls-ds-te-mib-00.txt	February 2001	Awaiting adoption by TE WG	Differentiated Services–aware traffic engineering
draft-ietf-mpls-tc-mib-*.txt	June 2001	Under IESG review	Common textual conventions for MPLS MIBs
draft-nadeau-gmpls-te-mib-*.txt	March 2002	Adopted by CCAMP WG	GMPLS TE
draft-nadeau-gmpls-tc-mib-*.txt	March 2002	Adopted by CCAMP WG	Common textual conventions for GMPLS MIBs

Table 1.2 enumerates many of the MIB modules that are available at the time this text was written. More will surely become available as time goes on. However, we provide this table to illustrate those that are currently available so that you might see the progression from essentially no standard management interfaces for MPLS, to the near dozen available today.

The remainder of this text will focus on SNMP MIB-based solutions for managing MPLS networks. To that end, Figure 1.10 illustrates how each MIB fits in with the others as well as how each one depends on the others. This illustration should be a guide for you on how you wish to pursue each chapter. The book is presented in an order such that the later MIBs depend on the earlier ones. The MIBs presented earlier generally do not have dependencies on the ones that come after them.

This order presents a road map of sorts regardless of whether you are reading the text from an operational or a developmental viewpoint. That is, we recommend that you examine the MIBs presented first before tackling those presented later in the text, as they form a foundation not only for inherited or reused MIB objects, but also for concepts that are used throughout the MIBs, but may be introduced in one of the earlier MIBs.

The MIBs are organized as follows. The MPLS-TC MIB describes textual conventions that are used by all MPLS-related MIBs (even ones too new to be covered at this time). A version of the TC MIB that existed at the time of publication of this book is included in Appendix B. Note that it may have changed to an RFC version, as it was being reviewed at that time. The remaining four MIBs are shown as having dependencies on the MPLS-TC MIB, as well as the IF-MIB (RFC 2863). Since nearly all of the MIBs are related to the IF-MIB, a specific icon has not been included for it in the figure; instead, a gray triangle in the corner of each MIB indicates this dependency.

The MPLS-LSR MIB (see Chapter 3) describes the basic label forwarding operations of an LSR. The MPLS-LSR MIB also exposes which interfaces the LSR has MPLS enabled on by cross-referencing each MPLS-enabled interface that appears in the IF-MIB. This MIB presents a foundation of actual objects (as opposed to TCs in the MPLS-TC MIB) that are used in many other MIBs; thus it is viewed as the base MPLS MIB by many.

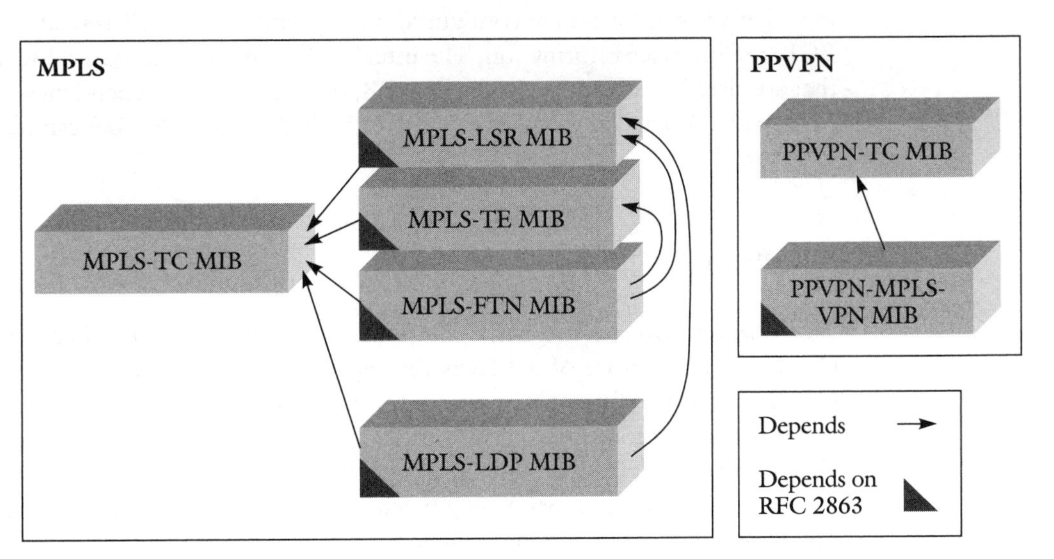

Figure 1.10 MIBs for MPLS network management discussed in this text.

The MPLS-TE MIB provides the operator with a view of which traffic engineering tunnels are configured, signaled, or presignaled (for backup). If a tunnel is also represented as an interface in the IF-MIB, an entry will exist there as well. The MPLS-TE MIB (see Chapter 8) depends on the MPLS-LSR MIB in that the system software in a device can be programmed to associate the active LSP with a tunnel when such a relationship exists.

Next, the MPLS-LDP MIB (see Chapter 4) provides insight into what the LDP protocol is doing on an LSR, assuming that LDP is enabled and in use. The MPLS-LDP MIB depends on the MPLS-LSR MIB for its mapping tables that are used to associate LDP sessions with active LSPs. The MPLS-LDP MIB also depends on the IF-MIB in that it exposes which label ranges are configured on an MPLS-enabled interface. Finally, the MPLS-FTN MIB (see Chapter 5) presents the operator with a view of how IP traffic is entering the MPLS network and how that IP traffic is being mapped onto MPLS LSPs or traffic-engineered tunnel interfaces.

The MPLS-FTN MIB depends on the MPLS-LSR and MPLS-TE MIBs because the way that it associates incoming IP traffic is to point at the associated LSP or traffic engineering tunnel head as represented in the MPLS-LSR and MPLS-TE MIBs, respectively. The MPLS-FTN MIB depends on the IF-MIB because it allows an operator to configure FEC-to-NHLFE mapping rules on a per-interface basis.

The PPVPN-MPLS-VPN MIB is shown to possess only dependencies on the PPVPN-TC MIB. This MIB contains common textual conventions used by the PPVPN-MPLS-VPN MIB as well as other MIBs defined by the IETF PPVPN Working Group. The PPVPN-MPLS-VPN MIB provides an operator with a view of which VPN instances are configured on a specific PE, as well as related statistics, BGP, and interface information. The interface information extends those interfaces that are already represented in the IF-MIB; thus yet another dependency on the IF-MIB exists. A detailed discussion of the PPVPN-MPLS-VPN MIB can be found in Chapter 11.

1.5 Summary

This chapter covered the basic components of Multi-Protocol Label Switching (MPLS). At the heart of MPLS is the separation of the control and forwarding planes. There are distinct advantages to this approach, as we have seen. The forwarding plane is composed of various IP-based protocols such as BGP and RSVP. The forwarding plane is based on switching a short, fixed-length label. This method is based on the forwarding mechanisms of several layer-2 forwarding technologies such as ATM and Frame Relay.

The remainder of the chapter introduced some of the reasons why it is crucial for world-class MPLS deployments to provide robust and comprehensive network management capabilities. We discussed fault and configuration management and how these two components of a management solution alone were critical if the network was to be deployed effectively, especially on a large scale. We then discussed performance measurement within the context of service level agreements. Monitoring and verifying the quality of a network connection is especially important when required by a service level agreement.

Finally, we presented an overview of how the MPLS MIBs fit together to provide the reader with a pictorial "30,000 view" of the MIBs that will be discussed later in the text. A current overview of how the MPLS MIBs interrelate can be found in Nadeau et al. (2001). We will refer to this figure throughout the book. Due to the large number of MIBs covered, we will repeat Figure 1–10 at the beginning of each chapter covering a particular MIB. In the opening of each chapter we will highlight the MIB that will be discussed to refresh your memory of where it fits into the larger picture of MPLS management.

It was not our goal for this chapter to be an exhaustive introduction to MPLS, since that is not the focus of this text. A comprehensive introduction is not in the scope of this book. For a more all-inclusive understanding of the basics of MPLS, please refer to the books listed in the Further Reading section. You are also invited to follow the ongoing work in the IETF via their Web site listed in the Further Reading section. Another way to follow the IETF work closely is by attending its meetings or by following the relevant discussions on its various mailing lists.

Further Reading

Davie, B. S., and Y. Rekhter. *MPLS: Technology and Applications.* First edition. San Francisco: Morgan Kaufmann Publishers. 2000.

Gray, E. W. *MPLS: Implementing the Technology.* Reading, Mass.: Addison-Wesley Professional. 2001.

To find out more about the IETF, visit their Web page at *www.ietf.org/*.

For more information about the Internet Assigned Numbers Authority (IANA), check out their Web site at *www.iana.org/*.

To locate information about the ITU: *www.itu.int/*.

To locate information about the MPLS Forum: *www.mplsforum.org/*.

To locate additional information about the OIF: *www.oiforum.com/*.

Cisco's Web site also provides a great deal of information regarding MPLS: *www.cisco.com*.

George Swallow

is the co-chair of the IETF's Multi-Protocol Label Switching Working Group. He is also a technical leader at Cisco Systems, where he is a member of the architecture team for label switching. He is the chief architect for MPLS traffic engineering. He is now involved in extending the MPLS control plane into the optical space. He is also Cisco's technical voting representative to the Optical Internetworking Forum.

George, you have been an active member of the IETF community since the early days of MPLS. At that time, people were concerned with early deployment issues such as just getting the technology working. However, today MPLS is a mature technology and has been deployed in numerous production networks. What do you see as the issues and hurdles (and possible solutions) for current and future deployments of MPLS?

As you point out, MPLS is a mature technology. Like any mature technology, its future deployment is driven largely by economics. As a technology, MPLS is not an end unto itself; the economics of MPLS are driven by the applications MPLS enables.

Let's speak of four kinds of applications:

1. Virtual private networks
2. Transport of legacy services
3. Traffic engineering
4. Quality of service

Note that these are not mutually exclusive and are often used in combination.

Virtual private networks: As IP comes to dominate data networks, it becomes economical for service providers and large corporate networks to combine their offering on a single backbone. VPNs offer a means of not only moving the customer base from a Frame Relay or ATM network, but also enable the service provider to offer directly what most customers desire, namely, secure outsourced IP connectivity.

Transport of legacy services: The economics here are similar to those for VPNs, except that one is offering only legacy service, Frame Relay, ATM, and in some cases SONET/SDH. Again, as IP dominates, it will at some point cease to be economically viable to maintain separate network(s) to offer the above service. MPLS allows these services to be offered over the IP backbone.

Traffic engineering: On a strategic basis, traffic engineering is an adjunct to network design. You design a network in anticipation of future traffic loads. In IP networks, you only have so much control over which path through a network a particular flow will take. Traffic engineering allows much tighter control over the mapping for flows to paths. Like any optimization problem, such as packing boxes, you can reach more optimal solutions if you have smaller items to put in the boxes. Serious bandwidth savings can be achieved. But in a time where there is a surfeit of bandwidth, the operational cost of engineering traffic cannot be justified in many cases. On a tactical basis, traffic engineering can be used to engineer traffic around network "hot spots" and to optimize the use of expensive transmission facilities such as transoceanic links. This allows a provider to allow a better level of service with the current network design. Tactical traffic engineering is particularly useful in dealing with unanticipated traffic growth or changes in usage patterns and with unplanned outages of network facilities. It seems to me that, even in the current economic climate, tactical use of traffic engineering makes sense.

Quality of service: I hesitate to guess here. QoS has been talked about for a long time. What I will say is that I believe it will be application-driven. At some point, it will make sense to push a lot of voice onto campus and wide-area backbone networks. I think that this will be the time when we really see QoS come to the fore. Note that MPLS can provide not only the QoS needed by voice (and other real-time applications such as video), but also fast restoration. This is a feature closely related to traffic engineering that we haven't spoken about. Essentially MPLS fast restoration offers IP the kind of restoration times found in SONET, allowing critical applications to suffer no more than a blip in the face of a link or node failure.

Some people feel that offline traffic engineering is the only way to correctly engineer and maintain an MPLS network. What is your take and advice in this area? Where do you see offline TE as fitting into a complete strategy for managing an MPLS network today, and in the future for a GMPLS network?

First, traffic engineering is not necessary for an MPLS or any other network. Congestion problems can always be solved by throwing more bandwidth at the problem. As I said above, you need to differentiate between strategic and tactical traffic engineering. In strategic traffic engineering you seek to essentially engineer all of the traffic in the network. The process is very tightly coupled to the process of network design. Like network design, good offline tools are needed in order to tackle a network of considerable size. In the tactical case, you try to engineer as little traffic as is necessary to adequately relieve some network hot spot. Here offline tools, though still useful, are not necessary.

2

Management Interfaces

A computer terminal is not some clunky old television with a typewriter in front of it. It is an interface where the mind and body can connect with the universe . . .

—**Douglas Adams,** ***Technology Computers***

Introduction

This chapter introduces several different types of management interfaces that may be used to manage MPLS deployments. In particular, we will introduce you to XML, CORBA, SNMP, and the command-line interface (CLI). We will investigate and explain why operators might or might not wish to utilize one, none, or all of these interfaces to manage their MPLS networks, as well as to hopefully provide device vendors with reasons for why they should or should not implement them on their MPLS devices. The end of the chapter will focus particularly on the SNMP interface by introducing it in such a way as to prepare you for the remainder of the book, which contains many sections on using SNMP for managing MPLS networks.

2.1 The Basics of Management Interfaces

Management interfaces allow network operators to manage the devices in their networks by providing access to each device's control, configuration, and status information. Many different types of management interfaces exist, but in general, a management interface is composed of two parts: a protocol describing the communication rules between the operator and the device, and the format of the information that will be exchanged using that protocol.

The basic features of a management interface are depicted in Figure 2.1. Notice how the management interface provides a unified external view of the managed device. It should be noted that devices might support more than one management interface, but all typically support at least one. Some management interfaces provide additional functions such as secure authentication, control of transactions, reliable or unreliable network transport options, and even functions that allow for the translation between other management interfaces. This collection of features, functions, and protocol comprise what is generally referred to as a management interface.

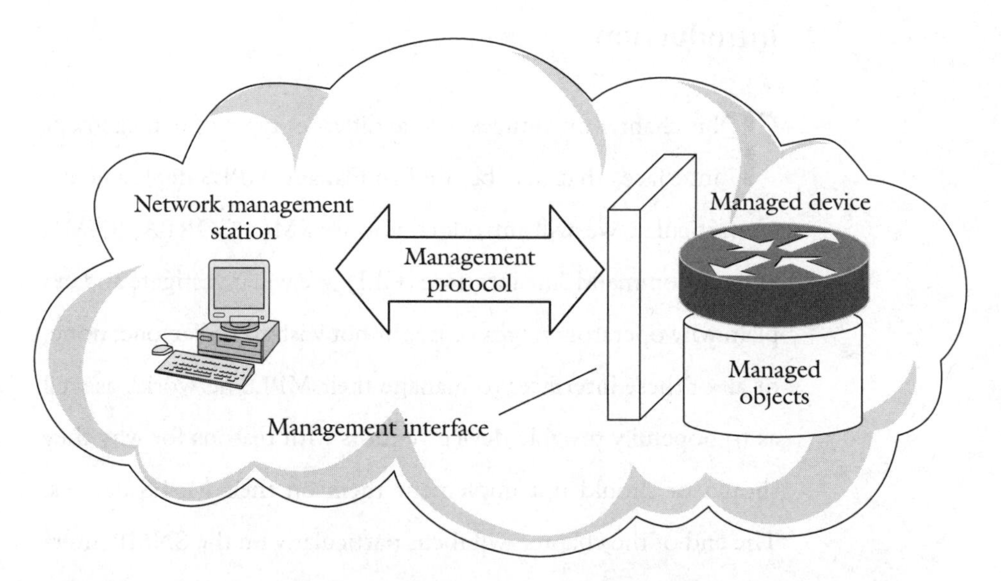

Figure 2.1 In general, a management interface provides two things: a protocol between the manager and the device and a consistent external representation of the managed device.

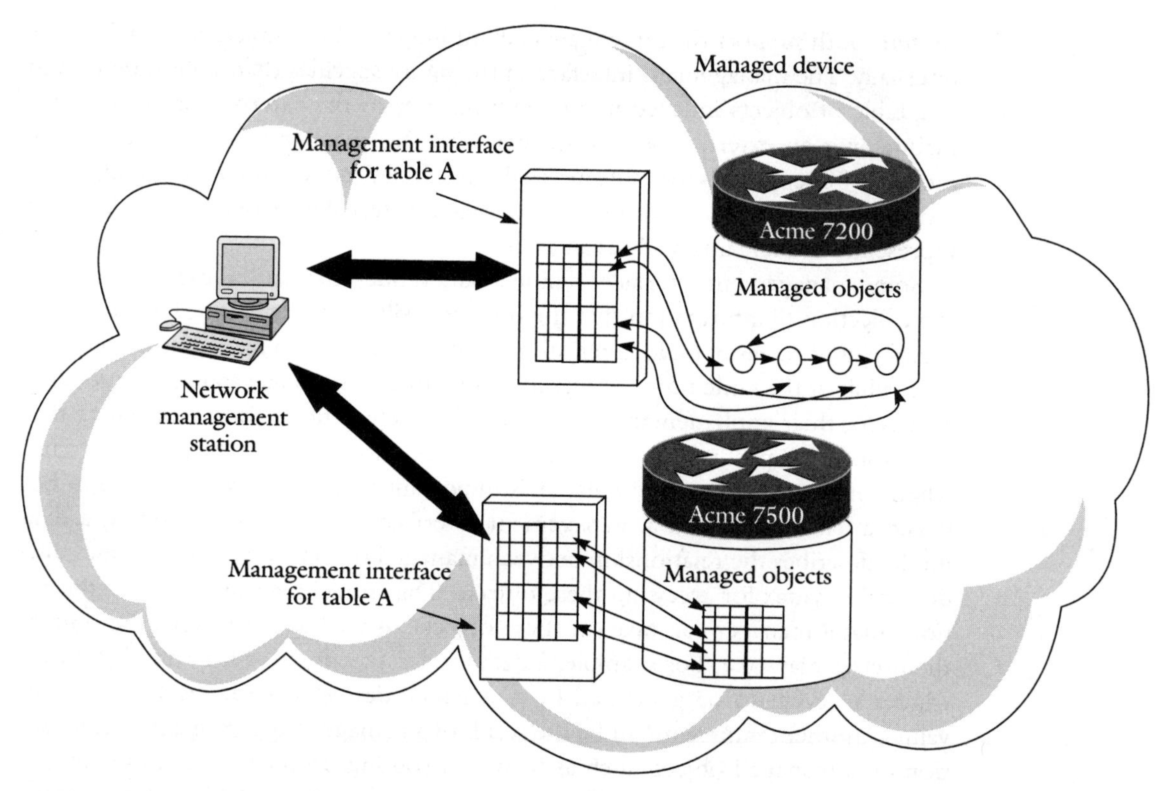

Figure 2.2 Two models of Acme's routers, both supporting the same management interface. The management interface specifies that table A be viewed as a table externally. While preserving the consistent external view, Acme model 7500 represents the table internally as an array. Acme model 7200 represents the table internally as a linked list, while preserving the same external representation.

The management interface provides a consistent external view of the manageable objects in a device. This view ideally remains consistent across all devices supporting the management interface, thus providing the same interface for the operator or the operational software used to manage these devices. To be more specific, the format of manageable objects is consistent. For example, the management interface may specify that the total time that the system has been running since it was last initialized be represented as a 64-bit integer. However, some devices may only be able to maintain 32-bit integers natively depending on the specific hardware used. Therefore, devices that are unable to support the wider data type natively must simulate it in order to support the management interface. Another example is objects that are stored in tables. A device may store the objects internally as an array of objects or as a linked list; however, the external representation will always be that of a table. Figure 2.2 demonstrates two models of Acme Corporation's routers; both support the same management interface but implement it differently internally. The management interface in the figure specifies that table A be viewed as a table of objects indexed in a certain manner to the external manager. While maintaining an external view consistent with this management interface, Acme model 7200 represents the table internally as a linked list. On the other hand, Acme model 7500 is still able to maintain a consistent external view of the same data, but represents the table internally as an array.

Some management interfaces define a data model that can be used as a map of the collection of information that the operator will have access to. This model can then be used to easily build applications that can be used to manage this information, while at the same time, devices implementing this model can use the model as a basis for their implementations. Still others simply define a format or syntax that operators can expect to view the managed information in, or will be required to use when configuring that same data. It is important to make a clear distinction between a data model and a management interface. As was just described, a data model describes the relationship between managed objects in a system. It may also define the syntax for accessing these objects. This model may also describe the actions that a manager can take on these objects, as well as the actions the objects themselves may take. For example, a data model may describe how an object may trigger an event to be generated by a managed device when it reaches a certain value. Another example might be the result of a manager triggering a particular action on a managed object, such as to start a routing protocol. For those familiar with object-oriented programming methodologies, a data model used for management purposes has the same meaning as one defined for a program. As crude as some may regard them, the SNMP management interface defines documents called a Management Information Base (MIB). Each document can be thought of as a form of a data model. These documents define the syntax, maximum access rights to the objects defined therein, and object interactions between those objects defined in that and other MIBs. On the other hand, more sophisticated data models can be constructed using other management interfaces such as CORBA IDLs. Not only can these other data models be used to describe what the MIBs can describe, but they can also be used to model the objects running inside of a network management application.

The most common form of a management interface is a proprietary command-line interface (CLI). This interface typically allows an operator to connect to a device using a remote telnet session. The format of the data viewed over this telnet session is specified as being in ASCII format and must be entered in the command-line syntax defined by the vendor. However, the format of the output can (and sometimes does) vary depending on the version of software running on the device. Other management interfaces, such as SNMP, define stricter rules for how the managed information will both be accessed and offered to the manager.

In addition to the raw tools and protocols provided by management interfaces, the actual use of the management interface can be arranged and deployed in many different ways depending on the network or the operational philosophy of those running the network. Some of these different approaches will be outlined in later chapters in this book, but there are far too many to investigate in this text. Instead, we will focus on showing you the tools that management interfaces provide.

There are at least a dozen management interfaces in use today, some useful to those managing MPLS networks, others that are less so. In general, all of the well-known management interfaces can be applied at least in some ways to MPLS networks. In addition to using a single management interface, many operators have chosen to use a combination of two or more interfaces for their networks. In the following sections, we will focus on and describe some of the management interfaces that are most prevalent in large operational MPLS networks today. Furthermore, the remainder of the book will focus on the specific tools that are made available within these more widely used management interfaces. While we recognize that management interfaces other than the ones we will discuss exist, and that they may be useful for the management of some MPLS networks, we will not focus on these at this time either because they are too proprietary, or because these management interfaces are simply not in enough use to interest a wide audience. We may choose to cover one or all of these in a future revision of this text.

2.2 The Command-Line Interface

Most if not all network devices since the early days of networks have provided the operator with some sort of character-based command-line interface. The command-line interface typically provides the user with screens of information for viewing specific device functions or configuration, as well as a structured syntax for interacting with it. The command-line interface provided a console screen similar to that of early mainframe computers. Early implementations of the CLI were as simple as a paper-based teletype that was wired directly to the device via a serial cable. Since then, the prevalent method of connecting to a network device is to use a telnet network connection, although other means exist that are still popular, including the good old hardwired serial connection; however, the sophistication of the input and output of this interface has not changed substantially. Figure 2.3 shows an example of a command-line interface from a well-known router vendor.

In general, a vendor will specify a command-line interface syntax that governs how an operator may interact with it. This syntax is typically broken into two areas: display screens and configuration entry. Display screens are used by the operator to view the information stored within the device. The information shown on these screens is typically status or configuration information. The display of information

```
Cisco Internetwork Operating System Software
IOS (tm) 7200 Software (C7200-JS-M), Experimental Version 12.2(20011220:212756) [tnadeau-
ldp_mib_122s_pi 101]
Copyright (c) 1986-2001 by cisco Systems, Inc.
Compiled Fri 21-Dec-01 10:43 by tnadeau
Image text-base: 0x60008960, data-base: 0x61738000

ROM: System Bootstrap, Version 11.1(13)CA, EARLY DEPLOYMENT RELEASE SOFTWARE (fc1)
BOOTLDR: 7200 Software (C7200-BOOT-M), Version 12.0(2)XE2, EARLY DEPLOYMENT RELEASE SOFTWARE
(fc1)

tagsw7200-43 uptime is 2 weeks, 5 days, 11 hours, 19 minutes
System returned to ROM by reload at 07:04:33 UTC Fri Dec 21 2001
System image file is "tftp://UNKNOWN/tnadeau/c7200-js-mz"

cisco 7206 (NPE200) processor (revision B) with 114688K/16384K bytes of memory.
Processor board ID 16065231
R5000 CPU at 200Mhz, Implementation 35, Rev 2.1, 512KB L2 Cache
6 slot midplane, Version 1.3

Last reset from power-on
Bridging software.
X.25 software, Version 3.0.0.

SuperLAT software (copyright 1990 by Meridian Technology Corp).
TN3270 Emulation software.
8 Ethernet/IEEE 802.3 interface(s)
2 ATM network interface(s)
125K bytes of non-volatile configuration memory.
4096K bytes of packet SRAM memory.

20480K bytes of Flash PCMCIA card at slot 1 (Sector size 128K).
4096K bytes of Flash internal SIMM (Sector size 256K).
Configuration register is 0x0

tagsw7200-43# show mpls forwarding

Local  Outgoing    Prefix             Bytes tag  Outgoing    Next Hop
tag    tag or VC   or Tunnel Id       switched   interface
1000   Pop tag     10.0.0.5 12 [72]   0          Et1/1       10.1.2.1
1001   Untagged[T] 55.55.0.0/32       0          Tu43003     point2point
1002   Pop tag     10.0.0.1/32        0          Et1/5       10.21.22.21
1003   Untagged[T] 10.3.5.0/24        0          Tu43003     point2point
1004   Pop tag [T] 10.0.0.3/32        0          Tu43003     point2point
1005   Pop tag [T] 10.0.0.5/32        0          Tu13        point2point
1006   26          10.0.0.1 1 [77]    0          Et1/2       10.2.3.3
1007   27          10.0.0.1 2 [76]    0          Et1/2       10.2.3.3
1008   28          10.0.0.1 3 [76]    0          Et1/2       10.2.3.3
1009   29          10.0.0.1 11 [76]   0          Et1/2       10.2.3.3
Local  Outgoing    Prefix             Bytes tag  Outgoing    Next Hop
tag    tag or VC   or Tunnel Id       switched   interface

[T]     Forwarding through a TSP tunnel.
        View additional tagging info with the 'detail' option
```

Figure 2.3 Sample command-line interface output from a popular LSR vendor.

on the screen either can be triggered as the result of an operator query or may be the result of an asynchronous display made by the device. Configuration information can contain all or part of the device's configuration.

For example, the commands "show router version" and "show mpls forwarding" were used to generate the output shown in Figure 2.3. The "show" command displays general system configuration information such as the version of software image that is currently executing the device, how much memory is installed in the device, or the version of the ROM code present. Notice that the command used to trigger the output has certain form or syntax. Every time the operator enters the command "show router version," the screen shown will appear with the same syntax. Variables in the fields may be different, however. Furthermore, if the just-mentioned command were entered as "show routerbbb version," it would have resulted in an error being reported to the operator, since this constitutes an illegal command.

Unfortunately, the command-line syntax specified by any two vendors is typically different even though they are used to manage identical abstract objects. In fact, sometimes it is even the case that different products from the same vendor use different screen formats and syntax to manage the same feature. Some newer vendors have tried to copy the syntax of older vendors, but even this inevitably results in some discrepancies when the vendor being copied decides to change their CLI without the other noticing. Unfortunately, no standard CLI syntax is defined by any standards body that might better help the situation. This is a large disadvantage for service providers who have to manage networks with disparate devices and corresponding CLIs, since it means that managing these types of networks will probably be much more difficult and expensive than a homogeneous network.

One important feature of a command line is the ability to display asynchronous notifications without any operator intervention. Some devices even provide a separate CLI session that allows an operator to more easily capture and recognize these messages. Notifications, or "alarms" as some call them, can be used to alert the operator to critical or fault situations. As with the syntax of the CLI, the format of the on-screen alarms will vary from vendor to vendor. Some vendors have chosen to specify a format for what is displayed on the screen, while others will allow the software to display a free-form string that may change from version to version of the system's software.

2.2.1 CLI Security

Other features of command-line interfaces include security functions such as authentication to verify the identity of the operator accessing the CLI, or encryption of the actual data stream over which the CLI data flows. Operators have options in terms of authentication, ranging from the simplest clear-text password authentication, whereby the operator must specify a user name and a password that are checked for authenticity within the system, to a complete RADIUS system for managing encrypted passwords. Different vendors will offer different security features with their CLIs.

One very common security option used by operators is the secure shell (SSH). Operators will run a secure shell session between their operation station or NMS and the network device rather than traditional clear-text telnet. SSH provides the user with transparent, strong encryption, as well as reliable public-key authentication that is quite easy to configure. Implementations are freely available, so vendors have little excuse not to offer this solution to operators. Furthermore, since SSH is a popular and rather robust TCP/IP-based solution that solves many network security and privacy concerns for operators, vendors are encouraged to implement this as an option for accessing their CLI. In addition to securing the CLI session, SSH also supports secure file transfer between management stations and network devices, so those vendors offering bulk file transfer of management information or configuration files can easily integrate this into their device software as well. Further, SSH provides the capability of "tunneling," which can be used to easily add additional encryption to otherwise insecure network applications. As has been mentioned, there are many feature-filled freeware versions of SSH available; therefore operators will find that many device vendors have already adopted SSH in their devices.

We recommend strongly that operators that have access to secure CLI implementations use them whenever possible. This added level of security aids in completely locking intruders out of devices, or at least making it much more difficult for them to access the devices. This is important not only in preventing the viewing of such sensitive information as a device's active configuration and the activity of a device, but also in preventing unwanted and unauthorized changes to a device's configuration.

2.2.2 Using Scripts with the CLI

One common means by which operators have effectively utilized the CLI is to use UNIX shell scripts or Perl scripts to interact with the various devices in their networks. Scripts are programmed to connect to remote devices using telnet or other means, authenticate, and then read data from what would be displayed on the user's screen, or send commands to the device to alter its configuration. Another terminology for scripts reading information from the CLI is called "screen scraping," which implies that the scripts are culling data from the characters they "scrape" off the screen.

Using scripts to scrape the CLI and configure devices can be a simple and effective means of managing a device and, in some cases, can approximate the efficiencies and ease of use provided by other management interfaces such as SNMP; however, with simplicity comes several problems. First, the syntax of the management interface that scripts are programmed to understand may change between versions of the vendor's software. Therefore, the operator needs to be keen on the changes made. This can be compounded by the fact that an operator may have to manage more than one vendor's equipment. Second, the volume of the data that is read back to the script over the network is relatively large as compared to that of other management interfaces. The amount of fixed information printed on a screen versus the characters that are used to display variable fields is generally significant, and therefore results in superfluous network traffic, since the entire screen is read back to the script via its telnet session. Other management interfaces provide a much more compact representation of variable data because they do not have to transmit the entire meaning of the fields being retrieved or modified. However, some of these fields may not be available via another management interface, so having the information in albeit a not-so-compact way may be better than not having it at all. This is again why the CLI is considered the lowest common denominator by many.

2.3 CORBA

The Common Object Request Broker Architecture (CORBA) defines a distributed object computing infrastructure that has several uses (OMG 1995a). In particular, CORBA provides for an architecture that automates many common network-programming tasks such as object registration, location, and activation of network objects. Network objects can reside anywhere in the network, including on a traditional management station or even in embedded networking devices such as routers and switches. In addition to locality control, CORBA facilitates parameter marshalling and demarshalling, and operation dispatching among objects, as well as request demultiplexing and multiplexing for specific objects.

The CORBA standard is maintained by the Object Management Group (OMG). The primary components of the OMG CORBA reference model architecture are listed in Table 2.1.

Table 2.1 OMG reference model architecture description.

Component name	Description
Application Interfaces	Application-specific interfaces are generally not standardized; rather, they are developed specifically for a

continued

Table 2.1 continued

Component name	Description
	type of application and used as such. However, if over time the service becomes widely applicable, it is possible that the OMG will standardize the proprietary application interface.
Domain Interfaces	These interfaces have roles similar to the Object Services and Common Facilities, but have specific application with a particular application domain such as telecommunications, medical, and finance.
Common Facilities	These interfaces are oriented toward end user applications. An example of such an application may be for the exchange of embedded objects within electronic documents. For example, objects originate within a spreadsheet application, but can then be linked into a word processor document.
Object Services	Object Services are domain-independent interfaces used by many distributed object programs. Two examples of Object Services that fulfill this role are a naming service that allows clients to find objects based on globally unique names or based on their properties. Another example of such a service might be life cycle management, security, transactions, and event notification, as well as many others (OMG 1995b).

The OMG reference model is composed of three general layers, each of which is assigned a variety of specific responsibilities. The bottommost layer in this model is called the *Object Services layer* and is composed of Object Services that are domain-independent. The functions and responsibilities provided in this layer can be applied to many differing types of objects. An example of such a service is a generic object location service. This service can be applied across most if not all objects within a system. The topmost layer in the model provides domain-specific services, which are outlined in Table 2.2. The architecture is also illustrated in Figure 2.4.

The middle layer of the OMG reference model is referred to as the Object Request Broker (ORB). In essence, the ORB provides the middle layer of abstraction that "glues" the Object Services layer to the upper Application Interface, Domain Interface, and Common Facilities functions.

Table 2.2 Description of OMG CORBA reference model ORB elements.

Component name	Description
Object	This is a CORBA programming entity. Each object has an identity, an interface, and an implementation. An implementation is also known as a servant.
Servant	A servant is an implementation programming language entity. Servants define operations that support a CORBA IDL interface.
Client	This is a CORBA programming entity that is capable of triggering operations on an object. When services are accessed, the specifics of the service should remain transparent to the client. Invocation of services is typically performed in the same way that an object method is invoked in order to keep the operation as simple as possible for the caller.
Object Request Broker (ORB)	The CORBA ORB provides a mechanism for transparently communicating client requests to target object implementations. The ORB decouples the client from the details of the method invocations, which results in client requests appearing to be local function calls. When a client triggers an operation, the ORB must find the correct object implementation to invoke the action, and then transparently activate it if necessary. The ORB is also responsible for delivering the request to the object and returning a response (if any) to the caller.
ORB Interface	The CORBA specification defines an abstract interface for an ORB that decouples applications from the details of their implementation. This interface achieves transparency by providing various functions that help hide the internal details of the interface.
CORBA IDL Stubs and Skeletons	CORBA IDL stubs serve as a layer between the client and server applications, and the ORB. Transformation from CORBA IDL definitions into the target programming language such as C or C++ is typically performed by a CORBA IDL compiler.
Dynamic Invocation Interface (DII)	The Dynamic Invocation Interface allows a client to access the underlying request mechanisms provided by an ORB directly. Applications utilize the DII interface in order to dynamically issue requests to objects without requiring an IDL interface-specific stub to be linked with its code. This is in contrast to IDL stubs that function as RPC-style requests. The DII interface also allows clients to invoke nonblocking, deferred, and synchronous (separate send and receive operations), which may be necessary for certain operations to function correctly. This interface also provides for one-way

continued

Table 2.2 continued

Component name	Description
	calls that are made to objects that act as events (they are sent only, and no response is sent).
Dynamic Skeleton Interface (DSI)	This interface is similar to the client-side DII interface except that it runs on the server side. The DSI interface allows an ORB to deliver requests to an object implementation that does not have compile-time knowledge of the type of the object it is implementing. Specifically, it allows the client issuing the operation request to have no specific knowledge of whether the implementation is using type-specific IDL skeletons, or if it has employed dynamic skeletons.
Object Adapter	The purpose of the object adapter is to assist the ORB in delivering request objects. It also assists the ORB with activating the specific object. An object adapter can be used to associate a specific object implementation with the ORB by smoothing out differences between the two. Furthermore, object adapters can provide support for certain *types* of object implementations. For example, OODB object adapters can be provided to support persistent library objects.

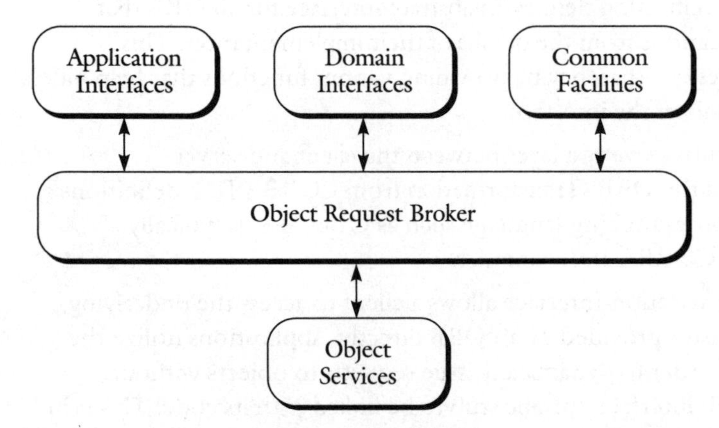

Figure 2.4 OMG reference model architecture.

Figure 2.5 illustrates the primary components in the CORBA ORB architecture. The ORB architecture is comprised of several components. The object (servant) is a programming language implementation component that defines the operations used to support a CORBA IDL interface. IDL interfaces can be implemented in several programming languages including C, C++, and Java. The client component is a program construct that invokes an operation on an object. This operation will potentially access the services of a remote object. When this occurs, this action should be transparent to the caller. The Object Request Broker provides a mechanism for communicating client requests to other object implementations transparently. The architecture of the ORB greatly simplifies distributed programming in

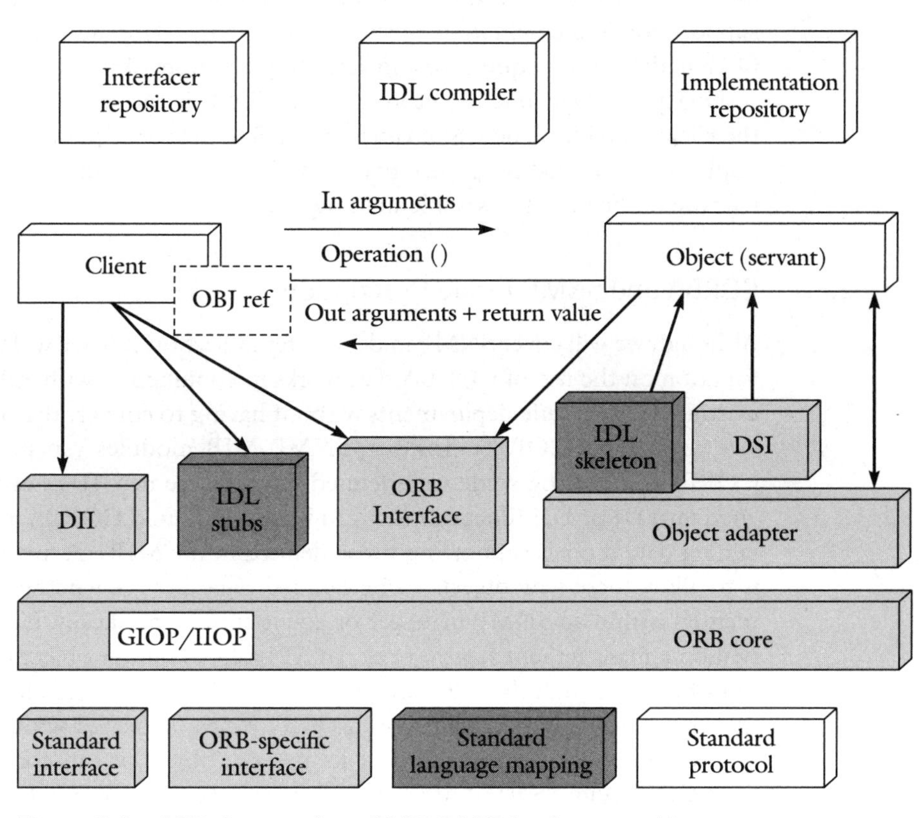

Figure 2.5 ORB elements from OMG CORBA reference architecture.

that it removes the client from the specific details of method invocation. This allows the client requests to appear to be local procedure calls, when in fact they may be remote calls. The ORB is responsible for locating the appropriate object implementation when a client invokes an operation on that object. The ORB must also deliver the request to the possibly remote object and return a response, if one is given, to the calling client. To decouple applications from their specific implementation details, the abstract interface for an ORB is called the ORB Interface. This provides various functions and routines that convert between internal implementation specifics and those provided in the standard CORBA API. The CORBA IDL provides various stubs and skeletons that "glue" the client and server applications together with the ORB. The conversion between the CORBA IDL definitions and the programming language used to implement the objects is usually automated using a CORBA IDL compiler. The object adapter associates object implementation specifics with the ORB. In doing so, an object adapter can be specialized to provide support for implementation-specific object styles. This function also assists the ORB in delivering requests to and activating a remote object.

Finally, the Dynamic Skeleton Interface (DSI) is the server-side counterpart to the client-side DII. The DSI enables an ORB to deliver requests to a specific object implementation that might not have compile-time knowledge of the specific internal type of the object it is implementing.

2.3.1 CORBA and SNMP Usage Description

Although we will cover SNMP in detail later in Section 2.6, we will touch upon one variation on the use of CORBA that works in conjunction with SNMP to leverage existing MIB module deployments without having to convert the code on each device to support CORBA. Instead, SNMP MIB modules can be translated into CORBA IDLs using some well-defined rules. These new IDLs can then be translated into DI or DII functions that can be used to build CORBA applications containing data models supporting those defined in the MIBs. A further modification is to allow these new objects to be queried regardless of whether they are implemented within an SNMP manager or agent entity. Specifically, if an object is queried on a management station (e.g., an API asks a counter object to display itself), this object, through the ORB architecture described earlier, can locate the value of its corresponding embedded side instance, which is located within an agent running on a network device. Once it locates the object, it can convert the request from CORBA into SNMP and send the managed device a request for its value using SNMP. This is illustrated in Figure 2.6 as the SNMP-aware CORBA management station.

One variation on this theme is to build applications that act as SNMP *proxy agents*, which are sometimes known as *midlevel managers*. These entities are

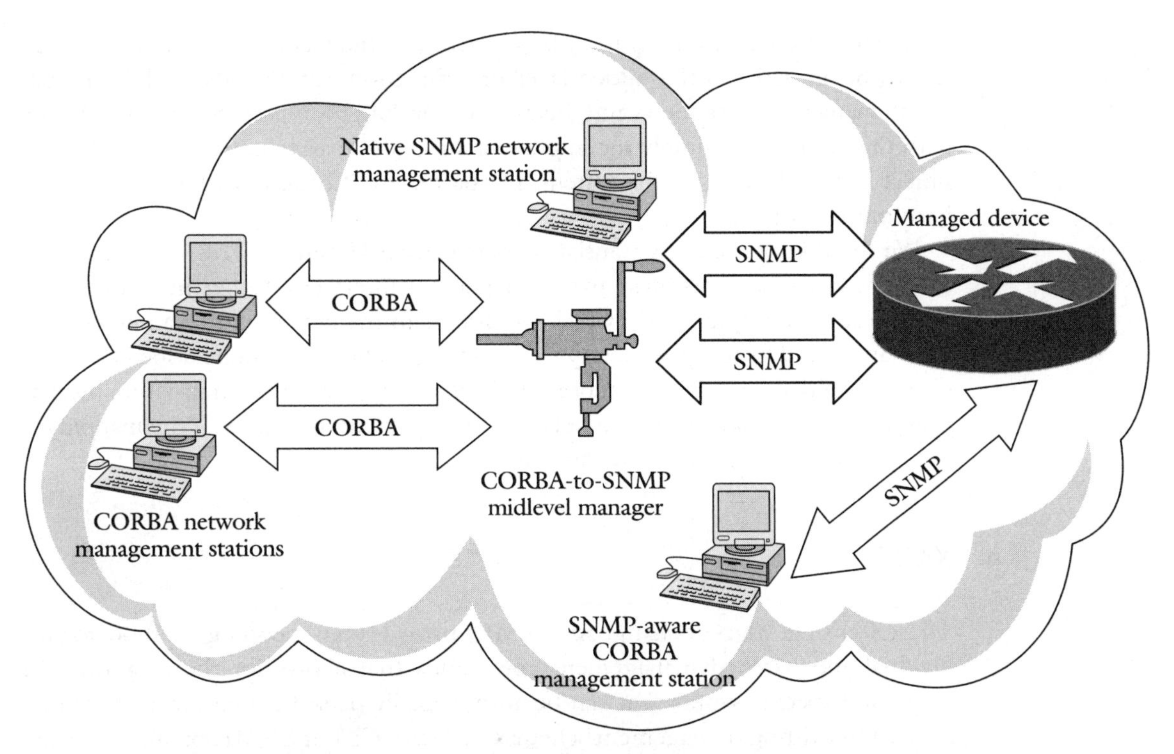

Figure 2.6 CORBA midlevel manager translates (SNMP protocol converter) between CORBA and SNMP for native CORBA management stations, while a native SNMP manager communicates directly with the managed device using SNMP. An SNMP-aware CORBA management station also can communicate directly with the managed device without the help of the midlevel manager.

responsible for translating SNMP requests and responses to and from CORBA. This allows existing CORBA applications to speak to SNMP devices without any modification. Instead, they speak to a CORBA ORB that does the translation externally. This ORB can also catch SNMP notifications and relay them to the appropriate application as CORBA events. Figure 2.6 demonstrates how a midlevel manager translates between CORBA and SNMP for native CORBA management stations that cannot speak SNMP. At the same time, a native SNMP manager entity communicates directly with the agent entity on the managed device using SNMP. Note that both the midlevel manager and the native SNMP management station can be configured to receive SNMP notifications from the managed device. The CORBA translator will translate these notifications into CORBA object messages that will be delivered to the appropriate objects, while the native SNMP NMS will process the events directly.

As with all translation mechanisms, we have witnessed that this method too suffers from performance issues, even if implemented correctly. Those that have used

CORBA implementations in the past can attest to the fact that they were not very fast to begin with, so this added layer of indirection simply adds to the time between management requests and their corresponding responses. A further difficulty with this approach is simply the implementation and maintenance. If a new SNMP object is introduced into the system, several pieces of code must be modified in a harmonious fashion.

We should note that the translation between SNMP and CORBA messages and events is not unique to these two management interfaces. Other translations are possible. For example, XML, which we discuss in the next section, can be used to translate to and from a proprietary CLI language. The same can be done between a CLI and SNMP. The important point to keep in mind about any translation mechanism is the data model mapping between management interfaces. If the mapping is not accurate and complete, inconsistencies may arise.

2.4 XML

The eXtensible Markup Language (XML) provides an encoding method for individual and batched management commands that allows for the creation of a command-specific syntax that can be automatically parsed and checked. It can be applied to existing management schemas such as a CLI or SNMP, or allows for the creation of a new set of commands and parameters specific to an XML client and server.

XML describes a class of data objects called *XML documents* and specifies the behavior of applications that are used to process these documents. XML represents a subset of the Standard Generalized Markup Language (SGML) (ISO 8879); thus XML documents are by nature conforming SGML documents.

To the untrained eye, the markup in an XML document may appear similar enough to an HTML document that the two may be confused with each other until some crucial differences are noticed. XML represents a metalanguage to markup text documents. XML offers the reality of both a truly cross-platform and long-term data format.

Data is embedded within XML documents as strings of text and is surrounded by special markers that describe the data so that it can be parsed and displayed. Each piece of data that is delineated by markup markers is referred to as an XML *element.* The exact syntax of how data is delineated by markup markers is detailed in the XML specification. The XML specification includes the details of such things as how each element is delimited by tags, what the precise definition of what a valid tag is, and the valid format of document names for elements, among others. XML documents are comprised of virtual storage bins called *entities.* An entity can contain parsed or unparsed data. While the contents of unparsed data are unknown, the contents of parsed data represent characters that form either markup or character data. Markup data is used to encode instructions indicating how the document is structured and how it is stored.

The markup used in an XML document not only describes the document's structure, but also allows you to determine which elements are associated with one another. Furthermore, good XML documents also include information about the semantics of the document so that you can understand which domain the document best applies in. For example, the semantics of a document might indicate that a data element represents model numbers for computer monitors, or automobile license plate numbers. An important distinction between XML and other document formats is that XML is not a document presentation language; rather, it is used to describe the structure of the document. For example, XML documents do not contain markers that specify that a particular word is to be displayed in a blue 12-point Times Roman font. This information can be embedded in the document, but it is not specifically XML data.

XML is a flexible metamarkup language. Specifically, XML does not have a rigid set of tags and elements; instead, XML allows elements to be defined as they are needed. This has fostered the development of domain-specific XML element sets that allow XML to be extended and expanded to meet the unique needs of a particular domain of use. Although XML allows user-defined sets of elements, it does not allow users to modify the grammar or syntax of XML documents. The grammar defined for XML is specific about the syntax for the placement and appearance of tags, as well as other points about the syntax of elements. This grammar is then used to build XML parsers that can interpret any standard XML document. Documents that follow the syntax defined in the grammar are known as *well-formed* documents, while documents that are not are referred to as *non-well-formed* documents.

The markup style permitted in a particular XML application is typically described in a *document type definition* (DTD). The DTD is a place where all markup and other specifics about the domain where this DTD applies are included. This is loosely analogous to a common header file in C that is included by all files that use a particular library. In this case, a particular instance of a document that is being applied to a specific domain will specify that it conforms to the specific DTD of that domain. XML parsers can then compare this document to the DTD to verify that it matches the DTD. If the document matches the DTD for the domain that it specifies, then it is considered a *valid* document. Bear in mind that it is possible to compare the same document to multiple DTDs and have the document be considered valid for all DTDs. This can happen, for example, if the DTDs are subsets of a common DTD. It is important to understand that DTDs are an optional element in XML; therefore, either they may not exist, or they may be inadequate to completely validate a document. Furthermore, documents that are considered invalid may still be useful under some conditions. This is simply because the syntax specified for DTDs is limited. In short, buyers beware!

It is important to understand that XML is *not* a programming language, nor is XML a database, or a network transport protocol. Many marketing departments may try to convince you that it is one or all of these things, but it is not. XML is simply a document description language. Some may be tempted to try to convince you that a configuration file for a network device contains XML programming, but it does not. In reality, this file is just formatted using XML and an XML DTD instead of a proprietary vendor's format. This configuration file cannot be written in XML; it is instead *described* in XML, which is an important distinction. It may contain configuration commands that are surrounded by XML tags, and these strings might be parsed by an XML-aware command parser. Actions such as configuration changes may even be taken as a result, but XML does not constitute a structured programming language for the file; it only describes its format (and perhaps its semantics). Similarly, the XML-formatted configuration file does not represent a database of information, even if stored within the network device in that format. The database must be implemented using operations and structure that are outside the scope of XML. An example may be an Oracle database format. SQL operations are used to access the database, even if the data is stored in XML format. Finally, XML does not represent a network transport protocol. XML, like HTML, cannot be used to send data across a network per se. Data can be sent formatted as XML across a network using some network transport protocol such as IP/UDP, IP/TCP, or IP/TCP/HTTP. ASN.1-encoded SNMP data can even be transported after being encapsulated within XML, but it still must be transported using a real network transport protocol such as UDP. In short, XML can be used to format data sent across a network, but software that is not part of the XML document must actually transmit the document.

XML documents are ASCII text files that are divided into logical pieces referred to as *records*. Therefore, any tool that is capable of reading ASCII text files can read an XML document. This is quite important, since most, if not all, document editors available today can read ASCII text. What most cannot do is read each others' proprietary formats. Several XML document fragments can comprise a single document. It is also possible for documents not to reside anywhere in particular. XML documents may reside in a device's memory after being dynamically generated, or the file fragments may be stored across multiple file systems. The temporal locality of an XML document does not have to be common among a set of document fragments that constitute its totality. XML represents a format that can be truly ubiquitous because the format is in clear ASCII text and provides all of the important information about the document's structure (and sometimes its semantics).

XML parsers are applications that interpret the contexts of XML documents and validate them. XML parsers can be contained within other parsers and be used to trigger actions that result from the successful interpretation of an XML document. The successful interpretation might include just a single line or a series of lines contained in an XML document. For example, XML-formatted CLI commands might first be transported to a command-line interface that understands XML-formatted commands. The command will first be validated by running it through the embedded XML parser. Once the command has been validated, the command is then passed to the command interpreter that reads the command and, if appropriate, triggers the appropriate action function. Similarly, the result of the command might be formatted internally and then sent to the command-line interface. The CLI might then, in turn, format the response using XML and return that to the operator. The operator's terminal or management application must then be able to interpret XML-formatted responses. This should not be difficult since the application had to send an XML-formatted command to the device in the first place.

2.4.1 XML-RPC and SOAP: XML Serialization and RPC over HTTP

Since its inception, XML has been applied in various ways. Originally, XML–Remote Procedure Call (XML-RPC) was developed to provide a simple RPC mechanism using XML and HTTP as a transport protocol. XML-RPC presented an interesting application of XML in that it positioned it as the basis for standards-based transactional computing. Several implementations of XML-RPC are still in use today. These implementations demonstrate that XML-RPC is platform-neutral and language-neutral while still being very useful.

SOAP, like XML-RPC, can be thought of as a Web-based abstraction of traditional distributed object communication. SOAP represents a lightweight XML-based protocol that can be used to exchange information in a distributed environment. SOAP consists of three principal components. First, it defines a container framework that is used to describe what the contents of a valid message are and how a parser should process it. Second, it defines a set of encoding rules that can be used to express instances of application-defined data types. Finally, it defines a representation for remote procedure calls and their corresponding responses.

SOAP itself does not address higher-level distributed object issues such as object activation, nor does it address object life cycle management. SOAP does not specify the messaging semantics of the XML transport encapsulation. That is, it does not define QoS, queuing, or other related issues. To the contrary, applications that process SOAP messages must provide the transport semantics used for the connection. Finally, issues do exist regarding the use of SOAP as an RPC mechanism over HTTP. Specifically, the issues of transaction control, replay protection, and encryption are in question. Due to these limitations, some have built true messaging models using SOAP as the base and addressed these issues in the new layer.

2.4.2 Encoding Managed Information Using XML

As was mentioned earlier, XML can be used to encapsulate or wrap managed objects from existing management interfaces such as proprietary CLI (see Section 2.2) or SNMP (see Section 2.6). When XML is used as a transport encapsulation, managed information can be encoded for display or storage at a network management station in several different ways. First, SNMP can be encoded within XML. In this way, XML tags can encode SNMP object names and values in clear text, allowing XML parsers to understand the information and display it. The same can be done for CLI data. Although allowing for easy display, this mechanism, however, has the disadvantage that the format of the data is quite verbose as compared to the standard data encapsulations, and thus results in significantly more network traffic.

2.5 Bulk File Transfer

Bulk file transfer is an option that some device vendors offer as a means of offloading large amounts of data from their devices via a file transfer protocol. In particular, deployments of devices that are required to maintain large configurations or large amounts of manageable information may find an advantage when exporting this data bulk form using a file transfer mechanism. Of course, other options such as the SNMP GET-BULK operation exist, but in some cases, even this optimized approach is still too inefficient for some networks.

Bulk file transfer generally results in one or more files containing the equivalent management data that would have been off-loaded using a more traditional management interface such as SNMP or CLI. Data can be exported to or imported from a device, just as files can be transferred to and from any networked computer. The choice of exporting or importing data depends on the goals of such activities. For example, statistics or an existing configuration might be exported from a device, while a new configuration might be loaded into a device by importing a bulk file. The actual details of how this process is configured and eventually triggered vary from vendor to vendor. For example, some vendors simply allow this to be engineered from the CLI, while other vendors may even provide a MIB that can be used to both configure and instigate the file upload/download process.

In all cases, the motivation for using such a mechanism is simple: efficiency of operations and a reduction in network overhead resulting from management protocol inefficiencies. In comparison to using any other management protocol, the amount of overhead when using a bulk file transfer of the same amount of manageable objects is very low. The reason is simple: the bulk file transfer generally requires one or two operations by the manager, and then the file is either sent or received using a very efficient file transfer protocol. The data being transferred can be highly compressed. By comparison, other management interfaces require several operations to achieve the same goals, as well as additional framing for each piece of managed data to identify it.

With this efficiency comes a low amount of flexibility. Essentially bulk file transfer is used for one or two types of operations: configuration upload/download or bulk statistical data off-load. In the case of bulk statistical transfer, potentially large volumes of data are off-loaded to an offline server for processing. Instead of the management station querying for each managed object, they are all packed a priori into the bulk file and are transferred in one operation. This is important also for the recording of historical data that may result in large volumes of data being transferred across the network to the management station. Instead, the device can cache the data and transfer it all together as a single file. This saves on network traffic and makes it more likely that sensitive statistical data will not be lost. In the case of configuration files, some network devices require very large configuration files. If each managed object must be transferred one at a time to the managed device, it may take several minutes (or hours!) to configure a device that has rebooted, or one that requires a large amount of reconfiguration. In these cases, it is preferable to simply transfer the configuration file to the device and allow it to read the data either from its memory or from an onboard disk drive.

Bulk files can either be "pushed" or "pulled" from a device. This describes how the manager extracts the file data. In the case of files being "pushed," an operator would specify a target machine to which the machine later transfers files using the file transfer protocol (FTP) or the UNIX Network File System (NFS). The machine either would then send the file at some specific time or might be triggered to send the file if some SNMP variable is set. The "pull" model is used when a device wishes to be the host of the files created. The operator would then use one of the aforementioned file transfer protocols to connect to the router as if it were a host computer, and then transfer the files from it. In either case, files, such as one containing the device's configuration, can be off-loaded or uploaded using the efficiency of a bulk file transfer rather than individual protocol configuration operations.

2.5.1 Encoding Bulk Data

The format of the bulk file can vary, as several encoding methods exist including XML, proprietary CLI, and SNMP. For example, the information in a file can be encoded using the SNMP SMI, type, and value. This information is then encoded using BER, ASN.1, for example. Other options are to simply include ASCII text of commands or to encode these in binary. Since the specific format is typically proprietary, the shortcoming of using such an approach is that it is not an open standard used by more than one vendor. The downside to this is that it is generally only provided by the vendor and the vendor's management applications, requiring decoding of multiple formats in a heterogeneous network. Sometimes the vendor will not even provide the format to third-party application developers or operators, requiring that they use that company's device management software. This may unfortunately result in a network comprised of devices from *N* vendors, requiring an operator to understand and manage as many as *N* different bulk file export/import applications. For most vendors we have interacted with, this is an unacceptable solution. This also applies to other management interfaces that we have already discussed or that we will discuss, including XML, CLI, and CORBA. Fortunately, SNMP does solve this problem. We will discuss how later in this chapter in Section 2.6.

Further compounding this problem for operators is the fact that in order for commercial parsers to convert proprietary formats that may have much of their semantic information "compressed" out, they may have to be uncompressed into more verbose representations that are more easily useable by applications that are generally unavailable. To alleviate this problem, some vendors provide applications to their customers that accomplish these things, but again, these applications are proprietary and only understand the format of the devices made by that vendor. This results in higher operating costs for the operator since they are only left with the options of either writing their own parsers to decode bulk-formatted data or purchasing one from each vendor. For these reasons, some vendors are moving toward providing their customers with an XML-based representation of bulk data that can be parsed and understood by many different applications, including readily available off-the-shelf versions. Although this does not solve the problem of providing a ubiquitous format for the encoded data, it does allow an operator to build a single application using a common protocol.

An example of how you can reduce the complexity associated with interpreting SNMP data encoded into a proprietary bulk transfer format is to define a new bulk data format that is based on an XML schema. Let's call this new format "vendor-x-bulk-xml." This format will be similar to the existing proprietary "vendor-x-bulk-snmp," which formats SNMP data as a pair containing an SNMP OID that describes the object and an instance of that data, as well as the value (if any) associated with that object instance. The XML version of this formation will modify the format slightly by replacing the OIDs with the verbose representation of the object. Additional XML tags may also be needed to denote fields within the file, as well as an XML DTL. The tags used in this format will be relatively small, while still providing enough semantic meaning about the tagged data. By formatting data in the "vendor-x-bulk-xml" format, we lose some compression that the "vendor-x-bulk-

SNMP" format provides, but the data produced in the "vendor-x-bulk-xml" format will still be sufficiently compressed. A benefit of using the "vendor-x-bulk-xml" format is that we can provide DTD specifications that provide a well-understood format for describing well-formed XML documents. This same format is either not provided by proprietary formats, or when it is, it is almost completely different from any other vendor's format. In addition, we can provide style sheets that can be used to transform the "vendor-x-bulk-xml" formatted data into user-friendly representations or into other XML representations.

There are significant advantages to this "vendor-x-bulk-xml" format. Such data would be easy to interpret with existing XML parsers, saving customers from having to write their own parsers. Further, the structured tagging makes for easier debugging in customer environments. Furthermore, the advantages of bulk transfer are still present: large volumes of data can be off-loaded from a device in a very efficient (and now well-understood) format. This reduces the complexity of managing devices, as well as provides an efficient mechanism for doing so.

2.6 The Simple Network Management Protocol (SNMP)

The Simple Network Management Protocol (SNMP) was devised many years ago by the IETF to solve the problem of managing network devices remotely using a standard protocol, access methods, and a well-known format for representing managed data stored in network nodes. The standard for SNMP has been enhanced and extended over the course of its existence to include additional protocol operations, enhanced security, and additional standard management models. The first version, called SNMPv1 (RFCs 1155, 1157, and 1213), has since been surpassed by SNMPv2c (RFCs 1901–1906) and, most recently, SNMPv3 (RFCs 2571–2575). With each new revision came many new features including new protocol operations, security features, and modifications to the language SNMP uses to represent managed information. In addition, during this time, the acceptance of SNMP in the marketplace grew. Today most production networks use SNMP as at least a part of their overall network management strategy.

SNMP is composed of three basic components: the Structure of Management Information (SMI), the Management Information Base (MIB), and the Simple Network Management Protocol (SNMP). Let us now describe each of these components in detail.

2.6.1 Structure of Management Information

Management information is viewed as a collection of managed objects, residing in a virtual information store that is referred to in SNMP as a Management Information

Base (MIB) (see Section 2.6.2). The Structure of Management Information (SMI) is the data modeling language used to model the management data. The SMI (RFC 2578) is the language that is used to write, define, and specify a MIB module. The roots of the SMI are actually in an adapted subset of OSI's Abstract Syntax Notation One (ASN.1). This adaptation was first done in 1988. The SMI has since changed in the second version of the SMI, called SMIv2. The SMIv2 has been used for several years now as the standard language in which to define MIB modules. Furthermore, a new SMI is being worked on at the IETF, called the SMIng, that encompasses even more powerful modeling language and constructs. It is important to understand that it is much better not to call this ASN.1.

The SMI is divided into three parts: object definitions, MIB module definitions, and notification definitions. An SMI macro called the MODULE-IDENTITY is used to specify the semantics of an information module. This macro is always found at the top of the module definition. Object definitions are used to describe managed objects. The ASN.1 OBJECT-TYPE macro is used to specify the semantics and syntax of a managed object. Finally, notification definitions are provided to describe spontaneous transmissions of management information. The ASN.1 NOTIFICATION-TYPE macro is used to specify the syntax and semantics of a notification.

Textual Conventions and Basic Data Types

The SMI provides a number of basic data types that are used to specify the semantics and syntax of objects defined in MIB modules. When designing a MIB module, it is sometimes beneficial to define new types that are derived from those defined in the SMI. Each of these new types has a different name, a similar syntax, but a more precise semantics than the type it is derived from in the SMI. These newly defined types are termed *textual conventions* (TC) and are defined with the TEXTUAL-CONVENTION data type. Textual conventions do not add any new basic types to the SMI. This is very important because a TC is encoded within an SNMP packet—a protocol data unit (PDU)—using the same rules that define their derived type. That is, the actual underlying basic data that is transmitted in the SNMP PDUs (i.e., on the wire) remains the same as the basic data type that the TC is derived from.

Table 2.3 lists most of the basic types that are defined in the SMIv2 (RFC 2578). These objects do not, however, represent the totality of object types in SMIv2. As mentioned, these types can be and frequently are extended using SNMP textual conventions to adapt their syntax or semantics to different domains.

Table 2.3 Basic data types as defined in SMIv2 (RFC 2578).

Object type	Description
INTEGER	Signed 32-bit integer quantity with a range of (–2147483648 . . 2147483647).
Integer32	Has the same syntax as INTEGER except that it never needs more than 32 bits for a two's complement representation. Valid range is (−2147483648 . . 2147483647).
Unsigned32	Unsigned 32-bit quantity. Range of (0 . . 4294967295).
OCTET STRING	Contains a series of bytes that can be of values (0 . . 255).
GUAGE	Unsigned 32-bit quantity used for counters. Range of (0 .. 4294967295).
GUAGE32	Unsigned 32-bit quantity used for counters. Range of (0 . . 4294967295). This object latches onto a specific value, but does not wrap around as a normal counter would, so it should not be confused with the standard Unsigned32 or Counter32.
OBJECT IDENTIFIER	Contains a unique OID. When displayed, it is typically shown as a series of dot-separated unsigned integers.
IpAddress	Contains an IPv4 address as a sequence of 4 bytes. Note that this type is only present in the SMIv2 for backward compatibility. It is no longer used in the specification of new MIBs.
Counter32	The Counter32 type is used when a nonnegative integer that monotonically increases is necessary. The Counter32 type increases in value monotonically until it reaches the maximum value of a 32-bit integer (4294967295 decimal), and then "wraps around" to 0, where it starts increasing again.
Counter64	64-bit counter used for counters that wrap in less than one hour with 32-bit counters. Valid range is (0 . . 18446744073709551615).
BITS	The BITS construct represents an enumeration of named bits. The collection of named bits is assigned nonnegative, contiguous values starting at zero. Only those named bits that are enumerated by the definition may be present in a value. Therefore, enumerations of bit positions must be assigned to consecutive bits (i.e., there cannot be holes in the enumeration).

continued

Table 2.3 continued

Object type	Description
Opaque	The Opaque type supports the capability to pass arbitrary ASN.1 syntax. A value is encoded using the ASN.1 Basic Encoding Rules (RFC 2578) into a string of octets. This, in turn, is encoded as an OCTET STRING, which in effect "wraps" the original ASN.1 value twice. The Opaque type is provided for backward compatibility and is no longer used in new MIBs.
TimeTicks	Represents an unsigned integer that represents the time modulo 2^{32} in hundredths of a second between two epochs. The DESCRIPTION clause defines both reference epochs.

SMI Versions

There have been two versions of the SMI. Let's briefly discuss them, since they are often confused or incorrectly used interchangeably. Back in the late 1980s, when SNMPv1 was first specified, SMIv1 was defined to facilitate the definition of SNMP MIB modules. The first version of the SMI consisted largely of the specifications in RFC 1155, RFC 1212, and RFC 1215. SMIv1 is currently a full standard (STD 16) within the IETF, although since 1995/1996 standard MIB modules have not been typically allowed to be defined using SMIv1. It should be noted that some corporations still define their enterprise MIB modules using SMIv1, although that is more of an exception to the norm. The SMIv2 consists of RFC 2578, RFC 2579, and RFC 2580 and is currently a full standard (STD 58) within the IETF.

It is important to understand that all of the data types defined in SMIv1 can be represented in SMIv2. It was the intent of the IETF when it defined the SMIv2 not to break existing implementations that used the SMIv1. It should be noted that although the data types in SMIv2 are a superset of those defined in SMIv1, some of the data types from SMIv1 are only in SMIv2 for backward compatibility, and so should not be used in new MIB module definitions. These exceptions are noted in the definitions for those types in SMIv2. Despite this backward compatibility between SMIv1 and SMIv2, translation in the other direction, from SMIv2 to SMIv1, is a bit more problematic. For example, the new type Counter64 is particularly difficult to convert because there is no direct mapping from this type to any other single type in SMIv1. For those readers interested in the specifics of how these conversions are performed or recommendations on this subject, RFC 2576 explains how to convert SMIv1 MIB modules into SMIv2 MIB modules and vice versa.

2.6.2 The Management Information Base (MIB)

Each managed system is composed of a collection of objects that are used to model system functions, concepts, or attributes. Objects are capable of representing many things and are only limited in power and flexibility by the data modeling language in which they are defined. For example, an object might represent a finite component within the system, such as an interface, and keep track of that interface's counters, state, or name. Another object might represent the state of a routing protocol on that same system. However, given the modeling language used, it is possible, for example, that some of the specific attributes or behaviors of these objects will not be possible.

The collection of managed objects can be thought of as a database of manageable information. This information can be used to form a data model of the system. In addition to the objects themselves, the database also must represent the type, behavior, and associated access policy for each object. It may also be necessary to represent the interaction between objects in the database. The objects in the model can be accessed using a variety of management interfaces to query or modify them. Examples of management interfaces include XML, CORBA, CLI, or SNMP, among others. However, maintaining this information in a database alone does not guarantee that all manager entities can access this information. Unless an external representation of the managed information such as an abstract data model is agreed upon by both manager and agent entities, an agent on one device may provide access to its information in a manner that is inconsistent with that of another device. This could easily happen, for example, if one manufacturer produced a representation of one variable as a string, while another represented it as an integer—perhaps because the management interface is nonstandard, as is the case with the command-line interface. In addition to these challenges, managed objects can be, and usually are, implemented in a variety of ways depending on the specific device, or even the software revision running on that device. Therefore, it is important for the management interface to provide a consistent view of the data model that agent entities can present to manager entities. SNMP accomplishes this using its Management Information Base.

The Management Information Base is the collective set of MIB modules that together make up the MIB that is being managed/monitored. The MIB is also sometimes referred to as the device's *virtual object store* because it defines the data model for a device. MIB modules are defined using the SMI, which provides a consistent data modeling language in which managed objects are defined. It also allows MIB modules to be parsed by MIB compilers that not only can check their syntax, but also can generate code that is used to implement the objects defined in that MIB module. An individual MIB module specification should not be confused with the entire collection of specifications implemented by any particular device, which is referred to as a device's Management Information Base proper. Many use the terms "MIB" and "MIB module" interchangeably. This is sometimes problematic, since the actual meaning must be distinguished based on the context in which it is used; therefore, we will henceforth use the term "MIB module" to describe a specific subset of the MIB, and "MIB" as the entire collection of MIB modules.

MIB module specifications are generally produced by two types of organizations: standards organizations and private entities such as corporations. In the case of standards-based definitions, standards bodies such as the IETF, ITU, and ATM Forum have produced, and continue to produce, MIB module specifications that can be used to manage various protocols and network device functions. The advantage to MIB modules that are produced by standards organizations is that they provide a common data model that can be adhered to by all vendors implementing the feature being managed. For example, the OSPF MIB module defines standard managed objects that can be used to manage a standards-based OSPF implementation. All devices implementing this MIB will represent the objects defined in that MIB externally in the manner in which they are specified in the MIB module, regardless of their actual internal implementation. Aside from standards-based MIB modules, company-specific versions called *enterprise* MIB modules are also produced by corporations. These MIB modules are typically produced to extend existing standards-based MIB specifications because they do not adequately reflect the entirety of managed information for that feature as implemented by a specific corporation's device. It is common that the standard MIB module contains a subset of managed objects required to manage a specific feature. Vendors typically augment the standard function, thereby adding value to it. These additional functions are then represented in that company's enterprise MIB module.

Each MIB module specification is composed of several components: a textual preamble, a MIB module defined using the SMI, and references to other related documents. Standards-based MIB module specifications rarely vary from this format, while proprietary MIB module specifications typically only contain the MIB module.

Each managed object in an agent's virtual object store must be modeled in a specific MIB module before it can be made available for access through the SNMP management interface. Each managed object is defined using the OBJECT-TYPE macro. This macro allows the specific syntax and semantics for an object to be defined. Two basic types of objects exist in MIB modules: *scalar* and *columnar.* If an object can only exist as a single instance, it is referred to as a scalar object. Management operations in SNMP apply exclusively to scalar objects; that is, the basic operations in SNMP always act on a scalar object. For example, the GET and SET operations may only act upon an instance of an object such as a scalar object or columnar instance. No specific construct exists within SNMP to organize together scalar objects; however, related scalar objects can be found grouped together at least conceptually within a MIB module. For example, scalars related to the OSPF protocol may be found in the OSPF MIB module. Organizing objects together in this way sometimes enhances the MIB module's readability and usefulness. Figure 2.7 illustrates a collection of scalar objects. Notice that the scalar objects are not organized in any particular way, and that each represents a single instance of an object (i.e., no two overlap in color).

We mentioned that two basic types of objects exist in SNMP. Let us now investigate columnar (i.e., tabular) objects. It is often a necessity to group managed objects together into a conceptual tabular structure to form an ordered collection of objects within the MIB module. Each conceptual table contains zero or more *rows,* each of which can contain one or more *columnar* objects. Thus, conceptual tables in SNMP contain a series of rows and columns. This conceptualization is specified using the OBJECT-TYPE macro. This macro can be used to define both an object that corresponds to a table and an object that corresponds to a row in that table. The intersection of a row and one or more columns represents a specific object *instance* and is how specific instances of objects are called out or indexed from within a conceptual table of objects. Figure 2.7 illustrates how a conceptual table in a MIB module might be configured with several rows and columns. The table represents horizontal rows, where the specific row is identified by the indexes for the table,

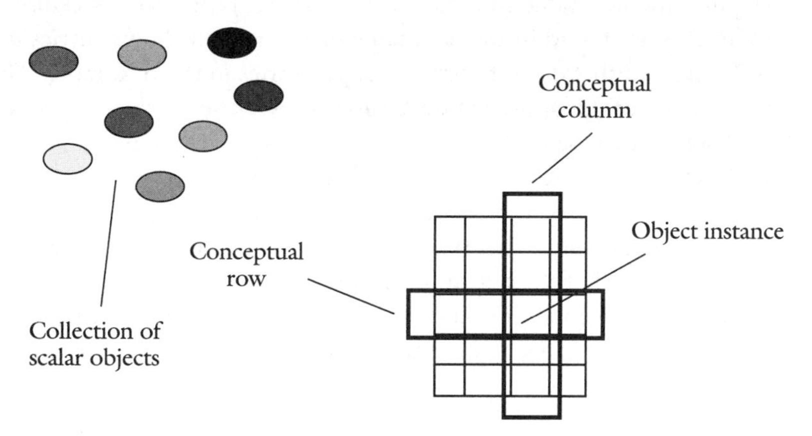

Figure 2.7 SMI conceptual table containing columns, rows, and object instances, and a collection of scalar objects.

and the specific instances of a columnar variable. Object instances are indicated by the cell within the table where both the columnar and row rectangles intersect.

Table Indexing

Each table defined in a MIB module is indexed by one or more objects. The indexing of a table is specified using either the INDEX or AUGMENTS clauses. The INDEX clause contains objects that either are present in the table or are found elsewhere that are used to index the rows in that table. The most common approach is to include objects defined in the table as indexes for the table. When this approach is used, the objects used as indexes must not be accessible.

Two additional choices exist when a table wishes to extend the entries of an existing table without having to add additional columns to that table. This may be necessary if, for example, you wish to extend a table defined in a standard MIB module with objects that are enterprise-specific. In this case, it would not make sense to add additional objects to the standard table. The choice of using one of the table extension mechanisms or another depends specifically on whether a one-to-one relationship exists at all times between the instances of the objects contained in both tables. If a row in the new table will exist for every row created in the base table, then the AUGMENTS clause can be used instead of the INDEX clause to define the indexing of this table. The AUGMENTS clause specifies the base table that the new table will augment, but does not specify any indexes. In doing so, the new table is implicitly indexed using *exactly* the same indexes as the base table. The other method of table extension is used if a sparse relationship exists between the objects in both tables. This method is called the *extends* relationship. This approach requires the new table to define its indexes with the INDEX clause using the same indexes as are found in the base table. This method allows entries in the base table to be created that do not correspond to entries in the new table. This is an important feature of this approach since this allows entries in the new table to correspond to those in the base table only if it makes sense to do so. For example, if a new table wishes to extend the IF-MIB's ifTable to include additional objects for a new type of interface, this new table would be implemented using an *extends* relationship because the possibility exists that the ifTable could contain interfaces that might not be related to all of the entries in this new table. However, if the new table was defined to contain counter objects that applied to *every* possible type of interface, then it would be appropriate to use the AUGMENTS clause to specify the indexing for this table.

2.6.3 Access to Objects

Each object defined in a MIB module must have a MAX-ACCESS clause associated with it. The valid values for MAX-ACCESS are ordered, from least to greatest:

"not-accessible," "accessible-for-notify," "read-only," "read-write," "read-create." The MAX-ACCESS clause indicates the *maximum* access that would make what is called "protocol sense" for the object. For example, if the MAX-ACCESS is defined as read-only, then a SET operation that attempted to modify that object would fail. If that object's MAX-ACCESS is set to read-write or read-create, then the SET operation may succeed. However, there may still be many reasons why a SET might fail. First, an implementation may not have implemented the object such that it could be modified. Therefore, the object's actual access is read-only, thus preventing modifications of the object. Similarly, the RowStatus for the conceptual row containing that object might be implemented or configured to disallow write access. Second, the SET request may be attempting to modify an object that is configured with noAuthNoPriv in the VACM. Third, the device's configuration may be set to not allow SETs for unsecured requests, and the SET request comes in unsecured. Fourth, the specific user requesting the modification of that object is not allowed to execute that operation on that object given the configuration in the VACM MIB (RFC 2575). Please note RFC 2576 also maps community-based (that is, SNMPv1 and SNMPv2c) access into the VACM access control. Table 2.4 enumerates the five specific access types in increasing order of permissions granted to the manager.

Table 2.4 Possible values of the MAX-ACCESS clause.

MAX-ACCESS type	Description
Not-accessible	Not allowed for scalar or columnar object types. Only for indexes.
Accessible-for-notify	This value indicates that an object is available only via a notification. That is, this object may only be contained within a notification, but cannot be accessed using the other SNMP operations (e.g., GET, SET, etc.).
Read-only	The object type may be an operand in only retrieval and event report operations.
Read-write	The object type may be an operand in modification, retrieval, and event report operations.
Read-create	Same as read-write, except specified for columnar objects that require a value to be set before a row in that table can be created. Once created, these objects can be read or written.

It is important to understand that the MAX-ACCESS clause specifies the maximum access that is required for an object, but it does not specify the minimum access possible. It is up to the implementation to choose this given how it has chosen to implement the object. It is permissible to implement an object using a MAX-ACCESS that is lower than the MAX-ACCESS specified in the MIB module. The implementation can specify how it has varied from the MIB module where the object is defined by noting this variation in the Agent Capability Statement (see RFC 2580 for more information) for that MIB module. For example, an object that is specified with a MAX-ACCESS as read-write might actually be implemented as read-only. Alternatively, this can be accomplished by specifying a MIN-ACCESS clause in the Conformance section of the MIB module.

2.6.4 Object Identifier

SNMP specifies a scheme by which all of the objects and instances of those objects present within a system can be uniquely identified. These items are called *object identifiers,* or more commonly, OIDs. OIDs are specified as an ordered sequence of nonnegative integers written from left to right and separated by a period (i.e., dot). This is referred to as the ***dot notation***. For example, "1.1" represents an OID. The OID space itself does not have any limitation as to how many branches (subIDs) are possible. For SNMP, however, a limit of 128 subIDs has been defined. Each consecutive integer is separated from the numbers around it by a period. The sequence must contain two integers at a minimum and does not have a maximum number (although all implementations will have a specific limit to this size).

OIDs are arranged and organized in a hierarchical tree structure. The topmost levels in the OID tree are controlled by the ITU and ISO standards bodies. These organizations delineate how new assignments are given. A portion of this OID tree is managed and maintained by standards organizations or corporations. A portion of the OID tree is shown in Figure 2.8. Note that the example contains the IETF subtree. Standard MIBs typically contain OIDs that use the prefix 1.3.6.1.2.1.

SNMP objects and object instances—sometimes referred to as SNMP *variables*—are assigned unique OIDs. As was noted earlier, SNMP variables are either scalar or tabular objects. SNMP MIB modules are also typically assigned an OID within the OID tree shown in Figure 2.8. This OID often represents the root OID for all objects present in that MIB module. The point where the MIB module is rooted depends on the status of the MIB specification as well as which organization owns and maintains the document. As mentioned earlier, two general types of MIB specifications exist: enterprise (or proprietary) and standard. Documents that are produced by a standards organization are typically rooted somewhere below that organization's node in the overall OID tree. For example, MIBs that are produced by the IETF are sometimes first placed under the experimental (3) OID. These MIB modules will be assigned OIDs with a prefix of 1.3.6.1.3. Once the MIB specification has been adopted for standards-track status, it is moved under a different portion of the IETF's OID subtree, typically mib-2. Similarly, private organizations

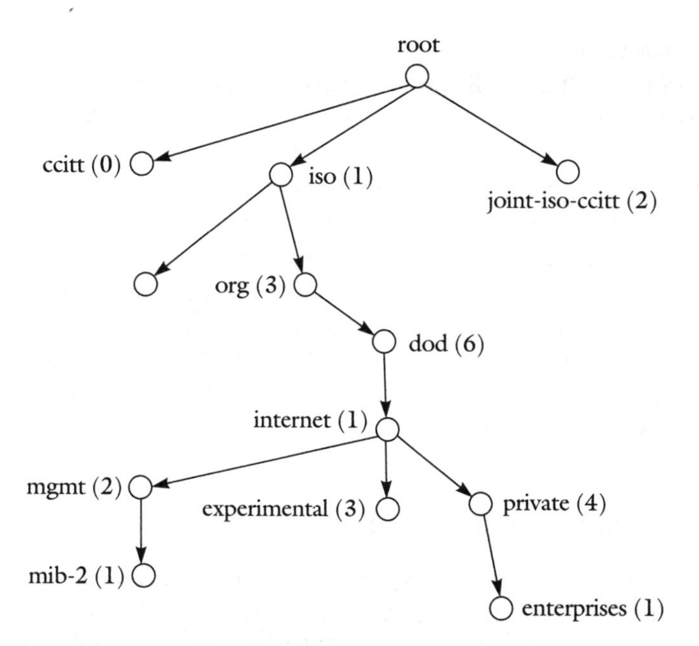

Figure 2.8 Sample portion of the OID tree containing IETF subtree OID assignments starting at internet (1).

maintain enterprise MIB modules. These modules are assigned OIDs that are rooted under the IETF organization's private (4).enterprises (1) subtree. For example, Cisco Systems maintains a subtree of OIDs under 1.3.6.1.4.1.9. Corporations sometimes further delineate their OID space in a manner that is consistent with the one used by the IETF. That is, corporations sometimes maintain additional subtrees that contain portions for experimental or prototype MIBs, as well as those that are officially released.

The general format of an OID is of the form <object identifier>.<instance id>. The first part of the identifier represents the OID of the object, as defined in the MIB module. The second portion identifies the specific object instance. In general, the OID representing a scalar object instance contains the object prefix and an instance identifier of 0. The format of a tabular object instance OID is comprised first of the table entry's OID. The second portion contains the columnar identifier. The last portion contains one or more identifiers that represent the index of the row. This is sometimes confusing, but remember that this must be the case because tables may be indexed by one or *more* object instances. Thus, the general format of an OID representing an object instance in a table is of the form

<table entry subOID>.<column>.<index$_0$> . . . <index$_n$>

We will investigate how OIDs are operated on later in Section 2.6.9.

MIB Module Versions

Since around 1996, MIB modules have, in general, been written using SMIv2. However, a small set of MIB modules are still written in SMIv1. The most visible and important one is MIB II, which is defined in RFC 1213 as well as Standard 17 (STD 17). It is also important to understand that much of MIB II has been split off into new MIB modules that replace these functions. These new modules are written in the newer SMIv2. For example, the system group from MIB II can now be found in the SNMPv2-MIB module (RFC 1907). This can sometimes be confusing for the novice (or someone who has been doing this for a while), so we recommend searching the IETF RFC archive and going over the v1 MIBs before implementing them, as newer SMIv2 versions of the MIB module (or the portion(s) you are interested in implementing) may exist.

2.6.5 SNMP Application Components

The Simple Network Management Protocol is the protocol that is used to send management information as is defined in MIB modules. This information is exchanged between SNMP entities. SNMP entities are traditionally referred to as SNMP *managers* and SNMP *agents*. However, in the current IETF SNMP architecture as defined in RFC 2571, managers and agents have been generalized into SNMP entities. The architecture of an SNMP entity is shown in Figure 2.9. All entities are comprised of an engine and applications. An SNMP engine provides services for sending and receiving messages (Dispatcher and Message Processing System), authenticating and encrypting messages (Security Subsystem), and controlling access to managed objects (Access Control Subsystem). There is a one-to-one association between an SNMP engine and the SNMP entity that contains it. All SNMP entities also contain SNMP applications. There are several types of SNMP applications, enumerated in Table 2.5.

Table 2.5 SNMP application components.

SNMP application	Description
Command Generator	Typically resides in an NMS or similar device.
Command Responder	Typically resides in a managed device such as a router.
Notification Originator	Typically resides in a managed device such as a router.
Notification Receiver	Typically resides in an NMS or similar device.
Proxy Forwarder	Translates between various SNMP versions.
Other	Other application types defined in the future.

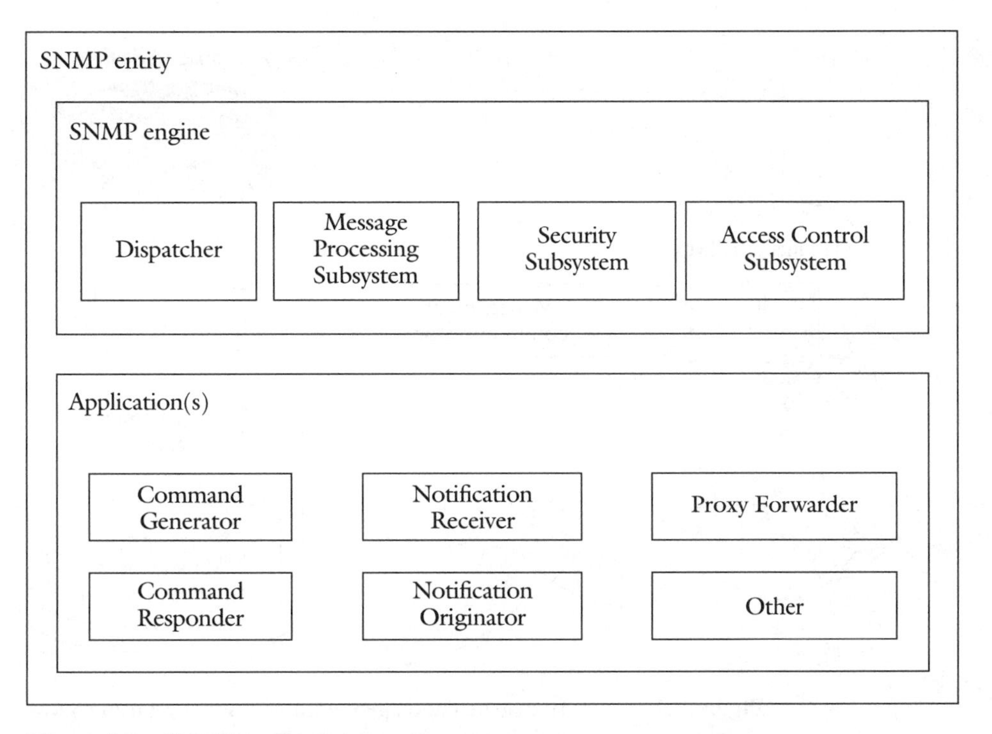

Figure 2.9 SNMP application components.

Previous versions of SNMP did not allow a Command Generator to reside within a managed device. Similarly, Notification Originators were not allowed to reside within a management station. However, the new architecture as defined in RFC 2571 does not make that distinction and instead only refers to an SNMP entity. This obviates the need for referring to the different components as an SNMP manager, agent, or a dual-role entity.

In Figure 2.10 an agent and a manager are shown conversing using the Simple Network Management Protocol. The agent maintains a database of managed objects in its MIB. When the agent accepts a request from the manager, it processes the requests using its Security and Access Control Systems. If all tests performed by these systems are successful, then the request is made to its database of managed objects to access the information specified in the request. Once the information is retrieved, it is encoded in a response message that is returned to the manager entity.

Dual-Role Entity

Until now, we have discussed the interaction between the manager and the agent as one of a manager requesting objects or modifying their values by sending protocol operations to an agent. However, a variation of this model of interaction is possible

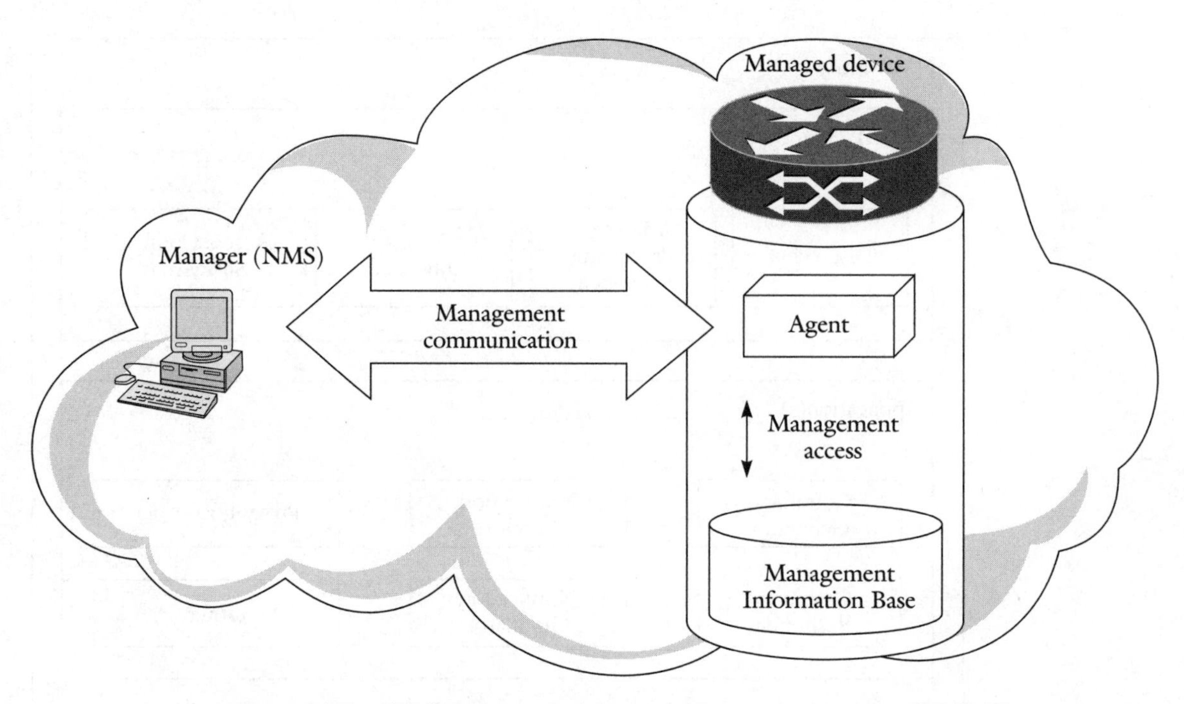

Figure 2.10 SNMP manager and agent entities conversing using SNMP. The agent maintains a database of managed information that a manager accesses to manage the device on which the agent resides.

that will allow an entity to act as a translator between SNMP and another protocol, including a different version of SNMP. Recall the scenario previously in the CORBA section (see Section 2.3.1) that explained how SNMP messages could be translated between SNMP and CORBA. This not only included protocol retrieval or modification messages, but also included SNMP notification conversion. For this configuration to work correctly, the entity responsible for translating between CORBA and SNMP needs to be implemented as an entity with a proxy forwarder application. Such an entity is commonly referred to as a *dual-role entity* or a *midlevel manager*.

2.6.6 SNMP Notifications

Network devices often maintain system software that is capable of operating in an asynchronous manner. Node software can sometimes be programmed to raise asynchronous alarms that are intended to alert the system's operator of some interesting condition or event. SNMP provides a means by which agents are able to issue asynchronous messages to managers (or midlevel managers). These messages are called

SNMP notifications. SNMP notifications are defined in a MIB module with the NOTFICATION-TYPE macro. Notifications can be sent from a notification originator to a notification receiver using one of two mechanisms: either a TRAP sent (a TRAPv1 or a TRAPv2, depending on the version of the Protocol Operations being used), or an INFORM (only available with version 2 of the Protocol Operations).

The notification message contains one or more pairs of OIDs and values. Each pair consists of an OID and a corresponding value that is informative for the notification the object is contained in. The Trap version of a notification message is sent either in a reliable (INFORM) or unreliable (TRAPv2) manner. Therefore, reception of a notification is not always guaranteed. SNMPv2 added a reliable notification called an INFORM. An INFORM notification contains similar semantics to the notification except that the agent continues to attempt delivery of the INFORM message until it receives an acknowledgment from the manager that it has received it (or it times out). The trap or inform destinations are specified either directly on a device via its CLI, or using SNMP's RFC 2573 MIB module. Retry counts and timeouts for INFORMs are specified in those MIB modules.

Special care should be taken when using traps to ensure that a manager is generally capable of both catching them and then reacting in a reasonable amount of time to those messages if necessary. The transmission of notifications in large amounts can actually exacerbate a failure condition by either overloading the network between the device entity and the manager entity, or overloading the manager entity such that it cannot take appropriate corrective action because it is busy processing notifications.

2.6.7 SNMP Security

SNMP provides varying degrees of security and security features depending largely on the version of SNMP used. SNMPv1 provides a very weak form of security called the ***community-based security model.*** In this model, a clear-text phrase is associated with certain access privileges. Unfortunately, this level of security has proven to be very weak because most implementations use the same phrases for access (e.g., "public" is used for read-only access, while "private" is often used for write access). When the IETF redesigned SNMPv1, one of the major hurdles to the new version being standardized was a new model of security. Unfortunately, a compromise could not be reached, so several versions of SNMP were released. At this time, the View Based Access Model (VACM) offered a stronger means of configuring access control than the community-based model of SNMPv1. The VACM allows for more constrained access to objects by allowing the per-object access policy to be configured. However, since this mechanism still used the community-based pass phrase token for authentication, it could still be easily compromised. When

SNMPv3 was standardized, one of the most significant new features it provided was a very robust and comprehensive security model. SNMPv3 security provides approaches for encrypting the SNMP message, replay detection/protection, and antispoofing mechanisms. These, coupled with the VACM, provide a very comprehensive security model that can be used as an effective means by which operators can allow full write access to their systems with the confidence that they are secure.

2.6.8 SNMP Transport Protocols

It is mandatory that an SNMP entity supports and implements SNMP over UDP/IP. However, SNMP can certainly be run over a variety of other network transport protocols. SNMP has been successfully run over IP/TCP, XML/HTTP, and IPX. It is difficult to say specifically whether one network transport is necessarily better than another, but it is clear that SNMP over UDP/IP is the most prevalent mode of operation, due to it being mandatory for compliance to the standard. We suggest that you determine what network and operational requirements exist and choose a transport protocol that best suits these requirements. From our experience, however, we have seen that although a small and dwindling number of vendors may implement some of the non-IP/UDP transports above for SNMP, the trend today is to use IP/UDP or IP/TCP for those requiring reliable transport services.

2.6.9 Protocol Operations

SNMP defines a simple[1] protocol that is used to carry out the interaction between the manager and agent. We will now discuss three of the most widely used operations, although please keep in mind that others such as GET-BULK exist and can be quite useful. These operations are used to access or modify the objects maintained by the agent. We will discuss the GET, GET-NEXT, and SET operations. We should first note that it is not possible to obtain or modify the value of anything other than an instance of an object; thus the GET and SET operations must include the OID of a valid object instance. However, the GET-NEXT operation may act on an invalid OID, as the agent will always attempt to find an object instance whose OID is lexically greater than the one specified.

The GET operation is issued by the manager entity when it wishes to know the value of a specific instance of a managed object. To do so, the manager entity inserts the OID of a specific object instance into an SNMP PDU and sends it to an agent entity. If the agent entity accepts the request, it will fill in the value portion of the

1 Some would argue that the "S" in SNMP does not necessarily stand for *simple*.

request with the value of the object instance and return it to the manager entity. This operation is very straightforward for scalar object instances. However, it is slightly more complicated for tabular object instances. To retrieve the value of an object instance that resides within a table, it is necessary to specify the index of that object instance in such a way as to precisely reference it within the table. Recall that conceptual tables represent rows and columns. The columns in the table represent objects and are specified with one or more indexes.

It is generally not possible to directly retrieve or modify indexes of a table using the GET, SET, or GET-NEXT operations. Columnar indexes defined in MIB modules are required to define their indexes as having a MAX-ACCESS of not-accessible, therefore an agent entity that receives a request for such an object will return an error. This restriction exists because it is not necessary to access the indexes of a table directly. This is because the indexes of a table must be specified in order to specify a unique object instance in the table.

The GET operation is demonstrated in Figure 2.11. The figure shows a simulated OID tree rooted somewhere under the enterprise OID. After a manager issues a GET operation using the OID 1.5.8.1.1.0, the agent accepts the request, queries its object database, and returns the value of this scalar object. All scalar

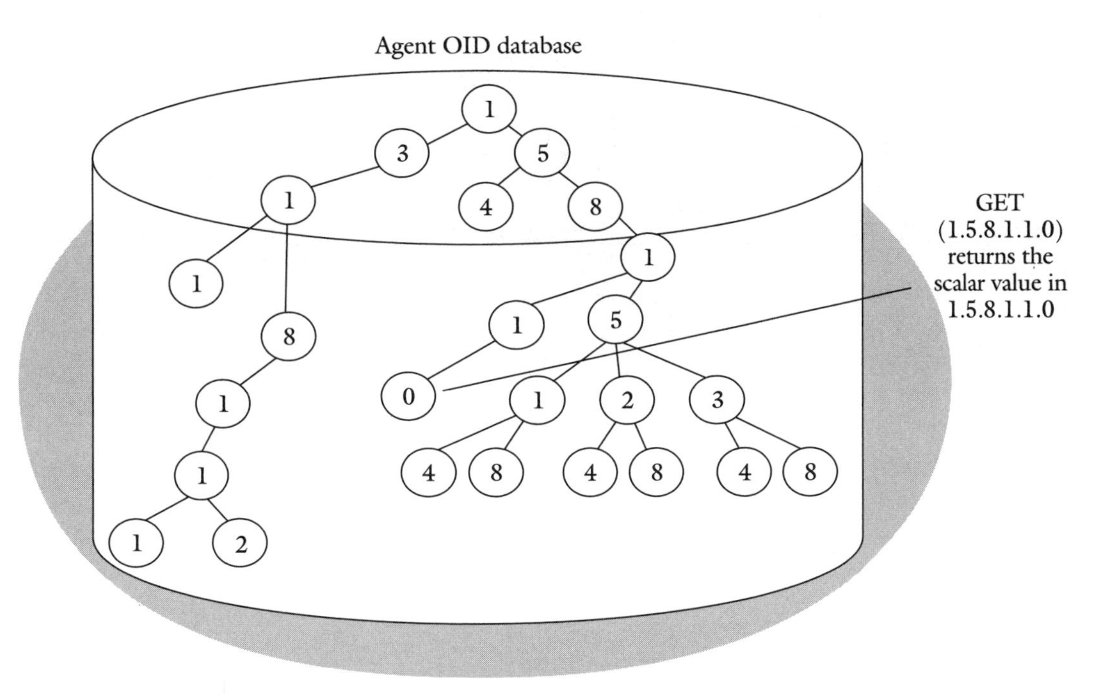

Figure 2.11 Example of an SNMP GET operation.

objects have associated instances that are conventionally identified with instance 0. In this case, the .0 at the end of the 1.5.8.1.1.0 OID specifies the instance of object 1.5.8.1.1. Notice that when the manager issues a GET operation containing the OID 1.5.8.1.5.3, the agent returns an error because 1.5.8.1.5.3 does not represent an instance of any object; it instead represents an actual object. This is demonstrated in Figure 2.12.

The second protocol operation defined is called the SET operation. SET represents the analog to the GET operation: it is used to modify the value of an object instance versus simply retrieving it. To this end, a SET operation must specify both an object instance *and* a value to assign to that instance. When an agent receives a SET request, it first checks the access specifics for that object. If the object is specified with a MAX-ACCESS of read-write (or read-create), the agent will modify the value of that object with the one specified in the request. If, however, write access to the object specified is not allowed, the agent will return an error informing the manager that the requested modification has not taken place.

The last operation we will cover is GET-NEXT. This operation is used to retrieve the value of the next *lexically* greater *object instance* in the OID database. Remember, the GET-NEXT operation still must retrieve the value of the next lexically

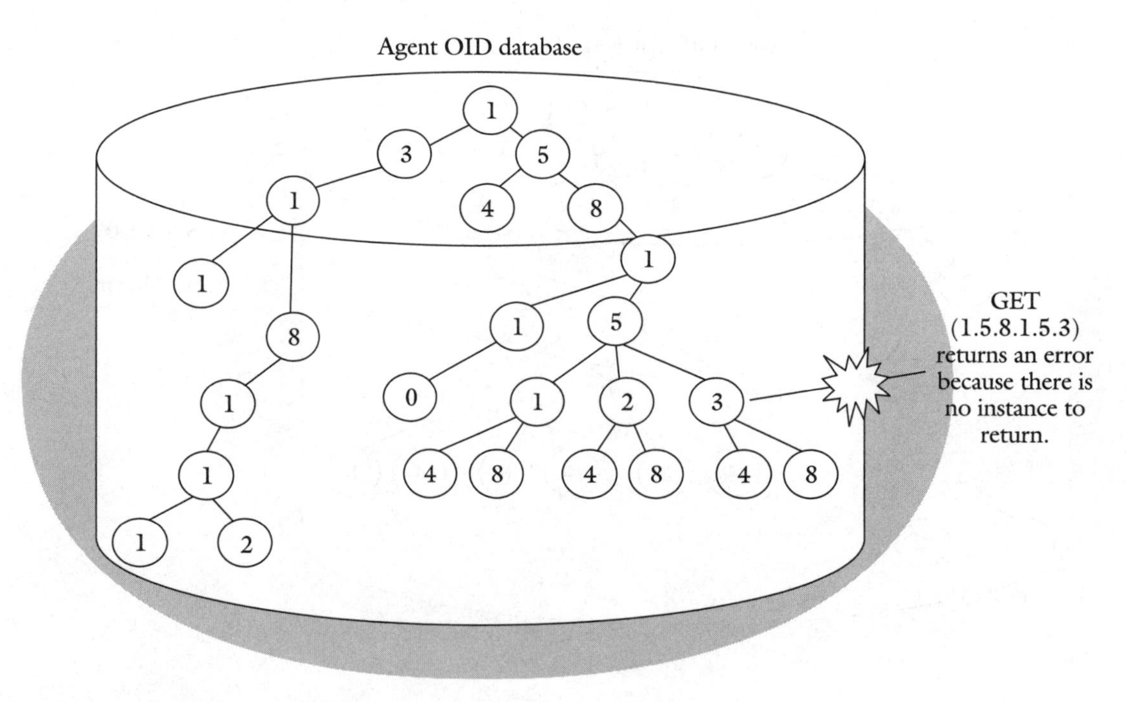

Figure 2.12 Example of invalid GET operation.

greater OID, so it must end up executing a GET operation on a valid object instance. By "lexically greater," it is meant that the object with the next largest OID should be retrieved. To find the next largest value, the agent traverses the OID tree from the point specified in the request using a depth-first search and will return either the next scalar object or instance of a tabular column. For example, using the OID tree shown in Figure 2.13, the object 1.3.1.8.1.1.2 is lexically greater than 1.3.1.8.1.1. Let us examine why. Begin at the root of the OID tree and traverse down it using a depth-first traversal, stopping when you arrive at the second-to-last node in the OID tree. If we do this, we arrive at the .1 node on the far bottom left of the figure (just above the 1 and 2 terminal nodes). At this point, we now need to find the OID of an object instance that is lexically larger than 1.3.1.8.1.1. What this means is that either we need to find the OID of an object instance that has the last number in its OID greater than 1.3.1.8.1.1, or we need to find the first OID with one more part to its OID (i.e., the OID is longer than the one we are using by at least one dotted decimal). In this case, since 1.3.1.8.1.1 possesses six parts to its OID, and since we have exhausted all of the available OIDs at this ply in the OID

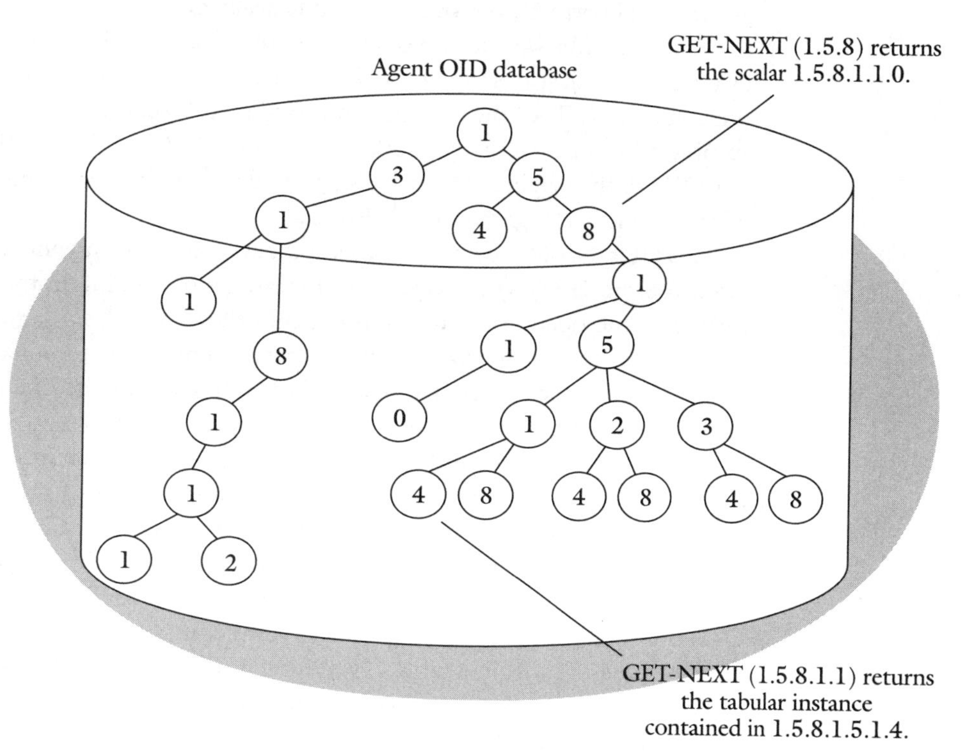

Figure 2.13 Examples of the GET-NEXT operation on both tabular and scalar objects.

tree, we need one with seven. Since no other nodes exist at this level, we will not be able to find an OID with a prefix of 1.3.1.8.1 that has six parts to its OID. Therefore, we must descend one more ply in the OID tree. If we do this, we find that the next level down contains two leaf nodes: 1 and 2. Appending either to the prefix of 1.3.1.8.1.1 will result in a lexically larger OID; however, the *next* lexically greater OID only results if we pick the first OID available—1. If we then choose this node and attach it to the current prefix of 1.3.1.8.1.1, we correctly arrive at the next lexically larger OID of 1.3.1.8.1.1.1.

Let's investigate a slightly more difficult OID. Let's examine how we discover the next lexically larger object instance OID to OID 1.5.4. If we begin with the depth-first search algorithm starting at node 1 in Figure 2.13 and traverse the tree down to node 1.5.4, we will first notice that we have run out of levels in the OID tree below this point. When this happens, we must use the same steps that the depth-first search algorithm would use. In this case, the algorithm would back up the tree until it found a place where it could descend down again. When the algorithm backs up one level in the tree, it will then try to descend down to a terminal leaf node. Since leaf nodes in the OID tree always represent object instances, the algorithm will correctly discover the next lexically greater instance. To continue with the example, after we back up one level in the tree, we will continue down .8.1.1.0 using the rules just described. We have arrived at a terminal leaf node. Examining the complete OID of the path just traversed reveals that we have found an OID for an object instance (actually a scalar object instance) of 1.5.8.1.1.0. If we compare this OID to the original one of 1.5.4, we find that we have satisfied all of the criteria of the next lexically greater OID. Eureka!

This GET-NEXT operation is useful for several tasks. In general, a manager will often use the GET-NEXT operation repeatedly by feeding in the OID that is returned by the operation into a subsequent call to "walk" some subset of the OID tree, or even the entire tree. The walk operation might be terminated when an object instance is encountered whose OID length exceeds the one used to start the operation. It might also be terminated when the OID of a valid object instance whose last dotted decimal that is greater than the one given is found. Some commercial SNMP toolkits provide tools with "walk" as part of their name, and they behave this way. Using these tools, several operations are available to the operator. First, it is possible to walk the entire OID tree contained in an agent by specifying 1.1 as the point of the walk. Second, it is possible for a manager to specify the "top" OID of a table and use this to walk all columnar instances, or even all rows and columnar instances within a table. This is often useful if either the manager does not know the indexes of the instances within a particular table, or the snapshot stored by the manager has grown stale.

In addition to the examples just shown for GET-NEXT, two examples of the GET-NEXT operation are illustrated in Figure 2.13. In the first example, the GET-NEXT operation is executed on OID 1.5.8. Using the algorithm we described earlier for locating the next lexically greater object instance reveals that this OID is 1.5.8.1.1.0. The value returned for this object is the value of the scalar object represented by the object instance. The type of this object is specified by the object 1.5.8.1.1. The second example attempts a GET-NEXT operation on OID 1.5.8.1.1. Using the algorithm described earlier again, we arrive at the OID 1.5.8.1.5.1.4. This OID represents an instance of a tabular object. The type of the value returned depends on the type of the columnar object.

Protocol Operation Versions

At the protocol level, SNMP has been specified as two versions. That is, two protocol specifications exist for the Protocol Data Units that are transported within SNMP messages. Version 1 of the Protocol Operations as specified in RFC 1157 specifies the version 1 Protocol Operations (PDU types) that can be used. These operations include GET, GET-NEXT, GET-RESPONSE, SET, and TRAPv1. It also specifies the remaining parts of the format of a PDU such as the request-id, the error-status and error-index, and the variable-bindings.

Version 2 of the Protocol Operations is specified in RFC 1905. This document is currently a draft standard, but a revision has been approved as full standard. This document specifies the version 2 Protocol Operations (PDU types) that can be used. These operations include GET, GET-NEXT, GET-RESPONSE, GET-BULKSET, INFORM, TRAPv2, and REPORT. It also specifies the remainder of the PDU format of a PDU that includes things such as the request-id, the error-status and error-index, and the variable-bindings. It is important to note that as of the second version of the Protocol Operations, the TRAPv1 no longer exists. The GET-BULK, INFORM, TRAPv2, and REPORT that are introduced in the specification as new PDU types represent what are essentially new SNMP protocol operations. The error-status value that is contained in the new message format has a much more detailed list of possible errors, which can be used to further isolate errors or recover from them. Lastly, the variable-bindings can now contain special values to indicate exceptions such as noSuchObject, noSuchInstance, and endOfMibView, which further enhance the functionality of SNMP.

2.6.10 SNMPv1, SNMPv2c, and SNMPv3

Several versions of SNMP have existed since the protocol's inception in the late 1980s. The original SNMPv1 was used until it was realized that it had many

SNMP message header	SNMP Protocol Data Unit (PDU)

Figure 2.14 Basic SNMP message format.

shortcomings. In particular, additional protocol operations were necessary as well as additional security features. It was at that time that work on SNMPv2 began. During that time, there were various iterations of SNMPv2. All but one of these intermediate versions has failed. The IETF has declared (or soon will) these versions as HISTORIC standards. The only version that is still used in the field is SNMPv2c as defined in RFC 1901. This version is an EXPERIMENTAL protocol as far as the IETF is concerned, and it too will soon be reclassified as HISTORIC. Since then, SNMPv3 has been approved as a full IETF standard.

When management applications and managed devices exchange management information, they do so by exchanging SNMP messages. The basic format of an SNMP message is shown in Figure 2.14. This basic format consists of an SNMP message header and an SNMP PDU.

Since three versions of the SNMP protocol exist, three versions of SNMP messages are also possible. The format of the SNMPv1 message version is described in the now full standard RFC 1157. An SNMPv1 message is composed of an SNMPv1 message header that includes a field to indicate it is an SNMPv1 message, a community string (a clear-text password that is included in the message), and the version 1 Protocol Operation (PDU). Version 1 Protocol Operations (PDUs) can only contain SMIv1 data types in the variable-bindings.

The format of the SNMPv2c message is described in RFC 1901. An SNMPv2c message is composed of an SNMPv2 message header (very similar to an SNMPv1 message header) that includes a field that indicates it is an SNMPv2c message, a community string, and a version 2 Protocol Operation (PDU). Version 2 Protocol Operations (PDUs) can only contain SMIv2 data types in the variable-bindings, plus three exceptions: noSuchObject, noSuchInstance, and endOfMibView. It is important to note that SNMPv2c messages have a very similar header as SNMPv1 messages, and as such, they are as (in)secure as SNMPv1 messages. The big difference is that an SNMPv2c message must carry a version 2 PDU and thus the data in such PDUs must be one of the SMIv2 data types.

The format of the SNMPv3 message is described in RFC 2572. This RFC is a draft standard, but a revision has been approved as full standard. An SNMPv3 message is composed of an SNMPv3 message header. This message header is much more extensive than the message headers defined in the earlier versions of the protocol, but does still carry a version 2 PDU, which implies that these PDUs can only contain SMIv2 data types. The SNMPv3 message header contains several new message fields. Specifically, the new header includes a field that indicates that it is an SNMPv3 message (instead of SNMPv1 or SNMPv2c), a message ID, the maximum message size, and other message flags. The header also contains a field indicating the security model in use. This field contains additional security-related fields that depend on the security model specified therein. For instance, if the User-based Security Model (USM) is specified, then the authoritative engineID, engineBoots, engineTime, username, authenticationParameters (MAC code), and privacyParameters are also included. This added header information provides for a high level of security for SNMPv3 messages.

To summarize, the difference between SNMPv1 and SNMPv2c message types is that the SNMPv1 message carries version 1 PDUs and thus SMIv1 data types, and an SNMPv2c message carries version 2 PDUs that use SMIv2 data types. SNMPv1 and SNMPv2c are similar in that they both use community-based security (i.e., plain-text passwords/pass phrases) and so both are equally insecure. The message formats of SNMPv2c and SNMPv3 differ in that SNMPv3 messages allow secure SNMP message exchanges, while SNMPv2 messages do not. SNMPv2c and SNMPv3 both carry version 2 PDUs that use SMIv2 data types. For the reader that wishes to pursue this topic further, we suggest reading RFC 2576. RFC 2576 explains the coexistence of the different SNMP versions and how to map from one to the other. This documentation is quite useful for those wishing to migrate from one version to the other.

2.7 Summary

This chapter introduced the concept of management interfaces. Management interfaces are useful because they provide a well-known and -understood method of both modeling managed objects within a device, as well as a protocol for accessing these managed objects. The differences between many of the management interfaces we discussed are largely a matter of object definition completeness using data modeling language used for that management interface. Other differences exist, however, such as security, efficiency under certain circumstances, and portability. Of course, with any technology that is deployed in the marketplace, the most important differences between it and its competitors are the *perceived* ones. These perceptions are largely related to cost and performance effectiveness of the interface. Unfortunately, in many cases, these perceptions are not grounded in technical reasoning, but rather on the marketing literature from one corporation or another.

The chapter began with a general introduction to management interfaces and what the advantages are to standards-based and proprietary versions. The discussion then focused on specific approaches to management interfaces. The first management interface discussed was the ever-ubiquitous command-line interface (CLI). We investigated how CLIs could be managed using various scripting languages, and how this was preferable to an operator accessing each device personally. We also discussed how this interface was the most pervasive in the industry today despite its shortcomings. We explained how the widespread use of the CLI, as the preferred management interface from device vendors, was not such a great achievement from the perspective of the network operators who operate a heterogeneous network. Having to manage multiple CLI languages and perhaps different data models for each type of device in a network is expensive and wasteful of resources.

Next, we investigated the Common Object Request Broker Architecture (CORBA). We gave an overview of the CORBA technology and then discussed how CORBA could be used to build management applications as well as agents in managed devices using CORBA ORBs. We investigated how you might translate between CORBA and another management interface. In particular, we showed how a CORBA translator could act as an SNMP midlevel manager and translate SNMP requests and notifications to and from CORBA, and how this would be advantageous for existing CORBA management systems that might be in use today without native SNMP support.

We then delved into a discussion of the eXtensible Markup Language (XML). We first investigated the basic definitions and properties of XML, in particular, how XML describes a class of data objects called *XML documents* and specifies the behavior of applications that are used to process these documents. We showed how data is embedded within XML documents as strings of text and is surrounded by special markers that describe the data so that it can be parsed and displayed, and how these delineated strings were called *XML elements.* One of the greatest advantages of XML documents is that the markup used in an XML document not only describes the document's structure, but also allows you to determine which elements are associated with one another. Furthermore, XML documents are written using plain ASCII text that can be read and written by a wide variety of word processors (including the popular one used to write this book). We showed how XML could be used as a general RPC mechanism, and potentially as a management protocol by building features on top of it. We then investigated how XML could be used as a general encapsulation of managed data stored natively in a variety of formats and, specifically, how CLI text could be encapsulated in XML for easy parsing and display by management stations. In addition, XML can be used to delineate SNMP data, both using HTTP as a transport and within bulk file transports.

Next, we discussed many forms of bulk file transfer. Bulk file transfer is a popular mechanism for off-loading large volumes of data from managed devices. Once transferred to a management station, it can be processed offline at the manager's convenience. This is important for management applications such as offline traffic engineering calculations that require large volumes of data with a high degree of integrity. That is, if the same offline traffic engineering application had fetched the same managed objects over the network as the data was available, it might have missed some of the data either due to network conditions causing the responses to be lost, or because the counters on the device changed too quickly. We also discussed the issues surrounding the format of these files, and how it could make a difference for an operator with a heterogeneous network.

Finally, we discussed the Simple Network Management Protocol (SNMP). We discussed the basic components of SNMP, which included the SMI, MIB modules, and the protocol operations. We first investigated the details of the SMI and why it is so important to the definition of a MIB module. We also explained the differences and similarities between the various versions of the SMI. We then defined a MIB module as containing object definitions using the SMI. MIB modules were collectively part of the larger MIB that constituted a data model for a device. The key elements of MIBs were discussed, including objects, object instances, tables, and tabular indexing. Next, we introduced and later demonstrated the various key protocol operations provided by SNMP. Specifically, we gave an example of how some of the protocol operations could be used to retrieve instances of managed objects from an entity's MIB. In particular, we gave detailed examples of how the GET, SET, and GET-NEXT operations would function under certain circumstances. We ended the section with an overview of the various versions of SNMP.

It is clear that management interfaces play an important part in the overall management of any network. Given this, you must weigh the relative benefits and weaknesses of each approach within the context of your network deployment to determine which interface or interfaces to deploy. We hope that our introduction has given you an even-handed look at many of the options for management interfaces. In many cases, one or more of the interfaces introduced in this chapter can be applied to an MPLS-enabled network. If you are a network operator, the question is how much work do you want to do to use a management interface? What are the benefits given your specific style of network operations management? If you are a device vendor, you must cater to the needs of your customers, who are in large part network operators if you are selling MPLS-enabled equipment. This probably means that you must implement at least two of the management interfaces described above—and sometimes more.

Further Reading

To find out more about the IETF, visit their Web page at *www.ietf.org/*.
For more information about IANA, check out their Web site at *www.iana.org/*.
To locate information about the ITU: *www.itu.int/*.

To locate information about the MPLS Forum: *www.mplsforum.org/*.

To locate additional information about the OIF: *www.oiforum.com/*.

Bray, T., J. Paoli, C. M. Sperberg-McQueen, and E. Maler. *Extensible Markup Language (XML) 1.0* (second edition). W3C Recommendation. October 2000. *www.w3.org/TR/2000/REC-xml-20001006.pdf*.

The Object Management Group maintains the standards for CORBA and can be located at *www.omg.org/*.

Arun Viswanathan

is Director of Protocol Development at Force10 Networks, Milpitas, California. Three years back, he joined Force10 Networks, where he provides leadership in development of protocols for a switch/router product. Prior to Force10 he was a distinguished member of the technical staff at Lucent Technologies, Holmdel, New Jersey. Earlier he worked at IBM in the Advanced Networking Laboratory in Hawthorne, New York. At IBM he co-invented the Aggregate Route-based IP Switching (ARIS) protocol, a precursor to MPLS, and worked on the NSFNET Milford router. He co-authored the MPLS architecture RFC 3031. He has several other MPLS Working Group drafts progressing toward standards. He holds an M.S. in mathematics and computer science.

Arun, having been an active member of the IETF in the early days of MPLS, how have the climate and activities of the standards body changed since then? Also, having been one of the people who got MPLS to be the successful technology that it is today, what do you see in the future of MPLS? In particular, what do you see as the future of MPLS applications?

The IETF started work on MPLS around the end of 1996. Since then significant progress has been made; several key protocol components have been standardized, making it possible to realize a true multivendor interoperable MPLS deployment. The scope of MPLS effort has certainly broadened since the initial days. In recent years, the Working Group has undertaken standardization of several new MPLS applications, ranging from virtual private networks, to TDM networks, to optical networks. Though early adoption in terms of deployment of the technology was slow, MPLS applications are now widely deployed, some in several large service provider networks. However, ratification of this technology has come about slowly. This we owe largely to MPLS's potential for being an omnibus protocol. This is quite evident from the current scope of the protocol. However, MPLS continues to be a promising technology for a multitude of applications, especially MPLS as a transport technology for the various VPN schemes. The scope being so wide today, it's even more critical that IETF ratify the various protocol components quickly to smoothen the deployment and operation of this neoteric technology.

Finally, as one of the original co-authors of the base MPLS MIBs, you have a keen insight into how these MIBs were formed and have progressed to the state that they

are today. What have been some of the challenges that the MIBs have undergone, and do you think that they are up to the task of being used in real production networks?

As MPLS standardization progressed it gathered complexity. It became quite evident that for a successful deployment and operation of an MPLS network a management application was necessary. Therefore, when I started MPLS protocol implementation I also designed the MIBs to enable management applications. That's how the MPLS MIBs came into being, part of an MPLS protocol implementation activity. The biggest challenge in designing the MIBs was in making them generic enough to suit any implementation and in covering all protocol aspects specified in the various drafts. The other challenge was in partitioning the MIB functionality properly. MPLS has multiple signaling protocols; therefore, the base LSR MIB had to be designed such that it would serve all the various protocol needs satisfactorily. The other aspect that took quite a bit of work was making the various conformance groups meet the needs of different vendor implementations. Managing all these requirements have added to the time it's taken to progress these MIBs toward ratification.

I think the MIBs are up to the task of being used in real networks. That's because they were developed off inputs from various vendors having deployed MPLS products. I believe these MIBs are already in use in production networks. Besides, MPLS is still a relatively new technology, and to find skilled and experienced manpower in a new technology area is difficult; a good management application will certainly ease the job.

3

The MPLS Label Switching Router Management Information Base (MPLS-LSR MIB)

> "See the world as it is, not how you want it to look."
>
> —John Chambers, CEO, Cisco Systems

Introduction

In this chapter, we examine the first of many standard tools used to manage MPLS networks. The MPLS Label Switching Router MIB (MPLS-LSR MIB) lies at the conceptual heart of the MPLS management framework and the standard MPLS MIBs. Accordingly, the MIB is also the basis of many management applications that are based on the MPLS MIBs. In short, this MIB provides the user with a snapshot of what the label forwarding (switching) database looks like at any moment in time. The label forwarding database is stored in the MPLS Label Forwarding Information Base (LFIB). This chapter will explain in detail the operation and usefulness of this MIB and present examples and tips for how it should be implemented by vendors, as well as how it should be utilized by operators.

This chapter will focus on the draft version 08 of the MPLS-LSR MIB, which is typically referred to as draft-ietf-mpls-lsr-mib-08.txt. This draft version may have been updated or replaced with an IETF RFC document after this book was published. Please keep this in mind when searching for the document on which this chapter is based.

3.1 Who Should Use It

The basic function of the MPLS-LSR MIB is to expose the active MPLS label switching of an LSR, as well as to allow for the configuration of some things such as static label mappings. Thus, all network managers wishing to monitor the basic label forwarding activities of a label switching router should monitor the tables provided in this MIB. Similarly, device vendors whose products are required to provide such information should implement this MIB. In doing so, both devices and network managers will have a common understanding of what should be provided by devices as well as what management stations should expect to manage.

Management stations wishing to monitor the behavior of other MPLS applications such as traffic engineering (see Chapter 8) or virtual private networks (see Chapter 11) should utilize the objects provided by this MIB in conjunction with the MIBs designed specifically for those applications. In this way, the basic label switching can be managed using the objects defined in this MIB, since they apply to all MPLS applications, and additional functions can be built around the MPLS-LSR MIB and/or separately. We should point out that although the MPLS-LSR MIB is not necessarily required to manage the other applications of MPLS (e.g., traffic engineering), it is certainly useful if a complete management picture of a label switching router is desired.

3.2 MPLS-LSR MIB at a Glance

The MPLS-LSR MIB contains several tables, including the MPLS Interface Configuration Table (mplsInterfaceConfTable), Interface Performance Table (mplsInterfacePerfTable), InSegment Table (mplsInSegmentTable), InSegment Performance Table (mplsInSegmentPerfTable), OutSegment Table (mplsOutSegmentTable), OutSegment Performance Table (mplsOutSegmentPerfTable), Cross-Connect Table (mplsXCTable), Label Stack Table (mplsLabelStackTable), and the Traffic Parameter Table (mplsTrafficParamTable). These tables interact together in a manner that provides a coherent view of the MPLS label switching router's label switching activity.

As is the case with many of the MIBs that you will encounter, it is often useful to first sit down at a board and illustrate each table contained in that MIB at a high level to show how each interacts with the others. It may also be useful to include the indexing of each of the tables. This may sometimes make some sense of how each table fits with the others in the MIB, as well as what the MIB's designers had in mind (or didn't!).

Figure 3.1 illustrates, at a high level, the interaction of the major tables found in the MPLS-LSR MIB, including the mplsInSegmentTable, mplsOutSegmentTable, mplsXCTable, and mplsInterfaceConfTable. The indexing of each table is shown

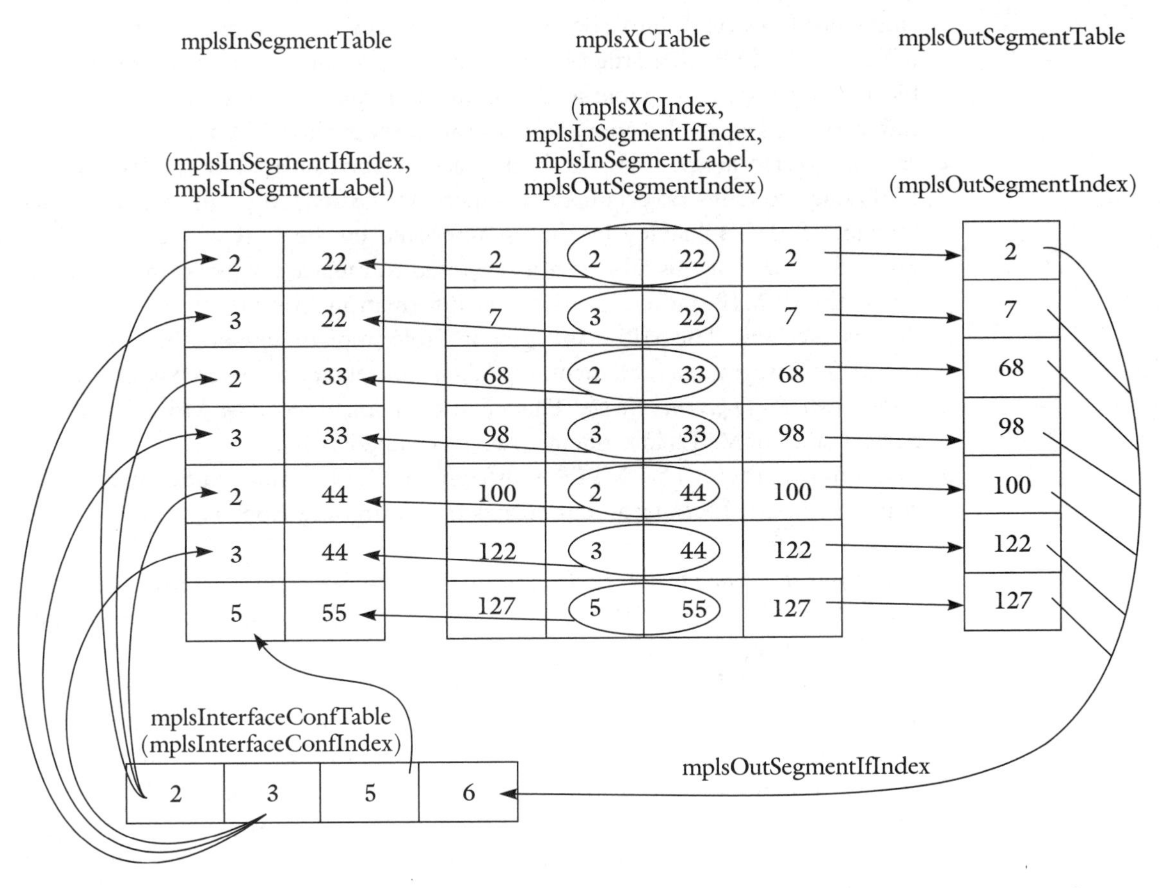

Figure 3.1 Relationship between major tables in the MPLS-LSR MIB. The indexes for each table are shown to indicate the relationships clearly.

with each table, as well as how this indexing relates to that of the other tables. In particular, the example shows how a fictitious LSR that has four interfaces (2, 3, 5, 6) is actively swapping labels from incoming interfaces 2, 3, and 5 to outgoing interface 6. Note that the relationship between the mplsOutSegmentTable entries and the outgoing interfaces as found in the mplsInterfaceConfTable are not explicitly implemented as indexes in the mplsOutSegmentTable; rather, they are referenced using that table's mplsOutSegmentIfIndex object. The pointers from the entries in the mplsOutSegmentTable in the figure have this relationship set up implicitly. Also, note that the label assignment shown here is arbitrary and only for the purposes of the example. We will show more complex and detailed examples of how these entries are configured and associated later in the chapter.

The following list is a summary of tables contained in the MPLS-LSR MIB.

- *MPLS Interface Configuration Table:* The MPLS Interface Configuration Table (mplsInterfaceConfTable) contains the interfaces that are enabled to support MPLS on the LSR where the MIB is queried. LSRs must create entries in this table for every MPLS-capable interface and indicate other interface-specific parameters. This entry also corresponds to an entry in the IF-MIB.
- *Interface Performance Table:* The Interface Performance Table (mplsInterfacePerfTable) contains objects used to reflect MPLS-related performance characteristics of MPLS-enabled interfaces supported by the LSR where the MIB is queried. Note that this table is not supposed to replace the performance counters in the IF-MIB; rather, it adds to the information found there.
- *InSegment Table:* The MPLS InSegment Table (mplsInSegmentTable) contains the MPLS insegments (i.e., incoming labels) and their associated parameters.
- *InSegment Performance Table:* This table contains objects used to measure the performance of MPLS insegments (i.e., incoming labels).
- *OutSegment Table:* The MPLS OutSegment Table (mplsOutSegmentTable) contains the MPLS outsegment entries (i.e., outgoing labels) and their associated parameters.
- *OutSegment Performance Table:* The MPLS OutSegment Performance Table (mplsOutSegmentPerfTable) contains objects used to measure the performance of MPLS outsegments (i.e., outgoing labels).
- *Cross-Connect Table:* The MPLS Cross-Connect Table (mplsXCTable) contains associations between in- and outsegments. When one or more insegments is combined with one or more outsegments, this notes that the LSR on which the MIB is viewed has been instructed to switch between the specified segments. This also indicates that an LSP has been constructed to support this configuration. Cross-connects may be administratively disabled using the associated administrative status (if supported). The associated operational status object indicates the actual status of the LSP at this LSR.

- *Label Stack Table:* This table contains a representation of the additional label stack imposed at the LSR where the MIB is queried. Specifically, this table contains the additional label stack entries that are replaced just under the topmost label on any labeled packet received on the associated LSP. Note that the topmost label is not found in this table.
- *Traffic Parameter Table:* The Traffic Parameter Table represents some traffic parameters that are commonly associated with an LSP.

3.3 Labels In, Labels Out

Let us preface the remainder of the discussion in this chapter with a simple explanation of the Label Forwarding Information Base. Put simply, an MPLS LSR has one basic function: to switch labeled (or unlabeled at the ingress) packets from one interface to another. This is illustrated in Figure 3.2, which shows graphically how incoming labels are associated with outgoing labels via a cross-connect object. Each label is associated with one interface. If a label is associated with more than one interface, it must be listed multiple times in the MIB, but may be associated with the same cross-connect object if it makes sense to do so.

Due to the fundamental nature of the MPLS label switching operation, it is important to expose this basic yet critical functionality to the operators not only so that they are able to examine the basic functionality of the LSR, but also as a means for debugging such problems as incorrectly routed (labeled) packets. The MPLS-LSR MIB provides the user with a snapshot of the label forwarding (switching) database at the time it is examined. The LSR's Label Forwarding Information Base (LFIB) is a proper subset of the MPLS Label Information Base (LIB). The LFIB contains the labels that are being actively used to forward/switch traffic, while the LIB contains the set of all possible valid MPLS labels on the specified router. It may

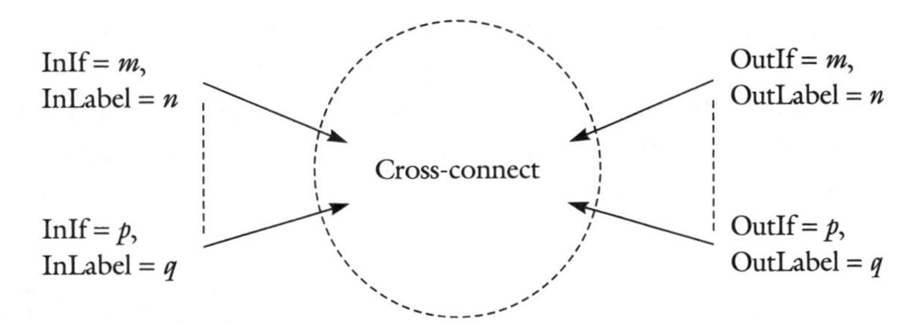

Figure 3.2 The MPLS-LSR MIB represents the Label Forwarding Information Base (LFIB) by associating incoming label and interface pairings with outgoing label and interface pairings.

be the case that all available labels are being used to switch traffic; thus the LFIB is a proper subset of the LIB.

The MPLS-LSR MIB typically displays labels that found their way into the LSR via one of the signaling protocols. The MPLS-LSR MIB can also be used to configure label bindings to form static LSPs without the assistance of a signaling protocol. This allows a user to preconfigure label allocations in a static fashion if such a requirement exists. For example, low-cost edge LSRs that serve one or more customers can be preconfigured with static LSPs to route their traffic into the MPLS core network, much in the same way that a static "default" IP route can be used to forward all of a user's traffic toward the core network. It should be noted that there are many drawbacks of using such an approach if not used correctly. These drawbacks include implementation difficulty related to the restoration of the static label bindings after a system reboot. Static label allocation must also be coordinated with any signaling protocols that might also use labels, since reuse of labels might result in misrouting of user traffic.

It is also important to mention that the MPLS-LSR MIB's cross-connection paradigm has the interesting property of allowing multiple LSPs to potentially share the same resources. This can be indicated by having the labels for both directions of the LSPs share the same cross-connect object. Although not mandatory, this can indicate resource sharing. This might be useful, for example, if two LSPs wish to carry traffic for what is, in reality, a bidirectional connection. In this case, it might be preferable to treat both directions of the path (i.e., both LSPs) in the same way (i.e., same queuing discipline) at each LSR. Another example might be a series of LSPs that travel between the same pair of LSRs. In this case, it might be desirable to group these LSPs together to share resources based on the source or destination of the LSPs. Despite its potential usefulness to indicate resource sharing, we strongly recommend that you instead use the mplsTrafficParamTable in much the same way as was just described. That is, point multiple mplsXCEntries at the same entry in the mplsTrafficParamTable to indicate resource sharing.

The LFIB is configured as the 4-tuple shown in Figure 3.3. The LFIB 4-tuple is read from left to right and is interpreted to mean that if a label is received on the specified interface, the label is replaced with the new one specified on the corresponding interface. It should be noted that some special cases exist, primarily at the ingress and egress, but also for load sharing. At the ingress, label imposition occurs, which results in the MPLS header being applied to an unlabeled packet. In this

Incoming interface	Incoming label	Outgoing label	Outgoing interface

Figure 3.3 Basic MPLS label switching 4-tuple.

case, the incoming label will not appear in this table since there is no label to receive; only an outgoing label will appear. In this case, the 4-tuple will have an empty "left-hand side"—incoming label and interface. When non-MPLS traffic is mapped into MPLS, the MPLS-FTN MIB can be used to show the missing left-hand side of the expression (i.e., the FEC). This consists of an IPv4 or IPv6 destination prefix that is mapped to the beginning of an LSP (i.e., outsegment and outgoing interface). The MPLS-FTN MIB is discussed further in Chapter 5.

At the egress of an LSP (or the LSP's penultimate hop) the outgoing label and interface will not appear in the 4-tuple since the MPLS header is stripped and no outgoing label will appear in the packet. However, the outgoing interface must be present in the LSR's LFIB, since the packet must still be transmitted to its ultimate hop, albeit lacking an MPLS shim header from that point on.

3.3.1 Load Sharing

When multiple paths exist through the MPLS domain to the same destination LSR, the forwarding LSR may choose to distribute traffic flows toward that destination across these paths in an effort to more fully utilize the resources available between the two LSRs. These paths may or may not represent equal-cost paths to the same destination. By sending some or all of its traffic across all possible equal-cost paths, an LSR can maximize more of the aggregate bandwidth available in a network between itself and the MPLS destination. For example, if two parallel paths exist that may reach the same destination LSR, the forwarding LSR can utilize both paths by sending the traffic from one destination prefix down one path, while sending the traffic for another down the other path. If the LSR does not do this, then essentially the bandwidth of one path goes unused—at least potentially. A parallel path may also include parallel links between two neighboring LSRs. If more than one link exists between two LSRs, it makes sense to send some traffic over one link, while other traffic is sent over other links. Doing so may realize the full bandwidth of all parallel links. In effect, what may be created is a point-to-multipoint LSP between the ingress and egress LSRs.

An LSR that wishes to load-share may forward traffic onto one or more same-cost next-hop paths based on the same destination prefix, based on its understanding of these paths from its routing database. Packets can be forwarded per packet,[1] per source flow, or per destination. The manner by which the traffic is spread across each parallel path is known as a *load-sharing algorithm*. Unfortunately, no standard exists for the load-sharing algorithm, so it differs from vendor to vendor.

1 Per-packet load sharing is dangerous, as it may result in a loss of packet ordering. Traffic flows that are based on TCP that experience a loss of packet ordering may experience degraded performance.

Consequently, the MPLS-LSR MIB does not note which algorithm is in use; we direct you to vendor-specific MIBs for this information.

It is important to note that multiple neighboring (next-hop) LSRs may be available to an LSR on the same interface when the interface is a multiaccess media such as Ethernet. Under this circumstance, the same incoming label may be switched to multiple next-hop neighboring LSRs. This may result in the LFIB having the same incoming label mapped to the same outgoing label, with possibly the same interface (in the multiaccess cases), but with different next-hop neighbor addresses. At first, this may appear strange until you realize what is happening. Simply put, the same outgoing label may be sent to different remote interfaces.

3.3.2 Load-Sharing Examples

The example in Figure 3.4 illustrates how load sharing can be employed by an LSR that might normally choose the same shortest path for each LSP that uses the same

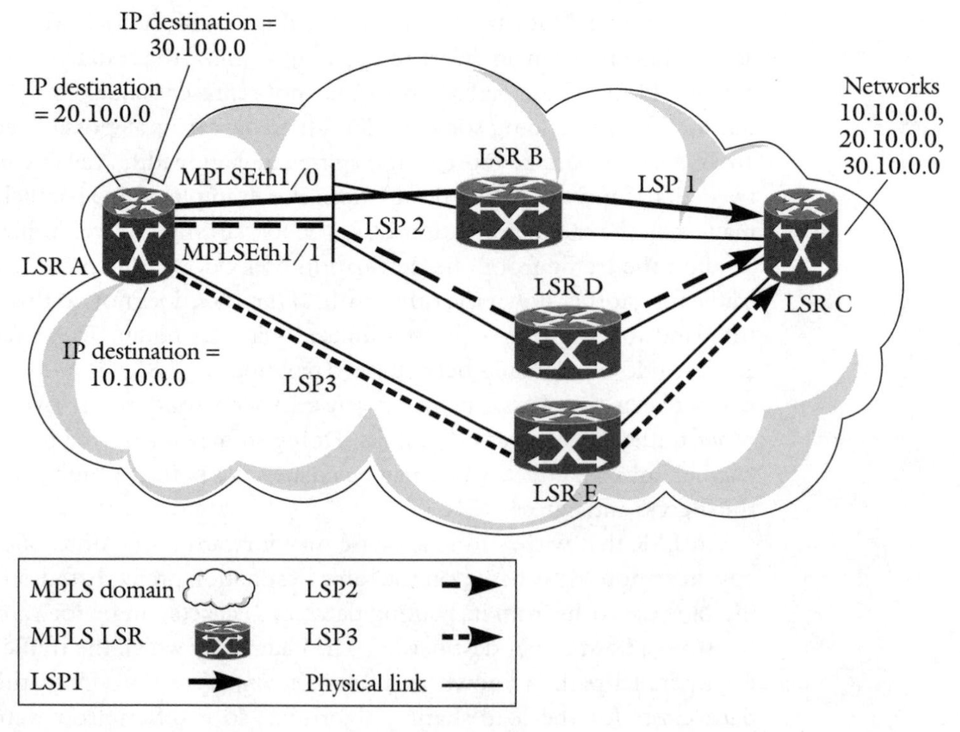

Figure 3.4 An MPLS network with three parallel label switched paths from LSR A to LSR C.

destination LSR. In particular, it shows how a single LSR can establish three LSPs along three parallel (equal-cost) paths to the same destination LSR (LSR C). This example is further complicated by showing how LSR A will send traffic for LSP 1 and LSP 2 using the same outgoing label on the same outgoing multiaccess interface. This will be done between LSRs A, B, and D. LSR A will also use the same outgoing label to forward traffic to LSR E on a different physical interface. Traffic for network 10.10.0.0 will be sent along LSP 1, while traffic with destination IPv4 prefix 10.20.0.0 will be sent along LSP 2. Traffic destined for network 30.10.0.0 will be sent along LSP 3. Note that given the earlier definition of load sharing, either per-packet or per-destination load sharing could be employed in this network, but we will illustrate per-destination load sharing, as this is the only method of load sharing that we recommend. Let us also assume that the LSRs in the figure are currently programmed to forward all traffic destined for networks 10.10.0.0, 20.10.0.0, and 30.10.0.0 toward LSR C, since it can reach all of the networks in question.

If we examine LSR A, any traffic it receives that is destined for networks reachable via LSR C can take any one of the three label switched paths that connect it to its immediate neighbors. These include either LSP 1, 2, or 3. LSR A has available to it two physical interfaces that connect to LSRs that can reach LSR C. These next-hops are LSR B and D out interface MPLSEth1/0, and LSR E via MPLSEth1/1. Assume that each of the three paths is assigned the same routing weight/cost, and thus if load sharing is not employed, only one of these paths will be chosen to carry all of the traffic to the aforementioned destination IP prefixes via LSR C.

Let us first examine the case where both next-hop LSRs are reachable via the same physical interface. Given the configuration described earlier, LSR A's LFIB will be programmed with entries to reach LSR C via both LSRs B and D. These entries will also note both LSRs as being reachable via the same next-hop Ethernet interface—MPLSEth1/0. Since both paths have equal cost, LSR A may choose to load-share traffic going to destination prefixes reachable via LSR C on a per-destination basis. To accomplish this, LSP 1 is assigned traffic from destination prefix 10.10.0.0. Traffic received containing this destination IPv4 prefix will have an MPLS header imposed containing the label 1050. These packets will be forwarded to LSR B via MPLSEth1/0. Similarly, traffic toward IP prefix 20.10.0.0 will be assigned label 1090 and forwarded to a next-hop of LSR D via MPLSEth1/0 as well. Finally, traffic toward 30.10.0.0 will be forwarded to LSR E via MPLSEth1/1 using the same label. Once LSRs B, D, and E receive the packets, they are free to forward the packets along the appropriate path that each LSP uses to reach LSR C. The details of the label forwarding for LSR A are given in Table 3.1.

Table 3.1 LFIB for LSR A in Figure 3.4.

Incoming interface	Incoming label	Outgoing label	Outgoing interface	Next-hop IP address
N/A	N/A	1050	MPLSEth1/0	LSR B
N/A	N/A	1050	MPLSEth1/0	LSR D
N/A	N/A	1050	MPLSEth1/1	LSR C

The first question that may come to your mind is how can the same label—1050—be assigned to the same LSP at LSR A? The reason is simple: labels are typically assigned on a per-neighbor basis, as they are significant only between two LDP neighbors (we will discuss the Label Distribution Protocol in further detail in Chapter 4). The second question that may come to mind is how are LSRs B and D distinguishable to LSR A given that they are both reachable via the same interface? They are distinguishable because they are kept track of as different next-hop neighbors. LSR A will thus keep track of their IP addresses. In addition, we should point out that LSR A must understand how to locate the MAC addresses of LSRs B and D given their IP addresses. This is typically kept track of internally on LSR A and/or by using the Address Resolution Protocol (ARP). If you wished to query the next-hop LSR MAC address using the MPLS-LSR MIB, you could locate the MAC address given the IPv4 address of the next-hop from the mplsOutSegmentTable entry and then use ipNetToMediaTable from STD17.

We should note that the label assignment given in the previous example is particularly pathological given the fact that the same outgoing label was used to carry the traffic from three different LSPs. Although this is perfectly valid, it is equally valid to use different labels for each destination prefix. It is good to know, however, that the MPLS-LSR MIB will handle both configurations.

3.4 A Simple Example

Figure 3.5 depicts a basic MPLS domain containing three LSRs and two LSPs traversing these LSRs. From left to right, these are LSR A, LSR B, and LSR C. LSRs A and C are each interconnected to LSR B via two Ethernet interfaces, which are noted next to each link. Furthermore, the leftmost and rightmost LSRs are also connected to other external networks via a serial interface and are noted as such. Traffic received at the leftmost LSR destined for networks reachable via the rightmost LSR will be forwarded down one of the two LSPs. This traffic will have

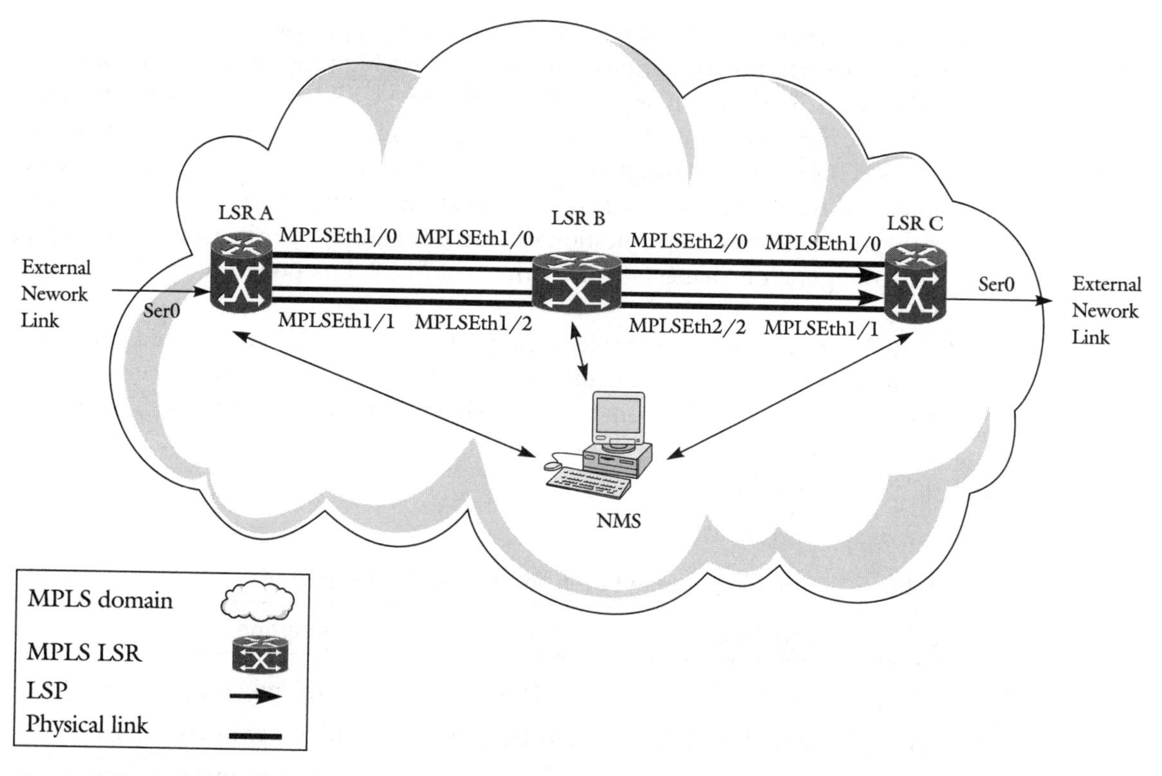

Figure 3.5 A simple MPLS network comprised of three LSRs. Two LSPs traverse the LSRs from left to right in order to carry traffic from LSR A destined to networks reachable via LSR C.

an MPLS header with a specific label assigned to it at the leftmost LSR. It will then be switched at LSR B and forwarded to LSR C, where the MPLS header will be removed and the traffic forwarded as native IP traffic. Note that the forwarding shown in the example only depicts the unidirectional flow of traffic from LSR A to LSR C. Any return traffic from LSR C to A would require additional LSPs to be established. The details of this configuration will be explored within the context of the MPLS-LSR MIB in the following sections.

One note about how such a network, albeit simple, might be managed. Typically, one or more management stations will reside within the MPLS domain and have access to all of the nodes within the domain. This will enable the management station or stations to manage the network directly. It is, however, possible that the management station resides elsewhere and its requests are relayed via a proxy entity. One additional possibility is for each node to be connected to a private "management" network that provides out-of-band network access to management stations.

In all cases, what is important is that SNMP access is possible between the network management station or stations and the label switching routers shown in the example network. In the example, we will assume there is a single NMS system that is responsible for monitoring the whole network. This station will have access to the nodes in the network through direct IP access. This assumption will simplify the example by removing the need for coordination among more than one management system. The NMS in question will provide fault and performance monitoring, and configuration management of the basic LSR functions via the MPLS-LSR MIB.

Let's now investigate a detailed example of how this MIB can be used to manage the network as illustrated in Figure 3.5. Before continuing with the example, let us first show how each LSR from Figure 3.5 has its interfaces configured. This will aid our discussion in the following sections. Each LSR has interface configurations as shown in Table 3.2.

Table 3.2 Interface configuration for routers in the example.

LSR	Interface	Speed	Description
A	Ser0	128 kbps	WAN link; ingress
A	MPLSEth1/0	100 mbps	100 mbps interconnect
A	MPLSEth1/1	100 mbps	100 mbps interconnect
B	MPLSEth1/0	1 Gbps	GigE interconnect
B	MPLSEth1/2	1 Gbps	GigE interconnect
B	MPLSEth2/0	1 Gbps	GigE interconnect
B	MPLSEth2/2	1 Gbps	GigE interconnect
C	MPLSEth1/0	1 Gbps	GigE interconnect
C	MPLSEth1/1	1 Gbps	GigE interconnect
C	Ser0	128 kbps	WAN link; egress

We will make a few assumptions in this example, the first of which is that the labels have been allocated from the global label space using LDP. Second, we will assume that all of the interfaces specified in the example have MPLS enabled and participate in the global label space (see Section 4.3). Third, given these assumptions, LDP (see Chapter 4) will then bind the two destination networks 1.2.0.0/16 and 2.3.0.0/16 to the labels shown in Tables 3.3–3.5. Note that these destinations are located on the other side of LSR C. We will allocate one outgoing label at LSR

A to map destination prefix 1.2.0.0/16 to LSP 1, and another to map destination network 2.3.0.0/16 to LSP 2.

It important to note that penultimate hop popping (PHP) is *not* in use in this example; therefore, LSR C will be popping the final label from the stack and transmitting the packet as an IP packet. If PHP were in use, then LSR B would pop the final label from packets traveling to LSR C and remove the MPLS header before forwarding traffic to it as native IP packets.

If all of the interfaces on LSRs A, B, and C are configured to participate in the global label space, then the Label Forwarding Information Base on each router might then appear as is shown in Tables 3.3–5. It is important to note that each valid incoming label may be accepted on any interface that participates in the global label space. Thus, the LSR may accept this label on all interfaces implicitly and only show a next-hop for it without the incoming interface, or it may explicitly list it out.

Table 3.3 Label Forwarding Information Base for LSR A.

Incoming interface	Incoming label	Outgoing label	Outgoing interface
See FTN MIB (Chapter 5)	See FTN MIB	1050	MPLSEth1/0
See FTN MIB	See FTN MIB	1050	MPLSEth1/0
See FTN MIB	See FTN MIB	1080	MPLSEth1/1
See FTN MIB	See FTN MIB	1080	MPLSEth1/1

Table 3.4 Label Forwarding Information Base for LSR B.

Incoming interface	Incoming label	Outgoing label	Outgoing interface
MPLSEth1/0	1080	2000	MPLSEth2/0
MPLSEth1/1	1080	2000	MPLSEth2/0
MPLSEth2/1	1080	2000	MPLSEth2/0
MPLSEth2/2	1080	2000	MPLSEth2/0
MPLSEth1/0	1050	2010	MPLSEth2/1
MPLSEth1/1	1050	2010	MPLSEth2/1
MPLSEth2/0	1050	2010	MPLSEth2/1
MPLSEth2/1	1050	2010	MPLSEth2/1

Table 3.5 Label Forwarding Information Base for LSR C.

Incoming interface	Incoming label	Outgoing label	Outgoing interface
MPLSEth1/0	2000	"pop"	Ser0
MPLSEth1/1	2000	"pop"	Ser0
MPLSEth1/0	2010	"pop"	Ser0
MPLSEth1/1	2010	"pop"	Ser0

The ingress destination to label mapping is not shown in LSR A's LFIB. Instead, we will explain how this mapping is exposed later in the section covering the FTN MIB.

3.5 The MPLS Interface Configuration Table

The MPLS-LSR MIB contains a table called the MPLS Interface Configuration Table (mplsInterfaceConfTable). The purpose of this table is to manage all interfaces on a device that are capable of running MPLS.

MPLS interfaces are allowed to have labels assigned from only one of two label ranges at a time; these include the global (or platformwide) and interface-specific label ranges (see Chapter 4 for more information regarding label ranges). In short, each label space is essentially a set of labels that are valid to be used on an interface. The manner in which these labels are distributed and assigned depends on the LSR configuration as well as which label distribution protocol(s) are in use. In most implementations, interfaces are assigned labels from the global label space in a consistent manner. That is, any interface may receive a valid label from the global label space, and thus all of these interfaces are configured with the same set of labels. In contrast to this approach, when an interface is assigned to an interface-specific label space, each interface is assigned a specific label space that is valid only for that interface. Although label spaces may overlap, this cannot be counted on; thus the specific label space must be consulted. The per-interface label space is commonly used when interfaces of the media types Frame Relay or ATM are used, but it is not a requirement. It should be noted that some implementations allow discrete subsets of the global label space to be assigned to specific interfaces using this label space. This results in an approach that is similar to the per-interface label spaces.

Given this description of global and per-interface label spaces and their relationship to MPLS interfaces, it is important to understand that several ways exist by which entries for interfaces can be represented in the mplsInterfaceConfTable. These distinctions are especially important to clarify, since they are often a point of common confusion for those reading the MIB for the first time.

Let us first begin with the simplest and most straightforward approach. When representing interfaces that participate in a per-interface label space, a single entry is created in the table for each interface. Its corresponding label space is represented as well as being specified as a per-interface label space.

The next methods apply to interfaces that utilize labels from the per-platform or global label space. If an implementation requires all interfaces that use labels from the global label space to support the exact same set of labels, it is easiest to implement a single entry in the table whose index is zero. This interface then applies to all current and future interfaces that use labels from the global label space. It is important to note that if the device chooses this implementation, it cannot represent the bandwidth values for those interfaces with the single entry.

The alternative to this approach is to follow the one just given for interfaces that use labels from the per-interface label space: create an entry in the table for each interface that participates in the global label space. The drawback to this approach is that many of the columns in the table will be duplicated. This may be more straightforward for some implementations, however, since it may be easier to iterate over all MPLS-enabled interfaces. It may also be desirable from a configuration perspective to allow this approach, since it allows a manager to specify which label space an interface should participate in. Finally, the one twist to this is for implementations that allow interfaces to use per-platform labels, but use a subset of this label space. In this case, each interface should be represented by its own entry in the table, but its label space type should be specified as per platform. An NMS should take care in this case to show the aggregated per-platform label space to the operator by taking the union of configured per-platform interface label spaces.

3.5.1 Creating Entries in the Table

Entries may be created in the mplsInterfaceConfTable to enable MPLS on an existing interface. When this occurs, an interface with ifType equal to mpls (166) must be created and stacked upon the underlying interface. This allows future applications, such as traffic engineering, that create additional interfaces for their services to stack them on the MPLS-type interface. This also allows an operator to view MPLS-only statistics for MPLS-labeled traffic on an interface. Any individual entry in the mplsInterfaceConfTable must have a corresponding entry in the IF-MIB before it may exist. This entry must have an interface type (ifType) set to mpls (166). Device agents that allow configuration of this table are advised to ensure this ordering behavior. Device agents should also take care when taking an MPLS-type interface out of service, since other interfaces (e.g., TE tunnels) may be carried over the interface in question, and thus the action will result in those interfaces being prevented from sending or receiving traffic.

Given this description of the mplsInterfaceConfTable, we recommend that device vendors implement each interface that uses labels from the global label space as its own entry in the table. There are several advantages to implementing the interfaces in this way over the approach that represents all as single entry. First, it seems more intuitive to access a subset of the IF-MIB ifTable and find a subset of those entries that are MPLS enabled. Furthermore, "walking" over the mplsInterfaceConfTable provides a convenient way in which an operator can quickly inspect all MPLS-enabled interfaces. Using the other approach, an operator would have to walk the entire IF-MIB's ifTable and pick out entries with ifType equal to mpls (166). Second, the advantage of aggregating all platformwide interfaces is lost when you wish to gather other information about those interfaces such as their bandwidth values. Again, the entire ifTable must be interrogated to find each MPLS-type interface before its corresponding statistics can be gathered.

An often-asked question is, Why did the co-authors of this MIB choose to include this alternative representation? The simple answer to this question is that there were some implementations that demanded this functionality, despite the fact that most did not. The reasoning was that this provides a bit of convenience for the operator in terms of being able to view a single entry in the table to view the label spaces being distributed on those interfaces. However, it is questionable that this efficiency is worth the confusion created. Simply put, the standards process is a series of compromises between vendor implementations and the operators who wish to use them. Therefore, the standard document generally results in the most common set of functionality between all of those implementations. In essence, the standard document is a great compromise and thus resembles a camel[2] with all of its imperfections.

One important reason for creating all MPLS-enabled interfaces in this table (as opposed to exclusively using entry 0) is that all MPLS applications may then use these interface entries as their underlying interface. That is, they may be "stacked upon" them using the Interfaces Stack Table (see Chapter 6 for more details about the IF-MIB and the ifStackTable). The stacking relationship, as well as the general

2 cam·el, *noun.*
1. A horse that was created by committee.
2. A humped, long-necked ruminant mammal of the genus *Camelus,* domesticated in Old World desert regions as a beast of burden and as a source of wool, milk, and meat.
3. A device used to raise sunken objects, consisting of a hollow structure that is submerged, attached tightly to the object, and pumped free of water. Also called *caisson.*
4. *Sports.* A spin in figure skating that is performed in an arabesque or modified arabesque position.
(From *Dictionary.com.*)

interface model for MPLS, are explained in more detail in Chapter 6. It is also useful to note that although some label space information is displayed in entries in this MIB, the user may wish to consult the MPLS-LDP MIB for much more detailed label space information if LDP is used to distribute the labels on the device in question. The MPLS-LDP MIB is outlined in Chapter 4.

In addition to the aforementioned objects, the mplsInterfaceConfTable includes objects that describe the bandwidth that is allocated to the interface, as well as what bandwidth might be available assuming that a protocol like RSVP-TE can allocate parts of this bandwidth. Additionally, some MPLS-specific performance information is available for each entry (detailed in later sections). Finally, since the MPLS-type interfaces are actually represented as ifTable entries, the same statistics counters (e.g., bytes, packets, errors, etc.) also apply to these interfaces (see Chapter 6 for more details about the IF-MIB). Table 3.6 enumerates the mplsInterfaceConfTable objects.

Table 3.6 Interface Configuration Table objects.

Object name	Definition
mplsInterfaceConfIndex	This is a unique index for an entry in the MPLS Interface Configuration Table. A nonzero index for an entry indicates the Interface Index for the corresponding interface entry in the MPLS layer in the Interface Table. Note that the per-platform label space may apply to several interfaces, and therefore the configuration of the per-platform label space interface parameters will apply to all of the interfaces that are participating in the per-platform label space.
mplsInterfaceLabelMinIn	This object is set to the minimum value of an MPLS label that this LSR is willing to receive on this interface.
mplsInterfaceLabelMaxIn	This object is set to the maximum value of an MPLS label that this LSR is willing to receive on this interface.
mplsInterfaceLabelMinOut	This object is set to the minimum value of an MPLS label that this LSR is willing to send on this interface.
mplsInterfaceLabelMaxOut	This object contains the maximum value of an MPLS label that this LSR is willing to send on this interface.
mplsInterfaceTotalBandwidth	This object indicates the total amount of usable bandwidth on this interface. Operators should note that this value is specified in

continued

Table 3.6 continued

Object name	Definition
	kilobits per second. This variable is not applicable when applied to the interface with index 0.
mplsInterfaceAvailable-Bandwidth	This value indicates the total amount of available bandwidth left on this interface. The bandwidth value is specified in kilobits per second. This value is calculated as the difference between the amount of bandwidth currently in use and that specified in the object mplsInterfaceTotalBandwidth. Note that this object does not apply to the interface with index 0.
mplsInterfaceLabel-ParticipationType	This object represents the label participation type of this interface. This value can only be set to one of either the **perPlatform (0)** or **perInterface (1)** bits, but never both. If the value of the mplsInterfaceConfIndex for this entry is zero, then only the **perPlatform (0)** bit has to be set, and the **perInterface (1)** bit is interpreted by network managers as meaningless. If the **perInterface (1)** bit is set, then the value of mplsInterfaceLabel-MinIn, mplsInterfaceLabelMaxIn, mplsInterfaceLabelMinOut, and mplsInterfaceLabelMaxOut for this entry will reflect the label spaces for this interface. If only the **perPlatform (0)** bit is set, then the value of mplsInterfaceLabelMinIn, mplsInterfaceLabelMaxIn, mplsInterfaceLabelMinOut, and mplsInterfaceLabelMaxOut for this entry will be identical to those objects found in the row corresponding to index 0.
mplsInterfaceConfStorageType	The SNMP storage type for this entry. Valid values are **other (1),** meaning that a storage type other than the ones defined below is available; **volatile (2),** meaning that the row will be stored in RAM and will disappear after the device reboots; **Nonvolatile (3),** meaning that the value is stored in some sort of nonvolatile RAM and will be preserved across reboots of the system; **permanent (4),** meaning that the row is stored partially in ROM; or **readOnly (5),** meaning that the row is stored completely in ROM. Operators will typically find this value to be set to either **Nonvolatile (3)** or **volatile (2).**

3.5.2 Example of InterfaceConfTable

The example shown in Figure 3.6 depicts how a simple interface configuration table might appear. Note that every mplsInterfaceConfEntry from the MPLS-LSR MIB has a corresponding entry in the IF-MIB. For example, mplsInterfaceConfEntry 100 corresponds to Interface Entry 100. Note that, since the InterfaceConfEntry's index is of type IfIndexOrZero, this index always must correspond to an entry in the IF-MIB unless it is set to zero. If this value is set to zero, it corresponds to the entry representing the global label space.

With this in mind, if we continue with the example from Figure 3.6, the MPLS-LSR MIB's Interface Configuration Table will appear as shown in Tables 3.7, 3.8, and 3.9 for each of the three LSRs. Note that the Interface Index (ifIndex) values included are hypothetical and can be any value chosen by the device at the time the interface is created. Also, notice that the ifIndex value corresponds directly with the mplsInterfaceConfIndex. This correspondence is intentional and is a requirement.

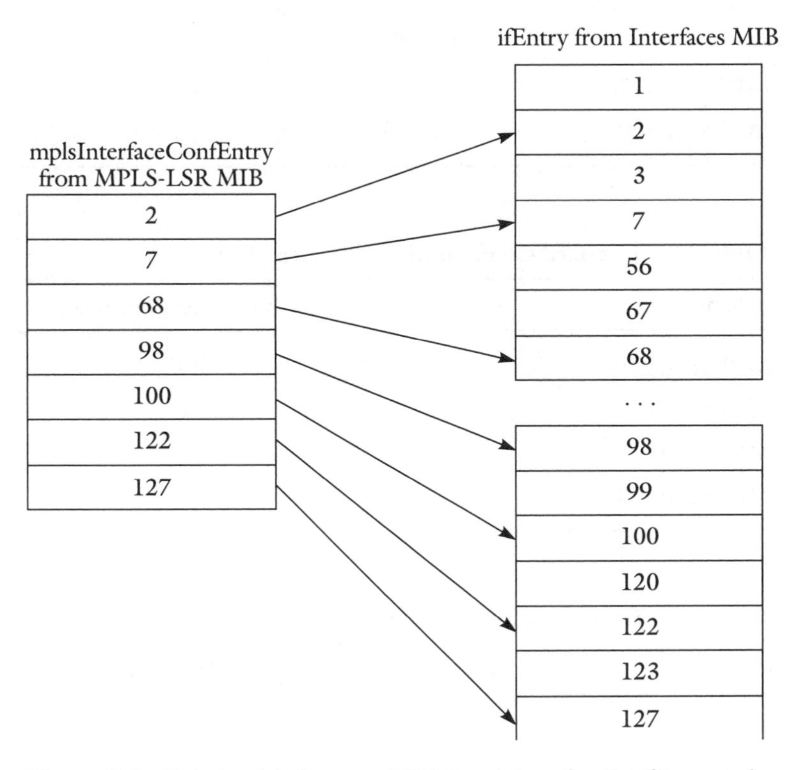

Figure 3.6 Relationship between MPLS mplsInterfaceConfEntry and RFC 2863's IfEntry.

It should be noted also that it is, in general, a good idea for the device to choose the ifIndex values for all of the interfaces that it supports; otherwise strange and odd behavior may arise from either the NMS and/or the agent. If devices wish to allow the NMS to allocate ifIndex values, which is sometimes common in provisioning systems, then great care must be taken when allocating these values so as not to collide with those chosen by the system or another instance of the provisioning system. Furthermore, devices allowing this behavior may also wish to implement the ifIndex values persistently, so as to make the provisioning or other management system's job easier by not having to remember what ifIndex values were assigned prior to a system reboot. This can sometimes be a complicated and error-prone task for an NMS to undertake, and it is quite often much simpler to leave it to a network node to do.

Table 3.7 Interface Configuration Table for LSR A.

ifName	ifIndex	mplsInterfaceConfIndex
Ser0	12	12
MPLSEth1/0	16	16
MPLSEth1/1	18	18

Table 3.8 Interface Configuration Table for LSR B.

ifName	ifIndex	mplsInterfaceConfIndex
MPLSEth1/0	44	44
MPLSEth1/1	46	46
MPLSEth2/0	55	55
MPLSEth2/1	57	57

Table 3.9 Interface Configuration Table for LSR C.

ifName	ifIndex	mplsInterfaceConfIndex
Ser0	78	78
MPLSEth1/0	99	99
MPLSEth1/1	109	109

3.5.3 MPLS Interface Configuration Performance Table

The MPLS Interface Configuration Table is augmented with a series of objects in the MPLS Interface Configuration Performance Table that provide some MPLS-specific statistical information about each MPLS interface. Other more general interface statistics can be found in the IF-MIB for the object referenced by the same Interface Index as the ones found in this table. The mplsInterfaceInLabelsUsed and mplsInterfaceOutLabelsUsed of these values provide the number of labels in use on the interface. The mplsInterfaceFailedLabelLookup object indicates the number of times the LSR was unable to switch an MPLS packet because it could not look up the label found in that packet. Note that if the media type of the interface is ATM, this object counts the number of cells that were received with incorrect label information.

The final object, mplsInterfaceOutFragments, indicates the number of times the LSR was required to fragment MPLS packets on this interface prior to their transmission. This counter might be most interesting for cases where IP packets are switched to ATM. It should be noted that some implementations might find it difficult or impossible to support packet-based counters because they support nonpacket media. ATM is such an example. In these cases, the counters should be set to zero, and the agent conformance statement for this device should make a clear note to this effect.

These statistics are in addition to those provided in the IF-MIB. There are four objects in this table that are listed in Table 3.10.

Table 3.10 Interface Configuration Performance Table objects.

Object name	Definition
mplsInterfaceInLabelsUsed	This object counts the number of labels that are in use just before or at the time this variable is interrogated. This value counts only those labels in use in the incoming direction. If the interface participates in the per-platform label space, then the value of this object will be identical to the instance in this table with index 0. If the interface participates in the per-interface label space, then this entry will represent the number of per-interface labels that are in use on this interface.
mplsInterfaceFailedLabelLookup	This object reveals the number of labeled packets that have been received on this interface, but were discarded because no corresponding cross-connect entry could be found. More

continued

Table 3.10 continued

Object name	Definition
	precisely, the MPLS software received an MPLS packet that it had no forwarding entry for and dropped the packet.
mplsInterfaceOutLabelsUsed	This object has the same syntactic and semantic meaning as mplsInterfaceInLabelsUsed except that it applies to labels used in the outgoing direction.
mplsInterfaceOutFragments	This object counts the number of outgoing MPLS packets that required MPLS fragmentation before transmission on this interface. This object *must* count on a per-interface basis regardless of which label space the interface participates in.

3.6 The InSegment Table

The MPLS-LSR MIB contains a table called the mplsInSegmentTable. This table is responsible for depicting the "left-hand side" of the LFIB 4-tuple as defined in Figure 3.7. That is, this table is responsible for depicting the incoming interface and incoming label portion of each switching entry. Accordingly, the table is indexed by both an interface index and an MPLS label, thus allowing easy retrieval. This table also contains a few other important elements, namely, the mplsInSegmentXCIndex and the mplsInSegmentTrafficParamPtr. If the switching entry has been configured correctly, the mplsInSegmentXCIndex should contain the cross-connect index of the cross-connect that is used to bind the left-hand side of the LFIB 4-tuple to the right-hand side. This value can thus be used to hop from the InSegment Table to the XCTable. The mplsInSegmentTrafficParamPtr is an OID that "points" at the first column of the mplsTrafficParamTable to indicate traffic-related parameters that are associated with this LSP. The reason why the type of this field is an OID versus the actual integer-based index used in the mplsTrafficParamTable is so that

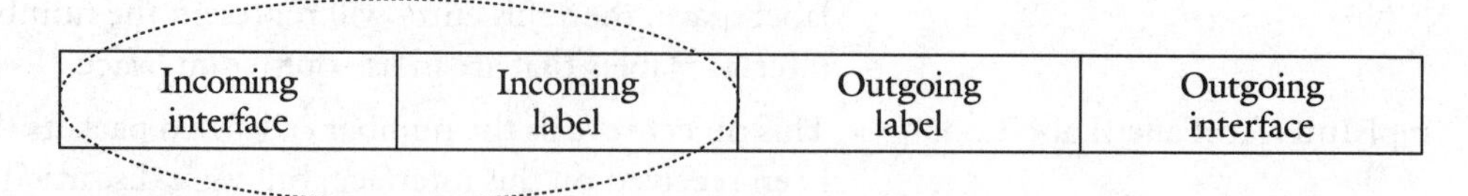

Figure 3.7 The InSegment Table contains the (incoming interface, incoming label) pair.

vendors can point the LSP at a vendor-specific traffic parameter table if desired. It is, however, recommended that the traffic parameter table included in the LSR MIB be used as often as possible, since the use of proprietary tables presents difficulties in terms of interoperability between two different implementations.

3.6.1 InSegment Table Example

To refresh your memory, take another look at Figure 3.5 repeated again here. Note that the LSRs are labeled from left to right: LSR A, LSR B, and LSR C.

Each LSR plays a specific role, that of ingress LER, core LSR, or egress LER. Tables 3.11–3.13 depict how the mplsInSegmentTable will appear on each of the LSRs in the example network. Note that the mplsInSegmentXCIndex in the table appears to be the same for groups of entries having the same incoming label. This is because the incoming label is the same, regardless of incoming interface. Once received, any packet containing that label will be switched to the same outgoing label.

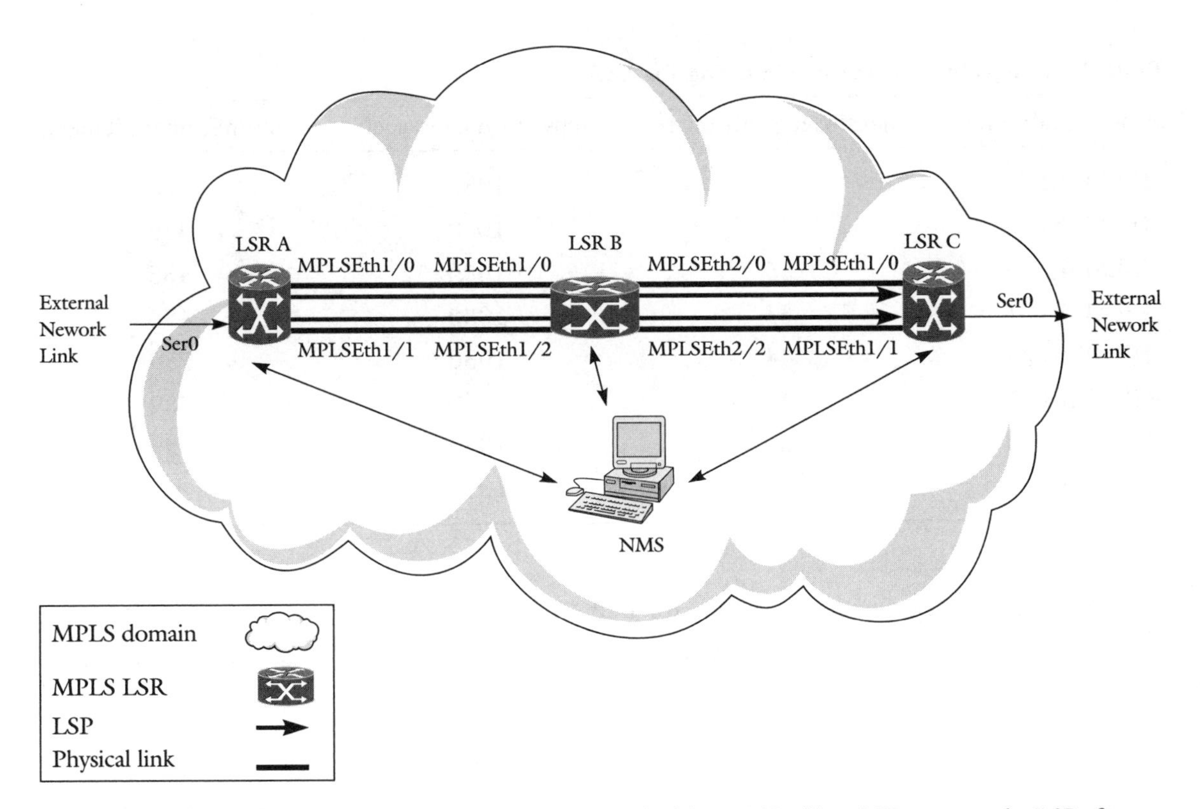

Figure 3.5 *(repeated)* A simple MPLS network comprised of three LSRs. Two LSPs traverse the LSRs from left to right in order to carry traffic from LSR A destined to networks reachable via LSR C.

For example, four entries exist with mplsInSegmentXCIndex equal to 456 for LSR B. Notice that, since we wish to switch any packet received with label 1080 onto the same outgoing interface, we will bind that label with any incoming interface to mplsInSegmentXCIndex 456. This will ensure that the packet will be switched out the correct interface, regardless of incoming interface. The actual value that is assigned for each mplsXCIndex is implementation-dependent. The values chosen in this example are merely there to suit the example. Also, note that care should be taken if configuring this table for static entries to avoid looping entries. If signaling protocols such as LDP are used to assign labels, there is little concern for this result.

Table 3.11 Empty mplsInSegmentTable indexing for LSR A.

Incominginterface	mplsInSegmentIfIndex	mpslInSegmentLabel	mplsInSegmentXCIndex
—	—	—	—

Table 3.12 mplsInSegmentTable indexing for LSR B.

Incoming interface	mplsInSegmentIfIndex	mpslInSegmentLabel	mplsInSegmentXCIndex
MPLSEth1/0	44	1080	456
MPLSEth1/1	46	1080	456
MPLSEth2/0	55	1080	456
MPLSEth2/1	57	1080	456
MPLSEth1/0	44	1050	789
MPLSEth1/1	46	1050	789
MPLSEth2/0	55	1050	789
MPLSEth2/1	57	1050	789

Table 3.13 mplsInSegmentTable indexing for LSR C.

Incoming interface	mplsInSegmentIfIndex	mpslInSegmentLabel	mplsInSegmentXCIndex
Ser0	78	2000	567
MPLSEth1/0	99	2000	567
MPLSEth1/1	109	2000	567
Ser0	78	2010	789

Table 3.13 continued

Incoming interface	mplsInSegmentIfIndex	mpslInSegmentLabel	mplsInSegmentXCIndex
MPLSEth1/0	99	2010	789
MPLSEth1/1	109	2010	789

3.6.2 Originating LSPs

Originating LSPs, regardless of how they are created (e.g., RSVP-TE, CR-LDP, LDP, or statically) are treated a little differently by the MPLS-LSR MIB than are typical label-to-label switching entries. When an LSP originates at an LSR, no corresponding mplsInSegmentEntry will be created because the incoming label to swap does not exist within the LFIB. Instead, a FEC-to-Next-Hop Forwarding Entry (FTN) will map a destination IP prefix (and perhaps other attributes) to an outgoing label (see Chapter 5 for a more detailed description of the MPLS-FTN MIB). Thus, there is no mplsInSegmentEntry present in the MIB.

Figure 3.7 illustrates how two LSPs originate at LSR A and terminate at LSR C. Upon examination of the mplsInSegmentTable for LSR A in Table 3.11, you will notice that no corresponding mplsInSegmentEntries exist for either LSP that originates at LSR A. As was discussed, this is because no incoming labels have been assigned to the outgoing labels, since IPv4 traffic that enters LSR A will have the MPLS header imposed with the outgoing labels shown only. This is demonstrated by the empty table shown in Table 3.11.

3.6.3 The MPLS InSegment Performance Table

The MPLS InSegment Table (mplsInSegmentTable) is augmented by a table called the mplsInSegmentPerfTable. This table contains objects used to indicate statistical values related to each InSegment present in the mplsInSegmentTable. The statistical objects are listed in Table 3.14 along with a description of how each object should behave. The objects in this table are provided in addition to those statistics provided on each MPLS interface in order to show per-label statistics. It is important to note that, although sometimes an important statistic, per-segment (per-label) counters are not always available on some implementations, specifically those that are implemented on older hardware or some that are cell-based such as ATM. Unfortunately, since MPLS came after certain hardware was designed, and since hardware is not easily modified to support new features, per-interface statistics may be the only traffic performance statistics available to the operator on those platforms. We encourage network operators to investigate the agent capability statements and/or vendor manuals for the MPLS-LSR MIB for the platforms they wish to manage this MIB on for further information on the availability of these statistics on their MPLS platforms of choice.

Table 3.14 InSegment Performance Table objects.

Object name	Definition
mplsInSegmentOctets	This value represents the total number of bytes that have been received using this segment (label).
mplsInSegmentPackets	This value represents the total number of packets that have been received using this segment (label).
mplsInSegmentErrors	This value represents the total number of packets that have been received using this incoming segment (label), but which have been discarded due to some error. These errors can include exhausted buffers or other resources. Operators should note that some switching devices could not properly implement this variable due to hardware limitations. This value represents the total number of errors experienced with labeled packets received with this label.
mplsInSegmentDiscards	This value represents the total number of packets that have been received using this incoming segment (label), but which have been discarded due to some error. These errors can include exhausted buffers or other resources. Operators should note that some switching devices could not properly implement this variable due to hardware limitations. Please inquire with the vendor of your hardware for specific details. This can happen, for example, if a Differentiated Services or buffer management function decides to actively discard some MPLS traffic.
mplsInSegmentHCOctets	This is the 64-bit version of mplsInSegmentOctets.
mplsInSegmentPerfDiscontinuityTime	This object contains the value of sysUpTime on the most recent occasion that one or more of this segment's (label's) counters suffered a discontinuity. This object will contain the value of 0 if no discontinuities have occurred since the last reinitialization of the local management subsystem.

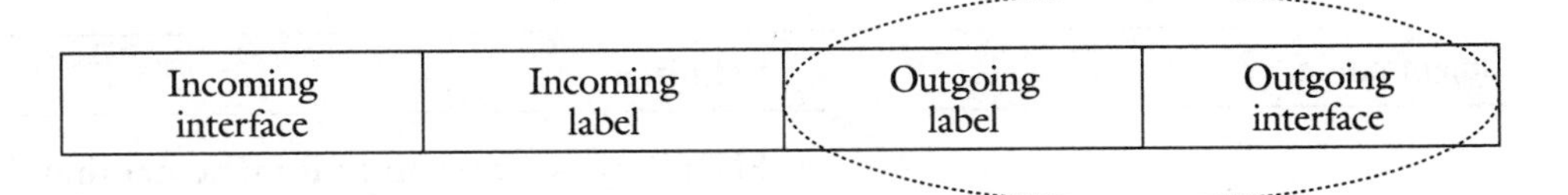

Figure 3.8 The OutSegment Table contains the (outgoing label, outgoing interface) bindings.

3.7 The MPLS OutSegment Table

The analog to the mplsInSegmentTable is the MPLS OutSegment Table (mplsOutSegmentTable). The purpose of this table is to show the label assignment for MPLS packets that will leave the LSR. In addition to simply showing the outgoing label portion of the switching 4-tuple shown in Figure 3.8, this table also shows the next-hop IP address (v4 or v6), the cross-connect pointer associated with this entry, as well as the traffic parameter pointer. As in the mplsInSegmentTable, the latter points at an entry in the traffic parameter table in order to show specific traffic parameters associated with this outgoing label. The objects in this table are listed in Table 3.15.

Table 3.15 MPLS OutSegment Table objects.

Object name	Definition
mplsOutSegmentIndex	This object contains a unique index for a specific row in this table. It should be noted that although the value of 0 is not a valid index for this table, it can be supplied as a valid value to index the MPLS XC Table in order to access entries that do not have an outsegment. This could occur if the MPLS XC Table has been configured for an entry that has not been fully configured.
mplsOutSegmentIfIndex	This object contains the Interface Index of the outgoing interface.
mplsOutSegmentPushTopLabel	This object shows if a top label should be pushed onto the outgoing packet's label stack. The value of this object will be **true (1)** if the outgoing interface does not support pop-and-go. This could be the case, for example, with an ATM interface or a tunnel origination.

continued

Table 3.15 continued

Object name	Definition
	SNMP software on an LSR implementing this MIB should note that the standard considers it an error if mplsOutSegmentPushTopLabel is set to false, but the cross-connect entry that refers to this outsegment has a nonzero mplsLabelStackIndex. Thus, this situation should be prevented.
mplsOutSegmentTopLabel	If the mplsOutSegmentPushTopLabel object is true, then this object represents the label that should be pushed onto the top of the outgoing packet's label stack.
mplsOutSegmentNextHopIpAddrType	This object indicates the next-hop address type. This can be set to **ipv4 (1), ipv6 (2),** or **unknown (0).** Note that the latter is valid only when the outgoing interface is of type point-to-point.
mplsOutSegmentNextHopIpv4Addr	If the object mplsOutSegmentNextHopIpAddrType is ipV4 (1), then this is the IPv4 address of the next-hop. Otherwise, implementations should return a value of 0 for this object.
mplsOutSegmentNextHopIpv6Addr	If the object mplsOutSegmentNextHopIpAddrType is ipV6 (2), then this is the IPv6 address of the next-hop. Otherwise, implementations should return a value of 0 for this object.
mplsOutSegmentXCIndex	This object contains a valid index into the MPLS XC Table. This index identifies the specific cross-connect entry using this segment. Implementations should return a value of 0 to indicate that an entry is not referred by any cross-connect entry. The SNMP agent software implementing this MIB should automatically update these values when a cross-connect entry is created that this outsegment is a part of.
mplsOutSegmentOwner	This object denotes the entity that created and is responsible for managing this segment. This can be set to many values, but most commonly, an operator will find it set to **snmp (2)** or **ldp (3).** Valid values for this

Table 3.15 continued

Object name	Definition
	object are **other (1),** indicating an owner entity other than one specified in the enumerated values in the standard; **snmp (2),** indicating that the SNMP process on the LSR created the entry or an NMS; **ldp (3),** specifying that the LDP process on the LSR created the entry; **rsvp (4)**, indicating that the RSVP process on the LSR created the entry; **crldp (5),** indicating that CR-LDP created the entry; and **policyAgent (6),** indicating that some form of policy agent such as COPS created this entry. Finally, the value of **unknown** (7) is used when this object is unsupported.
mplsOutSegmentTrafficParamPtr	This variable represents a pointer to the traffic parameter specification that corresponds to this outsegment. This value may point at an entry in the mplsTrafficParam-Table to indicate which mplsTrafficParamEntry is to be assigned to this segment, or it may point at an externally defined traffic parameter specification table. The latter case can be used by implementations that have additional proprietary traffic parameters that can be specified for a segment. A value of 0.0 indicates best-effort treatment for this segment. Implementations should note that it is possible and quite convenient for operators to allow the same value of this object to be specified by two or more segments to indicate resource sharing.
mplsOutSegmentRowStatus	This object contains the standard SNMP RowStatus that is used for creating, modifying, and deleting rows in this table.
mplsOutSegmentStorageType	This variable indicates the SNMP storage type for this object.

Tables 3.16–3.18 depict how the mplsOutSegmentTable will appear on each of the LSRs in the example network shown earlier.

Table 3.16 mplsOutSegmentTable for LSR A.

mplsOutSegmentIndex	mplsOutSegmentTopLabel	mplsOutSegmentXCIndex	mplsOutSegmentIfIndex
123	1050	123	16
987	1080	987	18

Table 3.17 mplsOutSegmentTable for LSR B.

mplsOutSegmentIndex	mplsOutSegmentTopLabel	mplsOutSegmentXCIndex	mplsOutSegmentIfIndex
2000	2000	456	55
2010	2010	789	57

Table 3.18 Empty mplsOutSegmentTable for LSR C.

OutSegmentIndex	mplsOutSegmentTopLabel	mplsOutSegmentXCIndex	mplsOutSegmentIfIndex
—	—	—	—

3.7.1 Indexing Tips

A note about the indexing of the mplsOutSegmentTable: if you had not already noticed, some of the earlier examples use the outgoing label as the mplsOutSegmentIndex. Since the index is simply an Unsigned32, at first glance it may be instinctive to assign it based on the outgoing label. However, this will not work if the LSR in question supports either load sharing or multicast operations. This is because in both cases it is possible to transmit the same label out to multiple destinations (i.e., neighbors). Thus, there exists the possibility that multiple entries of the form (outgoing label = L, outgoing interface = N_1), . . . , (outgoing label = L, outgoing interface = N_n) will exist in the MIB. Further compounding this difficulty is the case where the outgoing neighbors are reachable via the *same* outgoing interface. Such is the case if the outgoing interface is a multiaccess type such as an Ethernet interface. In these cases, it is important to consider that the following situations can arise: (outgoing label = L, outgoing interface = N, neighbor = P_1), . . . , (outgoing label = L, outgoing interface = N, neighbor = P_n). Therefore, a different indexing scheme is necessary from the simple mapping from the outgoing label.

There are many ways of choosing and maintaining a unique 32-bit integer: bit fields, arrays of used numbers, virtual address spaces, and so on. All of these methods, however, have intrinsic difficulties related to sparse index management, as well as other performance-related issues. In the end, one of the easiest and most straightforward methods is to simply use the memory address of a related data structure. This provides a 32-bit number, which is unique when in use, and which is maintained by the operating system. Thus, when the number is no longer needed, it is returned to the "free" pool of numbers, and when a new unique value is required, a new data structure is allocated. As with every method, this approach comes with some drawbacks. First, in-order iteration of the objects indexed by this scheme may be difficult if the data structures are not themselves arranged in a manner that lends itself to the indexing. Due to the random nature of memory address assignment in most systems today, it is generally an $O(n)$ problem to search for an index using this scheme. Second, if configuration is allowed through the MIB and operators are allowed to choose the indexes used for entries in the table, then the agent must ensure that the index relates to a valid memory address, or that it is managed correctly. Virtual address spaces may help in this case. However, as you can see, all problems and solutions come with their costs and benefits. You should weigh all of the pros and cons of a particular solution before making a decision to implement it.

3.7.2 Terminating LSPs

When an LSP terminates at a particular LSR, there will be no corresponding outgoing label because the MPLS header is stripped off the packet at that point, and the packet is then forwarded as whichever technology the underlying header represents. Therefore, the LSR will not create an outgoing label entry in the forwarding table, and thus no corresponding mplsOutSegmentTable entry will exist. This is demonstrated in Table 3.18. Figure 3.5, repeated here again, also demonstrates this in showing that the LSPs originate at LSR A, but end at LSR C. Note that an LSP may terminate at the penultimate hop if penultimate hop popping is enabled on the LSRs that the LSP is traversing. Within the context of Figure 3.5, this would mean that both LSPs originating at LSR A would in fact end at LSR B, where penultimate hop popping would occur. Packets from those LSPs would exit LSR B toward LSR C as IP-encapsulated packets instead of MPLS-encapsulated packets. However, note that the example does not implement penultimate hop popping, and thus does include an mplsOutSegmentTable entry.

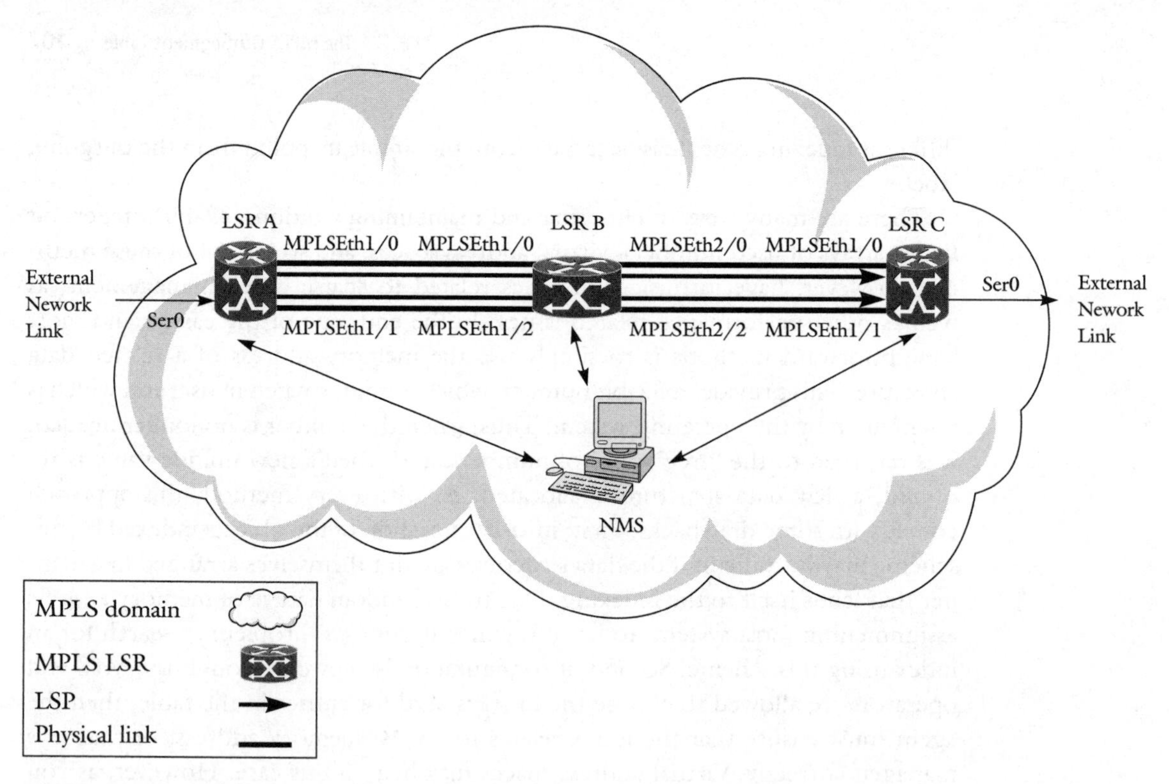

Figure 3.5 *(repeated)* A simple MPLS network comprised of three LSRs. Two LSPs traverse the LSRs from left to right in order to carry traffic from LSR A destined to networks reachable via LSR C.

3.7.3 The MPLS OutSegment Performance Table

The MPLS OutSegment Table (mplsOutSegmentTable) is augmented with a table called the mplsOutSegmentPerfTable. This table contains objects used to indicate statistical values related to each mplsOutSegmentTable entry. These objects are listed and defined in Table 3.19. The objects in this table are provided in addition to those statistics provided on the MPLS interface in order to show per-label statistics.

Table 3.19 OutSegment Performance Table objects.

Object name	Definition
mplsOutSegmentOctets	This object represents the total number of octets (bytes) sent on this segment.
mplsOutSegmentPackets	This object represents the total number of packets sent on this segment.

Table 3.19 continued

Object name	Definition
mplsOutSegmentErrors	This object represents the total number of packets that could not be transmitted due to errors using this label. This could occur, for example, if hardware errors are being encountered sending packets with this label.
mplsOutSegmentDiscards	This value represents the total number of packets that have been transmitted using this segment (label), but which have been discarded due to some error. These errors can include exhausted buffers or other resources. Operators should note that some switching devices could not properly implement this variable due to hardware limitations. Please inquire with the vendor of your hardware for specific details. This can happen, for example, if a Differentiated Services or buffer management function decides to actively discard some MPLS traffic.
mplsOutSegmentHCOctets	The 64-bit version of mplsOutSegmentOctets.
mplsOutSegmentPerfDiscontinuityTime	This object contains the value of sysUpTime on the most recent occasion that one or more of this segment's (label's) counters suffered a discontinuity. This object will contain the value of 0 if no discontinuities have occurred since the last reinitialization of the local management subsystem.

3.8 The Cross-Connect Table

The MPLS-LSR MIB contains a table called the mplsXCTable whose purpose is to associate the aforementioned mplsInSegmentTable and mplsOutSegmentTable entries together in such a way as to indicate how the LSR is switching the incoming labels to outgoing labels. Because of the flexibility required to facilitate the many different switching scenarios that an LSR may be required to accommodate, the indexing and structure of this table might seem a bit complicated at first. The cross-connect table is indexed by a 4-tuple consisting of the following objects:

- mplsXCIndex
- mplsInSegmentIfIndex

- mplsInSegmentLabel
- mplsOutSegmentIndex

The latter three parts of the index are taken directly from the mplsInSegmentTable and mplsOutSegmentTable entries. The first part of the index, however, is chosen by whoever created the row in the MIB to associate together the mplsInSegmentTable and mplsOutSegmentTable entries. In most implementations, this responsibility is filled by the agent implementing the MPLS-LSR MIB, but some implementations may allow a network manager to choose this index as well. For example, if we examine the switching entries that have already been demonstrated for LSR B earlier, we find that LSR B will switch two labels: 1080 and 1050. Let us first examine the first LSP represented by the switching entries with mplsInSegmentLabel equal to 1080 and mplsOutSegmentLabel equal to 2000. Label 1080 can be received on any of the valid MPLS interfaces, and in all cases must then be switched to outgoing label 2000 on interface MPLSEth2/0 (ifIndex = 55). This same switching relationship exists between incoming label 1080 and the MPLS interfaces MPLSEth1/0, MPLSEth1/1, MPLSEth2/1. Thus, it makes sense to associate these together in the relationship shown in Table 3.20.

Table 3.20 Cross-connect indexing for first LSP in LSR B.

mplsXCIndex	mplsInSegmentIfIndex	mplsInSegmentLabel	mplsOutSegmentIndex
2000	55	1080	2000
2000	57	1080	2000
2000	45	1080	2000
2000	47	1080	2000

As you can see from the table, any MPLS packets arriving on any of the MPLS interfaces with label 1080 will be switched to mplsOutSegmentIndex 2000. The LSR not only will replace the top label with 1080, but it will also transmit the packet out of the interface with ifIndex 55 (MPLSEth2/0). To complete the examples, in Tables 3.21–3.23 we will now demonstrate how each of LSR A's, LSR B's, and LSR C's cross-connect tables will appear given the original example in Figure 3.5.

Table 3.21 Cross-connect indexing for LSR A.

mplsXCIndex	mplsInSegmentIfIndex	mplsInSegmentLabel	mplsOutSegmentIndex
1050	0	0	1050
1080	0	0	1080

Table 3.22 Cross-connect indexing for LSR B.

mplsXCIndex	mplsInSegmentIfIndex	mplsInSegmentLabel	mplsOutSegmentIndex
2000	55	1080	2000
2000	57	1080	2000
2000	45	1080	2000
2000	47	1080	2000
2010	55	1050	2010
2010	57	1050	2010
2010	45	1050	2010
2010	47	1050	2010

Table 3.23 Cross-connect indexing for LSR C.

mplsXCIndex	mplsInSegmentIfIndex	mplsInSegmentLabel	mplsOutSegmentIndex
2000	78	2000	0
2000	99	2000	0
2000	109	2000	0
2010	78	2010	0
2010	99	2010	0
2010	109	2010	0

The MPLS Cross-Connect Table contains several objects. These objects are enumerated and defined in Table 3.24.

Table 3.24 The mplsXCTable objects.

Object name	Definition
mplsXCIndex	This object contains the primary index used to index this table. It is used to identify a group of (usually) related cross-connect entries.
mplsXCLspId	This object is used to identify the label switched path that this cross-connect belongs to.

continued

Table 3.24 continued

Object name	Definition
mplsXCLabelStackIndex	This object represents the primary index into the MPLS Label Stack Table. It identifies a stack of labels that are to be pushed beneath the top label of packets switched by this cross-connect entry. It is important to note that the topmost label identified by the corresponding outsegment entry (or entries) ensures that all the components of a multipoint-to-point connection will use the same outgoing label. Also, note that a value of 0 in this object indicates to the operator that no labels are to be stacked beneath the top label.
mplsXCIsPersistent	This object indicates whether the cross-connect entry and associated in- and outsegments should be restored automatically after the LSR has recovered from a failure. The standard states that this value is set to false if it is created dynamically by the LSR's switching or signaling software.
mplsXCOwner	This object denotes the entity that created and is responsible for managing this cross-connect. This can be set to many values, but, most commonly, an operator will find it set to **snmp (2)** or **ldp (3).** Valid values for this object are **other (1),** indicating an owner entity other than one specified in the enumerated values in the standard; **snmp (2),** indicating that the SNMP process on the LSR created the entry or an NMS; **ldp (3),** specifying that the LDP process on the LSR created the entry; **rsvp (4),** indicating that the RSVP process on the LSR created the entry; **crldp (5),** indicating that CR-LDP created the entry; and **policyAgent (6),** indicating that some form of policy agent such as COPS created this entry. Finally, the value of **unknown (7)** is used when this object is unsupported.
mplsXCRowStatus	This object contains the standard SNMP RowStatus that is used for creating, modifying, and deleting rows in this table.
mplsXCStorageType	The SNMP storage type for this entry. Valid values are **other (1),** meaning that a storage type other than the ones defined below is available; **volatile (2),** meaning that the row will be stored in RAM and will disappear after the device reboots; **Nonvolatile (3),** meaning that the value is stored in some sort of nonvolatile RAM and will be preserved across reboots of the system; **permanent (4),** meaning that the row is stored partially in ROM; or **readOnly (5),** meaning that the row is stored completely in ROM. Operators will typically find this value to be set to either **Nonvolatile (3)** or **volatile (2).**

Table 3.24 continued

Object name	Definition
mplsXCAdminStatus	This value indicates the desired status of this segment. Valid values for the objects are **up (1),** indicating that the operator desires that the cross-connect be enabled to pass traffic; **down (2),** indicating that the operator wishes for the cross-connect to stop forwarding traffic; and finally **testing (3),** indicating that the operator wishes that the cross-connect enter some test mode. In the case of the **testing (3)** mode, this mode is generally vendor-specific, so operators should consult vendor documentation to determine if such a mode exists. If such a mode does not exist, agents will reject requests to set this object to this state. Operators should also be aware that many implementations do not allow cross-connections to be disabled administratively, and thus this object may be implemented as read-only. Please consult the vendor documentation for specific details.
mplsXCOperStatus	This object indicates the operational (actual) status of the cross-connection. The valid states of this variable are **up (1),** indicating that the cross-connect is ready to pass packets; **down (2),** indicating that it is not forwarding packets; **testing (3),** denoting that the cross-connect (or the hardware interface that it runs over) is in some test mode; **unknown (4),** indicating that the status cannot be determined for some reason (usually there is an internal error preventing the SNMP task from querying the hardware); **dormant (5),** indicating that the cross-connect has been configured, but not used; **notPresent (6),** denoting that some component is missing, preventing the cross-connect from functioning (e.g., a valid outsegment has not been associated with the insegment to allow packets to be forwarded); and finally **lowerLayer-Down (7),** indicating that the cross-connect cannot forward traffic because it is down due to the state of lower-layer interfaces (i.e., the interface(s) over which the MPLS layer interface is running).

3.9 The Traffic Parameter Table

The MPLS-LSR MIB contains the mplsTrafficParamTable that allows the agent to demonstrate the traffic parameters associated with the insegments and outsegments of a particular LSP. An LSP's insegments and outsegments could "point" at entries in this table to indicate the traffic-related parameters they possessed. The mplsInSegmentTable and mplsOutSegmentTable accomplish this by providing an object defined using an SNMP construct called a RowPointer that can make this association. The mplsInSegmentTable contains an object called the mplsInSegmentTrafficParamPtr, and the mplsOutSegmentTable contains a similar object called the mplsOutSegmentTrafficParamPtr. By setting either one of these object's values to the OID that refers to the first accessible column of a row in the mplsTrafficParamTable, it is possible to cross-reference the appropriate traffic parameter table entry. It should be noted that, in cases where both the in- and outsegments share the same traffic parameter characteristics, it is possible to share the same traffic parameter entry by setting both pointers to the same mplsTrafficParamPtr OID. An example of this is demonstrated in Figure 3.9. The figure demonstrates how both insegments and outsegments can share the same traffic parameter pointer entry, or how they can be configured to use separate ones. In the example, mplsInSegmentEntry 17 and mplsOutSegmentEntry 88 both share traffic parameter entry 66. This indicates that they are both utilizing the same traffic parameter resources. This relationship can be used to denote, for example, that a bidirectional connection has been configured, and both directions of the connection should share the same traffic parameter resources.

The mplsTrafficParamTable contains three important objects. These are mplsTrafficParamMaxRate, mplsTrafficParamMeanRate, and mplsTrafficParamMaxBurstSize. To quote from the MPLS-LSR MIB, each is defined as an MPLSBitRate object, which is defined specifically as

> An estimate of bandwidth in units of 1,000 bits per second. If this object reports a value of 'n' then the rate of the object is somewhere in the range of 'n − 500' to 'n + 499'. For objects which do not vary in bitrate, or for those where no accurate estimation can be made, this object should contain the nominal bitrate.

This definition indicates that the bandwidth values reported by these objects can be within the range of +499 or −500 from the actual value recorded. For example, if the actual value recorded is 1000, the value reported by the agent may be 900 or 1200, since these values fall within the acceptable tolerance range. However, since these values should not vary with bitrate—meaning that the difference between the actual and recorded values should remain consistent regardless of bitrate—the agent should report the actual (nominal) value regardless of the bitrate.

3.10 A Note about SNMP RowPointer Use

RowPointer is a textual convention used to identify a conceptual row in an SNMP table by pointing to the first accessible column in that row. By "pointing," it is

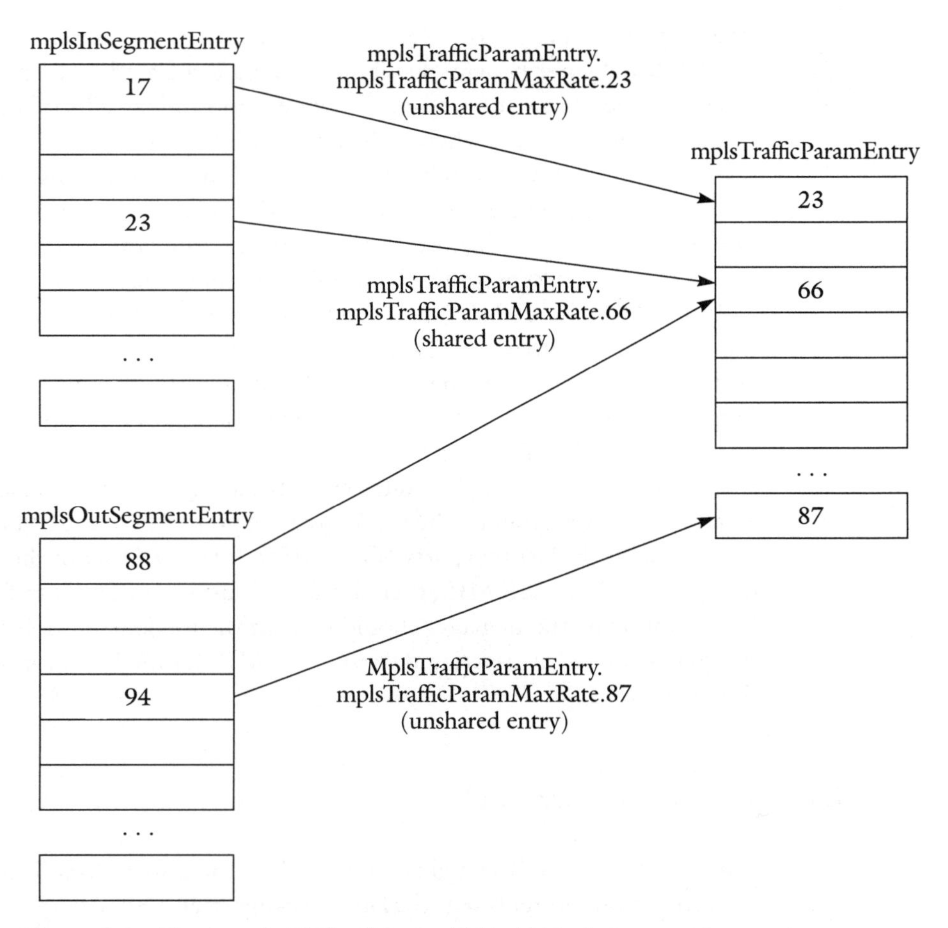

Figure 3.9 The use of mplsTrafficParamTable by the InSegmentEntry and outSegmentEntry.

meant that the RowPointer contains the OID of the columnar object. By doing so, you in effect have created a pointer to another table in the MIB through which the manager may quickly jump to the new table. One reason why RowPointers are used is to afford the agent with the flexibility of pointing at similar rows in different tables. For example, if a RowPointer in some MIB points at a standard table, but would like to extend that standard table, then it is possible to point instead to the new proprietarily extended table. For obvious reasons, it is advised the structure of the new table be similar (i.e., same indexing) to that of the original.

It is also common for a RowPointer to be associated with an object that indicates its type. This is somewhat analogous to a C programming language data type assigned to a *void** pointer in that language. Due to the opaque nature of a *void** pointer, it can possibly point at a variety of different data structures. Understanding what type of structure is pointed ahead of time is beneficial for a program. An SNMP RowPointer benefits similarly by being associated with an object that indicates the type of row it points at. To understand the benefits of an associated type object, imagine the case where a RowPointer is expected to point at rows in one of several tables, each with differing numbers of indexes. Knowing the type of the table ahead of time makes the task of the NMS easier when it needs to follow the pointer into another table, since it can quickly and easily infer the indexing contained in the RowPointer OID by simply examining the associated type value.

In the MPLS-LSR MIB, the RowPointer is used in both the mplsInSegmentTable and mplsOutSegmentTable for indicating a particular traffic parameter table entry. In either the mplsInSegmentTable or mplsOutSegmentTable, the trafficParamPtr should point to the first accessible column of the appropriate conceptual row in the mplsTrafficParamTable. However, it is possible that some implementations will point this value at a different table altogether. The manager can quickly ascertain this by examining the OID up to the point where the column and instance are indicated (the last two parts of the OID). If the OID is not the same as the one used in the MPLS-LSR MIB, then the agent is using a different traffic parameter table. In this case, the manager should consult the implementation's agent capabilities statement and an associated proprietary MIB for further information regarding this alternate table.

3.11 The Label Stack Table

The MPLS-LSR MIB contains a table called the Label Stack Table. This table is used to expose and represent the label stacking of an LSP when a specific label stack is imposed at the LSR supporting the MIB. The MIB only contains label stacks that are imposed since it is incorrect to display the label stack from LSPs that transition the LSR where the MIB is supported. LSPs always are associated with a label stack that has a minimum depth of one label; however, since this topmost label can be simply represented as the mplsInSegmentLabel or mplsOutSegmentLabel objects, this table is only used when the label stack depth is 2 or greater. The topmost label is never present in the labelStackTableEntry for this reason. Label stacking for a particular cross-connected set of insegments and outsegments is indicated by setting the appropriate mplsXCLabelStackIndex variable in the cross-connect entry with the correct index into the label stack table. The label stack table represents the MPLS label stack intuitively—as a stack of labels. That is, it is represented in an SNMP table where each entry in the stack is ordered by its position. This table has entries that contain the label itself as well. Since there can be many stack entries for many different LSPs, and since we only wish to have a single table in the MIB, the label stack table is indexed by two integers: mplsLabelStackIndex and mplsLabel-StackLabelIndex. The first variable indicates the label stack entry or group of label stack entries that are associated with a particular LSP or LSPs. The second index provides the relative position of each label stack entry within the group of entries. The lower the mplsLabelStackLabelIndex, the higher up the stack the label. Thus, an entry with index (1,1) would be used to represent the top of the label stack, whereas an entry such as (1, 3) would represent a label stack entry that is three deep into the stack. Figure 3.10 illustrates how this would actually be implemented. Notice that the indexes with lower values appear higher on the stack, and indexes with higher values appear lower in the stack. The indexing of the entries in the tables need not be contiguous. That is, gaps between entries are possible, since agents are free to assign the indexing of the entries therein as they find most convenient as long as they obey the rules for representing the stack.

It should be noted that although label stack entries may be reused by different cross-connect entries—multiple entries can "point" at the same entry—this practice is ill-advised due to the bookkeeping required to guarantee that if one label

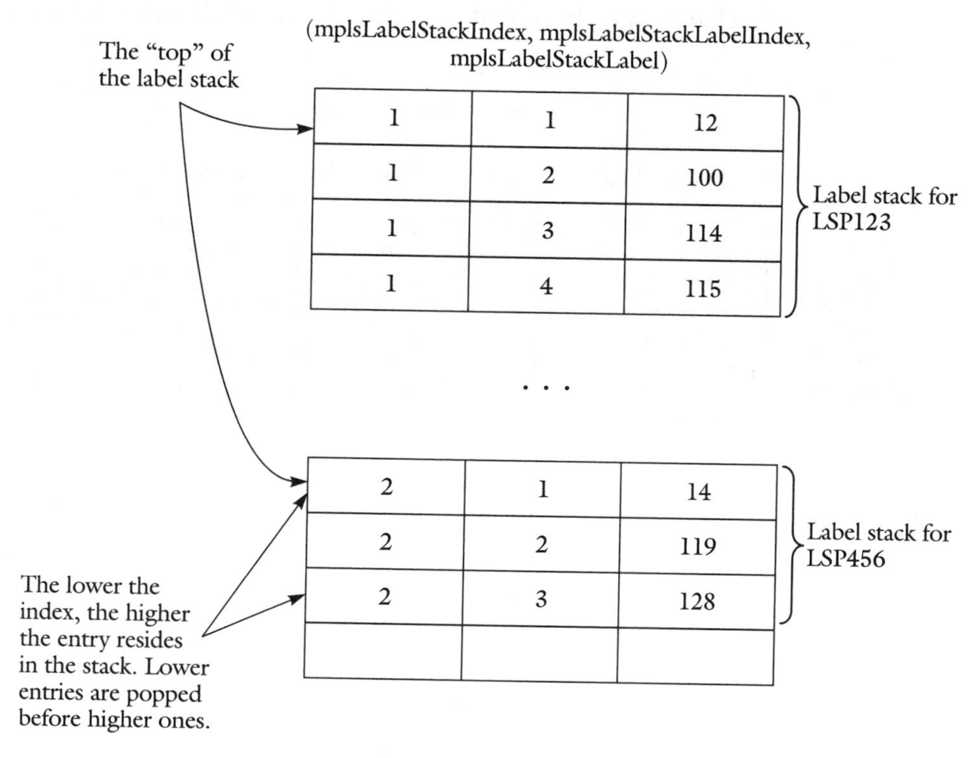

Figure 3.10 The conceptual label stacks for LSPs 123 and 456.

stack is changed, the remaining cross-connect entries are updated in a consistent manner. In short, save yourself some time and headache debugging some nasty synchronization problems by allocating a little more memory for each label stack entry. Memory is cheap these days, and it is probably cheaper than having irritated customers.

3.11.1 Label Stacking Example

The ordering of the stack is represented with the topmost stack having the lowest ranking index in the table. That is, the stack entry's precedence is inversely related to its index value. For example, index 3 will rank higher in the label stack than index 12. The example in Figure 3.11 demonstrates how label stack entries can be created for two sample LSPs, LSP123 and LSP456. In the first case, LSP123 contains a label stack that uses the labels 132, 100, 14, and 15. The example in Figure 3.11 demonstrates how label stacking is represented in the MIB for these LSPs. For example, the first block of entries correspond to label stack "1." This is ascertained from the mplsLabelStackIndex. The label stack entries are then represented within this block using the secondary indexes 1, 2, 5, and 8. These represent the label stack

mplsLabelStackEntry

labelStackIndex	labelStackLabelIndex	labelStackLabel	
1	1	132	Label stack for LSP123
1	2	100	
1	5	14	
1	8	15	
. . .			
9	1	144	Label stack for LSP456
9	4	18	
9	5	188	
. . .			

Figure 3.11 Label Stack Table Example.

entries 132, 100, 14, and 15, respectively. In this label stacking representation, the first entry happens to represent the label just under the "top" of the label stack. Remember, the topmost label is not present in the table; rather, it appears in either the mplsInSegmentTable or mplsOutSegmentTable entries.

3.11.2 Label Stacking Operations

The mplsXCLabelStackIndex is contained by each cross-connect entry in order to identify the stack of labels that are to be pushed beneath the top label. The LSR can take three actions with regard to the LSP's label stack. In the first case, the topmost label is simply replaced with the replacement label. For example, if an LSP has an incoming label *L1* that is replaced with an outgoing label *L2,* Table 3.25 results.

Table 3.25 LSP label stack top label replacement.

Variable	Value
mplsOutSegmentPushTopLabel	true
mplsOutSegmentTopLabel	*L2*
mplsXCLabelStackIndex	0

In the second case, the topmost label is popped and the packet is then forwarded using the underlying label stack (if it exists). An example of this might be an LSP with an incoming label *L1* that is popped and then the packet is forwarded with whatever label stack is underneath, as in Table 3.26.

Table 3.26 LSP label stack top label replacement.

Variable	Value
mplsOutSegmentPushTopLabel	false
mplsOutSegmentTopLabel	Next label in stack (if one exists). The label may be different for each packet processed, and therefore in some cases the value returned by this object is meaningless.
mplsXCLabelStackIndex	0

In the third and final case, the incoming label is replaced with a new stack containing more than one label. An example of this might be an LSP with an incoming label *L1*, which is replaced with a stack of more than one label: {*L2, L3, L4,* . . .}, where *L2* is the new top label (Tables 3.27 and 3.28).

Table 3.27 LSP label stack showing new top label *L2*.

Variable	Value
mplsOutSegmentPushTopLabel	true
mplsOutSegmentTopLabel	*L2*
mplsXCLabelStackIndex	x (entry into label stack table)

Table 3.28 The corresponding label stack entries.

Entry	mplsLabelStackLabel
(x,1)	*L3*
(x,2)	*L4*
. . .	. . .

The MPLS Label Stack Table contains several objects. These objects are enumerated and defined in Table 3.29.

Table 3.29 Label Stack Table objects.

Object name	Definition
mplsLabelStackIndex	This object contains the primary index for this table. It identifies a specific stack of labels that are to be pushed onto an outgoing packet, beneath the top label. The topmost label is indicated in the inSegment and outSegment tables. Only labels beneath this top label are indicated here.
mplsLabelStackLabelIndex	This object contains the secondary index for this table. It is used to identify a particular label in the stack of labels. It should be noted that entries in this table containing a smaller mplsLabelStackLabel-Index refer to a label that is higher up the label stack. Labels with a smaller index will be popped at a downstream LSR before a label represented by a higher mplsLabelStackLabelIndex.

Table 3.29 continued

Object name	Definition
mplsLabelStackLabel	This object contains the actual label.
mplsLabelStackRowStatus	This object contains the standard SNMP RowStatus that is used for creating, modifying, and deleting rows in this table.
mplsLabelStackStorageType	The SNMP storage type for this entry. Valid values are **other (1),** meaning that a storage type other than the ones defined below is available; **volatile (2),** meaning that the row will be stored in RAM and will disappear after the device reboots; **Nonvolatile (3),** meaning that the value is stored in some sort of nonvolatile RAM and will be preserved across reboots of the system; **permanent (4),** meaning that the row is stored partially in ROM; or **readOnly (5),** meaning that the row is stored completely in ROM. Operators will typically find this value to be set to either **Nonvolatile (3)** or **volatile (2).**

3.12 Notifications

The MPLS-LSR MIB provides two notifications: mplsXCUp and mplsXCDown. These notifications are emitted when the LSR either removes or installs a new cross-connect entry. This effectively notifies the operator whenever a new LSP is activated or is destroyed. Figure 3.12 demonstrates how an mplsXCUp notification is generated by the middle LSR after it has created the switching entry for the LSP. The notification is generated and delivered to the NMS.

3.13 Scalability Issues with Notifications

The MPLS-LSR MIB presents some interesting and potentially dangerous scalability issues for both the engineer who is saddled with the task of implementing them, as well as the network operator that must capture and process them. Put simply, the issue with these notifications is the sheer number that may be generated in a short period of time. If taken literally, the mplsXCUp or mplsXCDown notifications should be generated by an agent for *every* LSP that is created or deleted, respectively. The problem is that a typical core LSR may have tens of thousands of LSPs active or transitioning within its TFIB at any moment in time. Thus, it is

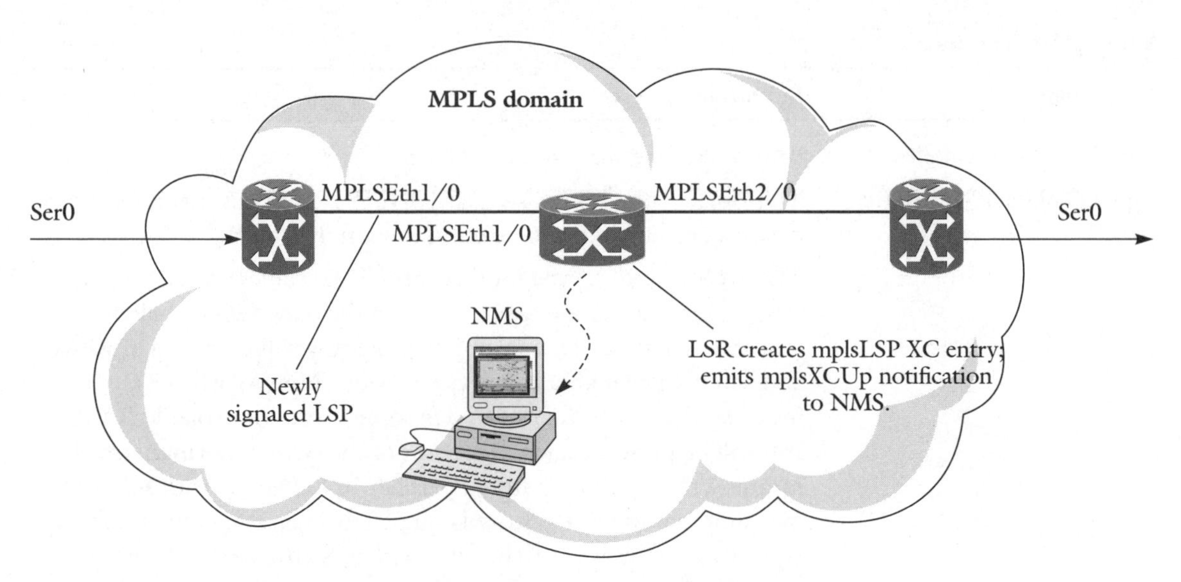

Figure 3.12 XCUp notification is emitted when a new LSP is created.

possible that many thousands of mplsXCEntries will change state suddenly, requiring the generation of a large number of notifications. In the worst case, it is possible to conceive of a situation where all or most MPLS interfaces become disabled either due to a physical fault (e.g., major fiber cut) or due to being shut down by the system's administrator. In a naive implementation, the result would be the generation of tens of thousands of notifications—one for each cross-connection entry in the MPLS-LSR MIB. Further compounding the problem would be a case where multiple managers were configured to receive the notifications. This would result in *N* copies of each notification being generated. A typical core router hosting, say, 30,000 LSPs and having perhaps three network management systems configured to receive notifications would require the device to generate 90,000 notifications within a very short period of time!

There is another problem to consider here as well. Once the notifications are generated, they need to go somewhere: onto the network. Once on the network, this large amount of traffic may result in congestion, which may exacerbate the fault condition. Even worse, additional notifications may be generated due to the worsening condition. Let's now consider the previous example where 90,000 notifications are generated. Assume that a scant 33% of those notifications do in fact make it out onto the network. That is still (90,000 * 33%) = 30,000 notifications! This number of notifications alone may result in saturation of the network connections between the device in distress and the NMS. Worsening this condition might be a situation where the device is configured to emit additional notifications when congestion occurs on a link, or when resource utilization exceeds some threshold, for example, when link bandwidth or transmit/receive buffer allocation exceeds some threshold, which may very well have resulted from 30,000 notifications being generated. Finally, compounding the scalability picture even further is the fact that someone needs to receive and process all of those notifications. If the NMS is not equipped to handle such a volume of information, it may take too long to process these messages, or worse, it may crash. As you can see, the problem of notification generation and processing becomes quite difficult in a hurry.

The co-authors of the MPLS-LSR MIB identified these scalability issues and modified the original versions of the notifications slightly in a subsequent version of the MIB. This modification, although simple, is quite effective in drastically reducing the number of notifications generated, transmitted, and processed by the entire system. In some cases, it is possible to reduce thousands of notifications to just a single one. The simple addition was to change the primary index of the notification from containing just the cross-connect index to a range of cross-connects. Thus, if the agent were implemented in such a way as to cache together all notification events occurring within some short period of time (e.g., a few seconds), it would be possible to send a single notification in place of potentially thousands. This new notification would contain the starting and ending range of the notification events. One caveat to consider is the fact that it is probably very idealistic to think that all events will be accepted over the event-sampling period from a contiguous range. More often than not, some of the notification events received during the sampling period will be within a contiguous range and can thus be compressed into a single one. Still, on average there is a good chance that the notifications can be compressed significantly. Even a 3:1 compression should suffice. Our experimental results show that, on average, this is roughly the compression achievable with this approach.

Figure 3.13 demonstrates how notification compression might be implemented. If an event queue is used to collect and buffer notifications within an agent, it is possible to then periodically go back over that queue of events and compress them together into a single notification, or sometimes several. Figure 3.13 shows how continuous ranges of the mplsXCDown notification are grouped together. These notifications can later be transmitted as a single notification with a range of indexes. For example, instead of transmitting three mplsXCDown notifications for indexes 11–13, single notification containing the index range of 11–13 could be transmitted, resulting in a reduction of notifications by one-third. Note that an appropriate interval of time must be used to buffer the notifications; otherwise, compression will not be possible. This value will differ depending on the system it is implemented on; therefore, we suggest experimentation to adjust this value. It may also be appropriate to allow the operator to adjust this value.

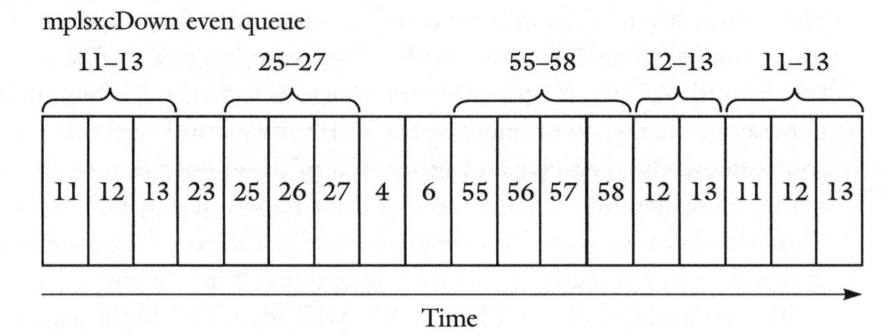

Figure 3.13 Compressing XCDown events into a single notification using an event queue.

3.14 Next Index

Each of the traffic parameter, outsegment, label stack, and cross-connect tables is associated with a scalar variable called <*table name*>IndexNext. For example, the traffic parameter table is associated with the variable mplsTrafficParamIndexNext. The intent of this variable is that it is to be used by the agent as a guide to the NMS to indicate which index is available to be used to create the next entry in this table. The intent is that this variable is incremented each time it is read so that an NMS wishing to create an entry will have a valid index that it is at least greater than any entries that currently exist or that might be in the process of being created by other managers (or the agent itself). This is important, since multiple managers may access this variable, each taking a value and creating a new entry with it. Agents that support creation of entries in this table should also utilize this variable to gauge how it creates new entries in the table that are consistent with those created by the NMS. The drawback to this approach is that managers do not have to create entries using a retrieved value right away; thus this can result in large gaps in the indexes, possibly making life more difficult for the agent. Another drawback to this approach is the fact that a manager that walks this MIB repeatedly will perhaps inadvertently increment this value each time it is read, which may not be an obvious side effect at first.

It is also important to note that two special cases exist in which this variable takes on the special value of 0. This variable will contain the value of 0 either if the agent is unable to create additional rows in this table (i.e., it is full and/or the agent has run out of memory), or if the table is accessible only for reading (i.e., read-only). In either case, the agent will disallow row creation operations on this table.

3.15 A Note about Indexing

One interesting question was raised recently with regard to those implementations that might implement both hand-routed and signaled LSPs (e.g., LDP, RSVP-TE, or CR-LDP) and wish to represent them as entries in the MPLS-LSR MIB's tables. The question asked specifically how the MIB might guarantee the uniqueness of the indexing of its tables given the fact that the signaling software could choose to use the same labels that an operator might have assigned for hand-routed or static LSP. As a matter of fact, the same situation exists for implementations that allow for multiple signaling protocols to be used simultaneously (e.g., RSVP-TE and LDP).

The simple answer is that if an implementation wishes to support such configurations, then the implementation *must* govern the allocation of labels such that it guarantees that the LSPs not overlap in their use of labels. It is important to understand that the MPLS-LSR MIB does not explicitly spell this out as a requirement; rather, it is implied. It is therefore important that implementations that do take care in these cases or disaster could result. Fortunately, most implementations that exist today do allow for mixed signaling protocols to be in used at the same time and do have label space management that ensures nonoverlapping utilization of labels.

3.16 Summary

The goal of the MPLS-LSR MIB is to expose the active MPLS label switching of an LSR, as well as to allow for the configuration of objects related to label switching such as static label mappings. The chapter began with an overview of the important characteristics of an LSR's label switching capabilities and why it was important to expose these things to the network manager. It was explained that network managers wishing to monitor the basic label forwarding activities of a label switching router should monitor the tables provided in this MIB. Management stations wishing to monitor the behavior of other MPLS applications such as traffic engineering or virtual private networks will benefit from understanding how to manage the objects defined in this MIB, since many of the other MPLS-related MIBs have been designed to work in conjunction with the objects defined in the MPLS-LSR MIB.

The components of an LSR's LFIB include labels that can be received or transmitted on its interfaces, as well as the associations between these labels and interfaces. The MPLS-LSR MIB exposes these salient components of a TFIB in its mplsInSegmentTable, mplsOutSegmentTable, mplsXCTable, and mplsInterface-ConfTables. In addition to the basic associations between these objects, the MIB also exposes label stacking, performance, and configuration characteristics of these components. The chapter presented each of the MIB's component tables and scalar objects, and gave examples of how each worked and could be used in concert with the other tables in the MIB to manage an LSR. Finally, we presented the notifications that are available in this MIB and how they can be used to complete the picture of managing an LSR.

The TFIB is an important function that is at the heart of an MPLS LSR. Label switching, imposition, and disposition are critical components of this function. These features should be exposed to network managers to enable them to effectively manage the LSR. Devices implementing the MPLS-LSR MIB can provide the network operator with an effective means for managing the device's LSR functions, as well as a consistent view of these features across all network nodes.

Further Reading

Srinivasan, C., A. Viswanathan, and T. Nadeau. "MPLS Label Switching Router Management Information Base Using SMIv2." IETF Internet Draft. January 2001. *www.ietf.org/internet-drafts/draft-ietf-mpls-lsr-mib-07.txt.*

McCloghrie, K., et al. "Management Information Base for Network Management of TCP/IP-based Internets: MIB-II." STD 17. March 1991. *ftp://ftp.isi.edu/in-notes/std/std17.txt.*

Davie, B. S., and Y. Rekhter. *MPLS: Technology and Applications.* First edition. San Francisco: Morgan Kaufmann Publishers. 2000.

Gray E. W. *MPLS: Implementing the Technology.* Reading, Mass.: Addison-Wesley Professional. 2001.

To find out more about the IETF, visit their Web page at *www.ietf.org/.*

For more information about IANA, check out their Web site at *www.iana.org/.*

The MPLS-LSR MIB can be found at *www.ietf.org/internet-drafts/draft-ietf-mpls-te-mib-08.txt.* Note that the document is an Internet Draft and is subject to change.

RFC 3031 specifies the MPLS architecture and is the basis for all MPLS documents. It can be located at *www.ietf.org/rfc/rfc3031.txt.*

Kireeti Kompella

is a Distinguished Engineer at Juniper Networks. His current interests are all aspects of MPLS, including traffic engineering, Generalized MPLS, and MPLS applications such as VPNs. Dr. Kompella is active at the IETF, where he is a co-chair of the CCAMP Working Group and the author of several Internet drafts in the areas of IS-IS, MPLS, OSPF, PPVPN, and TE. Previously, he worked in the area of file systems at Network Appliance and SGI.

Dr. Kompella received his B.S. in electrical engineering and M.S. in computer science at the Indian Institute of Technology, Kanpur, and his Ph.D. in computer science at the University of Southern California.

Being the chairman of the IETF Working Group that is chartered with advancing and extending MPLS to function as a general signaling mechanism for next-generation optical networks, what do you see as the key advantages of using this technology as compared to other competing technologies, both in its existing form and in its more generalized form (GMPLS)?

Let's start with the competing technologies. There are two main candidates: manual configuration and proprietary signaling mechanisms. Manual configuration is a nonstarter in *next-generation* optical networks. On the other hand, many vendors have put significant effort into their proprietary signaling mechanisms and have many features that come from their experience in the field and the feedback from customers; thus, proprietary mechanisms are a potential candidate.

However, there are two principal objectives that MPLS and GMPLS signaling try to achieve: multivendor interoperability and easy extensibility. The task of taking a proprietary signaling mechanism and making it a multivendor interoperable solution is a daunting one, fraught with intellectual property and competitive advantage hurdles. Furthermore, a proprietary mechanism will probably be tuned to the specifics of a particular vendor's product and architecture; making it general enough to cover multiple vendors and multiple architectures could also be a difficult problem.

Another approach is for a number of vendors to get together and design in common a new signaling protocol. However, that is exactly the genesis of MPLS; the signaling protocols that resulted were RSVP-TE and CR-LDP, among others. The advantage of continuing that effort, with incremental additions and changes to extend these signaling protocols to the optical domain, as opposed to designing a new protocol from scratch, is clear: the time to market is greatly reduced; vendors can leverage their current MPLS code; service providers can leverage their experience with and training in MPLS; GMPLS interoperability testing can leverage the fact that MPLS is interoperable. Finally, if you view the MPLS paradigm as virtual path creation aided by constraint-based routing, this paradigm is very well matched to the requirements of signaling for optical networks.

What do you see in the future of GMPLS and MPLS?

The first step is to complete the specification of GMPLS. These specifications are currently drafts, but are close to moving to the next stage and becoming Proposed Standards. In parallel, there are several vendors that are implementing these specifications, and over the next months, there will be several interoperability trials, both public and private. This will set the stage for the next step: testing in service provider networks and deployment.

Another direction that GMPLS will take is to add more "technologies." The initial GMPLS specifications are primarily for optical and SONET/SDH networks. There is already work in place to add mechanisms for the ITU-T's G.709 recommendation. This effort will test the extensibility of GMPLS, as well as validate that GMPLS is in fact "Generalized." And as new technologies come online, the hope is that GMPLS is well suited to provide signaling for those technologies.

As for MPLS, deployment is well under way. There are two directions for further development: (a) the feedback from deployment experiences will determine shortcomings and suggest remedies, and (b) the success of MPLS-based services will drive further services and applications. The success of a protocol is not measured by whether it did its job, but by whether it is flexible enough to adapt and change with changing needs.

4

The MPLS Label Distribution Protocol MIB (MPLS-LDP MIB)

"I may not have gone where I
intended to go, but I think I have
ended up where I intended to be."

—**Douglas Adams**, *Mostly Harmless*

Introduction

The Label Distribution Protocol was defined to allow MPLS LSRs to distribute FEC to label bindings in a variety of ways. In this chapter, we examine how this protocol is managed using the IETF MPLS-LDP MIB. In particular, we examine in detail each of the tables and how each fits together into the larger picture of LDP management. We also investigate in detail how the notifications defined therein can be used to detect and manage fault conditions. The MIB itself defines the management of the LDP protocol for any network that LDP runs over, including label controlled Frame Relay or ATM networks. The discussion is framed with one eye on the implementation of the MIB in an LSR and the other on how to use each component within the context of a network management system.

This chapter will focus on the draft version 08 of the MPLS-LDP MIB, which is typically referred to as draft-ietf-mpls-ldp-mib-08.txt. This draft version may have been updated or replaced with an IETF RFC document after this book was published. Please keep this in mind when searching for the document on which this chapter is based.

It was explained earlier that an MPLS label edge router acts as a bridge or conversion point between MPLS and non-MPLS traffic at the edges of an MPLS network. MPLS label edge routers (LERs) sit along the edge of an MPLS domain and impose MPLS headers onto non-MPLS packets containing a label based on a FEC that the packet is assigned to. These packets are subsequently forwarded along a label switched path that allows the packets to be switched through the MPLS domain so that they can exit the network at the appropriate place. However, before the packets can be forwarded along a label switched path, the path must first be constructed. Label switched paths can be established using a variety of mechanisms defined for MPLS, generally including hand specification or a signaling protocol. The Label Distribution Protocol (LDP) can be used to signal and maintain such an LSP with little operator intervention, but it can also be customized to suit specific operational needs.

4.1 The Label Distribution Protocol

When an interface on an LSR is configured to run LDP and has LDP enabled on at least one of its interfaces, the LSR will attempt to establish LDP *sessions* between itself and the LSRs connected to the other ends of the interface. A remote LSR that understands LDP is referred to as an LDP *peer*. When the LDP session is being established, LSRs exchange configured session attributes to understand each other's capabilities. Some of these parameters such as timer values are negotiated. Once the sessions are established between two LDP peers, these sessions can be used to exchange MPLS labels as well as their mappings to particular FECs. These label exchanges result in the establishment of LDP-signaled LSPs, which is the whole motivation behind using LDP.

4.1.1 LDP Neighbors

LDP neighbors—or *peers* in LDP terminology—are discovered dynamically when physically adjacent LSRs are configured to run the Label Distribution Protocol. Once configured, each LSR running LDP will attempt to contact other LDP peers by sending LDP Hello messages to the well-known (or specifically configured) TCP/IP or UDP/IP LDP port on other LSRs. It is also possible to contact indirectly connected peers via *directed* LDP Hello messages by specifying a distant LDP peer explicitly in the LSR's configuration. Directed LDP Hellos may result in a *targeted LDP session* to be established between two LSRs. Directed LDP sessions allow distant LSRs to contact each other in the same way as discovered when they are directly connected. In either case, once contacted, the LDP peers exchange various types of configuration information. During this time, certain configuration parameters are negotiated as well. A session is established once this phase is successful. Once configured, the peers will exchange label request and label binding messages that will allow them to construct and tear down LDP-signaled LSPs.

4.1.2 The Label Information Base and Label Spaces

Labels that are distributed between LDP peers are kept at each LSR in the LSR's *Label Information Base* (LIB). The LIB is distinct from the Label Forwarding Information Base (LFIB) discussed earlier in Chapter 3 in that it keeps a record of all labels distributed between LDP peers irrespective of whether or not those labels are actually in use. Recall that, in contrast, the LFIB only contains labels that are actually being used to forward MPLS traffic. The relationship between the LFIB and the FIB is demonstrated in Figure 4.1 and resembles a set relationship with labels

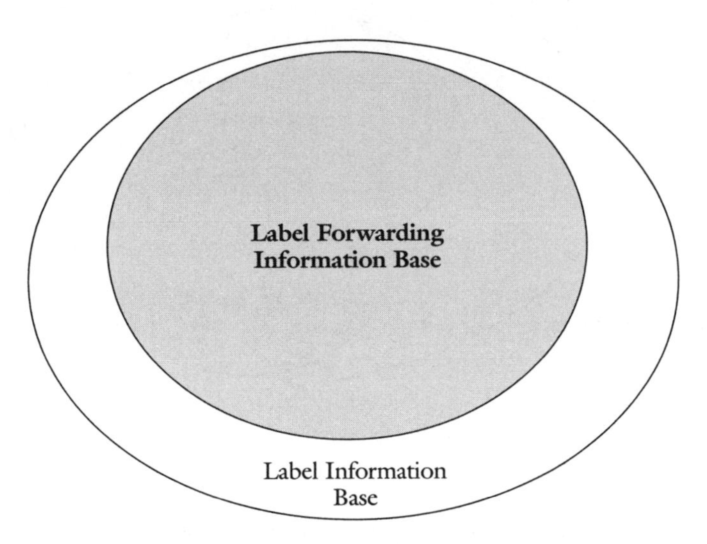

Figure 4.1 Label Forwarding Information Base as a proper subset of the Label Information Base.

being contained in one set or another depending on whether they are used for active switching or not. Note that the set of labels contained in the LFIB and the LIB can be identical under certain circumstances (see the later discussion of liberal versus conservative label retention modes). However, the databases are typically different at any given point in time for the simple reason that labels distributed to LDP peers may not be used immediately, or ever.

The LIB contains two types of label spaces that are used to maintain and partition the label-to-FEC mappings for LDP: the *global* or *platformwide* and the *per-interface* label spaces. The platformwide label space contains a pool of labels available to all MPLS-enabled interfaces except those supporting label switch controlled (LSC) Frame Relay or ATM interfaces. The per-interface label space contains a pool of labels that are significant only on a per-interface basis. Per-interface labels are typically assigned to label-controlled ATM interfaces that use virtual path identifiers (VPIs) and virtual circuit identifiers (VCIs) as labels or Frame Relay interfaces that use DLCIs that are mapped to MPLS labels. The LIB maintains each label space as a discrete proper subset of labels, and thus any set may overlap in part or completely with any other. This includes not only the interface-specific ranges, but the platformwide label range as well. An example of this is depicted in Figure 4.2. Notice that some of the labels from each label range overlap, which is legal since each subset is maintained individually for each label space.

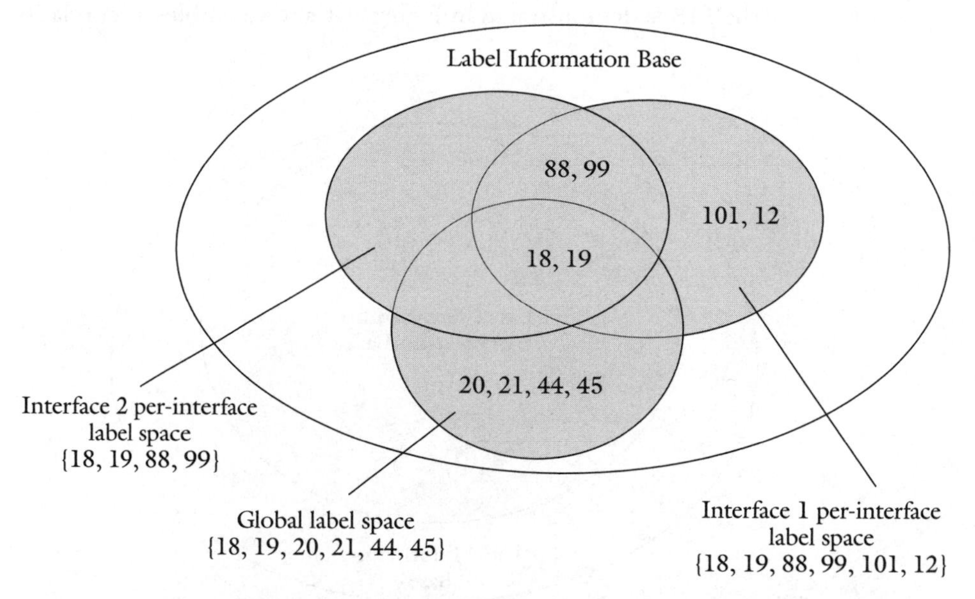

Figure 4.2 The Label Information Base contains overlapping subsets of label ranges that are subsets (possibly proper) of the overall set of labels contained by the LIB. Subsets of labels can be assigned to any supported label range, either per interface or per platform.

Interface-specific labels are maintained for several reasons. First, since ATM or Frame Relay labels possess local significance and can overlap—especially on a switch that supports multiple ATM and Frame Relay interfaces—it is important to maintain these separately. Second, it is important to closely govern the use of the labels on these types of interfaces since the method by which labels are allocated directly affects how ATM or Frame Relay–specific resouces are allocated (or overallocated) by a switch. Since there are a limited number of DLCIs for Frame Relay and VPI/VCIs for ATM, it is important to carefully control their allocation on a per-interface basis. This is accomplished in LDP through either *liberal* or *conservative* label retention modes. In conservative mode, LDP maintains only those label-to-FEC mappings that it needs at the current time. If it receives labels for which there are no current mappings, these labels are released. This saves on ATM VPI/VCIs or Frame Relay DLCIs being needlessly allocated because they are mapped to labels that are never used. In liberal mode, the LSR retains all mappings that have been advertised to it, even if some of them are not directly useful at the time of advertisement. The main advantage of liberal retention mode is that it generally results in quicker response to changes in routing. However, this comes at the cost of wasting labels that might never be used.

LDP offers several different modes of operation to distribute labels between peers, all of which are negotiated during LDP session establishment. These modes include *unsolicited* and *downstream on demand.* Both modes are available in *independent* (or *unordered*) *control mode* and *ordered control mode* flavors. The first mechanism allows the upstream LSR to explicitly request a label binding for a certain FEC from a downstream LSR. The downstream LSR must answer with a label-binding message to acknowledge the request. Under this operation, if independent control mode is in use, the label binding messages may occur at the same time. However, when using ordered control mode, these messages are distributed starting at the egress LSR and work their way back to the originating LSR. The second method, called *downstream on demand,* allows an LSR to send a label-binding message to another LSR for a specific LSP without being solicited. This mode is sometimes referred to as *downstream unsolicited,* or simply just *downstream.* The label binding message indicates the FEC that is associated with a particular LSP. When using ordered control mode, the first LSR to issue the label request message will act as the ingress LSR for the LSP. However, in independent control mode, any LSR along the LSP path associated with the FEC may act as the ingress point for the LSP.

Figure 4.3 depicts how label request and label mapping messages are distributed among LDP peers. In the figure, a FEC to label mapping is requested at the ingress and results in an LSP being signaled across the MPLS network. Different LDP signaling methods can be employed, but the eventual result is an LSP that carries traffic for the specified FEC.

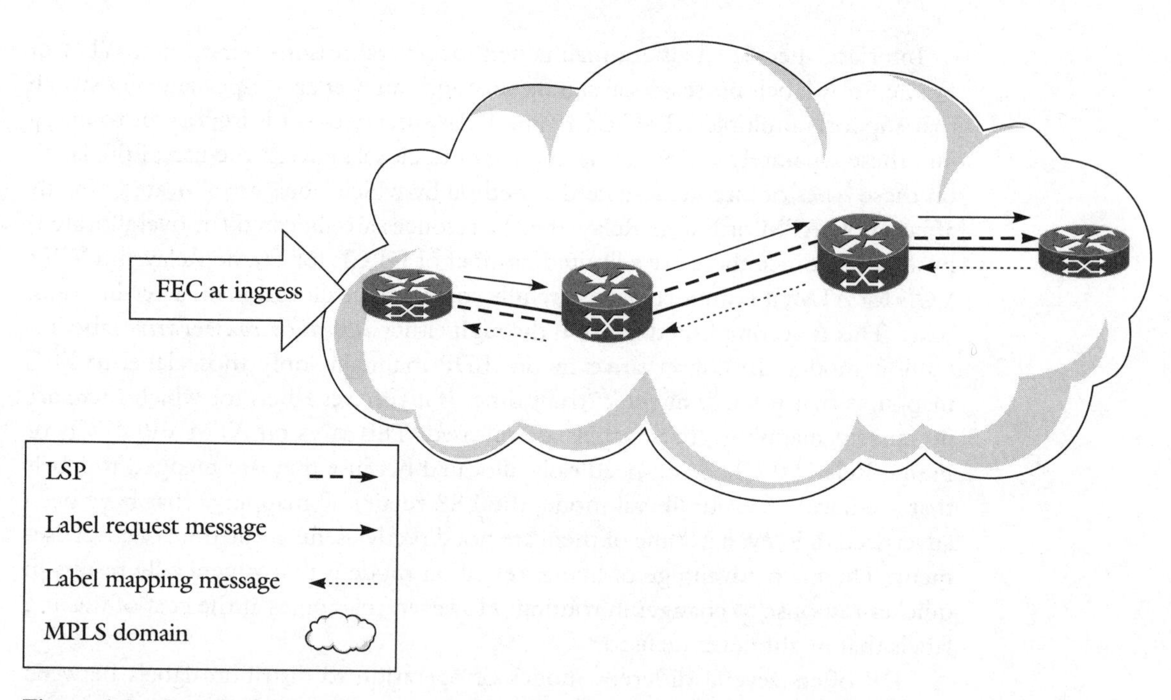

Figure 4.3 Signaling messages that are exchanged when signing an LSP using LDP. The label request and label mapping messages indicate the FEC to label mapping being requested by each LSR. Each LSR in the figure depicts an LSR running LDP. Each link between LSRs employs the Label Distribution Protocol; therefore each LSR is considered an LDP peer.

4.1.3 Label Switched Path Construction

Label switched paths are typically constructed in a control-driven manner using topology information found in the LSR's routing database. When converged, the routing database contains a list of shortest (or otherwise constrained) nonlooping paths within the MPLS domain. These paths are normally used by the routing process to determine the shortest path onto which to forward traffic. In the context of MPLS, this same information can be used to form the shortest path across the MPLS domain, except that instead of forwarding the traffic toward its destination using the IP encapsulation, for example, the traffic is instead encapsulated using an MPLS header and is forwarded along a label switched path using MPLS switching. This label switched path is first set up using the Label Distribution Protocol.

The example network shown in Figure 4.4 illustrates how LDP can be used to map traffic from a FEC containing several IP prefixes onto an LSP. Once each of the LSRs in the example has established LDP sessions with its adjacent neighbors, LDP can be used to construct control-driven LSPs that are based on information

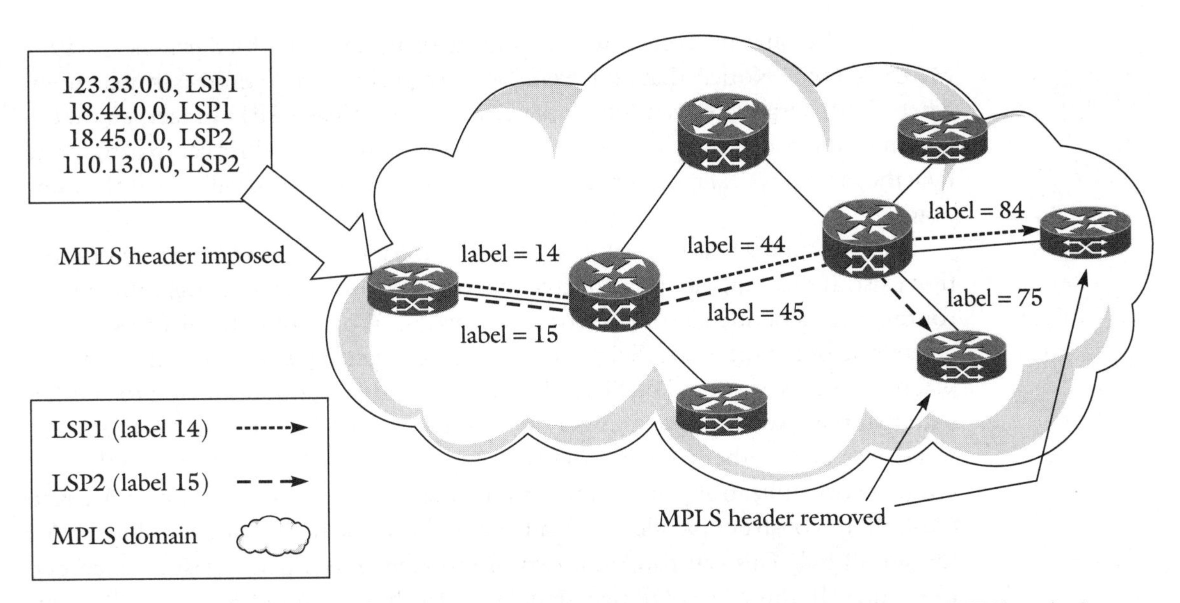

Figure 4.4 Four IP prefixes being mapped onto two MPLS label switched paths (LSP1/LSP2) at the ingress LER. Each LSP is an LDP-signaled LSP that was constructed using destination prefix information found in the ingress router's routing database. The IP packets have MPLS headers imposed at the ingress as they arrive, then traverse the MPLS domain using MPLS, and finally are stripped of the MPLS header at the egress LSR.

stored in the forwarding databases of each LSR. In the example, the leftmost LER accepts traffic for four IP destination prefixes that LDP will use the next-hop routing information from to create LDP-signaled LSPs across the sample network. Prefixes 123.33.0.0 and 18.44.0.0 may have similar MPLS headers imposed at this LER because they will both exit the MPLS domain from the same point and will traverse the same path across the network. Thus, in our example, traffic from these prefixes will be combined into the same FEC and forwarded using the same starting label of 14. Similarly, destination prefixes 18.45.0.0 and 110.13.0.0 exit the MPLS domain at the same point, and therefore will be forwarded with the same topmost label of 15. Note that it is also possible for individual FECs to be created for each IP prefix, and thus four labels and LSPs would be used to forward the traffic across the network. In the example, each LSP traverses the MPLS domain using locally significant labels until they reach the egress. At this point, the MPLS header is removed and the packets forwarded as IP encapsulated traffic. If MPLS were not enabled on any of the LSRs in Figure 4.4, the IP packets entering the network would still take the same path as the LSPs in the figure, but would be encapsulated using their native IP headers. This is the case because these paths represent the shortest, loop-free paths as calculated by the routing protocol for those destination prefixes. The only difference is that instead of using IP to forward the packets, MPLS is used. Notice that, for example, any packet that requires an exit point where LSP1 terminates can follow the exact same path as LSP1. In fact, an examination of the routing table entries for all of these prefixes will reveal that they will take the same physical path through the network and therefore will exit at the same point.

The only time when this will not happen is if parallel paths to the same destination exist. If this happens, it is possible that one path is chosen over the other by the routing protocol due to some metric comparison. When subsequent LSPs are signaled, one of the other parallel paths might be chosen. This can happen, for example, if two paths exist, and one has a higher bandwidth metric compared to another path that traverses the same number of hops. However, when the second LSP is signaled, one of the other parallel paths has a higher bandwidth metric due to the current network utilization, and so this path is chosen. A natural extension of this is to use LDP to construct parallel or "load-shared" paths across the MPLS domain for the *same* FEC. This can happen if two or more ***equal-cost*** paths exist for a certain FEC entry. In this case, LDP may signal multiple hop-by-hop LSPs that will terminate at the same egress LER, but that take different paths across the network. When LDP is used to construct such load-shared paths, it is up to the ingress LER to distribute the incoming packets among the multiple next-hop paths.

Finally, one interesting observation to make is to see how more than one (possibly hundreds or thousands) of IP destination prefixes can be collapsed into a single FEC and then mapped to the same outgoing label on the same interface. This can result in fewer switching entries at other LSRs within the core of the network. Some implementations have chosen to do this in order to save on labels, while others have taken the route of convenience and instead continue to add each IP prefix into its own FEC and thus to its own MPLS label. Both approaches work, and each has its benefits and drawbacks. It is also important for an operator who wishes to use the MPLS-LDP MIB to manage LDP to understand which approach a vendor has chosen to take because it may affect how the NMS displays the FEC to label mappings in the MIB tables. If two vendors' equipment employs different approaches, this could be potentially confusing for the operator managing the network.

4.2 Managing LDP

We now focus our attention on how the LDP protocol can be managed. We will now discuss the IETF MPLS-LDP MIB and how it can be used to effectively manage an LDP deployment. We will begin the discussion of managing LDP by first investigating the MPLS-LDP MIB and show how each object and table in this MIB can be used to manage a network. Along the way examples of how each table could be implemented by an LSR will be given, as well as an explanation of key concepts and features of the MIB. When appropriate, we will give illustrations of how an operator might use certain tables to monitor LDP not only when it is functioning, but also when it is malfunctioning. Particular attention will be paid to showing how particular tables might be viewed and used in collaboration with other tables in the MPLS-LDP MIB or in other MIBs as well.

LDP provides objects to configure potential, or monitor existing, LDP sessions on a specific LSR. The MPLS LDP Entity Table can be used to configure potential LDP sessions, where each row in the table represents an LDP session using a particular label space. This entry can be inserted by the LDP software or configured by the operator as a session the operator would like established in the future. For example, the operator may wish to configure targeted LDP sessions this way. Other entries in this table are added dynamically by the autodiscovery mechanism built into LDP. The MPLS LDP Peer Table is a read-only table that contains information learned from LDP peers via LDP discovery and the LDP session initialization message. Each row in the Peer Table represents an LDP peer relative to the LSR running the MIB. This table does not contain information about the local LDP entities per se. This is sometimes a subtle point made in the MIB that we would like to emphasize here for clarity. This table contains information that is specific to the peer-entity relationships that exist between the LSR running the MIB and those to which it has established LDP sessions, but which are not appropriate for the MPLS LDP Session Table. The MPLS LDP Session Table is used to represent the actual LDP sessions that are established between the LSR in question and its LDP peers. It is also particularly useful for monitoring those sessions that are in the process of being established, and of course, those specific sessions between an entity and a peer that already are established.

4.3 Definition of Terms Used in the MIB

The MIB uses some terminology either that is not included in the Label Distribution Protocol specification or that is difficult to understand—even for those who have used the MIB for some time—in particular, the terms *peer, entity, session,* and *hello adjacency.* The terms *cross-connect, insegment,* and *outsegment* were defined in Chapter 3. Table 4.1 enumerates and defines these terms in a manner that is hopefully clear and concise. Note that, for the purposes of this chapter, we will use the term *this LSR* to denote the LSR on which the MIB is being run (or queried).

Table 4.1 MPLS-LDP MIB terminology.

Term	Definition
LDP peer	An LDP peer is an MPLS LSR that has now, or has at some time in the past, established an LDP session with *this LSR*.
LDP entity	An LDP entity is an instance of the LDP protocol that controls a particular label space. This instance runs on *this LSR*. The entity is identified by the LSR ID of *this LSR* plus the label space identifier and is called the LDP ID. An LDP peer will never be an LDP entity.
LDP session	This is an LDP session run or running between *this LSR* and an LDP peer. The LDP Session Table will never contain an entry for any entry found in the LDP entity table since this refers to *this LSR*.
LDP hello adjacency	An LDP hello adjacency represents a relationship between *this LSR* and some LDP peer. Hello adjacencies do not exist between *this LSR* and itself.

4.3.1 MPLSLdpLsrId

This object reflects the LSR's LDP identifier. This object is defined as having a type of MPLSLsrIdentifier, meaning that it consists of 4 bytes that encode the label switching router ID (LSR ID). The LSR ID is typically the base IP address of the LSR. On some platforms, this address represents one of the loopback interfaces. This value should generally not change. If it does, it would potentially confuse an NMS, since NMS systems may wish to use this object to uniquely identify an LSR running LDP. Operators may wish to carefully regulate the configuration (or re-configuration) of this value on devices running in their network for the reasons just stated.

4.3.2 MPLSLdpLsrLoopDetectionCapable

This object is used to indicate whether the LSR supports LDP loop detection, and which loop detection modes the LSR in question has implemented. Table 4.2 enumerates and defines each possible value that an operator's NMS or OSS might encounter when interrogating this value. Implementations are encouraged not to use the other (2) value unless they have implemented some proprietary loop detection mechanism. This should be clearly indicated in their user documentation, as well as in their MPLS-LDP MIB agent capability statement.

Table 4.2 An LSR may indicate its LDP loop detection capabilities using the MPLSLdpLsrLoopDetectionCapable.

Enumeration name	Definition
None (1)	Loop detection is not supported on this LSR.
Other (2)	Loop detection is supported, but by a method other than those listed below. This may indicate that the LSR supports a vendor-proprietary or experimental loop detection mechanism.
HopCount (3)	Loop detection is supported by hop count only.
PathVector (4)	Loop detection is supported, but only with the path vector mechanism.
HopCountAndPathVector (5)	Loop detection is supported and both hop count and path vector methods are implemented.

Individual LDP sessions cannot be configured to run loop detection that differs from the mode of loop detection that is configured globally on the LSR; therefore, all sessions must run the same initial configuration of LDP loop detection. It is certainly possible to have two different sessions running different modes of loop detection after negotiation, since one LSR might be configured for HopCount-AndPathVector, but that LSR's peer is only configured for HopCount. In this case, the HopCount method will be employed. At the same time, another session might be capable of only PathVector. Since loop detection is negotiated during LDP session initialization, this value should be configured on the device *before* any sessions are negotiated.

4.4 The LDP Identifier

LDP entities are identified by a globally unique identifier called the LDP identifier or just the LDP ID. This identifier will be used throughout the MIB as a tabular index, making navigation through the MIB an easier task.

The LDP identifier is a 6-octet quantity used to identify an LDP entity that controls a specific label space on an LSR. This value is used to index several of the tables in the MIB and is therefore unique for any LDP sessions between *this LSR* and any LDP peers. Note that the entity table defined below contains a third index that can be used to disambiguate the case where an LSR has multiple label ranges defined for different interfaces using the same label space. This allows, for example, specific interfaces using the same overall label space to be confined to a subset of it. This also allows the agent to specify which interfaces are associated with a label space by creating multiple entries in that table for each.

The syntax for the LDP identifier is as follows:

<LSR ID> : <label space id>

The first 4 octets encode the LSR ID, which is typically the base IP address of the LSR. On some platforms, this address represents one of the loopback interfaces. The last 2 octets identify a specific label space within the LSR. The label space ID of 0x0000 (both bytes set to 0) indicates the global label space. Any other label space identifier indicates an interface-specific label space. An interface-specific label space is sometimes conveniently indicated using the ifIndex of the interface. Please note that this is a 2-byte quantity while the InterfaceIndex type is defined as a 4-byte value; thus, a direct mapping is not always possible. It was assumed that there will never be more than 2 bytes worth of parallel links between any two peers, so this approach should generally work as long as the value assigned never exceeds that of a 2-byte integer (0xffff). The following are examples of an LDP identifier: 10.1.1.1:0000, 195.0.0.5:0002. The first LDP identifier represents the global label space on LSR 10.1.1.1, while the second represents the interface-specific label space for interface with ifIndex = 02 and LSR ID 195.0.0.5.

Let's expand upon the label space example first shown in Figure 4.2 to show how those label spaces might be assigned LDP identifiers. This is shown in Figure 4.5. The base LSR ID for the LSR in the example is 12.14.5.1. This LSR maintains three label spaces; one platformwide label space maintains labels for interfaces IF5, IF6, and IF7. This label space is identified by 0x0000, and more completely, as the LDP Identifier {12.14.5.1:0000}. Note that all labels in this space may be received on or distributed to interfaces IF5, IF6, and IF7. The first interface-specific label range is maintained for IF1 and contains the labels 18, 19, 88, 99, 101, and 12. This label space is identified with the label space ID of {0x0001}, and when used in an LDP ID will appear as 12.14.5.1:0001. Similarly, the label range identifier maintained for interface IF2 is 0x0002, and when used as part of the LDP ID appears as {12.14.5.1:0002}.

With the discussion of the LDP ID in mind, we strongly recommend that operators or those constructing NMS systems used to manage networks running LDP consult the vendors who make the equipment that will reside in their networks to determine how their implementation assigns the label space ID component of the LDP ID. Some vendors may allow the operator to completely customize this value through configuration—some in a rather convoluted manner—while others may have an automatic algorithm that depends on another value such as the IP address of the first loopback interface on the device. It is also important to understand how

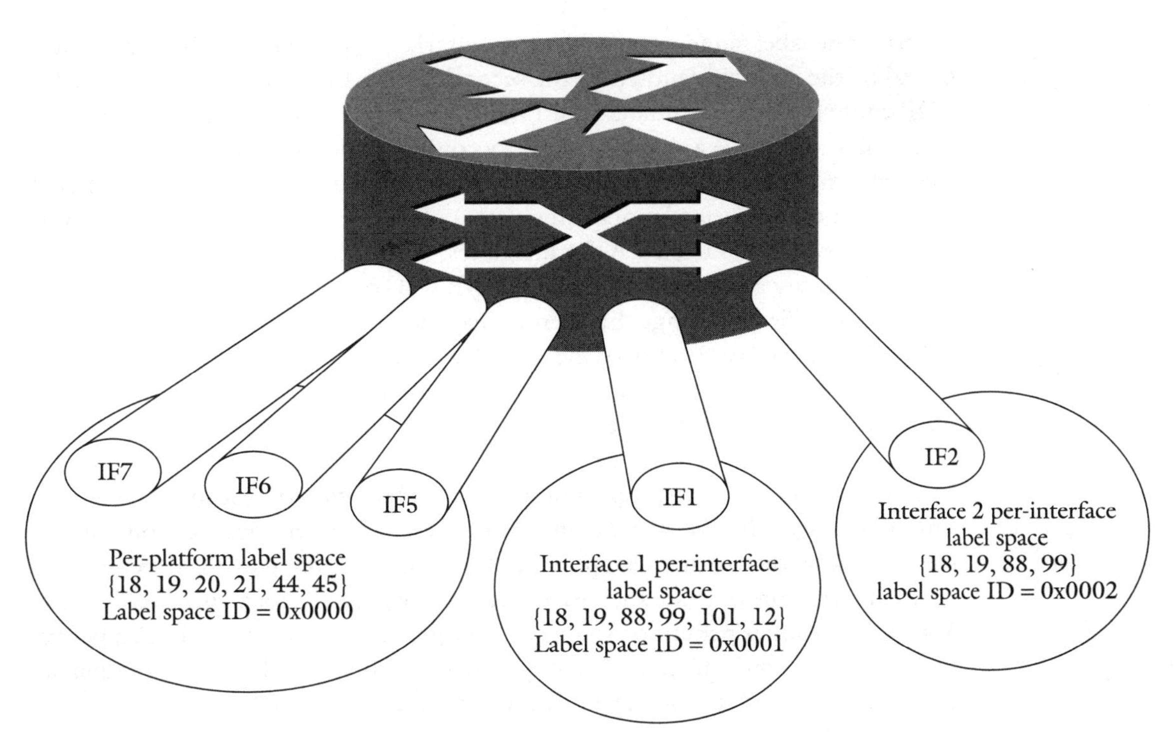

Figure 4.5 Label Forwarding Information Base for LSR 12.14.5.1.

the LDP Entity Instance ID is assigned (if it is used). In any event, the operator or NMS will need to know this information in order to understand how each device will identify itself. This is important because, for example, if the OSS or NMS systems an operator employs will be used to identify specific LSRs running LDP for tasks such as configuration, or detecting fault conditions when SNMP notifications are received, they will need to understand precisely which LSRs they are dealing with.

4.5 LDP Entity Table

The first table we will discuss is the LDP Entity Table. This table forms the foundation of the MPLS-LDP MIB by representing the LDP entities running on the LSR where the MIB is run as entries in this table. The term "entity" loosely correlates to a label space. This label space is typically managed by a device's LDP process or other software executing on the LSR where the MIB is running. This entity controls one label range over one or more interfaces. The specific label range maintained by the entity is denoted by the second 2 bytes of this identifier. At least one LDP entity must exist on the LSR if LDP is running. On some implementations, it is typical to only have a single LDP entity configured for all LDP sessions that utilize the same label space. However, other implementations will run separate LDP sessions for each interface connecting it to a peer. For example, if an LSR runs multiple label switch controlled ATM interfaces, each interface will typically have its own LDP entity assigned to it; thus the MIB will contain multiple entries for each interface-specific label range. Each entry may have the same LSR ID portion of the LDP ID, but the last 2 bytes of the LDP ID must be different to distinguish each interface-specific label range. Most implementations will have a single entity for the platformwide label range as well; however, some implementations will have multiple entities in this case: one for each interface connecting the LSR to its peer. If an implementation wishes to support different label retention or distribution methods simultaneously, different sessions and thus different LDP entities are required. Implementations that wish to support multiple signaling regimes for different purposes may separate out the label spaces accordingly, but are not actually required to. Session targeting is per entity, so only one targeted peer can be configured per entity. Thus, if an implementation wishes to share a label space between multiple targeted peers, it needs to use the entity index from the same LDP ID.

4.5.1 LDP Entity Table Indexing

The LDP Entity Table is indexed by two values: mplsLdpEntityLdpId and mplsLdpEntityIndex. The first value is used to identify the LDP ID. Typically, when an LDP session is established between two LSRs running LDP, each will identify itself using the aforementioned LDP ID. This is usually the case when LDP peers are connected via either a single link or multiple parallel links. In cases where only a single LDP session is run between peers, the mplsLdpEntityLdpId object is set to the entity's LDP ID and the mplsLdpEntityIndex will be set to 0 because it is unnecessary. This is illustrated in Figure 4.6, where two links exist between two LSRs. The LDP Entity Tables are shown in Table 4.3 for the leftmost LSR and Table 4.4 for the rightmost LSR.

Table 4.3 The LDP Entity Table for LSR 12.4.1.1 from Figure 4.6.

mplsLdpEntityLdpId	mplsLdpEntityIndex	Other columns . . .
12.4.1.1:00	0	...

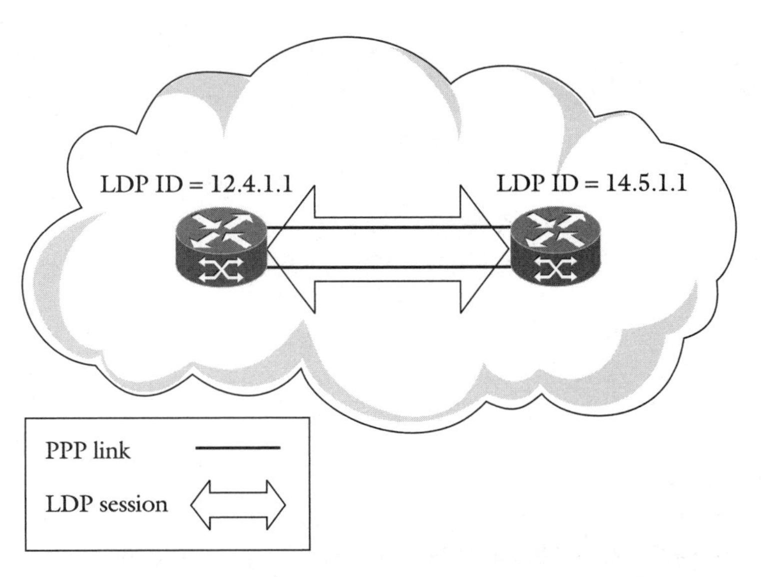

Figure 4.6 Parallel PPP links between peers running a single LDP session over both links. In this case, a single LDP entity will be used to represent the label space being exchanged between the peers. A single entry represents the LDP Entity Table on either LSR using the mplsLdpEntityLdpId as the primary index, and 0 as the mplsLdpEntityIndex in the LDP Entity Table.

Table 4.4 The LDP Entity Table for LSR 14.5.1.1 from Figure 4.6.

mplsLdpEntityLdpId	mplsLdpEntityIndex	Other columns . . .
14.5.1.1:00	0	...

Some implementations may wish to run individual LDP sessions for each link connecting two LDP peers. In particular, when an LSR has parallel connections between itself and another peer that will use the same label space, the same LDP ID must be used to identify the same entity running on each LSR. This poses a problem, as it may be difficult to identify which interface is being used for the session. What is done under these circumstances is to use an additional value that can be used to distinguish each interface being used by the same LDP entity, and thus the session. The MPLS-LDP MIB provides a secondary index for use under these circumstances; the secondary index mplsLdpEntityIndex will be set to a value that can distinguish one session from the other. An implementation may choose to simply start numbering the additional sessions by using a counting integer and

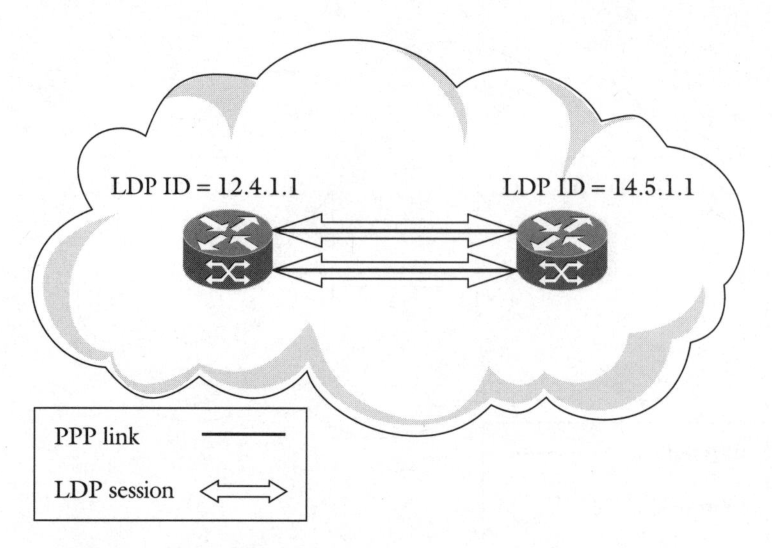

Figure 4.7 Parallel PPP links between peers running one LDP session per link. Each session will be represented as an (mplsLdp-EntityId, mplsLdpEntityIndex) entry in the LDP Entity Table. For example, the first may appear in 12.4.1.1's MPLS-LDP MIB as (12.4.1.1, 1).

increasing its value each time a new session is established. Other implementations may choose to assign the ifIndex of the link to the mplsLdpEntityIndex. For example, Figure 4.7 shows how two peers are connected via two links running PPP, with one LDP session running over each link. Both peers maintain a single platformwide label space, as is denoted by the "0x0000" portion of each entity's LDP ID. Since there are two links between each LSR, both running different sessions, but with the same label space, it is desirable to identify the sessions individually. The implementation in the example has chosen to assign the ifIndexes as the mplsLdpEntityIndex. The LDP Entity Tables are shown in Table 4.5 for the leftmost LSR and Table 4.6 for the rightmost LSR. In Table 4.5, the leftmost LSR has two interfaces, ifIndex = 1 and ifIndex = 2, that are used to carry the LDP sessions.

Table 4.5 The LDP Entity Table for LSR 12.4.1.1 from Figure 4.7.

mplsLdpEntityLdpId	mplsLdpEntityIndex	Other columns . . .
12.4.1.1:00	1	. . .
12.4.1.1:00	2	. . .

Similarly, the rightmost LSR's implementation uses two interfaces, ifIndex = 5 and ifIndex = 6, to support the LDP sessions; therefore, these appear in the LDP Entity Table in Table 4.6.

Table 4.6 The LDP Entity Table for LSR 14.5.1.1 from Figure 4.7.

mplsLdpEntityLdpId	mplsLdpEntityIndex	Other columns . . .
14.5.1.1:00	5	. . .
14.5.1.1:00	6	. . .

Table 4.7 lists and defines each object present in the LDP Entity Table.

Table 4.7 LDP Entity Table objects.

Object name	Definition
mplsLdpEntityProtocolVersion	This object indicates which LDP version number will be specified in the LDP session initialization message. This value will be requested during the negotiation of LDP parameters that takes place during LDP session establishment. The default protocol version is 1.
mplsLdpEntityAdminStatus	This value reflects the administrative status of the LDP entity. This object is set by default to **enable** so that LDP negotiation can take place as soon as this row is made active. When this object transitions from this state to the **disable** state ***and*** this entity has attempted to establish an LDP session with a peer, then any existing session with that peer is lost. All information pertaining to that peer also needs to be deprecated from the associated MIB tables. Note that this also implies that the mplsLdpEntityRowStatus object has been set to **down (2).** The user is only allowed to change values that are related to this entity when the value of this object is not **enabled.** When the administrative status is returned to **up,** then the LDP entity associated with this MIB entry should be instructed to establish new sessions with the specified peer.

continued

Table 4.7 continued

Object name	Definition
mplsLdpEntityOperStatus	The operational status of this LDP entity. The values of this are consistent with those defined in the SNMPv2 textual conventions for OperStatus.
mplsLdpEntityTcpDscPort	The TCP port used for discovering LDP peers on this entity. The default value is always set to the well-known value of 646.
mplsLdpEntityUdpDscPort	The UDP port used for discovering LDP peers on this entity. The default value is always set to the well-known value of 646.
mplsLdpEntityMaxPduLength	Indicates the maximum PDU length that is supported by this LDP entity. This value is transmitted as part of the Common Session Parameters portion of an LDP initialization message. Any value less than or equal to 255 indicates that the default maximum length of 4096 octets is desired.
mplsLdpEntityKeepAliveHoldTimer	This value contains the proposed keep alive hold timer for this LDP entity. This is a 2-octet quantity specified in seconds and has a default value of 40 seconds.
mplsLdpEntityHelloHoldTimer	This value contains the proposed LDP hello hold timer for this LDP entity. A value of 0 indicates that the default values of 15 seconds for link hellos and 45 seconds for targeted hellos are desired. A value of 65,535 indicates an infinite number of seconds—essentially no hello hold timer.
mplsLdpEntityInitSesThreshold	The LDP entity should send out the mplsLdpInitSes-ThresholdExceeded SNMP notification when the number of LDP session initialization messages sent when attempting to establish an LDP session exceeds this threshold. The transmission of this notification is important from an operational perspective, as it indicates a situation when this entity and its peer may be engaged in a cycle of messages resulting from one entity sending an error notification in response to the other's initialization messages. Setting this threshold appropriately allows the SNMP entity supporting this object to emit this notification to notify the operator. Setting this value to 0 indicates a desire not to have any notifications emitted.

Table 4.7 continued

Object name	Definition
mplsLdpEntityLabelDistMethod	This value specifies the method of label distribution employed by an entity. This value must be specified as either **downstreamOnDemand (1)** or **downstreamUnsolicited (2).**
mplsLdpEntityLabelRetentionMode	The LDP entity can be configured to use either conservative or liberal label retention mode. If the value of this object is **conservative (1),** then advertised label mappings are retained only if they will be used to forward packets immediately. If the value of this object is **liberal (2),** then all advertised label mappings are retained whether they are from a valid next-hop or not.
mplsLdpEntityPVLMisTrapEnable	Setting this object to **enabled (1)** instructs the SNMP agent supporting this MIB to emit the mplsLdpPVL-Mismatch notification. No notification will be sent if this value is set to **disabled (2).** The default value is **enabled (1).**
mplsLdpEntityPVL	This value indicates whether this entity supports loop detection for path vectors. The value of 0 indicates that this feature is disabled. A value greater than zero indicates both the Path Vector Limit and that the feature is supported. Note that if this object has a value greater than 0, the value of the object mplsLdpLsrLoopDetectionCapable must be set to either **pathVector (4)** or **hopCountAndPathVector (5).**
mplsLdpEntityHopCountLimit	This value indicates whether loop detection using hop counters is supported. This object is set to 0 if loop detection using hop counters is disabled; otherwise the value indicates both that it is supported as well as the entity's maximum allowable value for the hop count. The value of mplsLdpLsrLoopDetectionCapable must be set to either **hopCount (3)** or **hopCountAndPathVector (5)** if this feature is supported.
mplsLdpEntityTargPeer	If this LDP entity is configured to set up an LDP session with a targeted peer, then this value is set to true. The default value is false.

continued

Table 4.7 continued

Object name	Definition
mplsLdpEntityTargPeerAddrType	If mplsLdpEntityTargPeer is set to true, this value indicates the internetwork layer address type used for the extended discovery. The mplsLdpEntityTargPeerAddr object will contain an address using the format specified herein.
mplsLdpEntityTargPeerAddr	If mplsLdpEntityTargPeer is set to true, this value indicates the internetwork layer address used for the extended discovery. Note that the correct address type must also be specified in mplsLdpEntityTargPeerAddrType.
mplsLdpEntityOptionalParameters	This value is used to specify optional parameters for the LDP initialization message. If the value is set to **generic (1),** this indicates that no optional parameters will be sent in the LDP initialization message associated with this entity. However, if the value is set to **atmParameters (2),** then a row corresponding to this one must be created in the mplsLdpEntityAtmParms Table. Similarly, if the value is set to **frameRelayParameters (3),** then a corresponding row must be created in the mplsLdpEntityFrameRelayParms Table.
mplsLdpEntityDiscontinuityTime	The value of sysUpTime on the most recent occasion at which any one or more of this entity's counters suffered a discontinuity. Specifically, any Counter32 or Counter64 objects contained in the mplsLdpEntityStatsTable that encounter discontinuities should cause this value to be set. This object will contain the value of 0 if no discontinuities have occurred since the last reinitialization of the local management system.
mplsLdpEntityStorType	The storage type for this entry as defined in the SNMPv2 SMI's StorageType.
mplsLdpEntityRowStatus	An object that allows entries in this table to be created and deleted using the RowStatus convention. Once the mplsLdpEntityAdminStatus object has the value **up** and this object has the value of **active,** then the entity should instruct the LDP software on the device to contact the specified LDP peer. If the value of this object is changed to **notInService,** then the entity should instruct the LDP software on the device to drop the LDP session, thus losing

Table 4.7 continued

Object name	Definition
	contact with the LDP peer. All information related to that peer must be removed from the MIB. Note that doing so will have the same effect as changing mplsLdpEntity-AdminStatus from **enable** to **disable.** When this object is set to **active** and the value of mplsLdpEntityAdminStatus is set to **enable,** this entity will attempt to contact the peer and establish new sessions.

4.6 LDP Entity Configuration General Label Range Table

The MPLS LDP Entity Configurable Generic Label Range Table provides a mechanism by which contiguous ranges of platformwide labels can be specified. Labels are specified as a contiguous range that is associated with a specific LDP entity. Ranges in the platformwide label range can also be specified on a per-interface basis to allow for per-interface allocation of labels from the platformwide label space. Together, these ranges of labels form the definition of the label range associated with the LDP entity. Each entry in the table contains information about a single range of labels that are represented by the configured Upper and Lower Bounds pairs. Any LDP entity that uses labels from a platformwide label range must create at least one entry in this table to represent at least one *contiguous* label range. Some vendor equipment only allows for a single contiguous label range; thus operators may see only a single entry in the table. Each of the ranges specified for a particular entity must be unique and must not overlap with the other ranges specified for that entity. Therefore, if a noncontiguous label range needs to be specified, additional entries will need to be created to represent each nonoverlapping piece. An agent must not allow new rows to be created in this table unless a unique and nonoverlapping range is specified, so care should be taken by NMS systems that manage LDP not to allow them to be specified by the operator.

It is important to understand that there is no corresponding LDP protocol message that directly relates to the information stored in this table. That is, the entire label range is not distributed to any LDP peers directly as part of any LDP message; rather, it is done for each FEC to label mapping that is requested. This table does provide a useful way for an operator to view the configured label ranges in one single place, as well as for an NMS to visualize this information (see Figure 4.5). This table also provides the operator with a convenient location to configure reserved platformwide label ranges with minimal effort. It is important to note that this table is not to be used for the configuration of interface-specific label ranges; this is instead done in the mplsLdpEntityConfAtmLRTable (see Section 4.7.2) or mplsLdpEntityConfFrLRTable (see Section 4.8.2) for ATM or Frame Relay interface-specific label ranges, respectively.

4.6.1 LDP Entity Configuration General Label Range Table Indexing

The LDP Entity Configuration General Label Range Table is indexed by four objects: mplsLdpEntityLdpId, mplsLdpEntityIndex, mplsLdpEntityConfGenLRMin, and mplsLdpEntityConfGenLRMax. The first two objects identify the specific instance of the LDP entity. The last two objects specify the particular label range in question. This indexing allows for a manger to efficiently search for all label ranges associated with a particular LDP entity or LDP entity instance (if multiple instances are in use). This indexing facilitates the labels that are specified as a contiguous range that is associated with a specific LDP entity, or as a subset of labels that are assigned on a per-interface basis to allow for per-interface allocation of labels from the platformwide label space.

Table 4.8 lists and defines each object present in the LDP Entity Configuration General Label Range Table.

Table 4.8 LDP Entity General Label Range Configuration Table objects.

Object name	Description
MPLSLdpEntityConfGenLRMin	This object contains the minimum label value configured for this range.
MPLSLdpEntityConfGenLRMax	This object contains the maximum label value configured for this range.
mplsLdpEntityConfGenIfIndxOrZero	This object represents either the Interface Index of the IF-MIB layer where the generic labels are maintained by this LDP entity, or the value 0. The value of 0 indicates that the Interface Index is unknown. This can occur, for example, if the Interface Index is created after the entry in this table is created because the Interface Index would not be known. However, if the Interface Index is known, then the SNMP agent must fill in this field with the correct value. The standard also specifies that if a platformwide label is being actively used to forward data, then the value of this object must be that of the correct Interface Index.

Table 4.8 continued

Object name	Description
mplsLdpEntityConfGenLRStorType	The SNMP storage type for this entry. Valid values are **other (1),** meaning that a storage type other than the ones defined below is available; **volatile (2),** meaning that the row will be stored in RAM and will disappear after the device reboots; **Nonvolatile (3),** meaning that the value is stored in some sort of nonvolatile RAM and will be preserved across reboots of the system; **permanent (4),** meaning that the row is stored partially in ROM; or **readOnly (5),** meaning that the row is stored completely in ROM. Operators will typically find this value to be set to either **Nonvolatile (3)** or **volatile (2).**
mplsLdpEntityConfGenLRRowStatus	This object is configured with the object that allows entries in this table to be created and deleted using the SNMP RowStatus convention.

4.7 ATM Tables

Two tables exist to configure LDP when using ATM label controlled interfaces. These tables are the MPLS LDP Entity ATM Parameters Table and the MPLS LDP Entity Configuration ATM Label Range Table. The MPLS LDP Entity ATM Parameters Table provides a means by which an operator can configure information that would be contained in the Optional Parameter portion of an LDP PDU initialization message. The MPLS LDP Entity Configuration ATM Label Range Table provides a mechanism whereby an operator can configure information that would be contained in the ATM Label Range Components portion of an LDP PDU initialization message. Both tables can provide an operator with an effective means of managing LDP on label controlled ATM interfaces when used together and in collaboration with the aforementioned base entity tables.

4.7.1 MPLS LDP Entity ATM Parameters Table

This table contains ATM-specific information that can be configured in the LDP Optional Parameters. This table represents an *extends* relationship between it and the LDP Entity Table. That is, it augments those entries in the table, but only for those specific entries corresponding to label controlled ATM interfaces. Due to this potentially sparse relationship, the SNMP AUGMENTS tabular relationship is not used; instead, it is indexed using the same indexes as those of the base LDP Entity Table (see Section 4.5): mplsLdpEntityLdpId and mplsLdpEntityIndex. Thus, when this table is queried, it will only contain entries matching those in the mplsLdpEntityTable that represent LDP entities that are configured with ATM parameters. That is, entries in this table will not match one-for-one with those in the mplsLdpEntityTable.

Table 4.9 lists and defines each object present in the MPLS LDP Entity ATM Parameters Table.

Table 4.9 LDP Entity ATM Configuration Parameters Table objects.

Object name	Description
mplsLdpEntityAtmIfIndxOrZero	This object represents either the Interface Index of the IF-MIB layer where the ATM labels are maintained by this LDP entity, or the value 0. The value of 0 indicates that the Interface Index is unknown. This can occur, for example, if the Interface Index is created after the entry in this table is created because the Interface Index would not be known. However, if the Interface Index is known, then the SNMP agent must fill in this field with the correct value. The standard also specifies that if an ATM label is being actively used to forward data, then the value of this object should be that of the correct Interface Index, but it may also be zero.
mplsLdpEntityAtmMergeCap	This object denotes whether this entity supports ATM merge capability.
mplsLdpEntityAtmLRComponents	This object contains the number of label range components in the initialization message; that is, the number of entries in this and the MPLS LDP Entity Configuration ATM Label Range Tables that correspond to this entry.
mplsLdpEntityAtmVcDirectionality	This value is set to **bi-directional (0)** if the corresponding VCI and its associated VPI are used as the label for both directions independently. If the value is set to **unidirectional (1),** then the VCI and associated VPI designate one direction.
mplsLdpEntityAtmLsrConnectivity	If the peer LSR is connected to this LSR indirectly by means of an ATM VP, this value will be set to **indirect (2).** This can

Table 4.9 continued

Object name	Description
	happen so that the VPI values can be configured differently on either end point. If this happens, the label will be encoded entirely within the VCI field. If this is not the case, this value should be set to **direct (1),** which is the default value.
mplsLdpEntityDefaultControlVpi	This object contains the default VPI value for the non-MPLS control channel. The default value of this object is 0. Operators may modify this value to suit their specific deployment. This value must be configured as the same value on both ends of the ATM link.
mplsLdpEntityDefaultControlVci	This object contains the default VCI value for the non-MPLS control channel. The default value of this object is 32. Operators may modify this value to suit their specific deployment. This value must be configured as the same value on both ends of the ATM link.
mplsLdpEntityUnlabTrafVpi	This object contains the VPI of the VCC supporting unlabeled traffic (i.e., non-MPLS) for this interface. This non-MPLS connection is used to carry unlabeled packets that are typically IP. Note that the default value of this object is the same as the default value of the mplsLdpEntity-DefaultControlVpi. Operators may modify this value to suit their specific deployment. This value must be configured as the same value on both ends of the ATM link.
mplsLdpEntityUnlabTrafVci	This object contains the VCI of the VCC supporting unlabelled traffic (i.e., non-MPLS) for this interface. This non-MPLS connection is used to carry unlabelled packets that are typically IP. Note that the default value of this object is the same as the default value of the mplsLdpEntityDefault-ControlVci. Operators may modify this value to suit their specific deployment. This value must be configured as the same value on both ends of the ATM link.
mplsLdpEntityAtmStorType	The SNMP storage type for this entry.

continued

Table 4.9 continued

Object name	Description
mplsLdpEntityAtmRowStatus	This object is configured with the object that allows entries in this table to be created and deleted using the SNMP RowStatus convention. The SNMP agent implementation should make sure that this RowStatus object has the same value as that of the mplsLdpEntityRowStatus object related to this entry. NMS systems should prevent the user from changing this row status independently from that of the corresponding mplsLdpEntityRowStatus object; otherwise this could result in confusing errors being reported from the SNMP agent.

4.7.2 MPLS LDP Entity Configuration ATM Label Range Table

The MPLS LDP Entity Configuration ATM Label Range Table provides a mechanism for specifying a contiguous range of ATM VPIs and an associated range of contiguous ATM VCIs. Together, these represent an interface-specific label range for LDP entities that are used on ATM interfaces running the Label Distribution Protocol. Any implementation supporting this MIB on at least one ATM interface will have at least one entry in this table, but may have several if the entirety of the label range is noncontiguous.

This table is indexed in the same way that the LDP Entity Configuration General Label Range Table is. That is, it is indexed by four values: mplsLdpEntityLdpId, mplsLdpEntityIndex, mplsLdpEntityConfAtmLRMinVpi, and mplsLdpEntityConfAtmLRMinVci. The first two values indicate the specific LDP entity, while the final two values allow the table to be indexed by the specific minimum VPI/VCI label ranges defined for that LDP entity. Again, if the totality of the interface-specific label range is noncontiguous, multiple entries will appear in this table having different third and fourth indexes.

Each entry in this table contains information on a single range of labels represented by the configured upper and lower bound VPI/VCI pairs. The values specified in this table are the same used in the LDP session initialization message. It should be noted that operators are only allowed to specify unique and contiguous ranges for each specific LDP entity. No one range may overlap with any other entry configured for the same entity. In effect, each entry in the table represents a subset of the assigned label space that is analogous to a rectangle around a part of the label space. The collection of the subsets (i.e., rectangles) represents the entirety of the label space. The SNMP agent implementation should enforce this requirement. For example, take some LDP Entity Index 1.2.3.4:01, having its configured lower bound VPI/VCI configured as 0/32, and configured upper bound VPI/VCI set to 0/100. There might be an additional second entry with configured lower bound VPI/VCI configured as 0/101 and configured upper bound VPI/VCI set to 0/150. Both rows are legal, as the ranges do not overlap. However, a third entry with configured lower bound VPI/VCI configured to 0/150 and configured upper bound VPI/VCI set to 0/300 would be illegal because this label range overlaps with the values configured in the second entry. Specifically, both entries are configured with the label 0/150. One other important caveat specified in the standard for this MIB requires the agent to create at least one label range entry for a specific LDP entity that includes the default VPI/VCI values denoted in the LDP Entity Table.

As noted above, the SNMP agent should prevent the creation of rows in cases where an illegal (i.e., overlapping) label range is specified. The agent should return an error when an NMS attempts to create an illegal row. To facilitate easier row creation, the NMS should create rows in this table using "one-shot" row creation. This form of row creation requires that all the important OIDs be configured in a single set PDU. For this table, these objects would be the LDP EntityID, configured lower bound VPI/VCI, and configured upper bound VPI/VCI. Although recommended, it is not required, but this requires that the row not be active when the additional entries are added.

Table 4.10 lists and defines each object present in the LDP Entity Configuration ATM Label Range Table.

Table 4.10 LDP Entity Configuration ATM Label Range Table objects.

Object name	Object definition
mplsLdpEntityConfAtmLRMinVpi	This object is configured with the minimum VPI number for this range.
mplsLdpEntityConfAtmLRMinVci	This object is configured with the minimum VCI number configured for this range.
mplsLdpEntityConfAtmLRMaxVpi	This object is configured with the maximum VPI number configured for this range.
mplsLdpEntityConfAtmLRMaxVci	This object is configured with the maximum VCI number configured for this range.
mplsLdpEntityConfAtmLRStorType	This object is configured with the SNMP storage type for this entry.

continued

Table 4.10 continued

Object name	Object definition
mplsLdpEntityConfAtmLRRowStatus	This object is configured with the object that allows entries in this table to be created and deleted using the SNMP RowStatus convention. The SNMP agent implementation should make sure that this RowStatus object has the same value as that of the mplsLdp-EntityRowStatus object related to this entry. NMS systems should prevent the user from changing this row status independently from that of the corresponding mplsLdpEntityRowStatus object; otherwise this could result in confusing errors being reported from the SNMP agent.

4.8 Frame Relay Tables

Two tables exist in the MPLS-LDP MIB that can be used to configure LDP for using Frame Relay. These tables are the mplsLdpEntityFrameRelayParmsTable and the mplsLdpEntityConfFrLabelRangeTable. The mplsLdpEntityFrameRelay-ParmsTable provides a way for the operator to configure information that would be contained in the Optional Parameter portion of an LDP PDU initialization message. SNMP agents can also show the information that has been configured via other management interfaces in these tables for operators to view. This is especially important for those implementations that do not allow write access via the SNMP management interface, but do allow read access. The mplsLdpEntityConfFrLabel-RangeTable provides a means by which the operator can configure the information contained in the Frame Relay Label Range Components portion of an LDP PDU initialization message.

4.8.1 MPLS LDP Entity Frame Relay Parameters Table

This table contains Frame Relay–specific information that could be used in the LDP Optional Parameters and other Frame Relay–specific information. This table represents an *extends* relationship between it and the LDP Entity Table. That is, it augments those entries in the table, but only for those supporting Frame Relay interfaces. Due to this potentially sparse relationship, the SNMP AUGMENTS tabular relationship cannot be used; instead, it is indexed using the same indexes as those of the base LDP Entity Table (see Section 4.5) to *extend* the original table. Those indexes are the mplsLdpEntityLdpId and mplsLdpEntityIndex.

Table 4.11 lists and defines each object present in the MPLS LDP Entity Frame Relay Parameters Table.

Table 4.11 LDP Entity Frame Relay Parameters Table objects.

Object name	Description
mplsLdpEntityFrIfIndxOrZero	This object represents either the Interface Index of the IF-MIB layer where the Frame Relay labels are maintained by this LDP entity, or the value 0. The value of 0 indicates that the Interface Index is unknown. This can occur, for example, if the Interface Index is created after the entry in this table is created because the Interface Index would not be known. However, if the Interface Index is known, then the SNMP agent must fill in this field with the correct value. The standard also specifies that if a Frame Relay label is being actively used to forward data, then the value of this object must be that of the correct Interface Index.
mplsLdpEntityFrMergeCap	This object denotes whether Frame Relay merge capability is supported. Valid values are **notSupported (0)** and **supported (1).**
mplsLdpEntityFrLRComponents	This object is set to the number of Frame Relay label range components found in the LDP initialization message. It also represents the number of entries in the mplsLdpEntityConfFr-LRTable corresponding to this row in this table.
mplsLdpEntityFrLen	This object specifies the length of the DLCI bits in the Frame Relay label range portion of the LDP initialization message. There are two valid values: **tenDlciBits (0)** indicates that 10 bits are used, and **twentyThreeDlciBits (2)** indicates that 23 bits are used.
mplsLdpEntityFrVcDirectionality	This object indicates the directionality of the Frame Relay VC used. If the value of this object is set to **bi-directional (0),** then the LSR indicates that it supports the use of a DLCI as a label for both the transmitting and receiving directions

continued

Table 4.11 continued

Object name	Description
	independently. If the value of this object is set to **unidirectional (1),** then the LSR uses the given DLCI as a label for only one direction.
mplsLdpEntityFrParmsStorType	This object indicates the SNMP storage type for this entry.
mplsLdpEntityFrParmsRowStatus	This object is configured with the object that allows entries in this table to be created and deleted using the SNMP RowStatus convention. The SNMP agent implementation should make sure that this RowStatus object has the same value as that of the mplsLdpEntityRowStatus object related to this entry. NMS systems should prevent the user from changing this row status independently from that of the corresponding mplsLdpEntityRowStatus object; otherwise this could result in confusing errors being reported from the SNMP agent.

4.8.2 MPLS LDP Entity Configuration Frame Relay Label Range Table

The MPLS LDP Entity Configuration Frame Relay Label Range Table provides a mechanism for specifying a range of Frame Relay DLCIs that represent an interface-specific label range for LDP entities that are used on Frame Relay interfaces running the Label Distribution Protocol. Multiple contiguous ranges can be specified if the overall range is noncontiguous. Any implementation supporting this MIB on at least one Frame Relay interface will have at least one entry in this table, but may have several if the entirety of the label range is noncontiguous. The values specified in this table are the same used in the LDP session initialization message.

It should be noted that operators are only allowed to specify unique and contiguous ranges for each specific LDP entity. No one range may overlap with any other entry configured for the same entity. The SNMP agent implementation should enforce this requirement. For example, take some LDP Entity Index 1.2.3.4:01, having its mplsLdpConfFrMinDlci configured as 32, and mplsLdpConfFrMaxDlci set to 100. There might be an additional second entry with mplsLdpConfFrMinDlci configured as 200 and mplsLdpConfFrMaxDlci set to 250. Both rows are legal, as the ranges do not overlap. However, a third entry with mplsLdpConfFrMinDlci configured to 250 and mplsLdpConfFrMaxDlci set to 300 would be illegal because this label range overlaps with the values configured in the second entry. Specifically, both entries are configured with the label 250. One other important caveat specified in the standard for this MIB requires the agent to create at least one label range entry for a specific LDP entity that includes the default Frame Relay values denoted in the LDP Entity Table.

This table is indexed in the same way that the LDPEntity Configuration General Label Range Table is. That is, it is indexed by three values: mplsLdpEntityLdpId, mplsLdpEntityIndex, and mplsLdpConfFrMinDlci. The first two values indicate the specific LDP entity, while the final value allows the table to be indexed by the minimum DLCI value in a specific label range. Again, if the totality of the interface-specific label range is noncontiguous, multiple entries will appear in this table having different third indexes that are used to indicate the bottom of each range. The top of the range is specified by the mplsLdpConfFrMaxDlci object, but this object is not an index.

As noted above, the SNMP agent should prevent the creation of rows in cases where an illegal (i.e., overlapping) label range is specified. The agent should return an error when an NMS attempts to create an illegal row. To facilitate the easier row creation, the NMS should create rows in this table using "one-shot" row creation. This form of row creation requires that all the important OIDs be configured in a single set PDU. For this table, these objects would be the LDP EntityID, mplsLdpConfFrMinDlci, and mplsLdpConfFrMaxDlci.

Table 4.12 lists and defines each object present in the MPLS LDP Entity Configuration Frame Relay Label Range Table.

Table 4.12 LDP Entity Configuration Frame Relay Label Range Table objects.

Object Name	Description
mplsLdpConfFrMinDlci	This object indicates the lower bound DLCI value supported for the overall label range. This object should be set to the same value that is set in the Frame Relay Label Range Component's minimum DLCI field.
mplsLdpConfFrMaxDlci	This object indicates the upper bound DLCI value supported for this portion of the overall label range. This object should be set to the same value that is set in the Frame Relay Label Range Component's maximum DLCI field.
mplsLdpConfFrStorType	Indicates the SNMP storage type for this entry in the table.
mplsLdpConfFrRowStatus	This object is configured with the object that allows entries in this table to be created and deleted using the SNMP RowStatus convention. The SNMP agent implementation should make sure that

continued

Table 4.12 continued

	this RowStatus object has the same value as that of the mplsLdp-EntityRowStatus object related to this entry. NMS systems should prevent the user from changing this row status independently from that of the corresponding mplsLdpEntityRowStatus object; otherwise this could result in confusing errors being reported from the SNMP agent.

4.9 LDP Entity Example

Let us now investigate a detailed example of how the LDP Entity Tables just described might appear in the LDP-MIB. The example presented in Figure 4.8 depicts four LSRs connected together using various types of links. In particular, three

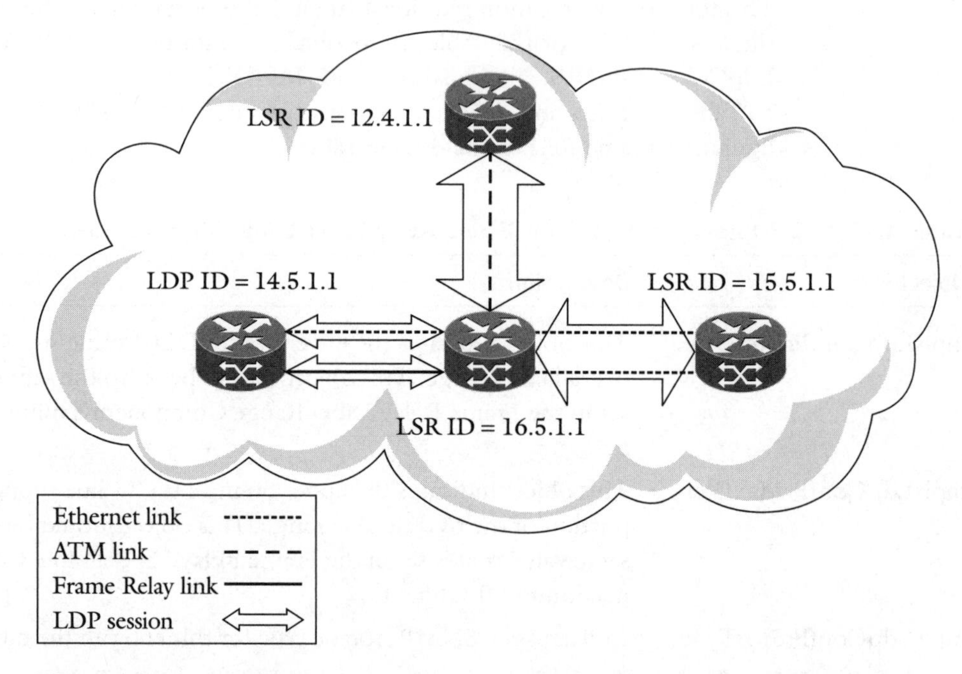

Figure 4.8 An example network running three LDP sessions between the LSR with LSR ID 16.5.1.1. Two parallel PPP links are run between this peer and the LSR with LSR ID 15.5.1.1, one Frame Relay and one Ethernet link between LSR ID 14.5.1.1, and one ATM link between it and LSR IDD 12.4.1.1.

LDP sessions are run between the LSR with LSR ID 16.5.1.1 and the other three LSRs. Two parallel Ethernet links are run between the LSR with LSR ID 16.5.1.1 and the LSR with LSR ID 15.5.1.1, with only a single LDP session running over both links, representing the platformwide label space. Additionally, two LDP sessions are run between LSR 16.5.1.1 and LSR 14.5.1.1, one over its Ethernet link and one over the Frame Relay link. In the case of the Frame Relay interface, an interface-specific label space is represented by the session utilizing that link. In the case of the Ethernet link, a new session using the platformwide label space is run over this link. The entity specified on 16.5.1.1 for the parallel PPP links will be the same one used for this link since all three links utilize the platformwide label space. Finally, a single ATM connection exists between LSRs 16.5.1.1 and 12.4.1.1. A single LDP session is run over this link supporting another interface-specific label range.

The LDP Entity Table for LSR 12.4.1.1 is shown in Table 4.13. Examining this table will reveal that it contains a single entry for the entity representing the single, interface-specific label range in use on the ATM interface. Let's assume that this interface possesses an Interface Index of 5. Thus, the entry in this table would be referenced by the indexes mplsLdpEntityLdpId = 12.4.1.1:0005 and mplsLdpEntityIndex = 0.

Table 4.13 The LDP Entity Table for LSR 12.4.1.1's ATM interface.

Columnar variable	Value of row with index (12.4.1.1:0005, 0)
mplsLdpEntityProtocolVersion	1
mplsLdpEntityAdminStatus	up (1)
mplsLdpEntityOperStatus	up (1)
mplsLdpEntityTcpDscPort	646 (well-known value)
mplsLdpEntityUdpDscPort	646 (well-known value)
mplsLdpEntityMaxPduLength	255 (default maximum length of 4096 octets in use)
mplsLdpEntityKeepAliveHoldTimer	40 (seconds)
mplsLdpEntityHelloHoldTimer	0 (default values of 15 seconds for link hellos and 45 seconds for targeted hellos)
mplsLdpEntityInitSesThreshold	5
mplsLdpEntityLabelDistMethod	downstreamOnDemand (1)
mplsLdpEntityLabelRetentionMode	liberal (2)

continued

Table 4.13 continued

Columnar variable	Value of row with index (12.4.1.1:0005, 0)
mplsLdpEntityPVLMisTrapEnable	enabled (1)
mplsLdpEntityPVL	hopCountAndPathVector (5)
mplsLdpEntityHopCountLimit	8
mplsLdpEntityTargPeer	false(0)
mplsLdpEntityTargPeerAddrType	0
mplsLdpEntityTargPeerAddr	0
mplsLdpEntityOptionalParameters	atmParameters (2)
mplsLdpEntityDiscontinuityTime	0
mplsLdpEntityStorType	readOnly (5)
mplsLdpEntityRowStatus	active (1)

Next, entries in both the LDP Entity ATM Configuration Parameters and the LDP Entity ATM Label Range Tables are required to support the ATM-specific parameters and label ranges required for the ATM interface. Both tables are indexed by the same indexing objects as the base LDP Entity Table, thus a single entry will appear in both tables corresponding to the LDP entity with LDP ID 12.4.1.1: mplsLdpEntityLdpId = 12.4.1.1:0005 and mplsLdpEntityIndex = 0. (See Tables 4.14 and 4.15.)

Table 4.14 LDP Entity ATM Configuration Parameters Table for the single LDP entity configured on LSR 12.4.1.1 in Figure 4.8.

Object name	Description
mplsLdpEntityAtmIfIndxOrZero	5
mplsLdpEntityAtmMergeCap	true (1)
mplsLdpEntityAtmLRComponents	1
mplsLdpEntityAtmVcDirectionality	bi-directional (0)
mplsLdpEntityAtmLsrConnectivity	direct (1)
mplsLdpEntityDefaultControlVpi	0
mplsLdpEntityDefaultControlVci	32
mplsLdpEntityUnlabTrafVpi	0

Table 4.14 continued

mplsLdpEntityUnlabTrafVci	32
mplsLdpEntityAtmStorType	readOnly (5)
mplsLdpEntityAtmRowStatus	active(1)

Table 4.15 LDP Entity ATM Label Range Table for the single LDP entity configured on LSR 12.4.1.1 in Figure 4.8.

Object name	Object definition
mplsLdpEntityConfAtmLRMinVpi	10
mplsLdpEntityConfAtmLRMinVci	15
mplsLdpEntityConfAtmLRMaxVpi	10
mplsLdpEntityConfAtmLRMaxVci	15
mplsLdpEntityConfAtmLRStorType	readOnly (5)
mplsLdpEntityConfAtmLRRowStatus	active (1)

Next, we list the LDP Entity Table and related Frame Relay tables for LSR 14.5.1.1. This LSR is connected to LSR 16.5.1.1 via one Frame Relay interface and one Ethernet interface. A separate LDP session will be run over each link due to the fact that the Ethernet interface will utilize the platformwide label range, and the Frame Relay interface will use an interface-specific label range. Since two label ranges are in use, two LDP entities need to be created, both of which will appear in the LDP Entity Table on this LSR. These entities will be designated as follows. The first mplsLdpLdpId = 14.5.1.1:0000 will be associated with the platformwide label range and the Ethernet interface. The other LDP entity for the Frame Relay interface's interface-specific label space will be designated as mplsLdpEntityLdpId = 14.5.1.1:0002. The Interface Index for the Frame Relay interface is assigned the value of 2, while the Ethernet interface has the Interface Index of 3. Since two LDP entities need to be represented, two entries will appear in the LDP Entity Table, indexed as follows. The interface-specific label range for the Frame Relay interface will be found in the row indexed by mplsLdpEntityLdpId = 14.5.1.1:0002 and mplsLdpEntityIndex = 0. The platformwide label range for the Ethernet interface will be found in the row indexed by mplsLdpEntityLdpId = 14.5.1.1:0000 and mplsLdpEntityIndex = 0.

Tables 4.16 and 4.17 illustrate possible configurations for these LDP entities on LSR 14.5.1.1 from Figure 4.8. Table 4.16 is associated with the LDP entity with the Ethernet interface. This LDP entity is indexed by mplsLdpEntityLdpId = 14.5.1.1:0000 and mplsLdpEntityIndex = 0.

Table 4.16 The LDP Entity Table for LSR 14.5.1.1's Ethernet interface.

Columnar variable	Value of row with index (14.5.1.1:0000, 0)
mplsLdpEntityProtocolVersion	1
mplsLdpEntityAdminStatus	up (1)
mplsLdpEntityOperStatus	up (1)
mplsLdpEntityTcpDscPort	646 (well-known value)
mplsLdpEntityUdpDscPort	646 (well-known value)
mplsLdpEntityMaxPduLength	255 (default maximum length of 4096 octets in use)
mplsLdpEntityKeepAliveHoldTimer	40 (seconds)
mplsLdpEntityHelloHoldTimer	0 (default values of 15 seconds for link hellos and 45 seconds for targeted hellos)
mplsLdpEntityInitSesThreshold	5
mplsLdpEntityLabelDistMethod	downstreamOnDemand (1)
mplsLdpEntityLabelRetentionMode	liberal (2)
mplsLdpEntityPVLMisTrapEnable	enabled (1)
mplsLdpEntityPVL	hopCountAndPathVector (5)
mplsLdpEntityHopCountLimit	8
mplsLdpEntityTargPeer	false (0)
mplsLdpEntityTargPeerAddrType	0
mplsLdpEntityTargPeerAddr	0
mplsLdpEntityOptionalParameters	generic (1)
mplsLdpEntityDiscontinuityTime	0
mplsLdpEntityStorType	readOnly (5)
mplsLdpEntityRowStatus	active (1)

Next, the entry representing one possible configuration for the LDP Entity General Label Range Table corresponding to the platformwide LDP entity is shown in Table 4.17. Note that this table is indexed by the same indexing objects as the base LDP Entity Table; thus a single entry will appear in both tables corresponding to the LDP entity with LDP ID 14.5.1.1: mplsLdpEntityLdpId = 14.5.1.1:0000, and mplsLdpEntityIndex = 0.

Table 4.17 LDP Entity General Label Range Configuration Table objects for the LDP entity with LDP ID 14.5.1.1: mplsLdpEntityLdpId = 14.5.1.1:0000, and mplsLdpEntityIndex = 0.

Columnar variable	Value of row with index (14.5.1.1:0000, 0)
MPLSLdpEntityConfGenLRMin	20
MPLSLdpEntityConfGenLRMax	1024
mplsLdpEntityConfGenIfIndxOrZero	3
mplsLdpEntityConfGenLRStorType	readOnly (5)
mplsLdpEntityConfGenLRRowStatus	active (1)

Finally, Tables 4.18–4.20 represent possible configurations of the LDP entity associated with the Frame Relay interface's interface-specific label range. This LDP entity is indexed by mplsLdpEntityLdpId = 14.5.1.1:0002 and mplsLdpEntityIndex = 0.

Table 4.18 The LDP Entity Table for LSR 14.5.1.1's Frame Relay interface.

Columnar variable	Value of row with index (14.5.1.1:0002, 0)
mplsLdpEntityProtocolVersion	1
mplsLdpEntityAdminStatus	up (1)
mplsLdpEntityOperStatus	up (1)
mplsLdpEntityTcpDscPort	646 (well-known value)
mplsLdpEntityUdpDscPort	646 (well-known value)
mplsLdpEntityMaxPduLength	255 (default maximum length of 4096 octets in use)
mplsLdpEntityKeepAliveHoldTimer	40 (seconds)
mplsLdpEntityHelloHoldTimer	0 (default values of 15 seconds for link hellos and 45 seconds for targeted hellos)

continued

Table 4.18 continued

Columnar variable	Value of row with index (14.5.1.1:0002, 0)
mplsLdpEntityInitSesThreshold	5
mplsLdpEntityLabelDistMethod	downstreamOnDemand (1)
mplsLdpEntityLabelRetentionMode	liberal (2)
mplsLdpEntityPVLMisTrapEnable	enabled (1)
mplsLdpEntityPVL	hopCountAndPathVector (5)
mplsLdpEntityHopCountLimit	8
mplsLdpEntityTargPeer	false (0)
mplsLdpEntityTargPeerAddrType	0
mplsLdpEntityTargPeerAddr	0
mplsLdpEntityOptionalParameters	FrameRelay (3)
mplsLdpEntityDiscontinuityTime	0
mplsLdpEntityStorType	readOnly (5)
mplsLdpEntityRowStatus	active (1)

Table 4.19 LDP Entity Configuration Frame Relay Label Range Table objects for the LDP entity with LDP ID 14.5.1.1: mplsLdpEntityLdpId = 14.5.1.1:0002, and mplsLdpEntityIndex = 0.

Columnar variable	Value of row with index (14.5.1.1:0002, 0)
mplsLdpConfFrMinDlci	15
mplsLdpConfFrMaxDlci	1024
mplsLdpConfFrStorType	readOnly (5)
mplsLdpConfFrRowStatus	active (1)

Table 4.20 LDP Entity Frame Relay Parameters Table objects for the LDP entity with LDP ID 14.5.1.1: mplsLdpEntityLdpId = 14.5.1.1:0002, and mplsLdpEntityIndex = 0.

Columnar variable	Value of row with index (14.5.1.1:0002, 0)
mplsLdpEntityFrIfIndxOrZero	2
mplsLdpEntityFrMergeCap	notSupported (0)
mplsLdpEntityFrLRComponents	1

Table 4.20 continued

Columnar variable	Value of row with index (14.5.1.1:0002, 0)
mplsLdpEntityFrLen	twentyThreeDlciBits (2)
mplsLdpEntityFrVcDirectionality	bi-directional (0)
mplsLdpEntityFrParmsStorType	readOnly (5)
mplsLdpEntityFrParmsRowStatus	active (1)

Let us now examine the LDP Entity Table for LSR 15.5.1.1. This LSR is connected to LSR 16.5.1.1 via two parallel Ethernet links. Notice that the implementation shown in the example has chosen to run a single LDP session over both links. This indicates that although the same LDP ID would be used to facilitate the platformwide label range for both connections, two LDP entities would appear in the MIB, representing both "instances" of the entity. These entries would be referenced by the indexes mplsLdpEntityLdpId = 15.5.1.1:0000 and mplsLdpEntityIndex = 0, and mplsLdpEntityLdpId = 15.5.1.1:0000 and mplsLdpEntityIndex = 1. Note that a simple counting integer is used to assign the mplsLdpEntityIndex to different instances of the same label space in this implementation. The implementation will distinguish instances of the same label space by assigning them the next value available and will "wrap around" when it reaches 0xffff. At this point, it should choose the next available instance index relative to this LDP entity. The topmost Ethernet interface is assigned Interface Index 2, while the other is assigned Interface Index 3. Tables 4.21 and 4.22 are examples of the tables that might be used to represent these LDP entities in the LDP MIB.

Table 4.21 The LDP Entity Table for LSR 15.5.1.1's first Ethernet interface.

Columnar variable	Value of row with index (15.5.1.1:0000, 0)
mplsLdpEntityProtocolVersion	1
mplsLdpEntityAdminStatus	up (1)
mplsLdpEntityOperStatus	up (1)
mplsLdpEntityTcpDscPort	646 (well-known value)
mplsLdpEntityUdpDscPort	646 (well-known value)
mplsLdpEntityMaxPduLength	255 (default maximum length of 4096 octets in use)
mplsLdpEntityKeepAliveHoldTimer	40 (seconds)

continued

Table 4.21 continued

Columnar variable	Value of row with index (15.5.1.1:0000, 0)
mplsLdpEntityHelloHoldTimer	0 (default values of 15 seconds for link hellos and 45 seconds for targeted hellos)
mplsLdpEntityInitSesThreshold	5
mplsLdpEntityLabelDistMethod	downstreamOnDemand (1)
mplsLdpEntityLabelRetentionMode	liberal (2)
mplsLdpEntityPVLMisTrapEnable	enabled (1)
mplsLdpEntityPVL	hopCountAndPathVector (5)
mplsLdpEntityHopCountLimit	8
mplsLdpEntityTargPeer	false (0)
mplsLdpEntityTargPeerAddrType	0
mplsLdpEntityTargPeerAddr	0
mplsLdpEntityOptionalParameters	generic (1)
mplsLdpEntityDiscontinuityTime	0
mplsLdpEntityStorType	readOnly (5)
mplsLdpEntityRowStatus	active (1)

Table 4.22 The LDP Entity Table for LSR 15.5.1.1's second Ethernet interface.

Columnar variable	Value of row with index (15.5.1.1:0000, 1)
mplsLdpEntityProtocolVersion	1
mplsLdpEntityAdminStatus	up (1)
mplsLdpEntityOperStatus	up (1)
mplsLdpEntityTcpDscPort	646 (well-known value)
mplsLdpEntityUdpDscPort	646 (well-known value)
mplsLdpEntityMaxPduLength	255 (default maximum length of 4096 octets in use)
mplsLdpEntityKeepAliveHoldTimer	40 (seconds)
mplsLdpEntityHelloHoldTimer	0 (default values of 15 seconds for link hellos and 45 seconds for targeted hellos)

Table 4.22 continued

Columnar variable	Value of row with index (15.5.1.1:0000, 1)
mplsLdpEntityInitSesThreshold	5
mplsLdpEntityLabelDistMethod	downstreamOnDemand (1)
mplsLdpEntityLabelRetentionMode	liberal (2)
mplsLdpEntityPVLMisTrapEnable	enabled (1)
mplsLdpEntityPVL	hopCountAndPathVector (5)
mplsLdpEntityHopCountLimit	8
mplsLdpEntityTargPeer	false (0)
mplsLdpEntityTargPeerAddrType	0
mplsLdpEntityTargPeerAddr	0
mplsLdpEntityOptionalParameters	generic (1)
mplsLdpEntityDiscontinuityTime	0
mplsLdpEntityStorType	readOnly (5)
mplsLdpEntityRowStatus	active (1)

Next, the entry representing one possible configuration for the LDP Entity General Label Range Table corresponding to the platformwide LDP entities is shown in Tables 4.23 and 4.24. Note that these tables are indexed by the same indexing objects as the base LDP Entity Table; thus two entries will appear corresponding to the LDP entity indexed by mplsLdpEntityId = 15.5.1.1:0000, mplsLdpEntityIndex = 1.

Table 4.23 LDP Entity General Label Range Configuration Table objects for the LDP entity with LDP ID 14.5.1.1: mplsLdpEntityLdpId = 15.5.1.1:0000, and mplsLdpEntityIndex = 0.

Columnar variable	Value of row with index (15.5.1.1:0000, 0)
MPLSLdpEntityConfGenLRMin	20
MPLSLdpEntityConfGenLRMax	1024
mplsLdpEntityConfGenIfIndxOrZero	2
mplsLdpEntityConfGenLRStorType	readOnly (5)
mplsLdpEntityConfGenLRRowStatus	active (1)

Table 4.24 LDP Entity General Label Range Configuration Table objects for the LDP entity with LDP ID 14.5.1.1: mplsLdpEntityLdpId = 15.5.1.1:0000, and mplsLdpEntityIndex = 1.

Columnar variable	Value of row with index (15.5.1.1:0000, 1)
MPLSLdpEntityConfGenLRMin	20
MPLSLdpEntityConfGenLRMax	1024
mplsLdpEntityConfGenIfIndxOrZero	2
mplsLdpEntityConfGenLRStorType	readOnly (5)
mplsLdpEntityConfGenLRRowStatus	active (1)

Let us now complete the investigation of the example shown in Figure 4.8 by showing how the entity tables might appear in LSR 16.5.1.1. LSR 16.5.1.1 is connected to LSR 15.5.1.1 via two parallel Ethernet links, and to LSR 14.5.1.1 via one Ethernet and one Frame Relay link. Since both of the links connecting to LSR 15.5.1.1 are Ethernet and use the platformwide label space, a single LDP session will run over these links. Two LDP entity entries can be created to represent the platformwide label space running over these links. Thus, two LDP Entity Table entries will be created to represent both with the following indexes: mplsLdpEntityLdpId = 16.5.1.1:0000 and mplsLdpEntityIndex = 5, and mplsLdpEntityLdpId = 16.5.1.1:0000 and mplsLdpEntityIndex = 6. LSR 16.5.1.1 is also connected to LSR 14.5.1.1 via an Ethernet connection. This link too participates in the platformwide label range, so a third LDP entity entry might be created to represent it: mplsLdpEntityLdpId = 16.5.1.1:0000 and mplsLdpEntityIndex = 7. The Frame Relay link connecting LSR 16.5.1.1 to LSR 14.5.1.1 utilizes the interface-specific label range for that interface; thus a separate LDP entity will need to be created for it with indexes of mplsLdpEntityLdpId = 16.5.1.1:0001 and mplsLdpEntityIndex = 0. Finally, an ATM link connects LSR 16.5.1.1 to LSR 12.4.1.1. This link utilizes a separate interface-specific label range for that interface; therefore, a separate LDP entity will need to be created for it with indexes of mplsLdpEntityLdpId = 16.5.1.1:0002 and mplsLdpEntityIndex = 0. The Interface Index assignments for LSR 16.5.1.1 are as follows: Frame Relay = 1, ATM interface = 2, topmost Ethernet connecting to 15.5.1.1 = 5, the bottommost Ethernet connecting to 15.5.1.1 = 6, and the Ethernet connecting to 14.5.1.1 = 7. Note that the mplsLdpEntityIndex has been assigned the Interface Index values. Special care has been taken with this example implementation not to exceed the 2-byte maximum for this value by not allowing ifIndexes with greater values to be assigned.

Table 4.25 represents the LDP entity that corresponds to the first Ethernet interface connecting to 15.5.1.1. The table represents the row indexed by mplsLdpEntityLdpId = 16.5.1.1:0000 and mplsLdpEntityIndex = 5 in the LDP Entity Table.

Table 4.25 The LDP Entity Table for LSR 16.5.1.1's first Ethernet interface connecting to 15.5.1.1.

Columnar variable	Value of row with index (16.5.1.1:0000, 5)
mplsLdpEntityProtocolVersion	1
mplsLdpEntityAdminStatus	up (1)
mplsLdpEntityOperStatus	up (1)
mplsLdpEntityTcpDscPort	646 (well-known value)
mplsLdpEntityUdpDscPort	646 (well-known value)
mplsLdpEntityMaxPduLength	255 (default maximum length of 4096 octets in use)
mplsLdpEntityKeepAliveHoldTimer	40 (seconds)
mplsLdpEntityHelloHoldTimer	0 (default values of 15 seconds for link hellos and 45 seconds for targeted hellos)
mplsLdpEntityInitSesThreshold	5
mplsLdpEntityLabelDistMethod	downstreamOnDemand (1)
mplsLdpEntityLabelRetentionMode	liberal (2)
mplsLdpEntityPVLMisTrapEnable	enabled (1)
mplsLdpEntityPVL	hopCountAndPathVector (5)
mplsLdpEntityHopCountLimit	8
mplsLdpEntityTargPeer	false (0)
mplsLdpEntityTargPeerAddrType	0
mplsLdpEntityTargPeerAddr	0
mplsLdpEntityOptionalParameters	generic (1)
mplsLdpEntityDiscontinuityTime	0
mplsLdpEntityStorType	readOnly (5)
mplsLdpEntityRowStatus	active (1)

Next, the entry representing one possible configuration for the LDP Entity General Label Range Table corresponding to the platformwide LDP entities is shown in Table 4.26. Note that this table is indexed by the same indexing objects as the base LDP Entity Table; thus two entries will appear corresponding to the LDP entity mplsLdpEntityLdpId = 16.5.1.1:0000 and mplsLdpEntityIndex = 5.

Table 4.26 LDP Entity General Label Range Configuration Table objects for the platformwide LDP entity with LDP ID 16.5.1.1 connecting over the first Ethernet interface between it and 15.5.1.1: mplsLdpEntityLdpId = 16.5.1.1:0000 and mplsLdpEntityIndex = 5.

Columnar variable	Value of row with index (16.5.1.1:0000, 5)
MPLSLdpEntityConfGenLRMin	20
MPLSLdpEntityConfGenLRMax	1024
mplsLdpEntityConfGenIfIndxOrZero	5
mplsLdpEntityConfGenLRStorType	readOnly (5)
mplsLdpEntityConfGenLRRowStatus	active (1)

Table 4.27 represents the LDP entity that corresponds to the second Ethernet interface connecting to 15.5.1.1. The table represents the row indexed by mpls-LdpEntityLdpId = 16.5.1.1:0000 and mplsLdpEntityIndex = 6 in the LDP Entity Table.

Table 4.27 The LDP Entity Table for LSR 16.5.1.1's second Ethernet interface connecting to 15.5.1.1.

Columnar variable	Value of row with index (16.5.1.1:0000, 6)
mplsLdpEntityProtocolVersion	1
mplsLdpEntityAdminStatus	up (1)
mplsLdpEntityOperStatus	up (1)
mplsLdpEntityTcpDscPort	646 (well-known value)
mplsLdpEntityUdpDscPort	646 (well-known value)
mplsLdpEntityMaxPduLength	255 (default maximum length of 4096 octets in use)
mplsLdpEntityKeepAliveHoldTimer	40 (seconds)

Table 4.27 continued

Columnar variable	Value of row with index (16.5.1.1:0000, 6)
mplsLdpEntityHelloHoldTimer	0 (default values of 15 seconds for link hellos and 45 seconds for targeted hellos)
mplsLdpEntityInitSesThreshold	5
mplsLdpEntityLabelDistMethod	downstreamOnDemand (1)
mplsLdpEntityLabelRetentionMode	liberal (2)
mplsLdpEntityPVLMisTrapEnable	enabled (1)
mplsLdpEntityPVL	hopCountAndPathVector (5)
mplsLdpEntityHopCountLimit	8
mplsLdpEntityTargPeer	false (0)
mplsLdpEntityTargPeerAddrType	0
mplsLdpEntityTargPeerAddr	0
mplsLdpEntityOptionalParameters	generic (1)
mplsLdpEntityDiscontinuityTime	0
mplsLdpEntityStorType	readOnly (5)
mplsLdpEntityRowStatus	active (1)

Next, the entry representing one possible configuration for the LDP Entity General Label Range Table corresponding to the platformwide LDP entities is shown in Table 4.28. Note that this table is indexed by the same indexing objects as the base LDP Entity Table; thus two entries will appear corresponding to the LDP entity mplsLdpEntityLdpId = 16.5.1.1:0000 and mplsLdpEntityIndex = 6.

Table 4.28 LDP Entity General Label Range Configuration Table objects for the platformwide LDP entity with LDP ID 16.5.1.1 connecting over the first Ethernet interface between it and 15.5.1.1: mplsLdpEntityLdpId = 16.5.1.1:0000, and mplsLdpEntityIndex = 6.

Columnar variable	Value of row with index (16.5.1.1:0000, 6)
MPLSLdpEntityConfGenLRMin	20
MPLSLdpEntityConfGenLRMax	1024

continued

Table 4.28 continued

Columnar variable	Value of row with index (16.5.1.1:0000, 6)
mplsLdpEntityConfGenIfIndxOrZero	6
mplsLdpEntityConfGenLRStorType	readOnly (5)
mplsLdpEntityConfGenLRRowStatus	active (1)

Table 4.29 represents the LDP entity that corresponds to the Ethernet interface connecting to 14.5.1.1. The table represents the row indexed by mplsLdpEntityLdpId = 16.5.1.1:0000 and mplsLdpEntityIndex = 7 in the LDP Entity Table.

Table 4.29 The LDP Entity Table for LSR 16.5.1.1's Ethernet interface connecting to 15.5.1.1.

Columnar variable	Value of row with index (16.5.1.1:0000, 7)
mplsLdpEntityProtocolVersion	1
mplsLdpEntityAdminStatus	up (1)
mplsLdpEntityOperStatus	up (1)
mplsLdpEntityTcpDscPort	646 (well-known value)
mplsLdpEntityUdpDscPort	646 (well-known value)
mplsLdpEntityMaxPduLength	255 (default maximum length of 4096 octets in use)
mplsLdpEntityKeepAliveHoldTimer	40 (seconds)
mplsLdpEntityHelloHoldTimer	0 (default values of 15 seconds for link hellos and 45 seconds for targeted hellos)
mplsLdpEntityInitSesThreshold	5
mplsLdpEntityLabelDistMethod	downstreamOnDemand (1)
mplsLdpEntityLabelRetentionMode	liberal (2)
mplsLdpEntityPVLMisTrapEnable	enabled (1)
mplsLdpEntityPVL	hopCountAndPathVector (5)
mplsLdpEntityHopCountLimit	8
mplsLdpEntityTargPeer	false (0)
mplsLdpEntityTargPeerAddrType	0

Table 4.29 continued

Columnar variable	Value of row with index (16.5.1.1:0000, 7)
mplsLdpEntityTargPeerAddr	0
mplsLdpEntityOptionalParameters	generic (1)
mplsLdpEntityDiscontinuityTime	0
mplsLdpEntityStorType	readOnly (5)
mplsLdpEntityRowStatus	active (1)

Next, the entry representing one possible configuration for the LDP Entity General Label Range Table corresponding to the platformwide LDP entities is shown in Table 4.30. Note that this table is indexed by the same indexing objects as the base LDP Entity Table; thus two entries will appear corresponding to the LDP entity mplsLdpEntityLdpId = 16.5.1.1:0000 and mplsLdpEntityIndex = 7.

Table 4.30 LDP Entity General Label Range Configuration Table objects for the platformwide LDP entity with LDP ID 16.5.1.1 connecting over the first Ethernet interface between it and 15.5.1.1: mplsLdpEntityLdpId = 16.5.1.1:0000 and mplsLdpEntityIndex = 7.

Columnar variables	Value of row with index (16.5.1.1:0000, 7)
MPLSLdpEntityConfGenLRMin	20
MPLSLdpEntityConfGenLRMax	1024
mplsLdpEntityConfGenIfIndxOrZero	7
mplsLdpEntityConfGenLRStorType	readOnly (5)
mplsLdpEntityConfGenLRRowStatus	active (1)

Next, let's investigate how the LDP Entity Table might be populated for the Frame Relay interface on LSR 15.5.1.1. An entity must be created to reflect an entity that represents an interface-specific label range, so a corresponding entity table entry must be created with the indexing mplsLdpEntityLdpId = 16.5.1.1:0001 and mplsLdpEntityIndex = 0 in the LDP Entity, LDP Entity Configuration Frame Relay Label Range, and LDP Entity Frame Relay Parameters Tables. Examples of how each table might be configured are given in Tables 4.31–4.33.

Table 4.31 The LDP Entity Table for LSR 16.5.1.1's Frame Relay interface.

Columnar variable	Value of row with index (16.5.1.1:0001, 0)
mplsLdpEntityProtocolVersion	1
mplsLdpEntityAdminStatus	up (1)
mplsLdpEntityOperStatus	up (1)
mplsLdpEntityTcpDscPort	646 (well-known value)
mplsLdpEntityUdpDscPort	646 (well-known value)
mplsLdpEntityMaxPduLength	255 (default maximum length of 4096 octets in use)
mplsLdpEntityKeepAliveHoldTimer	40 (seconds)
mplsLdpEntityHelloHoldTimer	0 (default values of 15 seconds for link hellos and 45 seconds for targeted hellos)
mplsLdpEntityInitSesThreshold	5
mplsLdpEntityLabelDistMethod	downstreamOnDemand (1)
mplsLdpEntityLabelRetentionMode	liberal (2)
mplsLdpEntityPVLMisTrapEnable	enabled (1)
mplsLdpEntityPVL	hopCountAndPathVector (5)
mplsLdpEntityHopCountLimit	8
mplsLdpEntityTargPeer	false (0)
mplsLdpEntityTargPeerAddrType	0
mplsLdpEntityTargPeerAddr	0
mplsLdpEntityOptionalParameters	FrameRelay (3)
mplsLdpEntityDiscontinuityTime	0
mplsLdpEntityStorType	readOnly (5)
mplsLdpEntityRowStatus	active (1)

Table 4.32 LDP Entity Configuration Frame Relay Label Range Table objects for the LDP entity with LDP ID 16.5.1.1:0001 and mplsLdpEntityIndex = 0.

Columnar variable	Value of row with index (16.5.1.1:0001, 0)
mplsLdpConfFrMinDlci	15
mplsLdpConfFrMaxDlci	1024

Table 4.32 continued

Columnar variable	Value of row with index (16.5.1.1:0001, 0)
mplsLdpConfFrStorType	readOnly (5)
mplsLdpConfFrRowStatus	active (1)

Table 4.33 LDP Entity Frame Relay Parameters Table objects for the LDP entity with LDP ID 16.5.1.1: mplsLdpEntityLdpId = 16.5.1.1:0001 and mplsLdpEntityIndex = 0.

Columnar variable	Value of row with index (16.5.1.1:0001, 0)
mplsLdpEntityFrIfIndxOrZero	1
MPLSLdpEntityFrMergeCap	notSupported (0)
mplsLdpEntityFrLRComponents	1
MPLSLdpEntityFrLen	twentyThreeDlciBits (2)
mplsLdpEntityFrVcDirectionality	bi-directional (0)
mplsLdpEntityFrParmsStorType	readOnly (5)
mplsLdpEntityFrParmsRowStatus	active (1)

Finally, we need to show how the LDP Entity Table might be populated for the ATM interface on LSR 16.5.1.1. An entity must be created to reflect an entity that represents an interface-specific label range, so a corresponding entity table entry must be created with the indexing mplsLdpEntityLdpId = 16.5.1.1:0002 and mplsLdpEntityIndex = 0 in the LDP Entity, LDP Entity Configuration ATM Label Range, and LDP Entity ATM Label Range Tables. Examples of how each table might be configured are given in Tables 4.34–4.36.

Table 4.34 The LDP Entity Table for LSR 16.5.1.1's ATM interface.

Columnar variable	Value of row with index (16.5.1.1:0002, 0)
mplsLdpEntityProtocolVersion	1
mplsLdpEntityAdminStatus	up (1)
mplsLdpEntityOperStatus	up (1)
mplsLdpEntityTcpDscPort	646 (well-known value)
mplsLdpEntityUdpDscPort	646 (well-known value)

continued

Table 4.34 continued

Columnar variable	Value of row with index (16.5.1.1:0002, 0)
mplsLdpEntityMaxPduLength	255 (default maximum length of 4096 octets in use)
mplsLdpEntityKeepAliveHoldTimer	40 (seconds)
mplsLdpEntityHelloHoldTimer	0 (default values of 15 seconds for link hellos and 45 seconds for targeted hellos)
mplsLdpEntityInitSesThreshold	5
mplsLdpEntityLabelDistMethod	downstreamOnDemand (1)
mplsLdpEntityLabelRetentionMode	liberal (2)
mplsLdpEntityPVLMisTrapEnable	enabled (1)
mplsLdpEntityPVL	hopCountAndPathVector (5)
mplsLdpEntityHopCountLimit	8
mplsLdpEntityTargPeer	false (0)
mplsLdpEntityTargPeerAddrType	0
mplsLdpEntityTargPeerAddr	0
mplsLdpEntityOptionalParameters	atmParameters (2)
mplsLdpEntityDiscontinuityTime	0
mplsLdpEntityStorType	readOnly (5)
mplsLdpEntityRowStatus	active (1)

Next, entries in both the LDP Entity ATM Configuration Parameters and the LDP Entity ATM Label Range Tables, shown in Tables 4.35 and 4.36, are required to support the ATM-specific parameters and label ranges required for the ATM interface. Both tables are indexed by the same indexing objects as the base LDP Entity Table; thus a single entry will appear in both tables corresponding to the LDP entity with LDP ID 16.5.1.1: mplsLdpEntityLdpId = 16.5.1.1:0002 and mplsLdpEntityIndex = 0.

Table 4.35 LDP Entity Configuration ATM Label Range Table objects for the LDP entity with LDP ID 16.5.1.1:0001 and mplsLdpEntityIndex = 0.

Columnar variable	Value of row with index (16.5.1.1:0002, 0)
mplsLdpEntityAtmIfIndxOrZero	2
mplsLdpEntityAtmMergeCap	true (1)

Table 4.35 continued

Columnar variable	Value of row with index (16.5.1.1:0002, 0)
mplsLdpEntityAtmLRComponents	1
mplsLdpEntityAtmVcDirectionality	bi-directional (0)
mplsLdpEntityAtmLsrConnectivity	direct (1)
mplsLdpEntityDefaultControlVpi	0
mplsLdpEntityDefaultControlVci	32
mplsLdpEntityUnlabTrafVpi	0
mplsLdpEntityUnlabTrafVci	32
mplsLdpEntityAtmStorType	readOnly (5)
mplsLdpEntityAtmRowStatus	active (1)

Table 4.36 LDP Entity ATM Label Range Table objects for the LDP entity with LDP ID 16.5.1.1:0001 and mplsLdpEntityIndex = 0.

Columnar variable	Value of row with index (16.5.1.1:0002, 0)
mplsLdpEntityConfAtmLRMinVpi	10
mplsLdpEntityConfAtmLRMinVci	15
mplsLdpEntityConfAtmLRMaxVpi	10
mplsLdpEntityConfAtmLRMaxVci	15
mplsLdpEntityConfAtmLRStorType	readOnly (5)
mplsLdpEntityConfAtmLRRowStatus	active (1)

4.10 Gathering Statistics for Entities

The MPLS-LDP MIB contains the MPLS LDP Entity Statistics Table (mplsLdp-EntityStatsTable) for maintaining counts related to errors when attempting to establish LDP sessions with a particular peer. This information can be used to alert the operator to misconfigurations in new or existing deployments, as well as LSR software that may have bugs. In particular, this table could be used to give insight into how to reconfigure values so that LDP sessions could be established successfully. For example, if the mplsLdpSessionRejectedLabelRangeErrors Counter object for a particular entity is viewed as increasing, then this would indicate that the label range might need to be adjusted. Another example might be when the hello timers are configured in an incompatible manner. In general, however, these counters should not increase on correctly configured deployments; therefore, if they are detected as increasing, the operator should give the LSR reporting them immediate attention. Note that the counters that keep track of errors should span all sessions past and present during the life of the entity.

The entries in this table are related to a specific LDP entity appearing in the LDP Entity Table (see Section 4.5). This table is associated with the LDP Entity Table using an SNMP AUGMENTS relationship. That is, every entry in this table corresponds to every entry in the LDP Entity Table. The mplsLdpEntityStatsTable is a read-only table that augments the mplsLdpEntityTable. The purpose of this table is to keep statistical information about the LDP entities on the LSR. Table 4.37 enumerates and defines each of the objects found in the mplsLdpEntityStatsTable.

Table 4.37 LDP Entity Statistics Table objects.

Object name	Description
mplsLdpAttemptedSessions	This object contains a count of the total attempted sessions for this LDP entity.
mplsLdpSesRejectedNoHelloErrors	This object contains a count of the Session Rejected/No Hello Error Notification Messages sent or received by this LDP entity.
mplsLdpSesRejectedAdErrors	This object contains a count of the Session Rejected/Parameters Advertisement Mode Error Notification Messages sent or received by this LDP entity.
mplsLdpSesRejectedMaxPduErrors	This object contains a count of the Session Rejected/Parameters Max PDU Length Error Notification Messages sent or received by this LDP entity.
mplsLdpSesRejectedLRErrors	This object contains a count of the Session Rejected/Parameters Label Range Notification Messages sent or received by this LDP entity.
mplsLdpBadLdpIdentifierErrors	This object contains a count of the number of Bad LDP Identifier Fatal Errors detected by the entity's past and present sessions.
mplsLdpBadPduLengthErrors	This object contains a count of the number of Bad PDU Length Fatal Errors detected by the entity's past and present sessions.

Table 4.37 continued

Object name	Description
mplsLdpBadMessageLengthErrors	This object contains a count of the number of Bad Message Length Fatal Errors detected by the entity's past and present sessions.
mplsLdpBadTlvLengthErrors	This object contains a count of the number of Bad TLV Length Fatal Errors detected by the entity's past and present sessions.
mplsLdpMalformedTlvValueErrors	This object contains a count of the number of Malformed TLV Value Fatal Errors detected by the entity's past and present sessions.
mplsLdpKeepAliveTimerExpErrors	This object contains a count of the number of Session Keep Alive Timer Expired Errors detected by the entity's past and present sessions.
mplsLdpShutdownNotifReceived	This object contains a count of the number of shutdown notifications sent by the entity's past and present sessions.
mplsLdpShutdownNotifSent	This object contains a count of the number of shutdown notifications received by the entity's past and present sessions.

4.11 LDP Peer Table

The LDP Peer Table (mplsLdpPeerTable) contains information pertaining to the LDP peers known to the LDP entities running on the LSR supporting the MIB. It is again possible to have two different flavors of implementations to support, keeping the examples shown earlier in mind. That is, if an implementation supports one LDP session per peer, regardless of the number of interfaces that connect it with the LDP entity, then only a single entity will exist in the LDP Entity Table regardless of the number of interfaces connecting the peer and entity together. Therefore, only a single peer entry corresponding to the entity will exist in the LDP Peer Table for each peer. However, as is the case with the LDP Entity Table, multiple entries are possible if multiple LDP sessions are connecting both peers since multiple entities will be connected to multiple (or possibly the same) peers.

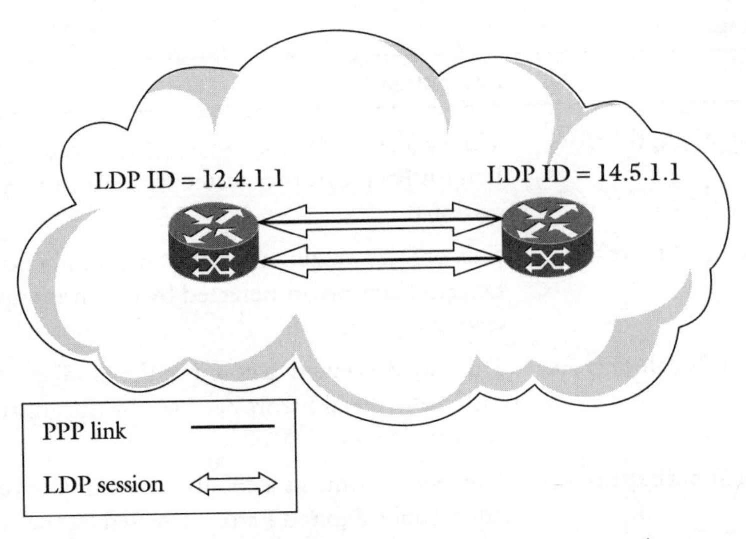

Figure 4.7 (*repeated*) Parallel PPP links between peers running one LDP session per link. Each session will be represented as an (mplsLdpEntityId, mplsLdpEntityIndex) entry in the LDP Entity Table. For example, the first may appear in 12.4.1.1's MPLS-LDP MIB as (12.4.1.1, 1).

Tables 4.38 and 4.39 demonstrate how LDP peers might be populated in the MPLS-LDP MIB's LDP Peer Table given the example from Figure 4.7, repeated here for your reference.

In Figure 4.7, a single LDP entity is connected on the leftmost LSR to a peer on the rightmost LSR. Both interfaces exchange labels from the platformwide label range; thus the label space ID used is 0x00. Because a single LDP session is run between each peer, despite multiple links connecting the two, a single LDP peer entry would appear in the MPLS-LDP MIB's Peer Table that corresponds with the single LDP entity entry. Notice that the indexing of Tables 4.38 and 4.39 reflects this.

Table 4.38 The LDP Peer Table for LSR 12.4.1.1 from Figure 4.7.

mplsLdpEntityLdpId	mplsLdpEntityIndex	mplsLdpPeerLdpId	Other columns . . .
12.4.1.1:00	0	12.5.1.1:00	. . .

Table 4.39 The LDP Peer Table for LSR 12.5.1.1 from Figure 4.7.

mplsLdpEntityLdpId	mplsLdpEntityIndex	mplsLdpPeerLdpId	Other columns . . .
12.5.1.1:00	0	12.4.1.1:00	. . .

Tables 4.40 and 4.41 demonstrate how LDP peers might be populated in the MPLS-LDP MIB's LDP Peer Table given the example from Figure 4.7. Recall that in this example, a unique LDP ID (LSR ID plus a unique Entity Instance ID) is created on each LSR for every interface connecting to the same LDP peer using the platformwide label range. Thus, in the example, two LDP entities are created, one for each interface. Table 4.40 demonstrates what the leftmost LSR's MPLS-LDP MIB Peer Table might look like. Notice that two LDP entities are created, each having the same LDP entity LDP ID, but differing in the mplsLdpEntityIndex, which corresponds to the Interface Index of each link. The mplsLdpPeerLdpId of the peer is the same for both connections, since both interfaces connect to the same entity that maintains the platformwide label range.

Table 4.40 The LDP Entity Table for LSR 12.4.1.1 from Figure 4.7.

mplsLdpEntityLdpId	mplsLdpEntityIndex	MPLSLdpPeerLdpId	Other columns . . .
12.4.1.1:00	1	12.5.1.1:00	. . .
12.4.1.1:00	2	12.5.1.1:00	. . .

Similarly, the rightmost LSR's LDP Entity Table contains two entries corresponding to interfaces 5 and 6 (see Table 4.41). Each interface denotes one LDP session connected to the LDP entity on the peer that maintains the platformwide label range.

Table 4.41 The LDP Entity Table for LSR 12.5.1.1 from Figure 4.7.

mplsLdpEntityLdpId	mplsLdpEntityIndex	MPLSLdpPeerLdpId	Other columns . . .
12.5.1.1:00	5	12.4.1.1:00	. . .
12.5.1.1:00	6	12.4.1.1:00	. . .

Rows in this table are read-only, as they reflect relationships that are learned via LDP discovery and the LDP session initialization messages. The values in this table are relevant to a peer and may or may not be the same values used in the session because some parameters are negotiated between peers. For example, the peer's Path Vector Limit information is learned from the session initialization phase. The actual value for the peer's Path Vector Limit may not be part of the session. There could also be a mismatch in this value between the entity and the peer. In the event of a mismatch, then the session will use the Path Vector Limit set by the entity (and not the peer). Finally, each row in this table is related to at least one row in the Hello

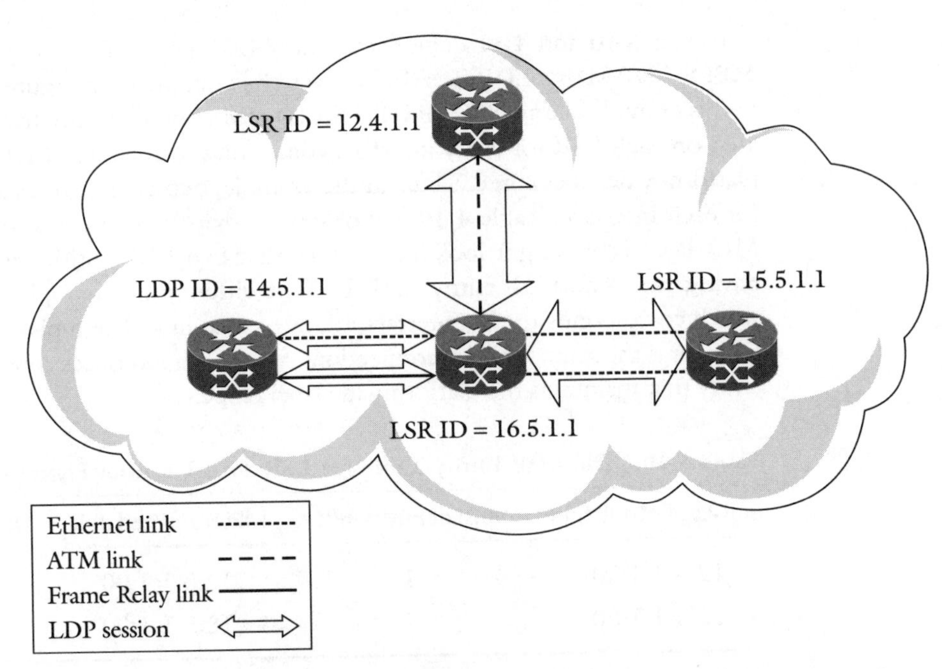

Figure 4.8 (*repeated*) An example network running three LDP sessions between the LSR with LSR ID 16.5.1.1. Two parallel PPP links are run between this peer and the LSR with LSR ID 15.5.1.1, one Frame Relay and one Ethernet link between LSR ID 14.5.1.1, and one ATM link between it and LSR IDD 12.4.1.1.

Adjacency Table. In addition, a row in this table is also associated with a row in the Session Table. Table 4.42 enumerates and defines the objects found in the mplsLdpPeerTable.

Table 4.42 LDP Peer Table objects.

Object name	Description
mplsLdpPeerLdpId	This object contains the LDP identifier of this LDP peer.
mplsLdpPeerLabelDistMethod	This object indicates the method of label distribution as either **downstreamOnDemand (1)** or **downstreamUnsolicited (2)** for any given LDP session.
mplsLdpPeerLoopDetectionForPV	This value indicates whether loop detection based on path vectors is disabled or enabled for this peer. If this object is set to **disabled (0)**, then loop detection is disabled; otherwise the object will be set to **enabled (1)** to indicate that loop detection based on path vectors is enabled.

Table 4.42 continued

Object name	Description
mplsLdpPeerPVL	This value represents that Path Vector Limit for this peer if mplsLdpPeerLoopDetectionForPV is set to **enabled (1).** SNMP agents implementing this object should set this value to 0 if mplsLdpPeerLoopDetectionForPV is set to **disabled (0).**

4.11.1 LDP Peer Table Example

Recall from the example in Figure 4.8, repeated again here for easy reference, that the LSRs in the example each had LDP Entity Table entries as shown in Table 4.43. Note that the Interface Indexes and descriptions are all relative to the LSR ID column in the table; thus some values such as Interface Index may be duplicated.

Table 4.43 Summary of LDP Entity Tables for LSRs in Figure 4.8.

LSR ID	LDP ID	LDP Instance ID	Interface Index	Interface description
12.4.1.1	12.4.1.1:0005	0	5	ATM
14.5.1.1	14.5.1.1:0000	0	3	Ethernet
14.5.1.1	14.5.1.1:0002	0	2	Frame Relay
15.5.1.1	15.5.1.1:0000	0	2	Topmost Ethernet
15.5.1.1	15.5.1.1:0000	1	3	Bottommost Ethernet
16.5.1.1	16.5.1.1:0000	5	5	Topmost Ethernet to 15.5.1.1
16.5.1.1	16.5.1.1:0000	6	6	Bottommost Ethernet to 15.5.1.1
16.5.1.1	16.5.1.1:0000	7	7	Ethernet to 14.5.1.1
16.5.1.1	16.5.1.1:0001	0	1	Frame Relay
16.5.1.1	16.5.1.1:0002	0	2	ATM

The example shows four LSRs interconnected in various ways. In particular, LSR 16.5.1.1 is depicted as being connected to the remaining LSRs using a variety of interconnection technologies. Let us now investigate what the LDP Peer Table might resemble on this LSR given the context of this example. This is shown in Table 4.44.

Table 4.44 Sample LDP Peer Table for LSR 16.5.1.1 from the example in Figure 4.8.

LDP ID	MPLS LDPEntity Index	LDPPeer LDPId	LDPPeer Label Dist Method	MPLS LDPPeer Loop Detection For PV	MPLS LDPPeer PVL
16.5.1.1:0000	5	15.5.1.1:0000	downstreamOnDemand (1)	enabled (1)	5
16.5.1.1:0000	6	15.5.1.1:0000	downstreamOnDemand (1)	enabled (1)	5
16.5.1.1:0000	7	14.5.1.1:0000	downstreamOnDemand (1)	enabled (1)	5
16.5.1.1:0001	0	14.5.1.1:0002	downstreamOnDemand (1)	enabled (1)	5
16.5.1.1:0002	0	12.4.1.1:0005	downstreamOnDemand (1)	enabled (1)	5

4.12 LDP Hello Adjacency Table

This is a table of all adjacencies between all LDP entities and all LDP peers. A session may have one or more adjacencies between it and an LSR if more than one parallel link between the two exists. This table is indexed by the mplsLdpEntity-LdpId, mplsLdpEntityIndex, mplsLdpPeerLdpId, and mplsLdpHelloAdjIndex. Table 4.45 enumerates and defines the objects contained in the LDP Hello Adjacency Table.

Table 4.45 LDP Hello Adjacency Table objects.

Object name	Description
mplsLdpHelloAdjIndex	This object contains an identifier that uniquely identifies this specific LDP hello adjacency.
mplsLdpHelloAdjHoldTimeRem	This object contains the value of the timer counting the remaining seconds for this hello adjacency. Operators monitoring this variable should note that this interval would change after each subsequent hello message corresponding to this hello adjacency is received.

Table 4.45 continued

Object name	Description
mplsLdpHelloAdjType	This object contains the type of hello adjacency that was received. This value is set to **link (1)** for nontargeted hello messages; otherwise, it is a result of a targeted hello message and is set to **targeted (2)** accordingly.

4.12.1 Hello Adjacency Table Example

Table 4.46 depicts how the Hello Adjacency Table might appear given the configuration from the example shown in Figure 4.8.

Table 4.46 Sample LDP Hello Adjacency Table for LSR 16.5.1.1 from the example in Figure 4.8.

LDP ID	MPLS LDPEntity Index	LDP Peer LDPId	MPLS LDP Hello Adj Index	MPLS LDPHello Adj Hold Time Rem	MPLS LDPHello Adj Type
16.5.1.1:0000	5	15.5.1.1:0000	0	4	link (1)
16.5.1.1:0000	6	15.5.1.1:0000	0	5	link (1)
16.5.1.1:0000	7	14.5.1.1:0000	0	3	link (1)
16.5.1.1:0001	0	14.5.1.1:0002	0	4	link (1)
16.5.1.1:0002	0	12.4.1.1:0005	0	5	link (1)

4.13 LDP Session Table

The LDP Session Table contains entries representing each session between an LDP entity on the LSR supporting the MIB and a peer. This table's objects must be implemented with read-only access since they only represent *active or activating* sessions; targeted sessions can be configured in the LDP Peer Table (see Section 4.11). This table is associated with the MPLS LDP Peer Table via the SNMP AUGMENTS relationship. That is, each entry in this table exists to correspond with each entry in the LDP Peer Table.

The following are a few important points about the content of some of the attributes in this table. Understanding these subtle points is very important, as we have seen them bring confusion to many an operator and device vendor's implementation! First, the Path Vector Limit for the session is the value configured in the corresponding local entry in the MPLS LDP Entity Table. The peer's Path Vector Limit is not located in this table; rather, it can be found in the MPLS LDP Peer Table. Second, in general, values found in this table may differ from those that have been configured depending on whether they have been negotiated with an LDP peer. If the values in this table are the same as those found in the configuration tables, several possibilities exist. First, negotiation might not have taken place or may be in progress, but values have not been exchanged. Second, it is also possible that negotiation will have resulted in the same values being used. Third, an entry may represent a session that was once established, but is now inactive, in which case the values in the table should contain the configured values. These tables cannot be used to make the distinction; what should be checked is whether the LDP session has been established between the peers. If so, then it is the case that the same values that were configured are now in use. Any value differing from the one configured in the corresponding configuration tables are noted in the objects of this table, the mplsLdpAtmSessionTable and the mplsLdpFrameRelaySessionTable. In summary, the values in these tables reflect the LDP parameters that have been *negotiated* during LDP session establishment if session negotiation has taken place; otherwise they will be equivalent to the configured values. SNMP agents implementing this MIB should make sure that they follow these guidelines. Table 4.47 enumerates and defines the objects contained in the mplsLdpSessionTable.

Table 4.47 LDP Session Table objects.

Object name	Description
mplsLdpSesState	This object indicates the current state of the LDP session. Valid states are **nonexistent (1), initialized (2), openrec (3), opensent (4)**, and **operational (5).** All states returned by this object are based on the state machine for session negotiation behavior. Please see RFC 3036 for specific definitions of each state.
mplsLdpSesProtocolVersion	This object indicates the version of the LDP protocol negotiated during session initialization.
mplsLdpSesKeepAliveHoldTimeRem	This object indicates the keep alive hold time remaining for this session.

Table 4.47 continued

Object name	Description
mplsLdpSesMaxPduLen	This value indicates the maximum allowable length for LDP PDUs negotiated during session initialization for this session.
mplsLdpSesDiscontinuityTime	The value of sysUpTime on the most recent occasion at which any one or more of this entity's counters suffered a discontinuity. Specifically, any Counter32 or Counter64 objects contained in this table that encounter discontinuities should cause this value to be set. This object will contain the value of 0 if no discontinuities have occurred since the last reinitialization of the local management system. Note that it is possible for an NMS to distinguish when a session is no longer functioning and then is reestablished; therefore, the agent should change this value in this case to indicate to the NMS that this is a different session instance. This also implies that NMS systems should monitor this value closely in addition to the mplsLdpSesState and the LdpSessionUp/Down notifications.

4.14 LDP ATM Session Table

The MPLS LDP ATM Session Table (mplsLdpAtmSessionTable) contains session information specific to an LDP session that runs over an ATM interface. This table is read-only, as it only reflects the active state of the session. Configuration information is stored elsewhere in the MIB. This table relates an ATM LDP session with its corresponding entry in the LDP Session Table. There could be one or more label range intersections between an LDP entity and LDP peer using ATM as the underlying media; thus the label range intersections are noted in this table as well. Each entry in this table represents a single label range intersection. The indexing of this table is that of an *extends* SNMP tabular relationship with the base LDP Session Table. This relationship is required instead of the SNMP AUGMENTS relationship because there is not necessarily a one-to-one mapping between all entries in this table and all entries in the LDP Session Table. The table contains five indexes: mplsLdpEntityLdpId, mplsLdpEntityIndex, mplsLdpPeerLdpId, mplsLdp-

SesAtmLRLowerBoundVpi, and mplsLdpSesAtmLRLowerBoundVci. Note that this indexing allows for the use of a subrange of label spaces on negotiation with a peer. Table 4.48 defines and lists each of the objects contained in the mplsLdpAtmSessionTable.

Table 4.48 LDP ATM Session Table objects.

Object name	Description
mplsLdpSesAtmLRLowerBoundVpi	This object contains the minimum VPI number for this range.
mplsLdpSesAtmLRLowerBoundVci	This object contains the minimum VCI number for this range.
mplsLdpSesAtmLRUpperBoundVpi	This object contains the maximum VPI number for this range.
mplsLdpSesAtmLRUpperBoundVci	This object contains the maximum VCI number for this range.

4.15 LDP Frame Relay Session Table

The MPLS LDP Frame Relay Session Table (mplsLdpFrameRelaySessionTable) contains session information specific to an LDP session that runs over a Frame Relay interface. This table is read-only, as it only reflects the active state of the session. Configuration information is stored elsewhere in the MIB. This table relates a Frame Relay LDP session with its corresponding entry in the LDP Session Table. There could be one or more label range intersections between an LDP entity and LDP peer using Frame Relay as the underlying media; thus the label range intersections are noted in this table as well. Each entry in this table represents a single label range intersection. The indexing of this table is that of an *extends* SNMP tabular relationship with the base LDP Session Table. This relationship is required instead of the SNMP AUGMENTS relationship because there is not necessarily a one-to-one mapping between all entries in this table and all entries in the LDP Session Table. This table is indexed by the following four objects: mplsLdpEntityLdpId, mplsLdpEntityIndex, mplsLdpPeerLdpId, and mplsLdpFrSesMinDlci. Table 4.49 shown below enumerates and defines the objects found in the mplsLdpFrameRelaySessionTable.

Table 4.49 LDP Frame Relay Session Table objects.

Object name	Description
MPLSLdpFrSesMinDlci	This object contains the lower bound of DLCIs that are supported.
MPLSLdpFrSesMaxDlci	This object contains the upper bound of DLCIs that are supported.
MPLSLdpFrSesLen	This object specifies the DLCI bits.

4.16 The LDP Session Statistics Table

The MPLS LDP Session Statistics Table (mplsLdpSessionStatsTable) is a read-only table that contains statistical information for each LDP session represented in the LDP Session Table. This table is associated with the base LDP Session Table using an AUGMENTS SNMP tabular relationship. More precisely, an entry in this table exists for every entry in the LDP Session Table. Table 4.50 enumerates and defines the objects found in the mplsLdpSessionStatsTable.

Table 4.50 LDP Session Statistics Table objects.

Object name	Description
mplsLdpSesStatsUnkMesTypeErrors	This object counts the number of Unknown Message Type Errors detected during this session. Discontinuities in the value of this counter can occur at reinitialization of the management system and at other times as indicated by the value of mplsLdpSesDiscontinuityTime.
mplsLdpSesStatsUnkTlvErrors	This object counts the number of Unknown TLV Errors detected during this session. Discontinuities in the value of this counter can occur at reinitialization of the management system and at other times as indicated by the value of mplsLdpSessionDiscontinuityTime.

4.17 The LDP Session Peer Address Table

The MPLS LDP Session Peer Address Table (mplsLdpSessionPeerAddressTable) stores addresses learned after LDP session initialization via the address message advertisement. An entry in this table represents information on a session's particular next-hop address that was advertised in an address message and received from an LDP peer. This table *extends* the MPLS LDP Session Table in that it is indexed using the same indexes as that table, but it does not necessarily correspond to each entry on a one-to-one basis because the address message advertisement may not have been used during each session initialization. All objects in this table are implemented by SNMP agents with a maximum level of access of read-only because entries in this table represent dynamically learned information. Nothing in this table is configurable.

Agents implementing this table should update it whenever LDP label withdraw address messages are received. Specifically, rows should be deleted as soon as they are withdrawn by the peer. It is important for implementations to note that since more than one address may be contained in a label address message, agents must be able to handle updates to this table under those circumstances. This table is indexed by four objects: mplsLdpEntityLdpId, mplsLdpEntityIndex, mplsLdpPeerLdpId, and mplsLdpSesPeerAddrIndex. The first three are the same indexes used in the LDP Session Table, and the last represents the address that was learned. Table 4.51 enumerates and defines the objects contained in the mplsLdpSessionPeerAddressTable.

Table 4.51 LDP Session Peer Address Table objects.

Object name	Description
mplsLdpSesPeerAddrIndex	This object contains an index that uniquely identifies this entry within a given session.
mplsLdpSesPeerNextHopAddrType	This object contains the internetwork layer address type of the next-hop address as specified in the label address message associated with this session. The value of this object indicates how to interpret the value of mplsLdpSessionPeerNextHopAddress.
mplsLdpSesPeerNextHopAddr	This object contains the network type used in mplsLdpSesPeerNextHopAddrType. However, operators and implementers alike should note that the LDP specification only defines the IPv4 type for use in the LDP Protocol Version 1.

4.18 Modification of Established LDP Sessions

LDP sessions are negotiated using the values found in the entries of the mplsLdpSessionTable. Negotiated values are then stored in this table once the session negotiation is complete. Therefore, entries in the mplsLdpSessionTable cannot be modified once session initialization starts. Special care should be taken when modifying the MPLS-LDP MIB objects that are used in the MPLS LDP session initialization. If the modification of any of these MIB variables takes place after the start of session initialization, the agent implementation must ensure that the entire session is stopped, and that any information learned by that session is discarded. This is can result in service disruption, as all of the LSPs using the labels from that session will also be torn down. Once this happens, the objects requested may be modified and the session initialization restarted. This is one of the major reasons why we have found that most implementations have chosen to implement the mplsLdpSessionTable as completely read-only.

Let's illustrate an example of how an operator might modify an LDP entity's objects. Assume that the operator wishes to change the configuration of a platform-wide label range that is currently being used by a session. The session in question was established some time ago. To accomplish this, the operator should first send the agent an SNMP SET request changing the mplsLdpEntityAdminStatus of the appropriate instance to **disable (2).** As a result, the session should be torn down and all LSPs established using that session should also be torn down (i.e., labels for these LSPs withdrawn). At this time, all nonconfigured information related to the session should be removed from the MIB. Further, if the agent has implemented the LSR MIB and the optional Mapping Table objects, all information pertaining to the LSPs using labels from this session should be removed as well. The session's mplsLdpEntityOperStatus should transition to the **down (2)** state, indicating that all of these things have happened and that the row is ready for modification. At this point, the operator may modify the label range associated with the specified LDP entity. Once modifications are complete, the operator should set the mplsLdpEntityAdminStatus to **enable (1)** for the index corresponding to this LDP entity. At this point, it may be appropriate for an NMS application to double-check with the user if their changes are acceptable, since changing them after the administrative status is returned to **enable (1)** will result in the LSPs that have just been resignaled to be torn down again. At this time, session initialization should occur, and the LSPs that were working previously should return to their normal function. Care should be taken by operators when making the decision to modify the attributes of a running LDP session, since the consequences can be far-reaching and potentially disastrous for customers using the affected LSPs.

4.19 Operational and Administrative Status

One subtle point to make about the mplsLdpEntityAdminStatus object used in the MPLS-LDP MIB is that it can be viewed as having a subset of the states defined in the mplsLdpEntityRowStatus object. More precisely, since the AdminStatus object has two states of **enable (1)** and **disable (2),** it is possible to set the mplsLdp-EntityAdminStatus object to **disable (2)** and achieve the same result as setting the mplsLdpEntityRowStatus object to **notInService (2).** This is because both will have the effect of shutting the session down, which will result in all LSPs signaled using those labels to be shut down as well. However, the co-authors of the MPLS-LDP MIB decided to keep the mplsLdpEntityAdminStatus object in the MIB because there are situations where a user might require the use of both objects, depending on the level of access that operator is given. For example, if the operator were given write permission to the mplsLdpEntityAdminStatus object, but not to the mplsLdpEntityRowStatus object, that would mean that the operator would have permission to start and stop LDP entities, but would not be able to create and delete them.

4.20 Mapping Tables

The MPLS Label Forwarding Information Base (LFIB) contains information about labels currently used by the LSR. The LFIB contains labels used by the three MPLS signaling protocols: LDP, CR-LDP, and MPLS-RSVP (see Chapter 8). The LFIB is exposed for operators in the MPLS-LSR MIB (see Chapter 3). The LFIB is represented by the interaction between the LSR MIB's MPLS Cross-Connect, InSegment, and OutSegment Tables. The MPLS Cross-Connect Table models the cross-connection of the ingress label (inSegment) with a specific egress label (outSegment). Rather than reinvent this table within the MPLS-LDP MIB, the co-authors decided to utilize the MPLS-LSR MIB's tables to represent the labels being used by LDP sessions by creating a mechanism that associated LDP sessions with the LSPs these sessions created. The mapping tables in this MIB (MPLS LDP Session InLabel Mapping Table, MPLS LDP Session OutLabel Mapping Table, and the MPLS LDP Session Cross-Connect Mapping Table) allow you to find all of the LSPs set up using the session given the session. Since the LSPs are represented in the MPLS-LSR MIB, these mapping tables map from sessions to entries in that MIB.

The mplsInSegmentTable, the mplsOutSegmentTable, and the mplsXCTable in the LSR MIB could contain rows that are created when an LDP LSP is created. Three mapping tables are contained in the MPLS-LDP MIB and are used to map

LDP sessions to the aforementioned tables in the LSR MIB. These mapping tables are described in the next few subsections. Operators should note that these mapping tables are optional, and therefore may find that they are not implemented by early implementations. Furthermore, these tables are indexed by many objects and can be not only difficult to implement, but also quite expensive to query; therefore, current implementations may choose not to implement them as well. If an implementation chooses to implement these mapping tables, then they must implement the MPLS-LSR MIB as well. We advise that operators check with their vendor(s) to verify that all of the appropriate tables are implemented. This information should be available via the vendor's agent capability statement for the MPLS-LDP MIB, if available. If they have not been implemented, we advise that you look closely at how useful the information is that the mapping tables provide to you and your particular OSS before insisting that they be implemented. The reasoning with this approach is that you may find that with some implementations these tables are quite costly to query, even infrequently. One suggestion is to limit the number of queries to this table if the information is deemed important.

4.20.1 The LDP Session InLabel Map Table

The LDP Session InLabel Map Table (mplsLdpSesInLabelMapTable) provides a way by which LSPs signaled by LDP can map their incoming labels to the MPLS-LSR MIB's mplsInSegment Table. For example, if an LSP is terminated on this LSR, then a corresponding entry will exist in the mplsLdpSesInLabelMapTable that associates the LDP session with the appropriate entry in the mplsInSegment Table. Implementations should be aware that many cleanup situations can be required because of the interrelationship between the two MIBs. This relationship may require complex cleanup scenarios. Some of these will be covered later in this chapter. Please also note that it is possible that entries in this table are also affected by LSPs being torn down by either disabling or removing entries in the MPLS-LSR MIB's mplsXCTable. Table 4.52 enumerates and defines the objects contained in the mplsLdpSesInLabelMapTable.

Table 4.52 MPLS-LDP MIB mplsLdpSesInLabelMapTable objects.

Object name	Description
mplsLdpSesInLabelIfIndex	This object contains the Interface Index of the mplsLdpSesInLabel. This value should be the same as the mplsInSegmentIfIndex object for the corresponding entry in the LSR MIB.

continued

Table 4.52 continued

Object name	Description
mplsLdpSesInLabel	This object contains the incoming label of this LSP. This value should be the same as the mplsInSegmentLabel object for the corresponding entry in the LSR MIB.
mplsLdpSesInLabelType	This object contains the layer-2 label type for mplsLdpInLabel.
mplsLdpSesInLabelConnType	This value contains the type of LSP connection. The possible values for this object are **unknown (1),** indicating that the LSP is in a state of flux (it is considered a temporary situation); **xconnect (2),** indicating that the mapping between the session and the insegment is associated with an LSP that is a true cross-connection; **terminates (3),** indicating that the mapping between the session and the insegment is associated with an LSP that terminates on this LSR and is not a cross-connection (i.e., this is the egress of an LSP).

4.20.2 The LDP Session OutLabel Map Table

The LDP Session OutLabel Map Table (mplsLdpSesOutLabelMapTable) provides a way by which LSPs signaled by LDP can map their outgoing labels to the MPLS-LSR MIB's mplsOutSegment Table. This table also provides a means by which these entries can be easily removed from both the MPLS-LSR MIB's mplsOut-Segment Table and from this table if the session is torn down. Specifically, if an LSP is originated on this LSR, then a corresponding entry will exist in the mplsLdpSes-OutLabelMapTable that associates the LDP session with the appropriate entry in the mplsOutSegment Table. Implementations should be aware that many cleanup situations can be required because of the interrelationship between the two MIBs. This relationship may require complex cleanup scenarios. Some of these will be covered later in this chapter. Please also note that it is possible that entries in this table are also affected by sessions being torn down by either disabling or removing entries in the MPLS-LSR MIB's Cross-Connect Table. Table 4.53 enumerates and defines the objects contained in the mplsLdpSesOutLabelMapTable.

Table 4.53 MPLS-LDP MIB mplsLdpSesOutLabelMapTable objects.

Object name	Description
mplsLdpSesOutLabelIfIndex	The ifIndex of the mplsLdpSesOutLabel.
mplsLdpSesOutLabel	The outgoing label of this LSP.

Table 4.53 continued

Object name	Description
mplsLdpSesOutLabelType	The layer-2 label type for mplsLdpOutLabel.
mplsLdpSesOutLabelConnType	The type of LSP connection. The possible values are **unknown (1),** indicating that the LSP may be in a state of flux (it is considered a temporary situation); **xconnect(2),** if the mapping between the session and the outsegment is associated with an LSP that is a true cross-connection; and **starts (3),** if the mapping between the session and the insegment is associated with an LSP that starts on this LSR and is considered an ingress to the LSP (i.e., this is the ingress of an LSP).
mplsLdpSesOutSegmentIndex	This value should contain the same value as the mplsOut-SegmentIndex in the LSR MIB. Note: This value will never be 0 because this table only maps from sessions to true outsegments.

4.20.3 The LDP Session Cross-Connect Mapping Table

The LDP Session Cross-Connect Mapping Table (mplsLdpSesXCMapTable) provides a way by which LSPs signaled by LDP can map their outgoing labels to the MPLS-LSR MIB's mpslXCTable. This table also provides a means by which these entries can be easily removed from both the MPLS-LSR MIB's mplsXCTable and from this table if the session is torn down. Specifically, if an LSP is originated on this LSR, then a corresponding entry will exist in the mplsLdpSesXCMapTable that associates the LDP session with the appropriate entry in the mplsXCTable. Implementations should be aware that many cleanup situations can be required because of the interrelationship between the two MIBs. This relationship may require complex cleanup scenarios. Some of these will be covered later in this chapter. Please also note that it is possible that some entries in this table are affected by sessions being torn down by either disabling or removing entries in the MPLS-LSR MIB's mplsXCTable. Table 4.54 lists and defines the managed object contained in the mplsLdpSesXCMapTable.

Table 4.54 MPLS-LDP MIB mplsLdpSesXCMapTable object.

Object name	Description
mplsLdpSesXCIndex	This object indicates the mplsXCIndex from the MPLS-LSR MIB. Implementations should never set this value to 0 because this table only maps from LDP sessions to true cross-connects.

4.21 Cross-Connects FEC Table

The Cross-Connects FEC Table (mplsXCsFecsTable) contains FEC (Forward Equivalency Class) information for LDP-signaled LSPs. Each entry in this table represents a single FEC element. The mplsXCsFecsTable maps FECs to their associated cross-connects. Each row in this table represents a single cross-connect to FEC association. This table is implemented only as read-only because it represents active information that cannot be configured here. The table is indexed by eight objects: mplsLdpEntityLdpId, mplsLdpEntityIndex, mplsLdpPeerLdpId, mplsLdpSesInLabelIfIndex, mplsLdpSesInLabel, mplsLdpSesOutLabelIfIndex, mplsLdpSesOutLabel, and mplsFecIndex. The importance of the indexing of this table is that it shows which FEC-to-NHLFE relationships are being constructed by LDP. For some operators, this is quite useful information. However, for some, a more complete FEC-to-NHLFE relationship is required showing all FEC-to-NHLFE relationships that are active on the device, as well as the ability to configure them directly. This can be done with the MPLS-FTN MIB (see Chapter 5). As with all of the mapping tables in this MIB, we recommend that you carefully examine how critical gathering this information is to your deployment because queries to this table will be quite expensive. One suggestion is to limit the number of queries to this table if the information is important to your OSS strategy. One approach to take might be to query the table only upon notification that an LSP has been established (i.e., via the mplsLdpSessionUp notification). Again, due to their complexity, some vendors may have chosen not to implement them. Furthermore, please check with your device vendor(s) to determine if they have implemented these mapping tables. Table 4.55 enumerates and defines the managed objects contained within the mplsXCsFecsTable.

Table 4.55 MPLS-LDP MIB mplsXCsFecsTable objects.

Object name	Description
mplsXCFecOperStatus	This object provides the operators with an indication of the operational status of the FEC associated with this cross-connect. Valid values are **unknown (1),** indicating that the LSP-FEC association may be in a state of transition and is typically a temporary state; **inUse (2),** indicating that the FEC associated with the cross-connect is currently being applied; and **notInUse (3),** indicating that the FEC associated with the XC is not being applied.
mplsXCFecOperStatusLastChange	This object contains the value of sysUpTime when the mplsXCFecOperStatus last changed state.

4.22 Notifications

Currently, there are several notifications that are specific for LDP that can be very useful for operators wishing to manage LDP with the MPLS-LDP MIB. These are described in this section. With these notifications, a device can alert an operator or NMS as to a change in the state of the LDP protocol. In some cases, an NMS or operator might wish to act upon these notifications to correct a fault condition.

4.22.1 MPLSLdpInitSesThresholdExceeded

This notification is generated when the value of the mplsLdpEntityInitSes-Threshold object is not zero, and the number of session initialization messages exceeds the value of the mplsLdpEntityInitSesThreshold object.

The mplsLdpInitSesThresholdExceeded notification indicates to the operator that there may be a misconfigured mplsLdpEntityEntry because the session associated with this entity is not being established, and the entity keeps trying to establish the session. A side effect of this situation is that a row in the mplsLdpSessionTable may not be reaching the operational state as indicated by the mplsLdpSesState object. If the value of mplsLdpEntityInitSesThreshold is 0, then this is equivalent to specifying the value of infinity for the threshold, and the mplsLdpInitSes-ThresholdExceeded trap will never be sent. As of the version of the MIB covered in this chapter, no explicit mechanism for disabling this notification other than the one just described exists. We, however, suspect this will be added in a later version of the MIB.

4.22.2 MPLSLdpPVLMismatch

This notification is generated only when the value of the mplsLdpEntityPVLMisTrapEnable object is **enabled (1)** and can be prevented by setting this object to **disabled (0).** If this notification is enabled, it is emitted when there is a mismatch in the Path Vector Limits between the entity and peer during session initialization between that entity and that peer. In particular, it is generated when the mplsLdpEntityPVL does not match the value of the mplsLdpPeerPVL for a specific entity. In this situation, a session could still be established between an entity and the specified peer if the session uses the value configured as the entity's Path Vector Limit. However, a notification should still be sent to indicate to the operator that a mismatch in parameters has occurred in case a modification to the LSR's configuration is appropriate.

4.22.3 MPLSLdpSessionUp

This notification is generated when the mplsLdpSesUpDownTrapEnable object is set to **enabled (1).** If this notification is enabled, and the value of mplsLdpSesState transitions into the **operational (5)** state, this notification is emitted. In particular, this notification is sent when there is an appropriate change in the LdpSesState object from any state to the **operational (5)** state by a specific session represented in the MPLS LDP Session Table.

4.22.4 MPLSLdpSessionDown

This notification is generated when the mplsLdpSesUpDownTrapEnable object is set to **enabled (1).** If this notification is enabled, and the value of mplsLdpSesState transitions out of the **operational (5)** state, this notification is emitted. In particular, this notification is sent when there is an appropriate change in the LdpSesState object from the **operational (5)** state into any other state by a specific session represented in the MPLS LDP Session Table.

4.23 What the MIB Does Not Support

At the time of publication, the MPLS-LDP MIB did not include support for many of the new applications of MPLS. These include MPLS/BGP virtual private networks (enterprise, carrier-of-carriers, etc.), VP merge configuration, multicast for LDP, or pseudo-wire emulation functions. This is because the Label Distribution Protocol specification (RFC 3037) upon which this MIB was based did not specify the use of these technologies. It in fact currently does not. The management of

LDP-related features within the context of these additional uses of LDP objects can sometimes be found in MIBs that were written specifically to manage those functions. For example, the MPLS/BGP VPN MIB contains some provisions for the configuration of LDP features that are not configurable via the base MPLS-LDP MIB.

4.24 How the MIB Varies from the LDP Specification

The MPLS-LDP MIB varies slightly from the specification, largely due to requirements of standard MIBs. There are currently three differences between this specification and the LDP specification. As previously mentioned, the MPLS-LDP MIB is almost entirely based on the LDP specification. The differences are documented here in the hope of avoiding any confusion between the two documents.

The first difference is that the LDP Entity Table contains some DEFVAL clauses that are not specified explicitly in the LDP specification. These values, although not documented in the LDP specification, are widely used by existing LDP MIB implementations and thus have been adopted within this MIB. Please note: they can certainly be changed during row creation or a subsequent set request.

The second difference is the mplsLdpEntityConfGenericLabelRangeTable. This table, although provided as a way to reserve a range of generic labels, does not exist in the LDP sspecification. It was added to the MIB due to a request from the Working Group and because this table was considered useful for reserving a range of generic labels.

The third difference is documented by the textual convention MPLSAtmVc-Identifier that is in the MPLS-TC MIB. This TC was added to restrict vci values to be greater than 31 as described in RFC 3032.

The ways in which the MPLS-LDP MIB varies from the LDP specification are summarized in Table 4.56.

Table 4.56 Summary of how the MPLS-LDP MIB differs from the LDP specification.

Difference from LDP specification	Description
DEFVAL clauses	Default values for some variables are used that are not specified in the LDP specification.
mplsLdpEntityConfGenericLabelRangeTable	This table is provided as a way to reserve a range of generic labels. This table does not appear in the LDP specification.

continued

Table 4.56 continued

Difference from LDP specification	Description
MPLSAtmVcIdentifier	This TC was added to restrict ATM vci values to be greater than 31.

4.25 Using the MPLS-LDP MIB with TDP

The Label Distribution Protocol is based on the combination of the (control-driven) Cisco Tag Distribution Protocol (TDP) and the IBM ARIS protocols. No known deployed implementations of ARIS currently exist, but some implementations of TDP remain deployed in the field. In these situations, it is important to understand how to manage these deployments. First, the IETF standard MPLS-LDP MIB was not designed to handle the TDP protocol; it was only designed to manage the IETF Label Distribution Protocol. Therefore, it is necessary for a vendor to take certain measures to assure that an operator can manage a network where both TDP and LDP coexist, especially on devices where some interfaces may continue to run the older TDP, while still others run LDP. In these cases, it is important for the vendor to provide the operator with access to the TDP function via one or more management interfaces. This may include a proprietary TDP MIB, specific CLI commands, XML schema, CORBA IDL, or all of the above.

Operators might be interested in knowing a few pieces of information about any hybrid implementation. First, it is important to discern whether LDP, TDP, or both are supported on a particular device. Second, it is interesting to understand which interfaces of a particular device are configured to run LDP, TDP, or no Label Distribution Protocol. In general, either TDP or LDP, but not both, may be run as the Label Distribution Protocol on any particular interface. Notifications from TDP are important for events such as those that have been defined in the LDP MIB and described earlier. We recommend that these be duplicated for the Tag Distribution Protocol. Finally, any additional implementation-specific events may be important for an operator. For example, it may be important to notify an operator when TDP exhausts its pool of tags to distribute. In the end, the vendor needs to work closely with the operators using these hybrid implementations to assure that they are manageable.

4.26 Summary

This chapter introduced the Label Distribution Protocol and explained its basic uses. The MPLS-LDP MIB was then introduced within this context. Each of the many tables in the MIB was then introduced and the details of each were explained. Each subsection pertaining to a specific table included a discussion of the table's attributes, the table's indexing, and how it related to the other tables in the MIB. Where appropriate, we included discussions of scaling issues related to using and implementing each table. Then the notifications defined in the MIB were explained in detail. To further enhance the discussion of the MIB's usefulness, sufficiently complex examples of ATM and Frame Relay networks were given to show how the MIB could be used to successfully manage these types of MPLS deployments. Finally, we gave a description of which MPLS applications of LDP the MIB was and was not intended to manage, and how to potentially manage these features. Some applications of LDP were so new that they are not covered by the MPLS-LDP MIB, nor are they included in this text. These will be covered in a subsequent update of the MIB and perhaps of this book in the near future.

Further Reading

"Definitions of Managed Objects for the Multiprotocol Label Switching, Label Distribution Protocol (LDP)," IETF Internet Draft, can be found at *www.ietf.org/internet-drafts/draft-ietf-mpls-ldp-mib-08.txt.* Note that the version of the draft (i.e., the last two digits of the hyperlink) is subject to change.

RFC 3036, "LDP Applicability," can be found at *www.ietf.org/rfc/rfc3036.txt.* This document provides some insight as to when the use of the LDP protocol itself is useful.

RFC 3037 specifies the standard definition of the Label Distribution Protocol and is the document on which the MPLS-LDP MIB is based. It can be found at *www.ietf.org/rfc/rfc3037.txt.*

RFC 3031 specifies the MPLS architecture and is the basis for all MPLS documents. It can be located at *www.ietf.org/rfc/rfc3031.txt.*

Davie, B. S., and Y. Rekhter. *MPLS: Technology and Applications.* First edition. San Francisco: Morgan Kaufmann Publishers. 2000.

Gray, E. W. *MPLS: Implementing the Technology.* Reading, Mass.: Addison-Wesley Professional. 2001.

To find out more about the IETF, visit their Web page at *www.ietf.org/.*

For more information about IANA, check out their Web site at *www.iana.org/.*

Joan E. Cucchiara

is a principal software engineer at Crescent Networks, a leading equipment provider for scalable virtual routed networks and services. She has over 10 years experience in developing embedded network management solutions for networking devices for companies including Cabletron Systems and Wellfleet/Bay Networks/Nortel Networks. She has authored several MIBs in different standards bodies including the IETF and the ATM Forum. Most recently, she authored the IETF MPLS-LDP MIB. This MIB evolved from an LDP implementation that used IP and ATM.

Being the principal author of the MPLS-LDP MIB and one who has been involved in the management of MPLS networks, how have you seen the state of MPLS and the management of those deployments progress over the past few years?

MPLS and the interest in management of MPLS networks have grown tremendously over the past few years. Initially, deployments were nailed up LSPs across the network, but more recently, deployments are turning to the use of RSVP-TE with MPLS for traffic-engineered networks. Additionally, there is a growing amount of interest in using MPLS to provide VPNs and QoS capabilities.

What do you think are the most important hurdles to the complete management of MPLS deployments?

This would depend upon what is meant by "complete management." Today, LSPs are provisioned from the head end, and then monitored at the transit nodes. Areas, which are tougher to address with MPLS, are on-demand services with QoS, usage-based accounting, traffic engineering services, security, and verification of services. Customers are requesting more flexibility with the services and scalable solutions. We see the next hurdles in network management as being able to provide for the needs of these areas.

Do you think that MPLS will ever be completely manageable, or is this simply something that is not attainable?

Again, this depends on what is meant by "completely manageable." The technology is still evolving. This is particularly true in areas like GMPLS and traffic engineering. Vendors are providing new and better equipment all the time. In addition, customers are requesting more and more MPLS services. They are also interested in a high degree of verification of those services. From a network management perspective, these aspects are certainly quite challenging. There are many questions that still need to be addressed. For example, "What are acceptable metrics for determining if a customer is receiving the service they have contracted?" If you consider that this is part of making MPLS "completely manageable," then we would say we are not yet there. However, we do believe the industry is responding in very positive ways to these requirements.

For example, recent advancements in the Operational Support Systems (OSSs), specifically the Enterprise Application Integration model, and efforts like bundling services, will help provide manageability and flexibility of services. Again, the customer requests are being responded to by the OSS industry, and due to deregulation and convergence of IP, we see possibilities for advancements that were not there until quite recently.

IETF technologies such as DiffServ, PPVPN, Policy, and SnmpConf are opening the doors for new and enhanced services with MPLS. These technologies are still evolving, and deployments are not yet widespread. However, we do believe that as these technologies mature, MPLS is likely to be more manageable and provide more services and flexibility for the customers than it does today. In particular, there is a huge level of interest in areas such as the creation of scalable virtual routed networks made possible through MPLS and straightforward MPLS management techniques. Such areas are being addressed by new product offerings and are the subject of healthy discussions in the various standards bodies involved in the application of MPLS as a technology. All of these aspects will contribute to greater MPLS manageability.

5

The MPLS Forward Equivalency Class to Next-Hop Label Forward Entry MIB (MPLS-FTN MIB)

"Sour, sweet, bitter, pungent, all must be tasted."

—Chinese proverb

Introduction

In this chapter, we examine one of the many standard tools used to manage MPLS networks. The MPLS Forward Equivalency Class (FEC) to Next-Hop Label Forwarding Entry (NHLFE) MIB (MPLS-FTN MIB) was defined with the specific function of exposing the destination prefix to next-hop label mapping for all ingress points of LSPs on an MPLS LER. In short, the MPLS-FTN MIB provides the user with a snapshot of how any particular LSR has been programmed to accept traffic and assign this traffic to an LSP, and then which MPLS next-hop to forward it to. This MIB is meant to be used in conjunction with the MPLS-LSR MIB as well as the soon-to-be-discussed MPLS-TE MIB. This chapter will explain in detail the operation and usefulness of this MIB and will present examples of how it could be implemented by vendors, as well as how it could be utilized by operators.

This chapter will focus on the draft version 04 of the MPLS-FTN MIB, which is typically referred to as draft-ietf-mpls-ftn-mib-04.txt. This draft version may have been updated or replaced with an IETF RFC document after this book was published. Please keep this in mind when searching for the document on which this chapter is based.

5.1 Who Should Use It

All management stations wishing to monitor or configure the basic label forwarding capabilities of a label edge router should monitor the tables provided in this MIB. This MIB is of particular interest to those who wish to understand the ingress prefix-to-label mapping being performed by the LER at any given time. Management stations wishing to monitor the core behavior of MPLS while monitoring other MPLS applications such as traffic engineering or virtual private networks should also utilize the objects provided by this MIB for a clearer picture of what a particular LSR is doing.

5.2 IP Traffic In, MPLS Labels Out

Let us begin the discussion of the FTN MIB with a simple explanation of the Label Forwarding Information Base (LFIB) that is present at MPLS label edge routers (LERs). Put simply, an MPLS LER has one basic function: to accept IP packets on non-MPLS interfaces, determine if there is an appropriate label switched path that will eventually reach the destination desired by the packet, impose the appropriate MPLS shim header, and forward that packet into the MPLS domain. The latter would be either an LSP or a traffic-engineered tunnel head interface. Thus, the basic function of an LSR is to accept unlabeled packets and to forward them onto an LSP or TE tunnel. These activities describe the operations of the Forward Equivalency Class to Next-Hop Label Forwarding Entry (FEC-to-NHLFE) mapping.

5.3 Forward Equivalency Classes

The forwarding function of a router is responsible for forwarding traffic toward its ultimate destination. The information in the forwarding table is programmed based on information from the control plane. If a packet is not delivered via a local interface directly to the destination, the router must forward the packet toward the ultimate destination using a port that will steer that traffic on a path considered most optimal by the routing function. For this reason, a router must forward traffic toward its destination via a next-hop router. This next-hop router may be the next-hop along the most optimal path for more than one destination subnetwork, so many packets with different network layer headers may be forwarded to the same next-hop router via the same output port. The packets traversing that router can then be organized into sets based on equivalent next-hop network nodes. We call such a set a Forward Equivalency Class (FEC). Thus, any packet that is forwarded to a particular next-hop is considered part of the FEC and can thus be forwarded to the same next-hop.

One important feature of the FEC is the granularity of the classification of traffic it can encompass. Since the FEC is based on a routing next-hop, it can include different classifications of packets. For example, since the routing information for a particular next-hop classification can be based on a destination prefix, it might include every packet traveling toward that destination. In this way, the granularity of packets classified by that FEC is quite coarse. However, if the routing database has programmed some next-hops for some traffic based on an application layer, for example, the traffic granularity might be much finer.

Each FEC is assigned an MPLS Next-Hop Forwarding Entry. This is fancy jargon for an MPLS label. The label assigned to a FEC is used to carry the FEC's traffic to the next-hop LSR, where it may continue along the label switched path. What this does is in effect forward all of the traffic from a FEC along the same LSP. When a FEC is mapped to a label, this represents the FEC-to-NHLFE (FTN). The basic FEC-to-NHLFE operation is demonstrated in Figure 5.1.

The FEC-to-NHLFE operation is critical to the behavior of a correctly functioning LER. In some cases, this relationship is established manually, but more often it is established automatically by the LSR's control plane. It is for this reason and the important nature of the FEC-to-NHLFE relationship that this relationship needs to be exposed to network managers. Incorrect behavior of this function can result in misrouting or discarding of traffic. This is the basic premise behind the MPLS-FTN MIB.

As was noted earlier, the MPLS-LSR MIB provides the user with a clear picture of what the label forwarding (switching) database (LFIB) looks like at any moment

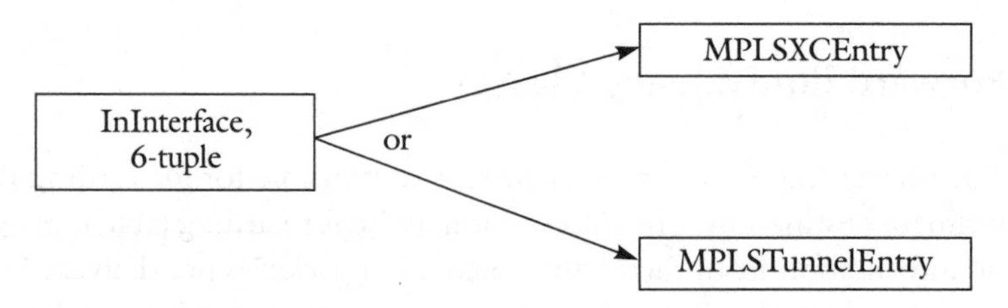

Figure 5.1 The basic functionality of the MPLS-FTN MIB: 6-tuple in, cross-connect/tunnels out.

in time. It is thus possible to map incoming traffic to the LSPs that are modeled in this MIB. Figure 5.1 illustrates this. In Figure 5.1, traffic from a certain interface matching a 6-tuple used to uniquely describe an incoming packet is mapped to either an MPLS LSP or TE tunnel interface. The 6-tuple consists of the IP source and destination addresses, the source and destination ports, the layer-4 protocol identifier, and the EXP bits. Once a match is found within one of the 6-tuples, the traffic is then mapped to an MPLS LSR or TE tunnel interface present in the MPLS-LSR MIB or MPLS-TE MIB, respectively.

5.4 A Simple Example of FEC-to-NHLFE

Let us now investigate a simple but detailed example of how the MPLS-FTN MIB can be used to manage an MPLS LER in a simple MPLS network. This is illustrated in Figure 5.2. The example network presented in Figure 5.2 is composed of several label switching routers configured in a simple MPLS topology. The MPLS network is denoted by the cloud. Notice that several label edge routers are located around the edges of the network, as well as several internal or core LSRs. Let us focus on the leftmost LSR in Figure 5.2. Notice that this router is accepting traffic from sources that are destined for networks 123.33.0.0, 18.44.0.0, 18.45.0.0, and 110.13.0.0, and thus it needs to impose MPLS labels onto packets it receives with these destination prefixes before forwarding them into the MPLS domain. Notice that not all of the traffic is treated in the same manner. In the case of the first two prefixes, any traffic received destined for network 123.33.0.0 or network 18.44.0.0 will be pushed into the traffic-engineered tunnel “tun123.” In contrast, traffic destined for networks 18.45.0.0 or 110.13.0.0 is pushed into the MPLS LSP “LSP1.”

In the second case, traffic is being mapped onto an LSP. In general, this mapping is done automatically by the LSR’s autorouting function. This function takes what would be routing protocol destination routes and, instead of using IP to traverse

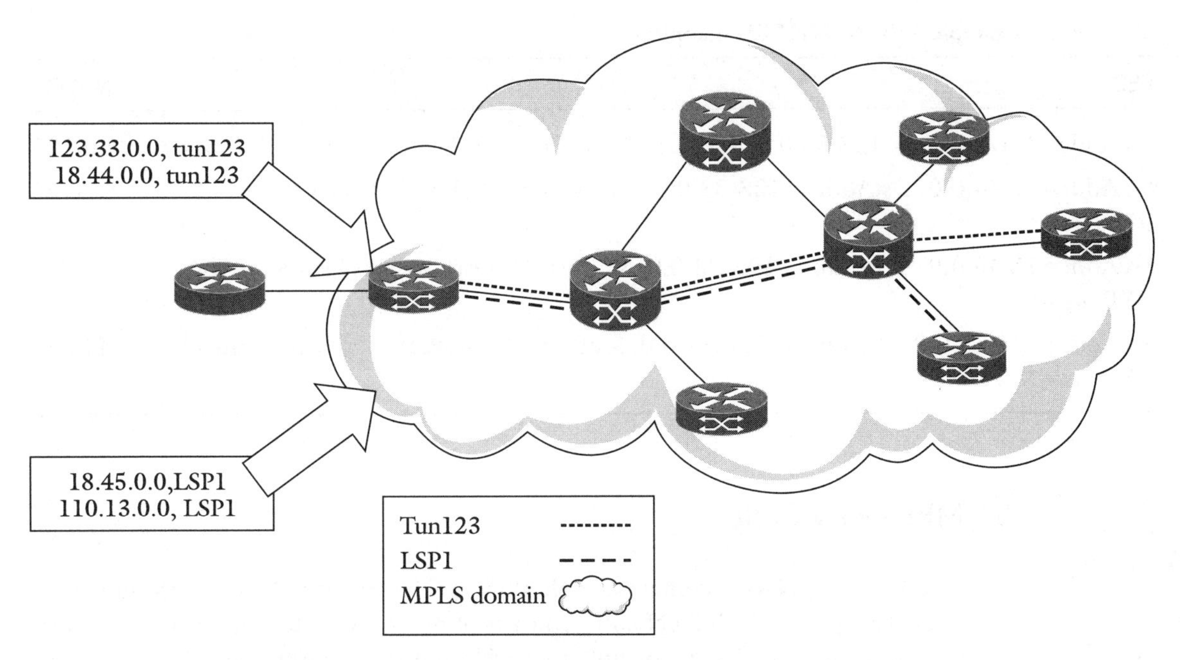

Figure 5.2 FEC-to-NHLFE mapping in an MPLS network.

the network to reach those destinations, the LSR uses one or more MPLS LSPs. There are cases where LSPs can be set up statically or "by hand," and these cases are similar to the next scenario, but are called E-LSPs and are present in the LSR MIB. The first example depicts traffic that is being explicitly forwarded onto a preconfigured traffic-engineered tunnel. This tunnel is generally statically configured and remains until it is removed from the LSR's configuration. Thus, it is desirable to also statically map any traffic that would be forwarded into this tunnel in a static manner. This is typically done by configuring a static route on the LSR with a next-hop interface of the traffic-engineered tunnel. It is also possible to allow the routing protocols to forward traffic into TE tunnels. In particular, if the tunnel allows any traffic to its destination to be carried by the tunnel, traffic may be forwarded into it via mechanisms other than static routes.

To be more concise, the FEC-to-NHLFE mappings just described appear in Table 5.1. Notice that to be precise, the 6-tuple must be completely specified in order to fully describe the FEC.

Table 5.1 Example FEC-to-NHLFE mapping.

FEC	NHLFE
SrcAddr=*, DstAddr = 123.33.0.0, SrcPort=*, DstPort =*, ProtocolId=*, EXPbits=*	Tun123
SrcAddr=18.44.0.0, DstAddr = 123.33.0.0, SrcPort=*, DstPort =*, ProtocolId=*, EXPbits=*	Tun123
SrcAddr=18.45.0.0 , DstAddr = 123.33.0.0, SrcPort=*, DstPort =*, ProtocolId=*, EXPbits=*	LSP1
SrcAddr=110.13.0.0, DstAddr = 123.33.0.0, SrcPort=*, DstPort =*, ProtocolId=*, EXPbits=*	LSP1

5.5 MPLS FTN Table

The MPLS-FTN MIB contains a table allowing the operator to configure and monitor FTN entries. Each FTN entry consists of many attributes, but its primary purpose is to fully specify the 6-tuple that will uniquely identify the FEC entry. This entry also contains a pointer to a next-hop entry in the MPLS-LSR MIB or MPLS-TE MIB. This second portion completes the "N" or NHLFE portion of the FTN name. Let us now look at this table in detail.

5.5.1 FTN Table Indexing

The FTN Table is indexed by a single index called the mplsFTNIndex. This value represents an arbitrary but unique unsigned integer representing an FTN entry. There are several reasons for employing this arbitrary index. First, it allows for the definition of a single FTN entry that is applied to multiple interfaces (see FTNMappingTable). This arrangement also allows the implementation flexibility in how it allows these entries to be configured. These reasons aside, the only other option would have been to use six or seven indexes for this table. Such a scheme would have been burdensome on both the implementation in terms of engineering and performance, as well as on the operator or network manager who would have to decipher the indexing scheme. On the other hand, the simplicity of a single index always carries with it a price. In this case, that price is paid primarily in the fact that an implementation is free to order the FTN entries any way it pleases. So, for example, an implementation may have two FTN entries that are identical for all but, say, the EXP bits, yet the entries not only are noncontiguous, but are separated by thousands of indexes! Remember, the implementation is free to order (or misorder) the entries since there is no order implied by this arbitrary index. One bright note to this indexing is that implementations that allow configuration through SNMP allow the operator to configure these entries in the same flexible manner as was just described. Thus under these circumstances the operator may order the objects any way it wishes.

5.5.2 FTN Objects

The remainder of the FTN Table is composed of objects that are used to describe the FTN entry. The FEC entry can be thought of as a filter against which to match incoming traffic. These filters are only applied in the positive sense. That is, if they match, the rule will fire and forward the incoming traffic to the specified MPLS next-hop entry (i.e., an LSP or a TE tunnel head).

If a packet is received that does not match the filter, no action is taken. Actually, the packet is most likely dropped. It is also important to understand that due to the way the rules work, the counters associated with a specific rule will be modified only when a match occurs. It is prohibitive in terms of device performance and implementation effort to tell how many times a rule has not been matched using these tables.

The objects in this table include a description (FTNDescr), a bit denoting whether or not the FTN entry has yet been applied or enabled for service (FTN-Applied), a RowStatus variable that allows for configuration of the entries (mplsFTNRowStatus), a description of the rule (mplsFTNDescr), and finally, the storage type of the table (mplsFTNStorageType). The latter object provides the operator with an indication of the persistent status of the row in the table. The remaining objects are provided specifically as a means for matching incoming data. The first object in this group is the mplsFTNMask. This object is used to specify which of the remaining attributes are to be applied as the rule. This bit mask contains a corresponding bit position for each of the following attributes. The mplsFTNMask is defined in Table 5.2.

Table 5.2 FTNMask values.

Attribute name	Bit position
Source address	0
Destination address	1
Source port	2

continued

Table 5.2 continued

Attribute name	Bit position
Destination port	3
Layer-4 protocol type	4
EXP bits	5

Table 5.3 lists the remaining objects in the FTN Table with a brief description of each object.

Table 5.3 FTN entry matching parameters.

Object name	Object description
mplsFTNAddrType	Denotes which address objects to use—IPv4 or IPv6.
mplsFTNSmyceIpv4AddrMin	If mplsFTNAddrType is set to IPv4, this contains the minimum source address range to use.
mplsFTNSmyceIpv6AddrMin	If mplsFTNAddrType is set to IpV6, this contains the minimum source address range to use.
mplsFTNSmyceIpv4AddrMax	If mplsFTNAddrType is set to IPv4, this contains the maximum source address range to use.
mplsFTNSmyceIpv6AddrMax	If mplsFTNAddrType is set to IPv6, this contains the maximum source address range to use.
mplsFTNDestIpv4AddrMin	If mplsFTNAddrType is set to IPv4, this contains the minimum destination address range to use.
mplsFTNDestIpv6AddrMin	If mplsFTNAddrType is set to IPv6, this contains the minimum destination address range to use.
mplsFTNDestIpv4AddrMax	If mplsFTNAddrType is set to IPv4, this contains the maximum destination address range to use.
mplsFTNDestIpv6AddrMax	If mplsFTNAddrType is set to IPv6, this contains the maximum destination address range to use.
mplsFTNSourcePortMin	The minimum source port to use.
mplsFTNSourcePortMax	The maximum source port to use.
mplsFTNDestPortMin	The minimum destination port to use.
mplsFTNDestPortMax	The maximum destination port to use.

Table 5.3 continued

Object name	Object description
mplsFTNProtocol	The layer-4 protocol ID as defined by IANA.
mplsFTNActionType	Describes whether the MPLS next-hop entry is a TE tunnel entry or an MPLS-LSR MIB XCEntry.
mplsFTNActionPointer	The RowPointer that is represented as an OID of the next-hop entry. This will be the first accessible column of the row corresponding to the entry in either the MPLS-LSR MIB or the MPLS-TE MIB or 0.0, depending on whether or not the mplsFTNActionType is set to Lsp or TeTunnel. If the value is not configured, it will contain 0.0.
mplsFTNExpBits	MPLS EXP bits to match against incoming traffic.

A few important notes about the objects described in these tables. With the exception of the EXP bits and protocol ID, all objects are specified as ranges of port numbers or addresses. In the case of the addresses, this may be a bit confusing at first for those who are used to expressing these things as Classless InterDomain Routing (CIDR) addresses and mask lengths (e.g., "1.2.0.0/12"). It was felt by those implementing this MIB that it would be more efficient to represent the addresses as ranges because many of the CIDR ranges overlapped and contained the same EXP bits, protocol ID, and port numbers. Thus, it was thought that, rather than having multiple entries that differed only in an address, it would be more efficient to specify a single address range that encompassed several entries. Furthermore, this arrangement does not preclude the expression of the addresses as single entries; operators and implementations are able to do this as a range consisting of a single address range.

5.6 MPLS FTN Map Table

The MPLS FTN Map Table (mplsFtnMapTable) was designed to allow the operator or implementation to apply FTN entries to one or more interfaces. Due to the way that FTN entries are represented as a linked list of entries, it is also possible to map more than one FTN entry to a single interface. This list of FTN entries is assigned a priority or order of application. The order of application of FTN entries on an interface reflects the order in which the rules specified therein will be compared against incoming packets. When applying multiple FTN entries to an interface, it is advised that the most-specific FTN entries be applied first and the least-specific ones last. Reversing this order introduces the possibility that the more-specific FTN entries will never be matched.

Each entry in the FTN Map Table is indexed by three specific values: the interface index that the FTN entry is applied to, the FTN entry's index, and the index of the next FTN entry in the chain of FTN entries being applied to the interface specified by the first index. This structure implements what is commonly referred to as a *linked list*. The linked-list structure employed in this table allows FTN entries to be inserted into arbitrary positions in the list of FTN entries being applied to the specific interface. This affords the implementation and operator a degree of freedom when configuring FTN entries.

It should be noted that an entry in this table containing a primary index of 0 represents a filter that is applied to all interfaces. This is provided as a convenience in cases where the same rule might need to be applied to all interfaces. This allows the operator to configure a single rule and apply it many times.

Let us now investigate some of the important caveats about this table. First, agents must not allow the same FTN entries to be applied multiple times to the same interface. Doing so would present the possibility of recursive chaining of the rules, which could cause the agent to crash or malfunction. This is especially a concern in deployments that have multiple manager entities accessing this table at the same time. Unless they are coordinated, they may accidentally create chain loops. Furthermore, agents cannot allow the creation of rows in this table until the corresponding rows are created in the MPLS FTN Table. Again, doing so would create the possibility for inconsistencies or malfunctions by the agent. Finally, when rows in the FTN Table are destroyed that correspond to FTN entries that are named in this table, the agent must destroy the corresponding entries in this table as well. Not doing so will result in hanging or "dangling" references to the FTN Table. Even worse, if new entries are created using the old FTN entry indexes, unpredictable behavior could ensue on the agent. Table 5.4 enumerates and defines each of the objects specified in the mplsFtnMapTable.

Table 5.4 MPLS FTN Map Table objects.

Object	Description
mplsFTNMapIfIndex	This object contains the Interface Index that this FTN entry is associated with. The value of 0 indicates that this entry represents all Interface Indexes active on the system.

Table 5.4 continued

Object	Description
mplsFTNMapPrevIndex	This object contains the index of the previous FTN entry that was applied to this interface. The value 0 indicates that this should be the first FTN entry in the list.
mplsFTNMapCurrIndex	This object contains the Interface Index of the FTN entry that this interface is applied to.
mplsFTNMapRowStatus	This object contains the standard SNMP RowStatus that is used for creating, modifying, and deleting rows in this table.
mplsFTNMapStorageType	The SNMP storage type for this entry. Valid values are **other (1),** meaning that a storage type other than the ones defined below is available; **volatile (2),** meaning that the row will be stored in RAM and will disappear after the device reboots; **Nonvolatile (3),** meaning that the value is stored in some sort of nonvolatile RAM and will be preserved across reboots of the system; **permanent (4),** meaning that the row is stored partially in ROM; or **readOnly (5),** meaning that the row is stored completely in ROM. Operators will typically find this value to be set to either **Nonvolatile (3)** or **volatile (2).**

5.7 MPLS FTN Performance Table

The MPLS-FTN MIB contains one final table that is used to provide the operator with statistics related to the operation of the FTN entries. This table provides four simple yet effective counters that the operator can use to monitor the FTN behavior of the LSR. These are enumerated and defined in Table 5.5.

Table 5.5 MPLS FTN Performance Table objects.

Object	Description
mplsFTNMatchedPackets	The number of packets matched on this FTN entry.
mplsFTNMatchedOctets	The number of octets (bytes) matched by this FTN entry.
mplsFTNMatchedHCPackets	64-bit version of mplsFTNMatchedPackets.
mplsFTNMatchedHCOctets	64-bit version of mplsFTNMatchedHCOctets.

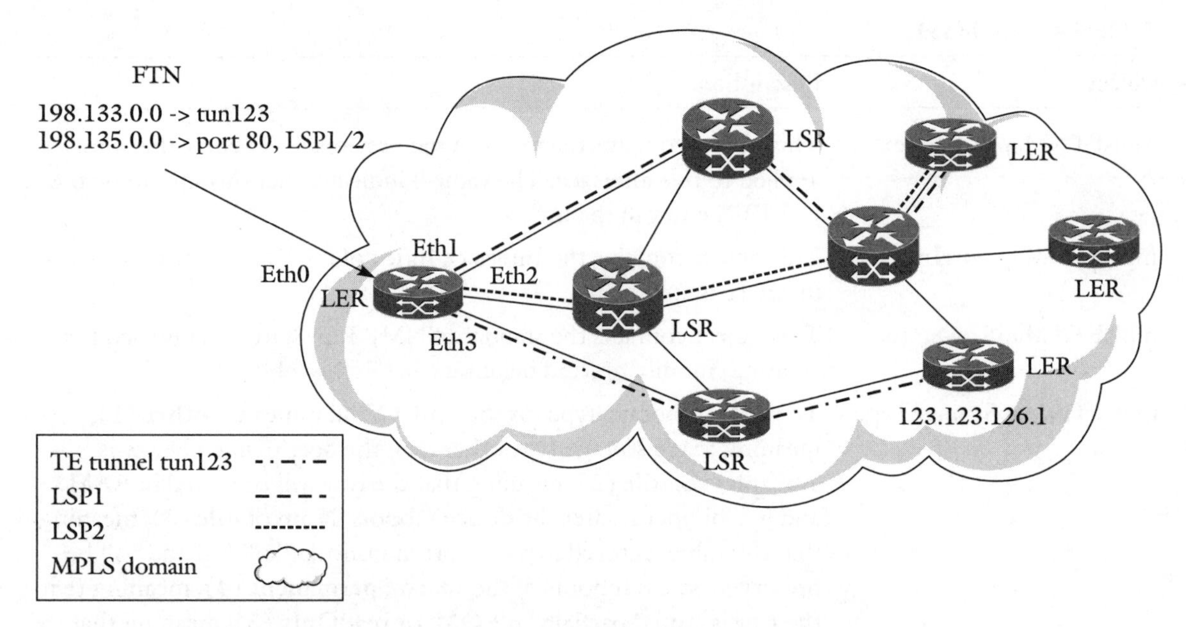

Figure 5.3 FEC-to-NHLFE mapping for an LSP using load sharing and a TE tunnel.

5.8 Another FTN Example

The following example will build upon the first example, but will significantly complicate it by presenting a realistic configuration, and then go through how to configure not only the FTN MIB, but the corresponding tables in the MPLS-LSR and MPLS-TE MIBs as well. This example will demonstrate how two prefixes will be handled at the ingress of the MPLS network. The first prefix is a prefix to Cisco.com. Because we want Web traffic to Cisco.com to travel at a high priority with low latency, the routing protocol has calculated just such a connection via the path shown in Figure 5.3. In order to meet these requirements, the traffic to 198.133.0.0 with destination port 80 will be mapped onto two LSPs that traverse the network in parallel along these paths. Note that the layer-4 port for HTTP is 80. Since both LSPs are considered equal-cost paths to the egress LER, the ingress LER will utilize per-packet load sharing across those LSPs. The second incoming prefix maps sensitive voice-over-IP traffic onto a specifically configured traffic engineering tunnel. This tunnel is named "tun123."

5.8.1 Configuring the FTN MIB for LSP1 and LSP2

Let us first consider the two LSPs noted in Figure 5.3 as LSP1 and LSP2. Notice that these LSPs originate from the leftmost LSR and continue across the network in parallel to the same terminating LER. The traffic with destination IP prefix for network 198.135.0.0 and port 80 is mapped onto LSP1 and LSP2 at the first LER. This is done on a per-packet basis. In order to implement this in the FTN MIB, one FTN entry containing this IP prefix must first be created in the FTN Table. Table 5.6 shows an example of how this entry could be created.

Table 5.6 FTN entry for sample destination prefix 198.135.0.0, destination port 80 mapped to LSP1 and LSP2 via cross-connect pointer.

FTN entry variable	Value
mplsFTNIndex	45 (arbitrary index chosen by NMS)
MPLSFTNRowStatus	active (1)
mplsFTNDescr	"Traffic to www.cisco.com"
mplsFTNApplied	True (1)
mplsFTNMask	0x50 (binary: 0101 0000)
mplsFTNAddrType	IPv4
mplsFTNSourceIpv4AddrMin	0.0.0.0
mplsFTNSourceIpv6AddrMin	0.0.0.0.0.0.0.0
mplsFTNSourceIpv4AddrMax	0.0.0.0
mplsFTNSourceIpv6AddrMax	0.0.0.0.0.0.0.0
mplsFTNDestIpv4AddrMin	198.135.0.0
mplsFTNDestIpv6AddrMin	0.0.0.0.0.0.0.0
mplsFTNDestIpv4AddrMax	198.135.0.0
mplsFTNDestIpv6AddrMax	0.0.0.0.0.0.0.0
mplsFTNSourcePortMin	0
mplsFTNSourcePortMax	0
mplsFTNDestPortMin	80
mplsFTNDestPortMax	80
mplsFTNProtocol	0
mplsFTNActionType	redirectLsp (2)
mplsFTNActionPointer	mplsXCIndex.2.1.16.2
mplsFTNExpBits	0x0
mplsFTNStorageType	Nonvolatile (3)

Table 5.6 demonstrates the configuration that is necessary to capture the FEC for traffic with a destination prefix 198.135.0.0 and destination port number of 80. Notice that the entry contains ranges for the destination address that begin and end with the same address prefix. This is how single prefixes should be present in the table. Note, however, that it would be quite easy to augment this entry's matching of IP destination prefixes with additional (contiguous) addresses by simply extending the value set for mplsFTNDestIpv4AddrMax or lowering the value of mplsFTNDestIpv4AddrMin. This gain in efficiency is possible with any of the fields that are specified using ranges. Also notice that the mplsFTNActionType is set to "redirectLsp" and that an OID value is present in mplsFTNActionPointer. The value of mplsFTNActionPointer points at the primary index of an mplsXCEntry. In doing so, this pointer may refer to one or more entries in the MPLS-LSR MIB's mplsXCTable. In this case, we will need to refer to two entries: one for each outgoing label used to support load sharing. This will be covered in more detail later in this chapter.

5.8.2 Configuring the FTN MIB for "tun123"

Let's now examine the FTN entry required to capture the second destination prefix that is being mapped onto the TE tunnel "tun123." Table 5.7 demonstrates how the FTN entry would be configured to define the FEC for this entry.

Table 5.7 FTN entry for sample destination prefix 123.33.0.0 mapped onto TE tunnel "tun123."

FTN entry variable	Value
mplsFTNIndex	55
MPLSFTNRowStatus	active (1)
mplsFTNDescr	"VoIP Traffic for customer X."
mplsFTNApplied	true (1)
mplsFTNMask	0x80 (binary: 0001 0000)
mplsFTNAddrType	IPv4
mplsFTNSourceIpv4AddrMin	0.0.0.0
mplsFTNSourceIpv6AddrMin	0.0.0.0
mplsFTNSourceIpv4AddrMax	0.0.0.0
mplsFTNSourceIpv6AddrMax	0.0.0.0

Table 5.7 continued

FTN entry variable	Value
mplsFTNDestIpv4AddrMin	198.133.0.0
mplsFTNDestIpv6AddrMin	0.0.0.0
mplsFTNDestIpv4AddrMax	198.133.0.0
mplsFTNDestIpv6AddrMax	0.0.0.0
mplsFTNSourcePortMin	0
mplsFTNSourcePortMax	0
mplsFTNDestPortMin	0
mplsFTNDestPortMax	0
mplsFTNProtocol	0
mplsFTNActionType	redirectTunnel (3)
mplsFTNActionPointer	mplsTunnelEntry.12.1.4.123.123.125.1.4.123.123.126.1
mplsFTNExpBits	0x0
mplsFTNStorageType	Nonvolatile (3)

The example FTN entry shown in Table 5.7 demonstrates how the FEC for IP destination 123.33.0.0/16 could be configured to map traffic onto TE tunnel "tun123." This mapping is configured by setting up mplsFTNDestIpv4AddrMin and mplsFTNDestIpv4AddrMax to be equal to the desired destination of 198.133.0.0/16. Because it is desired to map this traffic onto "tun123," it is therefore necessary to link this traffic with the MPLS-TE MIB's mplsTunnelEntry that represents tunnel "tun123." This tunnel's index in the table is used as the OID pointer found in mplsFTNActionPointer. The latter portions of the index are deciphered by realizing the syntax of an mplsTunnelEntry. The entry is indexed by four values: mplsTunnelIndex, mplsTunnelInstance, mplsTunnelIngressLsrId, and mplsTunnelEgressLsrId. Notice that these values are set as follows: MPLSTunnelIndex = 12, MPLSTunnelInstance = 1, MPLSTunnelIngressLSRId = 123.123.125.1, MPLSTunnelEgressLSRId = 123.123.126.1. Without going too deep into the semantics of the values of each portion of the index, suffice it to say that together they uniquely identify an mplsTunnelEntry in the mplsTunnelTable. Further explanation of these values, their syntax, and semantics will be deferred for now and will be covered in detail in MPLS-TE MIB (Chapter 7).

5.8.3 Applying FEC Entries to Interfaces

Now that the mplsFTNMapTable entries have been set up to accommodate the desired FECs, let us configure the mplsFTNMapTable for both destinations. Recall from the introduction that the FTN Mapping Table is used to apply FTN entries to particular incoming interfaces on which the filter should be used. This is accomplished by creating entries in the mplsFTNMapTable. Before we do this, we need to specify which Interface Index (ifIndex) values are assigned to which interface. The ingress LER in Figure 5.3 has four interfaces configured. These are named Eth0, Eth1, Eth2, and Eth3. We will use the ifIndex assignment in Table 5.8 for the remainder of this example.

Table 5.8 Interface Index value assignment for this example.

Interface name	ifIndex
Eth0	12
Eth1	13
Eth2	14
Eth3	15

At this point, we have configured FTN entries for the FECs and now need to consider which interfaces to apply them to. Given what we have covered in the earlier examples, it is logical to conclude that we only care about the destination of the incoming traffic and not where it came from, that it is desirable to apply the FTN entries to *all* interfaces. In applying the FTN entries to all interfaces on the ingress LER, we have two options for assigning them. The first option would be to define individual FTN map entries for each interface to FTN entry mapping. To do this, we would create an mplsFTNMapEntry for each interface and map it back to both FTN entries. This would result in eight FTN map entries. This is demonstrated in Figure 5.4.

However, recall that the MIB defines a shortcut for the FTN Mapping Table that accommodates such configuration with ease. That is, the mapping table allows for the mapping of an entry to ifIndex = 0 to apply to all interfaces on the device. This is the second configuration that may be chosen to map the FTN entries to interfaces. In this configuration both FTN entries are assigned to 0 for mplsFTNMapIfIndex. This is demonstrated in Tables 5.9 and 5.10. This will be the configuration used to continue the example to completion.

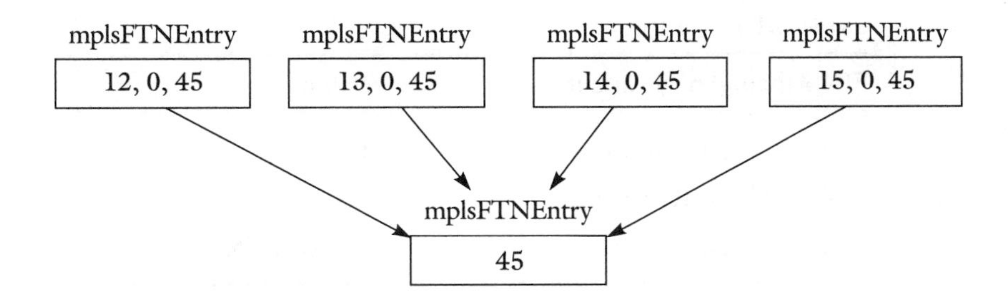

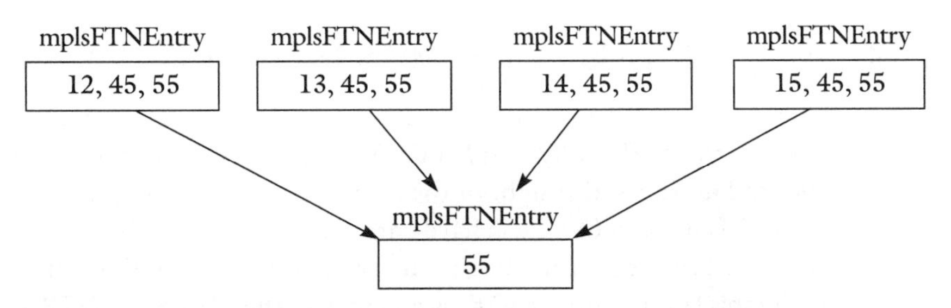

Figure 5.4 Creating mplsFTNMapEntries for each interface that map to each interface shown in Figure 5.3.

Table 5.9 Simplified FTN mapping entry for all interfaces being mapped to FTN entry 45.

FTNMappingEntry variable	Value
mplsFTNMapIfIndex	0
mplsFTNMapPrevIndex	0
mplsFTNMapCurrIndex	45
mplsFTNMapRowStatus	active (1)
mplsFTNMapStorageType	Nonvolatile (3)

Table 5.10 Simplified FTN mapping entry for all interfaces being mapped to FTN entry 55.

FTNMappingEntry variable	Value
mplsFTNMapIfIndex	0
mplsFTNMapPrevIndex	45

continued

Table 5.10 continued

FTNMappingEntry variable	Value
mplsFTNMapCurrIndex	55
mplsFTNMapRowStatus	active (1)
mplsFTNMapStorageType	Nonvolatile (3)

5.8.4 Pointing FTN Entries at the MPLS-LSR or MPLS-TE MIBs

Now that the MPLS-FTN MIB has been configured, let us investigate the configuration of the MPLS-LSR and MPLS-TE MIBs to support the desired FECs. We will refer to Figure 5.4 throughout the remainder of this example.

First, LSP1 and LSP2 will have entries in the MPLS-LSR MIB. These entries will be created to support the beginning of both LSPs. The FEC entry will be pointed at an mplsTunnelEntry, and thus an entry in the MPLS-TE MIB's mplsTunnel Table must exist.

The MPLS-LSR MIB must contain two sets of entries, one to support each LSP. As was explained earlier, the MPLS-LSR MIB contains three primary tables: the mplsInSegmentEntry, mplsOutSegmentEntry, and mplsXCTables. Because these LSPs are considered originating LSPs, there will be no corresponding mplsInSegmentTable entries. However, cross-connect and mplsOutSegmentTable entries must be created. Figure 5.5 demonstrates how the MPLS-FTN MIB's mplsFTNEntry with index 45 is pointed at the cross-connects in the MPLS-LSR MIB. In turn, each cross-connect represents a parallel or load-shared outgoing path for the labeled traffic to take. In this case, traffic will exit to two LSR neighbors located via interfaces with ifIndex equal to 13 and 14. Notice that the cross-connect is primarily indexed by the integer 2. Conceptually, this represents a single cross-connection to multiple outsegments. Practically speaking, by only using the primary cross-connect index, the MPLS-FTN MIB is allowed to address many cross-connects. In particular, in this case where multiple outsegments are associated together via a common cross-connect, it is possible to point the mplsFTNEntry at a single "cross-connect." In reality, several cross-connect entries must be created, but they are associated together nicely via the first index. It is also possible to create several mplsFTNEntries and point them at several unrelated cross-connect entries to achieve the same goal. This approach just requires more configuration and effort on the part of the agent to keep track of where the actual outsegments are. It is also potentially more confusing for the operator.

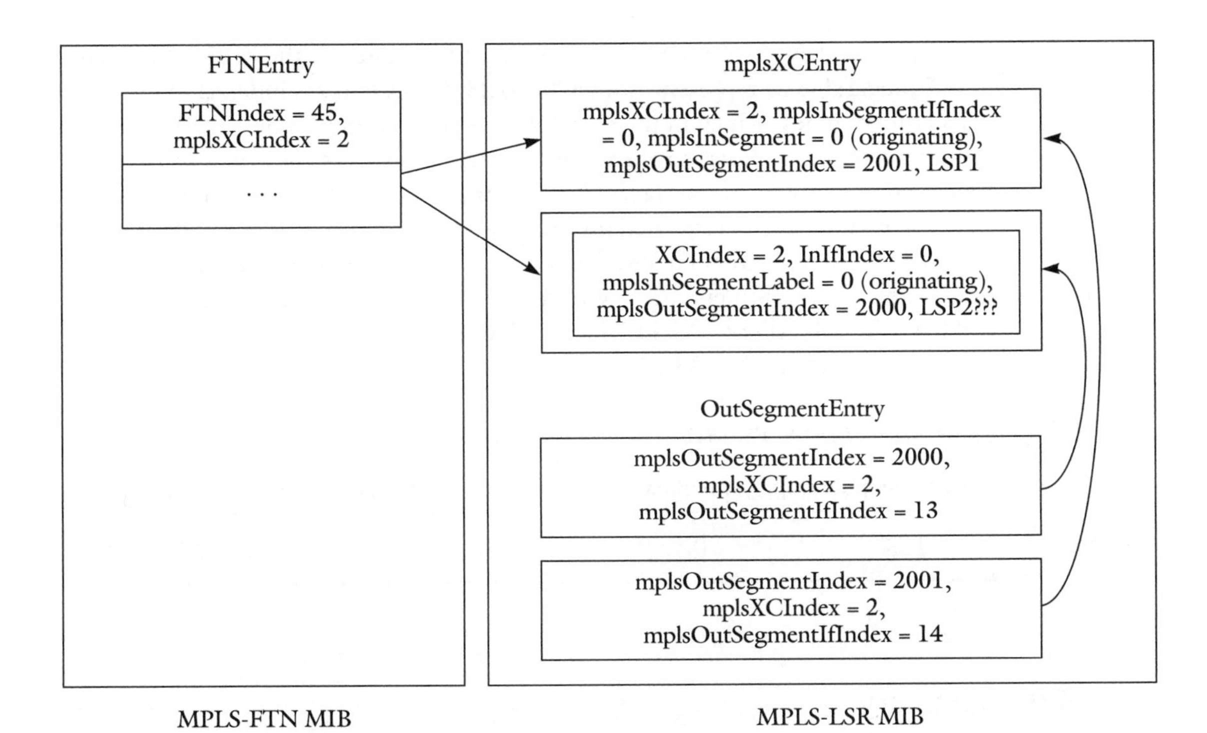

Figure 5.5 MPLS-FTN and MPLS-LSR MIB interaction to support LSP1 and LSP2.

Table 5.11 demonstrates how the cross-connect entries will be created for LSP1 and LSP2 in the MPLS-LSR MIB. Bear in mind that in the interest of simplicity, the LSP for "tun123" will not be created in the LSR MIB. Technically this will suffice, since this will actually be the case if no LSP(s) have been signaled for the tunnel head. If the tunnel head has not yet been signaled, no LSPs exist to carry the tunnel traffic, and thus no corresponding LSPs are represented in the MPLS-LSR MIB. When the tunnel is signaled, one or more LSPs should appear in the LSR MIB and must be pointed to by the TE MIB.

Table 5.11 MPLS-LSR MIB XCEntry to support LSP1 and LSP2.

mplsXCIndex	mplsInSegmentIfIndex	mplsInSegmentLabel	OutSegment
2	0	0	2000
2	0	0	2001

Table 5.12 demonstrates how an entry must be created in the MPLS-TE MIB's mplsTunnelTable to represent tunnel "tun123." This entry is indexed by the tunnel's primary index, which in the original example is chosen to be 123. The tunnel instance is set to 0, indicating that this instance represents the configured tunnel head, and not a signaled interface. If the instance were a value other than 0, this instance would represent a signaled instance. See Chapter 7 for more information on the indexing of this table. Finally, the tunnel instance is indexed by the source and destination LSR identifiers. For the purposes of this example, the originating LER will have the LSR ID of 123.123.125.1. The terminating LER for this tunnel will have the LSR ID of 123.123.126.1, which is noted in Figure 5.6.

Table 5.12 MPLS-TE MIB TunnelEntry to support TE tunnel "tun123."

mplsTunnelIndex	mplsTunnelInstance	mplsTunnelIngressLSRId	mplsTunnelEgressLSRId
123	0	123.123.125.1	123.123.126.1

5.9 Summary

It should be clear that the MPLS-FTN MIB is a powerful and essential tool for those deploying MPLS. It exposes the FEC to NHLFE mapping at ingress LERs and, in conjunction with the already-mentioned MPLS-LSR MIB and the to-be-discussed MPLS-TE MIB, can be used to provide a very complete picture of MPLS label switching within the MPLS network.

It has been explained that entries must be created in either the MPLS-LSR MIB's cross-connect or MPLS-TE MIB's mplsTunnelEntry (or both), as well as in the mplsFTNTable to facilitate the connection between the FEC and the NHLFE.

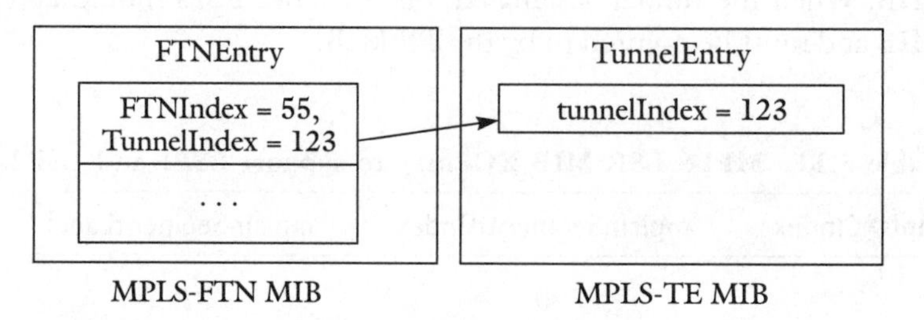

Figure 5.6 MPLS-FTN and MPLS-TE MIB interaction to support tunnel "tun123."

Furthermore, multiple entries in these tables may be required to implement multiple exit paths on the ingress LSR in order to support load sharing. Engineers implementing this MIB may choose cross-connect entries with similar indexes to represent the load-shared entries, or they may use unrelated ones. This has at least been our experience. One final tip to the engineer implementing these MIBs is to pay special attention to the indexing used above. Notice that the same index is carried for all three tables. Since these indexes are arbitrary and are designated by the agent (unless you allow writable tables), it is therefore easy to just carry the same index along. This allows for easier implementation when cross-referencing things internally, as well as quicker GET-NEXT functions that are easier to implement. Finally, note that it is legal to use the memory address of the XC as the index, which both guarantees its uniqueness and is easy to retrieve and store. Be aware that use of a memory address as an index on embedded platforms implemented with a distributed architecture (e.g., AgentX) may not necessarily work.

Further Reading

MPLS FEC-To-NHLFE (FTN) Management Information Base: *www.ietf.org/internet-drafts/draft-ietf-mpls-ftn-mib-04.txt.*

RFC 3031 specifies the MPLS architecture and is the basis for all MPLS documents. It can be located at *www.ietf.org/rfc/rfc3031.txt.*

Srinivasan, C., A. Viswanathan, and T. Nadeau. "MPLS Label Switching Router Management Information Base Using SMIv2." IETF Internet Draft. January 2001. *www.ietf.org/internet-drafts/draft-ietf-mpls-lsr-mib-07.txt.*

Srinivasan, C., A. Viswanathan, and T. Nadeau. "Multiprotocol Label Switching (MPLS) Traffic Engineering Management Information Base." IETF Internet Draft. August 2001. *www.ietf.org/internet-drafts/draft-ietf-mpls-te-mib-08.txt.*

Davie, B. S., and Y. Rekhter. *MPLS: Technology and Applications.* First edition. San Francisco: Morgan Kaufmann Publishers. 2000.

Gray, E. W. *MPLS: Implementing the Technology.* Reading, Mass.: Addison-Wesley Professional. 2001.

To find out more about the IETF, visit their Web page at *www.ietf.org/.*

For more information about IANA, check out their Web site at *www.iana.org/.*

To locate information about the ITU: *www.itu.int/.*

To locate information about the MPLS Forum: *www.mplsforum.org/.*

To locate additional information about the OIF: *www.oiforum.com/.*

Cisco's Web site also provides a great deal of information regarding MPLS: *www.cisco.com.*

Bruce Davie

works at Cisco Systems in Chelmsford, Massachusetts, where he is a Cisco Fellow. From 1988 to 1995 he worked at Bellcore on a variety of networking research projects. Since 1995 he has been at Cisco, where he was part of the team that created the initial Tag Switching architecture, which directly led to the MPLS effort. He now leads a group working on the development of MPLS and quality-of-service capabilities for IP networks. He is the author of three books on networking, an active participant in the IETF, and a senior member of the IEEE.

Bruce, having been an active member of the IETF in the early days of MPLS, how have the climate and activities of the standards body changed since then? Also, having been one of the people who got MPLS to be the successful technology that it is today, what do you see in the future of MPLS?

MPLS attracted a lot of attention as soon as it was introduced to the IETF in 1996. And there has always been plenty of controversy, beginning with the attempts to reconcile the differences between the two major contributing efforts, IBM's Aris and Cisco's Tag Switching, followed by the RSVP versus CR-LDP debate.

The most noticeable recent change has been the growth of the MPLS effort both in terms of the number of people and companies actively involved and in the number of technical areas in which work is being done. As a result of the second factor, it's become necessary to break the work effort up into multiple Working Groups, and there has also been some attempt on the part of the IESG to figure out how MPLS should fit into the broader efforts of the IETF. I think this hasn't been entirely successful, for a number of reasons. MPLS is not easily pigeonholed as a layer-2 or layer-3 technology—indeed the argument has been made (by Yakov Rekhter, among others) that MPLS illustrates the difficulties of viewing the world from a strictly layerist perspective. Also, MPLS affects a large number of control protocols, only one of which (LDP) is purely focused on MPLS—it also affects routing protocols and RSVP, for example. And finally there remains the whole question of how much of MPLS belongs in the IETF. For example, if you want to run layer-2 or layer-1 protocols over MPLS, can you justify standardizing that in the IETF, which is primarily focused on things to do with IP? I would say you can, because of the close relationship MPLS has with IP and its control protocols, but some people would disagree.

As for the future of MPLS, I have been consistently impressed with the new applications of MPLS that keep appearing. In the early days, we predicted that the simple forwarding paradigm, augmented with a range of different control paradigms, would provide the basis for a rich and evolving set of capabilities. This has turned out to be very accurate—that prediction was made before any of the VPN capabilities of MPLS had been thought about, for example. So I think we'll continue to see new applications of MPLS, some of which haven't yet been thought of. Emerging MPLS applications that look promising at this stage include its use to control a wide range of optical devices (GMPLS), restoration and protection at the network layer, and tunneling layer-1 and layer-2 traffic over IP backbones.

6

The Interfaces MIB and MPLS

"Success seems to be largely a matter of hanging on after others have let go."

—William Feather

Introduction

This chapter provides a detailed examination of how interfaces are represented and stacked in MPLS. Specifically, we will investigate how interfaces are represented in various MIB modules and how they are related to entries in RFC 2863. This chapter will explain, in general, how RFC 2863 interface entries apply to MPLS. We will investigate how they specifically are applied to general MPLS networks, as well as MPLS traffic-engineered and MPLS virtual private network deployments.

6.1 Who Should Use It

This chapter will explain how RFC 2863 interfaces operate and how they are specifically applied to general MPLS networks, as well as MPLS traffic-engineered and MPLS virtual private network deployments. This MIB module is possibly the most widely implemented and deployed MIB module. Since MPLS interfaces can and are being represented as Interfaces MIB (IF-MIB) entries in various capacities, there is a high probability that it will be available in many implementations by the time you are involved with one. It is therefore useful to explore the IF-MIB as an essential tool and show how it can be useful to the operator, as well as why it is important for vendors to support it in their MPLS implementations.

6.2 IF-MIB Overview

RFC 2863, "The Interfaces Group MIB Using SMIv2," or otherwise known as the IF-MIB, has played an important part in the standards-based network management solution provided by many different networking technologies over the years. The reason why has simply been that it defines a general model for network interfaces, as well as their layering and arrangement, that is applicable to a wide variety of data communications media technologies. In particular, it defines a model of an interface as a portal to and from a particular networking layer beneath the internetworking layer (i.e., layer 3 in the OSI model). For example, in the case of a router that supports Frame Relay virtual circuits, the IF-MIB would represent the circuits as interfaces. In addition, it is possible to determine what the hierarchical layering relationship is between any related entries. For example, if a Frame Relay virtual circuit utilizes a physical channelized interface, both the virtual circuit and the channelized interface would be represented in the ifTable. Furthermore, the Frame Relay circuit entry would be shown to be stacked above the channelized interface. This would indicate the layering relationship of the interfaces.

6.2.1 How to Number Interfaces

All interfaces represented in the IF-MIB are uniquely identified by a single index called the ifIndex. This object contains an unsigned integer with values ranging between one and the maximum value of an unsigned integer supported on that system. Agents implementing the IF-MIB and its ifTable generally assign ifIndexes in a monotonically increasing order. When the agent reaches the maximum value for

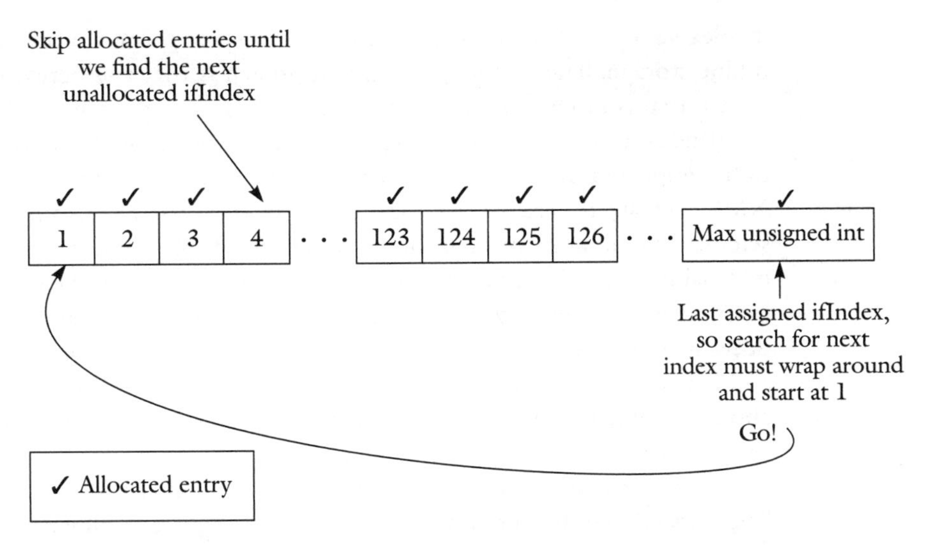

Figure 6.1 Example of the ifIndex allocation algorithm.

the index, it will wrap around to the next unused value starting at one.[1] The value of ifIndex must be unique in that a newly chosen value may not reuse an existing value. Note that an interface that is taken out of service (i.e., deleted from the ifTable) may be given the same ifIndex if it is re-created in the ifTable. The value of the ifNumber given to each interface must remain constant, at least until the device reinitializes. The algorithm for ifIndex assignment and allocation is illustrated in Figure 6.1. Note that once the value for the ifIndex reaches the maximum value supported on the agent, it wraps around to 1. At that point, unless the value of 1 is no longer assigned to an active interface, the agent must continue through the list of possible indexes until it finds one that is not currently assigned to an interface. Note that in the example, allocated entries are denoted with a check mark. Unallocated indexes are left unmarked.

Persistent Interface Index Assignment

It is possible that the device preserves its assignment of ifIndexes in a persistent manner. That is, the device remembers which interfaces were assigned which

---

1 The value of 0 is an invalid ifIndex value because it has special semantic meaning with regard to media-specific MIBs that extend the Interfaces MIB. When an entry is created in the media-specific MIB, but has not yet been associated with a valid ifTable entry, the ifIndex of that entry is typically assigned the value of 0 to indicate this fact.

ifIndex values by storing that information in some form of nonvolatile online or offline storage. The advantage of this approach is that any network management system that is monitoring this device need not reread the entire ifTable to relearn the ifIndex assignments in the event that the device reboots. This feature may be rather important if the NMS has a requirement that limits the amount of time the NMS may take to reread the ifTable and reinitialize itself accordingly. The NMS itself may have limitations on having to reset the ifTable if a device reboots, due to its internal bookkeeping of a device's interface objects. Due to the overhead involved in implementing persistent ifIndex assignment, the maintenance of Interface Indexes in a persistent manner is optional. Therefore, we recommend that operators and network managers interested in this feature consult with their device vendors about this functionality well in advance. This is especially important in the event that this feature must be implemented from scratch.

When an agent reboots, it may choose to remember the last ifIndex value and begin assignment from that point on, or it may begin assignment from one. When interfaces are deleted due to being disassociated within software or are physically removed, their ifIndexes may be reused by the agent. However, these values should not be reused for "a long time" so as to not confuse any NMS. That is, the value should not be reused for as long a time as an NMS might have been monitoring the interface and not received either a notification indicating the removal of the interface or a polling cycle that revealed that its RowStatus was destroyed. Due to the size of the ifIndex space (Unsigned32), typical systems will not reuse an ifIndex for hours, days, or until they are rebooted. If a system maintains indexes persistently, it may take many thousands of system reboots before it will reuse an ifIndex.

Dynamic Interface Creation and Deletion

Effective management of the ifIndex space is not possible unless the agent enforces the constancy requirements that have just been described. This becomes especially evident when considering the many modern networking devices that provide functionality allowing network interfaces to be removed or added dynamically (i.e., "hot swapped"). For example, a device that contains some number of media-specific interfaces on a modular board may have that board removed by the operator either due to a desire to change media types or because the board is malfunctioning. This obviates the need to power down the entire unit, thus allowing the remaining, presumably operational, interfaces to continue transmitting and receiving network traffic. This is demonstrated in Figure 6.2. The leftmost I/O board is removed and replaced with a similar one, whereas the rightmost board is removed but replaced with one of a different type of interface. The act of removing the modular board may result in the SNMP agent dynamically deleting all of the physical interface entries from its internal bookkeeping, the corresponding

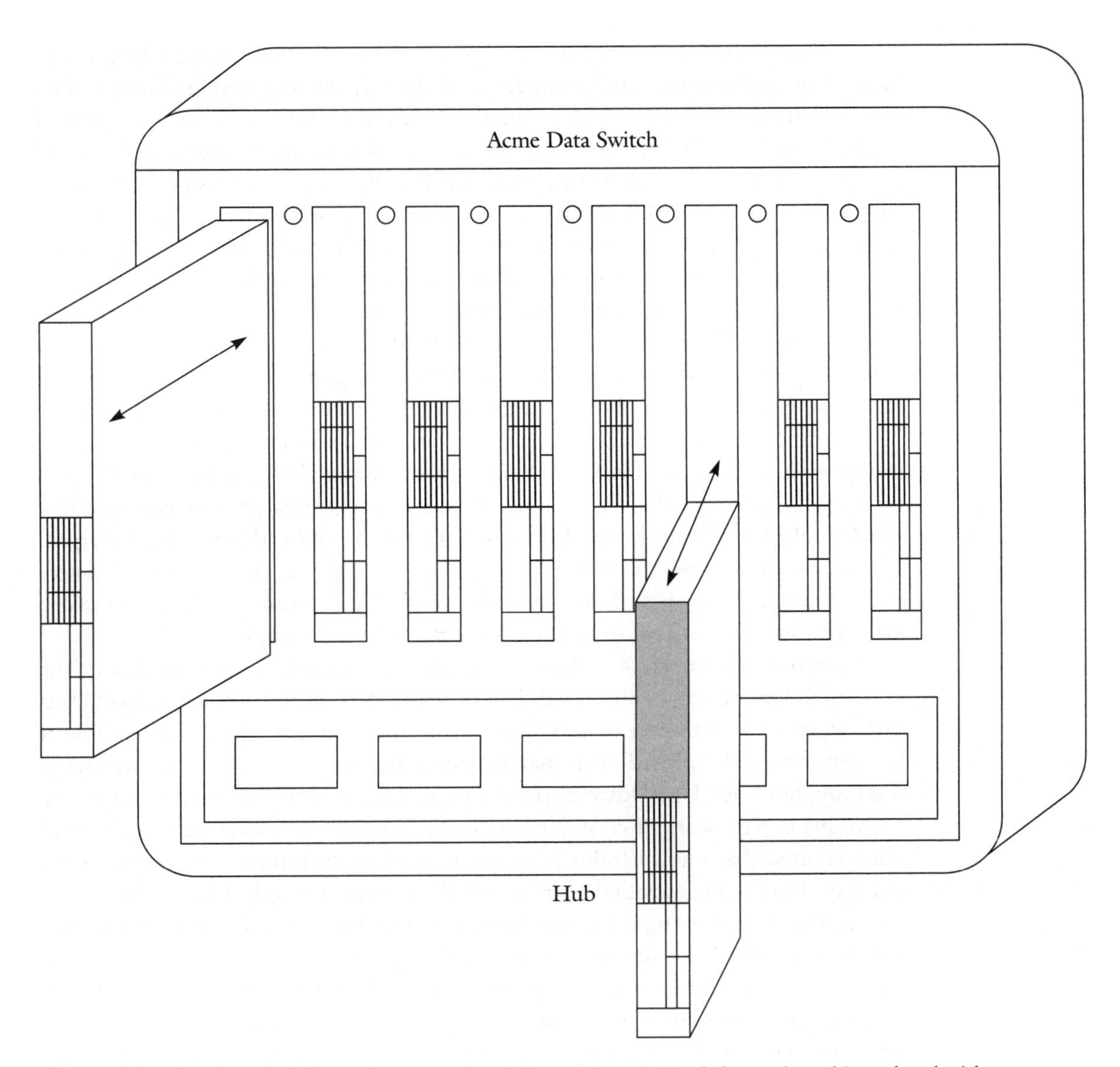

Figure 6.2 Switch with I/O cards removed and then replaced. The leftmost board is replaced with an identical board, but the other (shaded) is replaced with a different type of I/O board.

subinterfaces from the ifTable, and the stacking relationship for these interfaces as represented in the device's ifStackTable. Such dynamic additions/removals may result in the possibility of "holes" in the ifTable—that is, ifIndex values no longer being contiguous. This is, however, not a problem for managers, as the GET-NEXT and GET-BULK operations easily deal with sparse tables.

Consistency between reinitializations of an interface's ifIndex value is further enhanced by requiring that after an interface is dynamically removed, its ifIndex value not be reused by a *different* dynamically added interface until after the agent is reinitialized. This obviates the need for a priori assignment of ifIndex values for all possible interfaces that might be added dynamically. Clearly, this is not something that is easily achieved. At best, it imposes restrictions on the agent that result in a nonefficient allocation of the ifIndex number space. It is typically acceptable for an agent to reuse the ifIndex if the new interface is of the same ifType (i.e., same media). For example, in Figure 6.2, the I/O board that is removed but replaced with the same type of board containing presumably the same number of interfaces may have those interfaces reassigned the same ifIndex assignment as the previous board had. Conversely, the other board that is not replaced with the same type may not.

It is important to note that different implementations interpret the exact meaning of what a different board type is in different ways. For example, some implementations may reuse the ifIndex if the interface that is replacing the old one is of a *similar* subtype. For example, if the interface being removed is of type Ethernet, and the new one is of type Fast Ethernet, it is possible that the ifIndex will be reused. Again, the specifics of this behavior are implementation-specific, so network managers should check with their device vendors for more details.

When the new interface is the same as an old interface, a discontinuity in the value of the interface's counters will always result. A potential problem exists in this case because an NMS may be monitoring this interface and may not detect that it was removed and replaced. This may happen if the poll interval on the interface is too long, or if the ifUp/Down notifications are lost, or if they are never sent by the agent due to a network or configuration error. The only way to avoid this has until now required that a new ifIndex value be assigned to the interface that replaces the old one. That is, the agent either retained all counter values during the absence of an interface, thus making it appear that no discontinuity occurred even after assigning the same ifIndex value on the interface's replacement, or a new ifIndex value was simply assigned to the returning interface. Both of these options pose different problems to agent implementations. Namely, it may be difficult or impossible for an agent to maintain the counter values. This may be the case if, for example, the counters are maintained on removable hardware. In its absence, the agent will be required to either make a copy of the counters or forget about them. Neither case is easily implemented. A second problem occurs due to the use of a new ifIndex. Taking this approach results in additional bookkeeping and processing effort for management applications.

A solution to both of these problems is for the NMS to monitor the ifCounterDiscontinuityTime object. This object records the sysUpTime when the last discontinuity in any of an interface's counters occurred. An NMS can detect counter discontinuities by checking this value periodically. The advantage of doing so is that the NMS now does not need to keep track of whether or not the ifIndex of the interface has changed (or if it was silently replaced and the same one reused) because it can now detect either case simply by checking this value. Implicit is the assumption that agents will implement this variable correctly! Agents implementing this object and wishing to reuse that interface's existing ifIndex will need to update the interface's value of ifCounterDiscontinuityTime when appropriate. A management application will have to discard any calculated difference between successive polls of any of the objects on an interface for which the value of ifCounterDiscontinuityTime has changed. It is important that an NMS perform this test *in addition to* the normal checking it should be doing of sysUpTime to detect agent reinitialization.

6.2.2 Interface Layering

When managing media-specific interfaces on a device, you must distinguish between each sublayer beneath layer 3. It is therefore desirable that each sublayer is aware of which, if any, internetworking or intermediate protocols run either above or below it. Interfaces are distinguished by their ifType and ifIndex values. They are conceptually "stacked" in the ifStackTable, which allows identification of the protocol sublayer interactions. Note that the stacking may be arbitrary and is typically defined by the standards that specify the protocols.

Figure 6.3 shows an example of conceptual interface layering. The figure depicts a device that has several virtual PPP interfaces that utilize an HDLC link layer. This layer in turn uses an RS232 connector. In this case, it is important to the manager of this system to reveal this layering, as is shown in Figure 6.3. Note that interfaces are represented in Figure 6.3 by the tube segments and have their respective ifIndex values noted within. The layering just described is represented in the IF-MIB by creating an individual conceptual row in the ifTable to represent each sublayer interface. Each entry in this table is assigned a media-specific ifType that is maintained by the Internet Assigned Numbers Authority (IANA). In the example shown in Figure 6.3, all of the interfaces at a particular sublayer will have the same ifType value. For example, the interfaces at the PPP sublayer will all have the type for "ppp." Note again that interfaces are represented in Figure 6.3 by the tube segments and have their respective ifIndex values noted within each icon. The layering relationship between each entry that is also known as the interface layer is represented in the ifStackTable by creating an entry for each interface there as well, and then "pointing" that layer to the corresponding higher or lower layer or layers. This is illustrated conceptually in Figure 6.3 with the bidirectional arrow, which connects the appropriate interfaces to each other. Each arrow in the example indicates a flow of traffic between layers. Table 6.1 also illustrates this, but demonstrates how the ifStackTable would be configured to represent this scenario.

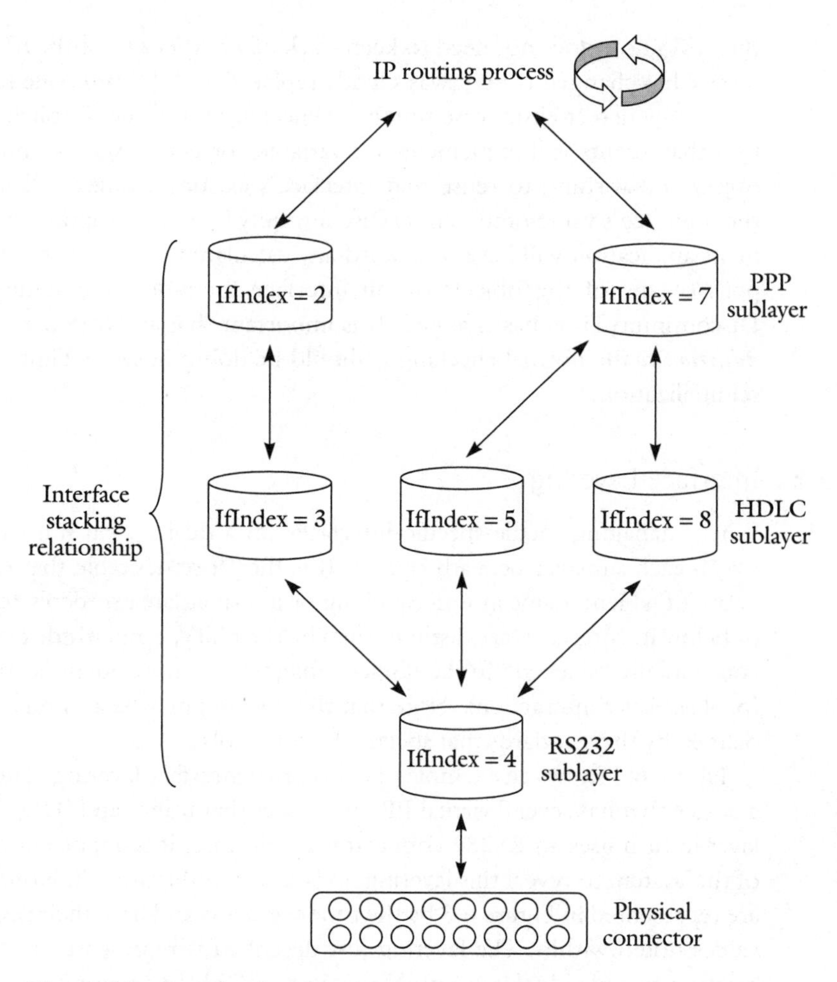

Figure 6.3 Interface layering.

Table 6.1 The ifStackTable representing the subinterface layering for Figure 6.3. Note that the 0 entries are required to indicate the upper and lower limits of the stack. For example, the first two entries indicate the top of the stacking relationship for both PPP interfaces, while the 0 as the lower layer indicates the bottom of the stack for all interfaces stacked above it.

ifStackHigherLayer	ifStackLowerLayer
0 (none)	2 (PPP)
0 (none)	7 (PPP)

Table 6.1 continued

ifStackHigherLayer	ifStackLowerLayer
2 (PPP)	3 (HDLC)
7 (PPP)	5 (HDLC)
7 (PPP)	8 (HDLC)
3 (HDLC)	4 (RS232)
5 (HDLC)	4 (RS232)
8 (HDLC)	4 (RS232)
4 (HDLC)	0 (none)

It is possible that more than one layer or entries within each layer are used to facilitate load balancing. That is, when an upper layer wishes to balance traffic between multiple lower-layer interfaces to improve throughput, it will use some sort of distribution discipline (e.g., round-robin, per-destination, etc.) to divide its traffic among multiple lower-layer interfaces. An example of load sharing is shown in Figure 6.3, where the interface with ifIndex = 7 load-balances its traffic between interfaces at the HDLC sublayer entries 5 and 8. It is also possible to repeat the traffic from a single lower layer to multiple upper layers as well. For example, notice that there is a single RS232 interface at the bottom of the interface stack in Figure 6.3. This interface is connected to interfaces 3, 5, and 8 at the next highest sublayer. In order to deliver traffic to higher layers, the interface must know how to demultiplex traffic to each subinterface stacked above it. It is important to realize that the interface-stacking scheme employed by the IF-MIB supports multiplexing in both the upward and downward directions. It further allows us to identify which media-specific MIB module can be used to fully manage any particular sublayer by associating a particular IANA ifType with a media-specific MIB module.

The counters of packets received on an interface are defined as counting the number of packets delivered by a "higher-layer" protocol to this interface. The higher-layer protocol simply indicates the interface(s) that are stacked above an interface that will accept traffic from the underlying interface(s). The actual higher-layer protocol is defined as a forwarding module that accepts packets, frames, or octets using the specified interface. This interface then passes the traffic to a protocol entity (e.g., OSPF) or to another interface. The delivery of an interface's data may be to an intermediate layer within the interface stack. For example, in Figure 6.3, data is received from the HDLC layer and is pushed up the stack to the PPP layer.

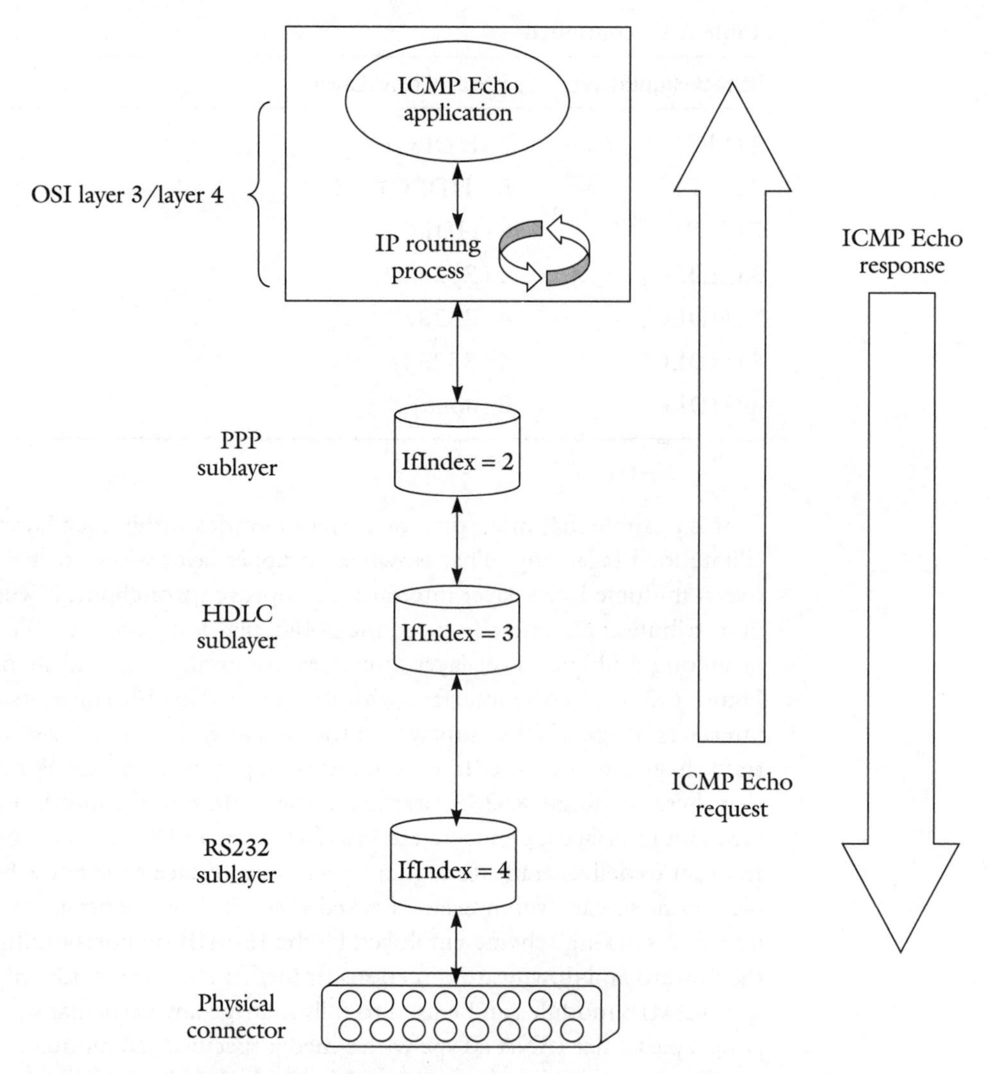

Figure 6.4 Delivery of ICMP traffic addressed to this router to the layer-4 ICMP Echo application on the router.

Finally, delivery may be to a higher-layer protocol that cannot do packet forwarding. That is, the ultimate layer on the stack may be the local IP module. For example, Figure 6.4 demonstrates how an ICMP Echo processing application module that is part of the IP routing process may handle ICMP Echo request messages directed to the routing device. In the figure, the router passes the received Echo request data up through the stack to the ICMP Echo application, processes it, and transmits an Echo response message down the stack. For output counters, the number of packets transmitted out an interface is defined as counting the number of packets or protocol data units that one or more upper-layer protocols requested that this interface transmit.

6.2.3 How Sublayers Are Defined in the IF-MIB

The author of a media-specific MIB module must decide two things when addressing a new media type. First, the decision of whether to add a new interface layer carries with it some important consequences. It is important to weigh the pros and cons of such a decision carefully, as a poor or ill-thought-out decision could result in a waste of agent, network, and manager resources. In particular, if the designer decides to create an interface at a layer where it is inappropriate, entries at this layer that add little or no value to the system must then be maintained. These superfluous entries require resources from the agent, the network, and the manager. Second, whether a layer exists, or a new one is created, the designer should strive to conserve system resources by keeping the number of entries in the ifTable that correspond to this layer and its related tables to the minimum required for successful and effective network management. Specifically, a new interface layer should not be created if an existing layer can be used to represent the layer, or if a new layer will add little or no value. Furthermore, if a new layer exists or is to be created, the designer should be sure that a single interface entry in the ifTable might not be used instead of creating multiple ones. Several conditions may be used to guide a MIB module designer on whether or not to create multiple interface entries or a single one. First, if none of the group of interfaces performs multiplexing for any other interface in the agent, it makes sense to create a single interface. Second, if there is a meaningful and useful way for all of the ifTable's information (e.g., the counters and the status variables) and all of the ifTable's capabilities (e.g., write access to ifAdminStatus) to apply to the group of interfaces as a whole, then a single interface should be used. Figure 6.5 illustrates each of the points just described pictorially. In cases where new interfaces or layers are superfluous, they are grayed out or marked with an "X."

If the decision is made by the designer of a new MIB module to create a new interface layer, it is then appropriate to assign a new media-specific ifIndex value to all interfaces that reside at this new layer. This is accomplished by requesting a new ifType from IANA. This can be done by sending email to iana@iana.org. Note that new media types must have corresponding standards-based documents that define them.

When managing various devices, either from the same vendor or from multiple vendors, it is possible to encounter situations where the sublayers are implemented slightly differently. This is also possible between interconnected devices when

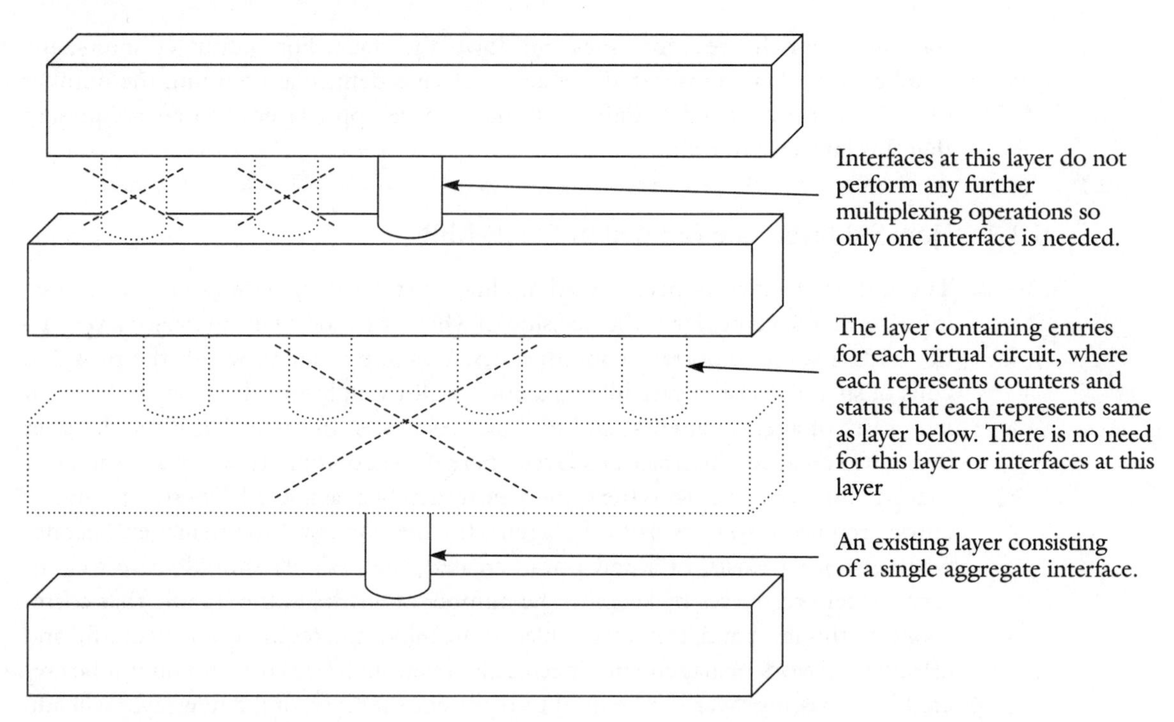

Figure 6.5 Examples of generally when and when not to create additional layers and corresponding interfaces for a new protocol sublayer.

examining the interface stacking of the interfaces on either side of the connection. This occurs since most interface stacking standards provide a certain amount of latitude to accommodate different implementations. Be sure to check the appropriate standards documents to verify that the interface stacking is within allowable guidelines.

6.2.4 ifName, ifDescr, and ifAlias

The IF-MIB contains three objects that are intended to aid the operator in identifying interface entries and distinguishing them from the many other entries in the table. The first of these objects, ifName, was added to aid the operator in distinguishing physical interfaces from the proliferation of virtual ones. Early deployments of the IF-MIB typically created a direct mapping of physical interfaces to ifTable entries. This mapping was simple and resulted in a relatively small number of entries that could be physically verified and corroborated with the entries in the ifTable by looking at the device and comparing these entries with those found in the table. However, with the advent of virtual interfaces came the need to create additional ifTable entries for these interfaces. These additional table entries no longer had the simple and intuitive one-to-one correspondence with physical interfaces on the device and thus had the potential of resulting confusion on the part of the operator. The purpose of the ifName is to aid the operator in identifying these new interfaces and distinguishing them from the physical interfaces on the device. This object contains a string representing the device's local name for that interface. This name should be the same as the one used on the other management interfaces on the device, including the command-line interface. As an example of how this object can be used, consider the router shown in Figure 6.3. The router in the example contains an interface stacking composed of PPP running over an HDLC interface running over an RS232 port. In this case, the router should assign names similar to "ppp1" to the PPP interface, "hdlc1" to the HDLC interface, and "ser1" to the serial port's interface. If the router in the agent simply assigns a name like "ser0" for this interface, then the ifName objects for the corresponding PPP and RS232 entries in the ifTable might both be assigned the same values of "ser0." This would result in operator confusion. In contrast, the existing ifDescr object is intended to contain a description of the interface rather than a canonical name. In contrast to ifName, the ifDescr object should describe the interface in sufficient detail, which may be similar to or the same as other interface entries on the device. For example, the PPP interface could be described as "Point to Point Protocol Interface." Another interface on that same device might very well have that same name.

Finally, the IF-MIB contains the ifAlias object, which is also used to describe the ifTable entry, but differs from the aforementioned ifName and ifDescr in that it should be configured by a management system or operator. This name is intended to be a shorthand for the interface that is perhaps more familiar or germane to the operator's context. For example, the PPP interface in Figure 6.3 might be assigned an ifAlias of "PPP interface for K-Mart home office to Burlington Satellite Office." The alias is irrespective of any interface-stack relationship. Because the network management system may count on this object remaining the same across system reboots, the value should be stored in nonvolatile storage by the agent so that the interface can retain this value. The agent should persist in associating this value with the interface, even if an agent chooses to assign a different ifIndex value for the interface.

6.2.5 Virtual Subinterfaces

Many media-specific MIB modules have been defined for connection-oriented media types. Such MIB modules have been defined for Frame Relay, X.25, ATM, and recently MPLS traffic engineering. Some of these MIBs have taken the approach of representing *all* virtual circuits with a single entry in the ifTable that is used to aggregate counter and status information together in one place. This is demonstrated in the bottommost sublayer in Figure 6.6. The dotted interface symbols would be represented as individual interface entries if this aggregation approach were not taken. However, some technologies such as MPLS have chosen to take a different approach—to represent their virtual circuits as separate entries in the ifTable. One major reason for the divergence from the previous approach of aggregating together all VCs into a single interface was that operators found it very useful to have access to counters and stacking for these circuits. The reasoning was that the counters and status of any single virtual circuit differed significantly enough from other circuits that it was important to model it separately. This is illustrated by the topmost layer in Figure 6.6. Notice that true interface entries are created for each VC in this example.

Despite the utility of such an approach, there is still a price to be paid for such convenience. MIB modules that specify this approach for agent implementations will require much more memory and processing power than those that choose to represent all of the circuits at a sublayer as a single entry. It is therefore still strongly recommended by most MIB module designers as well as the IETF's best current practice documents that connection-oriented sublayers not have a one-to-one relationship with entries in the ifTable for each virtual circuit.

6.2.6 How to Represent Non-Packet-Oriented Interfaces

Several networking technologies that are in use today are non-packet-based. The most commonly used are character-based serial (e.g., HDLC) or cell-oriented (e.g., ATM VC). Due to the packet-based nature of many of the objects in the ifTable

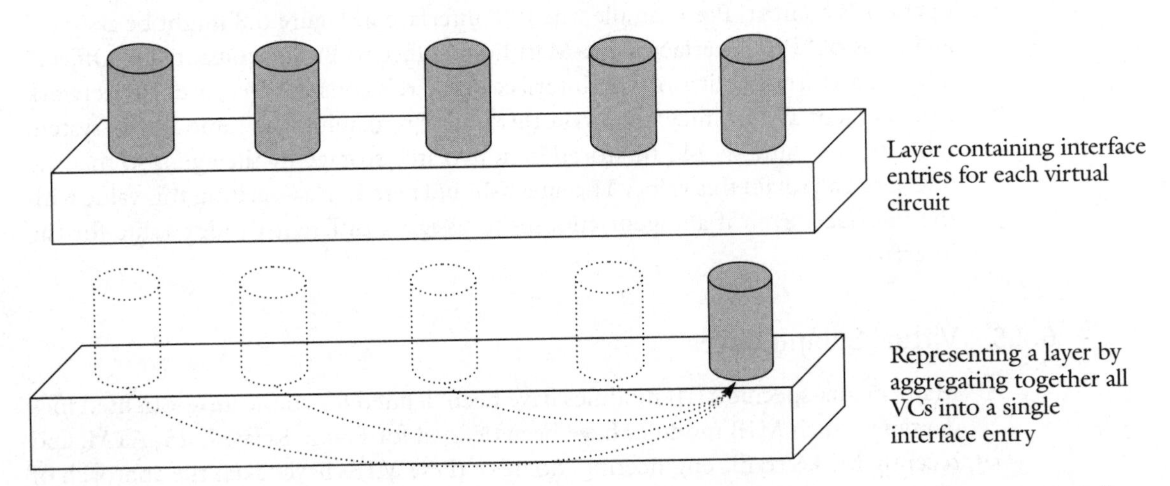

Figure 6.6 Example of a sublayer that aggregates all VCs into a single entry versus one that creates interface entries for each virtual circuit.

and ifXTable, it is generally inappropriate to represent character-oriented or fixed-length interfaces and their associated sublayers as individual interfaces that are stacked together in the IF-MIB. Instead, the physical interface typically represents the entire layer of interfaces as a single interface. An example of this is Asynchronous Transfer Mode (ATM). ATM constrains data transfer to fixed-length cells. Representation of ATM interfaces as packet-based interfaces may result in redundant information being kept by the agent. This can sometimes happen because it is possible for the number of packets and the number of octets in either direction to be fixed by the size of a cell; therefore in some cases, it may be appropriate to represent all of the interfaces as a single entry that aggregates the information from all layers together.

As with all rules, there are exceptions to them, including the rule just described regarding the representation of non-packet-oriented interfaces within the IF-MIB. It is sometimes desirable to have management information for non-packet-oriented interfaces represented as individual sublayers stacked above the physical interface. This is usually done if the previous rule does not apply. That is, if there is sufficient difference between the counters and status of the sublayer interfaces stacked above the physical interface, then it is appropriate to represent them this way. For example, when the DS0 channels of a DS1 circuit are used to carry packet-based data, or when the DS0 channels are bundled together into a DS0 bundle type interface, it is appropriate to represent them as subinterfaces in the interface's IF-MIB despite the fact that the underlying transport is bit-oriented. This model is demonstrated in Figure 6.7. Notice that the underlying DS1 interface is segregated into five DS0 channels. Three of these channels are grouped into a "DS0 bundle" interface, which is then utilized by the Frame Relay service (i.e., a router process) as a single interface. The remaining two DS0 channels are not bundled together into a DS0 bundle. This demonstrates the fact that although interfaces *may* be bundled

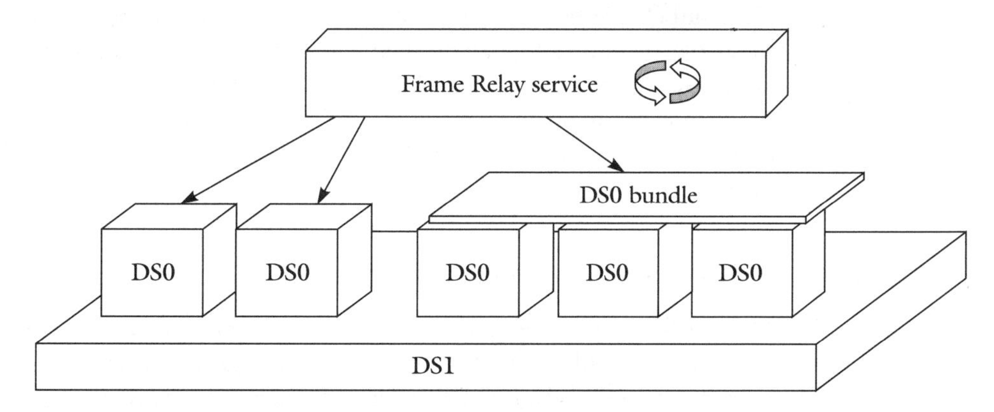

Figure 6.7 DSx interface stacking model with DSO link bundles.

together, it is up to the discretion of the device and/or system operator to decide on the bundling characteristics of the interface stacking. Finally, both the DS0 bundled interface and the unbundled DS0 interfaces are presented to the Frame Relay process as available interfaces.

6.2.7 Interface and Link Bundling

The example shown in Figure 6.7 not only broke the rule of representing byte-oriented interfaces as subinterfaces, but it also introduced a new layer referred to as a "DS0 bundle" layer. Layer bundling is desired for several reasons and is common in several other technologies; thus it is worth investigating.

First, a bundling layer is created to group or aggregate together similar lower layers into a new sublayer. In the case of Digital Subscriber type interfaces, a new layer type referred to as a "DS0 bundle" layer can be created. This additional layer has its own ifType assigned so that it can be distinguished from the other DS0 layers. Other technologies define their own bundling types. When the bundling layer is created, it is done for several reasons. First, channelized technologies typically divide a larger channel into smaller ones for the express reason of link sharing or multiplexing. However, when additional bandwidth is desired that is greater than a single channel, several channels are collected together and treated as a new aggregate channel. A second advantage of creating an aggregation layer is to present the upper routing layer with a smaller number of interfaces to keep track of. For example, if instead of presenting 100 56K DS0 interfaces to the routing layer, there is a single interface of 56K * 100 = 5600K, it should be clear that the bookkeeping and processing required by the routing protocol is greatly reduced. For example, rather than presenting OSPF with 100 interfaces over which it must transmit, receive, and keep track of hello messages, it now must only do this processing for a single interface. Finally, implementations bundle channels for redundancy. When multiple channels are bundled and presented to the routing layer as a single interface, they do protection and load balancing automatically without bothering the routing layer. This has the advantage of fewer routing updates being generated as opposed to when each channel is represented as an individual link. If a channel is reduced, only the link's aggregate bandwidth and throughput is reduced, but the link stays up and can still transmit data.

6.2.8 Creation and Deletion of Interfaces

Most devices prevent either the operator or the network management station from creating or deleting physical interfaces for logistical reasons. For example, if a router were built with one Ethernet and one serial port that cannot be removed, it would probably render the device useless if the operator removed those interfaces from the system. On the other hand, if a device contains modular interfaces (or removable I/O cards with a number of interfaces on them), it may make sense to allow the operator to remove existing interfaces from the system or create new ones. It is also possible—and sometimes preferable—for the interface management of physical interfaces on modular I/O boards to be created and deleted by the system. Therefore, operators and network management stations are typically only allowed to create and delete virtual subinterfaces. In cases where operators are allowed to create interfaces, the IF-MIB specifies that the interface must be first created in the media-specific MIB module rather than in the ifTable. The reason is that the ifTable contains generic information about interfaces and most likely does not contain enough information for an operator to sufficiently configure the interface. However, once the interface has been created in the media-specific MIB module, either the device should create the corresponding ifTable entry automatically, or the operator should create it manually. The method required depends largely on the device. Similarly, when the interface is deleted by either the operator or the device's software, the corresponding ifTable entry should also be deleted or be brought to a not-present state.

6.2.9 Addition of New IfType Values

IF-MIB entries are all assigned a specific value for ifType, which identifies when possible the type of media supported by this interface. The ifType is assigned to each interface so that operators and network management systems can identify their type, and both distinguish it from other media type entries in the table as well as identify other interfaces that support the same media type. It is also important to distinguish interfaces that are similar but of a different type. For example, simply identifying all Ethernet interfaces without identifying their transmission capabilities (e.g., half-duplex versus full-duplex, or 10 Mbps versus 1 Gbps) could be misleading to the operator. Differentiation of interfaces is particularly crucial when the total collection of interface types supported within a system goes beyond a simple few. Typical modern switches and routers support dozens of different media types. Identification of the interface type also assists the operator in understanding whether an interface's media type is physical, virtual, bundle, or perhaps a new and undefined type. Imagine attempting to figure out how interfaces are functioning on a system with 20 different types of media simply by looking at the stacking relationships! The importance of identifying the interface type is demonstrated in Figure 6.8. Notice that at the physical layer there are two Ethernet-like interfaces, but they differ in speed and duplex capabilities. The leftmost is defined as ethernet3Mbit (ifType = 26), and the next is defined as gigabitEthernet (ifType = 117). Both

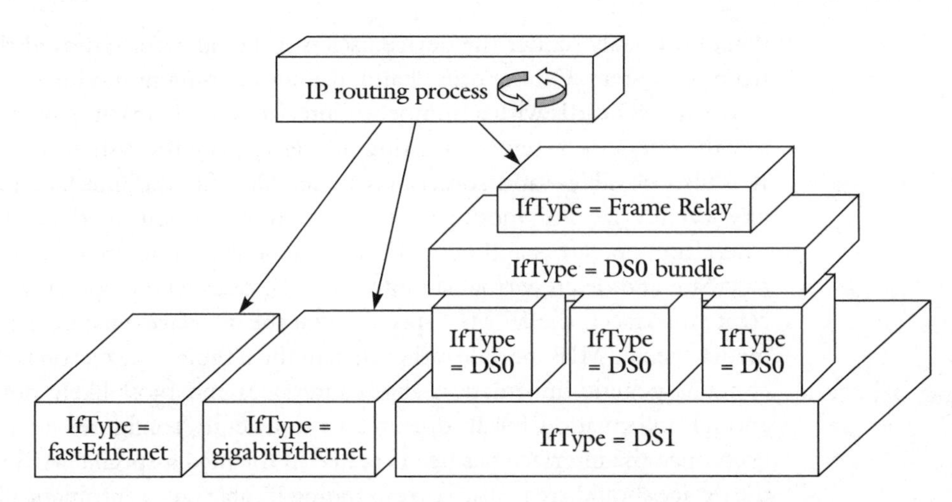

Figure 6.8 Importance of ifType when stacking heterogeneous subinterfaces.

could be loosely called "Ethernet" interfaces, but their capabilities are very different. In addition, in the case of the third physical interface, it is important to distinguish the DS channels from one another. Simply referring to each as a "DSx" channel would not be an accurate description. Therefore, the DS0 and DS1 channels have a specific ifType. Finally, note that even the Frame Relay interface has a specific type. At the topmost routing layer, it is sometimes important or desirable to understand which underlying types are connected or available.

It was once thought that the descriptive field of an interface could be used to define which type of media was supported by the interface. However, in a multivendor environment, this becomes increasingly problematic when you consider the fact that a single media type can be defined differently. For example, consider 10 Mbps 10-base-T Ethernet. This could be represented descriptively as "enet1 10mbps enet" or "10bT enet." Both strings identify the type of media, but in a manner making it sufficiently difficult for an NMS to identify. This is why the IETF, in cooperation with the IANA, has defined types that explicitly and precisely reveal the media type of an interface. Thus, all standard ifType values are maintained by the IANA as the IANAifType textual convention in the IANAifType MIB module. When there is the need to add a new ifType, a request must be made to IANA by sending email to them at iana@iana.org. Note that new ifTypes are not added unless they are sanctioned by an approved standards body. In the case of the IETF, a new ifType is typically assigned only if a MIB module has been accepted as an official Working Group draft. The reason is that this draft will probably become a standard at some point. Accepting ifTypes for individual submissions is problematic because it has the potential for ifType assignments that do not become standardized. One or all of the authors of the new MIB module should send the request to IANA for the new ifType assignment so that they can be assigned as the contact person(s) for the new ifType. Alternatively, the current Working Group chair can be assigned as the contact person.

The syntax of ifType is represented as a simple enumerated integer. When new values are added, this textual convention is modified by adding to the end of the enumeration. This is done so as not to affect existing implementations that assume prior assignments. IANA is responsible for the assignment of all Internet numbers, including various SNMP-related numbers. It is also responsible for maintaining the ifType values. We have included references for where to find the IANA ifType MIB module in the Further Reading section at the end of the chapter.

6.2.10 Interface Speed

The IF-MIB includes a variable for identifying the speed of an interface. As was described with the ifType, it is important for an operator to understand not only the type of interface, but also its current bandwidth. The reason why it is important to report this for each interface is twofold. First, the description or the naming of the specific ifType value (or both) may be insufficient to describe the speed of the interface. Furthermore, on interfaces where this value can vary depending on load, these values will incorrectly describe the interface's speed. Thus, a separate value for ifSpeed is necessary.

The ifSpeed object is defined as an estimate of the interface's current bandwidth in bits per second. The reason why it is defined as an estimate is that on interfaces where the value varies, it may not be possible or desirable to fetch the exact current interface speed. For example, if the average bandwidth is being maintained, the current bandwidth reported will be just that—an average, and not the actual value. For performance reasons, it may also be desirable to delay or buffer the bandwidth value and report the stored value. This value may be several seconds old, thus again not reflecting the exact current bandwidth. However, implementations should make every effort to return a value that is as close to the actual value as is possible. The nominal bandwidth should be returned when the value of the interface's speed is fixed. The nominal bandwidth might, for example, differ from the actual bandwidth if this value varies with the loading or utilization of the interface. For example, it is reasonable to return 10 Mbps on a 10 Mpbs 10-base-T Ethernet interface, despite the fact that the actual bandwidth on this interface varies at any point in time depending on how many multiaccess users are attempting to transmit on the LAN segment at that time, or over some period. This should also be done when it is not possible or desirable to actively measure the bandwidth of an interface. For example, this approach is sometimes taken on lower-end devices where

CPU horsepower is limited and is used only for critical operations such as PDU forwarding. Due to cost constraints, lower-end systems may also not be equipped with the appropriate hardware to measure interface utilization. One final case is possible. The concept of bandwidth is inappropriate for some sublayers. In these cases, the ifSpeed should be set to zero by the device. For example, in some deployments of MPLS, traffic-engineered tunnels are signaled with a bandwidth of zero, denoting that they will take whatever bandwidth is currently available. These interfaces will thus report a value of 0 when their ifType is queried.

Because modern interface speeds may exceed the maximum speed of $2^{31}-1$ bits/second, or approximately 2.2 Gbps, it is necessary to have a mechanism for representing these higher speeds. When the measured or actual bandwidth of an interface exceeds the largest possible value for a 32-bit integer, this value (4,294,967,295) should be returned and the ifHighSpeed object used. The ifHighSpeed object is intended to be used to report speeds that exceed 32 bits without incurring the overhead of a 64-bit integer. Instead of reporting the exact bandwidth, the ifHighSpeed object is designed to report the speed of the interface in 1 million bits/second units. As a result, the actual speed of the interface might be the value reported by this object, plus or minus 500,000 bits/second. For example, in the case of SONET the slowest interface is defined as an OC-48 interface. This interface is defined with a bandwidth in excess of 2.4 Gbps. If ifSpeed were only supported, an accurate representation of this interface's bandwidth could not be reported. However, it can if the ifSpeed is first set to 4,294,967,295, and then the actual bandwidth reported in ifHighSpeed as 2400 (1 million bits/second units) = 2400 * 1,000,000 bits/second = 2.4 billion bits/second.

6.2.11 Counters

The IF-MIB provides several standard counters for each entry in the ifTable. These counters include bytes and packets received, transmitted correctly, as well as in error. These counters provide the network management station and operator with a standard set of counters to expect from all interfaces represented in the table. However, not all counters will be utilized depending on the nature of the interface. For example, character-oriented interfaces that have no concept of packets may report a value of 0 for ifInUcastPkts.

The more recent versions of the IF-MIB define two flavors of counters: 32- and 64-bit. The latter were added to the IF-MIB to support high-speed interfaces that would wrap the older 32-bit counters around to zero before the NMS could poll those counters. This would result in an inaccurate measurement of the counters' values. For example, an Ethernet interface repeatedly accepting full-sized Ethernet frames at about 10 Mbps would wrap its associated ifInOctets value in just over 57 minutes. Increasing the rate of packets to 100 Mbps would reduce the time it took this same counter to wrap around to roughly 5.7 minutes. Both of these maintain a reasonable period of time during which an NMS could safely retrieve the counter. However, if we then increase the interface's speed to 1 Gbps and assume that it receives traffic at that rate, the time it takes the ifInOctets counter to wrap is decreased to 34 seconds! This is simply not enough time for an NMS to poll this counter considering both the network traffic required to poll the counter at very frequent intervals, as well as the fact that it most likely has some reasonably large number of counters to poll during this interval.

In order to support interfaces that wrap around too fast, the IF-MIB contains 64-bit versions of all 32-bit counters. These counters should only be used when the 32-bit counters *could* wrap too fast. Note that the fact that a counter *can* wrap around is sufficient to warrant the use of the higher-capacity counter because the NMS does not want to lose any data. It is better to be safe than sorry. The guidelines from the IF-MIB stipulate that any interface operating at 20 million bits per second or less has to implement 32-bit byte and packet counters and not implement the 64-bit version. When an interface operates faster than 20 million bits/second, but slower than 650 million bits/second, both 32-bit packet counters and 64-bit octet counters are to be used. This is because although the number of packets will not cause the 32-bit counters to wrap, the larger number of bytes per second could. Finally, for interfaces that operate at 650 million bits/second or faster, 64-bit packet counters as well as 64-bit octet counters should be used, and the 32-bit counters should not be used. Note that when a 64-bit counter is in use, if queried its 32-bit counterpart should continue to return a value containing the lower-order 32 bits of the counter. This is because the standard requires it to be backwards compatible with implementations that were available prior to the standard. Thus, it is possible that given the point in time that the variable is polled, it returns a reasonable value if the total value of the counter is less than or equal to 32 bits. However, when the counter exceeds 32 bits, this value will only return the maximum value that can fit into an integer. This behavior is also required because typical implementations can implement 64-bit counters natively or as two 32-bit words. In the case where two 32-bit counters are supported as a single 64-bit quantity, one word represents the lower-order 32 bits, and another the upper 32 bits. When the first 32 bits are exhausted (i.e., the value is greater than 0xffffffff), the remainder of the value is carried over into the high-order bits. This is sometimes the case when a MIB module is defined using SMIv1, or when it is defined in SMIv2 but the agent entities supporting the module do not support the 64-bit types defined in SMIv2.

6.2.12 IfAdminStatus and IfOperStatus

Each entry in the IF-MIB ifTable contains two objects called the ifAdminStatus and the ifOperStatus. These variables are used to allow the operator to modify the status of the interface, as well as to provide a place where the system can reveal to the operator what the actual status of the interface is. Each of these objects will now be discussed in detail.

ifAdminStatus

The ifAdminStatus object is provided so that the operator may modify the state of an interface. However, modification of this object is only possible on systems allowing write access to it. In cases where write access is not available, it must be permissible for a system to set this value appropriately, perhaps during initialization. The ifAdminStatus can be used by the operator to *request* that the system bring the specified interface into one of the predefined states. If successfully set, the result should eventually be indicated in the ifOperStatus variable. That is, if the ifAdminStatus is successfully changed from **down (2)** to **up (1),** then the ifOperStatus for the corresponding interface should eventually become **up (1).** This is also the case for the other two states of **down (2)** and **testing (3).**

The ifAdminStatus has three possible states. The first two states are **up (1)** and **down (2).** The definition of these states is straightforward. The operator sets the state to **up (1)** if the interface is desired to be enabled and ready to forward traffic. The down state is set when the opposite is desired. When a system is initialized, all interfaces are initialized to the **down (2)** administrative state unless the device contains persistent configuration with a different value. The third state—**testing (3)**—is set when the operator wishes the interface to transition into its testing state. This state may or may not be available depending on the type of interface. Even in the event that a particular interface is defined to have some testing state, it is possible that an agent does not support a testing mode. In either case, an agent may reject a request to enter the **testing (3)** state. If an agent allows an interface to enter the testing state, the ifOperStatus will reflect this new state as soon as the interface enters its testing mode.

The reason why this action is a request is simply because the system does not have to honor the operator's request depending on several variables including the type of interface, the stacking relationship of that interface relative to other interfaces, and the state of the system, among others. The agent may also delay the operation for a reasonable amount of time, so the result may not appear in the corresponding ifOperStatus for some reasonable period of time. For example, if the ifAdminStatus is set to down at the bottom of a stack of interfaces, the agent may choose not to set the ifOperStatus of this interface until such time as it has notified all stacked interfaces to go into the down state and verified that they have done so. If after setting the ifAdminStatus to **down (2),** the NMS checks the value of the corresponding ifOperStatus and finds that it is not set to **down (2)** for a particular interface, the NMS should wait a short time and check again. If the condition persists after a reasonable period, then the NMS should raise an error indication. However, the NMS should also ensure that ifLastChange has not changed for that interface during this interval, since this could indicate that the interface was removed and replaced, for example. One option to the explicit ifOperStatus checking done by the NMS is to enable notification generation by the agent entity. This may obviate the need for the NMS to explicitly check for a change in the operational status.

ifOperStatus

The IF-MIB includes a variable called the ifOperStatus. This variable is provided as a means for the device to expose the actual operational state of an interface. This variable always represents the *actual* state of the interface, in contrast to the ifAdminStatus that reflects the desired state. This variable can take on one of seven states. The first three states—**up (1), down (2),** and **testing (3)**—correspond to the three states defined for ifAdminStatus. That is, the interface enters the **up (1)** state when it can forward traffic, the **down (2)** state when it cannot enter the **up (1)** state to forward traffic due to some fault or as a result of the administrative status being set to **down (2),** and the **testing (3)** state if it is in some testing mode. Note that an interface cannot forward traffic, other than testing traffic, unless it is in the **up (1)** state. If an interface is stacked upon one or more lower layers, it should not enter the **up (1)** state unless all of those lower layers are also in the **up (1)** state. Further, if any of the lower layers enters a non–**up (1)** state, it should be set to the appropriate non–**up (1)** state as well. The state of **dormant (5)** is entered when the interface is waiting for external actions. For example, if a serial interface has been created to answer incoming modem calls, this interface would enter the dormant state until the call has been received. The **notPresent (6)** state is entered when components of the interface are missing. For example, if a set of interfaces requires an external clock for synchronization, and this clock is implemented as a separate daughter board that is applied to the main I/O board where the interfaces reside, all of the interfaces may enter this state until this daughter board is plugged in to provide the clock signal.

The **lowerLayerDown (7)** state is entered if the interface is stacked upon other interfaces that are either disabled or not present. This is illustrated in Figure 6.9. Notice that the leftmost DS0 interface has entered the **lowerLayerDown (7)** state because it does not have a DS1 interface stacked beneath it. Until this interface is created and the stacking relationship implemented, this interface cannot forward

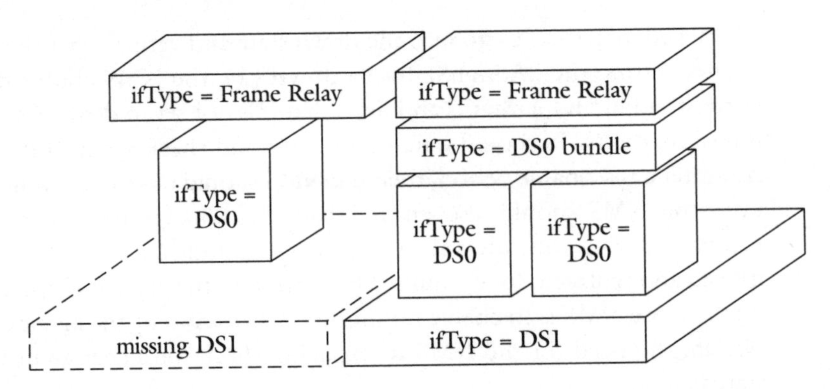

Figure 6.9 The leftmost DS0 interface is missing its DS1 lower layer.

traffic. Finally, the **unknown (4)** state is entered if the agent cannot determine the state of the interface.

6.2.13 Notifications

The IF-MIB contains two notifications that are useful for entries in the ifTable—the linkUp and linkDown notifications. These notifications are generated when an interface operational status transitions into the up state or out of the down state.

A linkDown notification is generated when a transition of operStatus into the down state (from a state other than notPresent) has just occurred. By "just occurred" this means that the agent should emit this notification in a reasonably short period of time following the actual event. Given the sometimes heavy loading of systems, this can take up to several seconds in real time, but should not take much more. Doing so will delay the NMS from raising an alarm to the operator. The linkDown notification is emitted when the first error is detected on the interface preventing it from forwarding traffic. However, an interface may transition into the down state simply because the operator requested that this interface do so. This can be accomplished, for example, by setting the ifAdminStatus to **down (2).** Once set, the device should perform any cleanup it needs to do prior to disabling the interface, and then bring it into the **down (2)** state. Interfaces typically depart the down state and transition to a state other than the notPresent state. This generally indicates that the interface will transition into either the **up (1)** or the **dormant (5)** state. Transitions to the notPresent state are typically utilized when the interface resides on a modular piece of hardware that has just been inserted or removed from the device. No notification is generated when transitioning out of the **down (2)** state into one of these states. However, if an interface is in the **up (1)** state and transitions to the notPresent state, an ifLinkDown notification may be generated by the interface. This can happen, for example, due to the modular I/O board on which the interface resides being removed from the system. In this case, it is very informative for the operator to know that the interfaces on this board have been removed. Note that if an interface is transitioning into the down state due to a failure condition such as CPU failure, the ifLinkDown notification might not be emitted because the subagent responsible for emitting the notification may reside on the same hardware.

An ifLinkUp notification is generated whenever the interface transitions into the **up (1)** state from any other state. This occurs once the interface is ready to forward traffic. For example, if the link is in the **down (2)** state and is enabled by the system's operator, the link would enter the **up (1)** state as soon as it was ready to forward traffic.

One important feature of each interface entry is the ifLinkUpDownTrapEnable variable. This variable allows the operator to suppress or enable ifLinkUp and ifLinkDown notifications from specific interfaces—or subinterface layers of interfaces. This is critical for devices that contain numerous physical interfaces that may have many sublayers of subinterfaces stacked thereon. For example, a multislot chassis device may contain several I/O boards. Each I/O board can, for example, contain several hundred physical interfaces. Depending on the media used on the interfaces, numerous subinterface layers may be stacked upon these physical interfaces in the IF-MIB. If the board is removed, the device's SNMP manager may receive requests for several hundred or even several thousand ifLinkDown notifications! This situation should clearly be avoided.

It is therefore recommended that agent entities suppress notifications when possible since interface state changes would tend to result in one or more notifications as the state changes propagate through the interface stack (from top to bottom, or bottom to top). This approach would likely result in only a few notifications being generated for each linkUp/Down occurrence.

It is also possible for management entities or network operators to control the generation of IF-MIB notifications using the ifLinkUpDownTrapEnable object. This object allows managers to squelch the generation of notifications by specific interfaces. Judicious use of this object on the appropriate interfaces will allow the operator to tailor notification generation by a sublayer or sublayers of interest. We note the judicious use of the ifLinkUpDownTrapEnable object because inappropriate use can result in notifications being missed. For example, if a higher layer experiences state transition into the **down (2)** state, but the underlying layers do not, this event might be missed. An example of this is shown in Figure 6.10. Notice that all but the lowest-layer subinterfaces have the ifLinkUpDownTrapEnable object set to false. In the example, notifications are enabled only at the lowest-layer subinterface

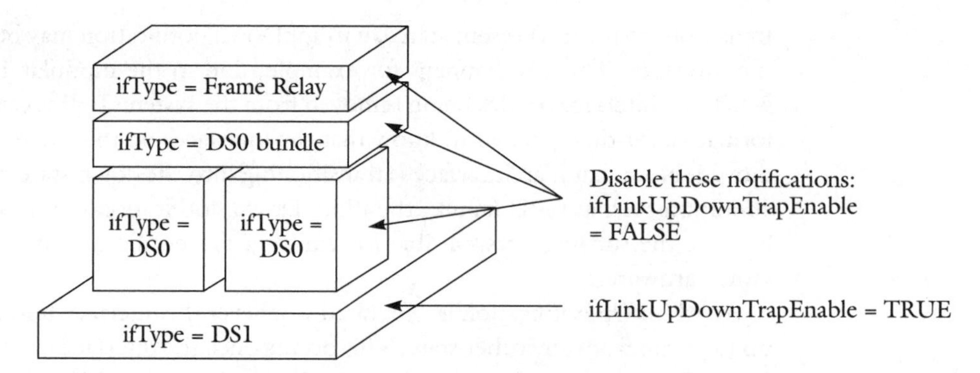

Figure 6.10 Disabling notifications of higher sublayers in a stack of interfaces to reduce the overall notifications generated when the lowest layer changes state.

because state changes at this level will result in all of the interfaces stacked on this layer's interface entries to transition into new states as well. If a notification from this layer is received, the NMS can assume that all of the interfaces stacked above it are also affected. This condition might be set temporarily because the upper layers are being tested and may transition many times to and from the **up (1)** state. This would not allow the emission of notifications from these layers, but would alert the operator of the fact that the DS1 layer had failed.

6.2.14 MIB Tables

This IF-MIB consists of four tables. Each table is briefly described in this section, and each table's objects enumerated in tabular form for quick reference.

The IfTable contains a list of interface entries. It was derived from the original table found in MIB II. The number of entries is given by the value of ifNumber. The objects contained in the IfTable are defined in Table 6.2.

Table 6.2 IfTable objects.

Object name	Object description
ifIndex	A unique value, greater than zero, for each interface. Values are typically assigned contiguously starting from one. Once assigned, the agent must maintain a constant value for each interface sublayer until the device reboots. Some devices can maintain this value across reboots. Check with the device vendor for specific details.

Table 6.2 continued

Object name	Object description
ifDescr	A human-readable string containing information about the interface. This typically includes, but is not limited to, the name of the manufacturer, the product name, and the version of the interface hardware/software.
ifType	The type of interface as assigned by the IANA. The enumeration of the IANAifType textual convention can be found at *www.iana.org*.
ifMtu	The size of the largest packet that can be transmitted or received on this interface. This value is specified in octets (i.e., bytes).
ifSpeed	An estimate of the interface's *current* bandwidth in bits per second. For interfaces that do not vary in bandwidth, or for those where no accurate estimation can be made, this value is usually set to the nominal bandwidth. This value can be set to zero in the rare circumstance where the concept of bandwidth does not apply to the interface. Note that in cases where the bandwidth of the interface is greater than the maximum value that will fit into an unsigned integer (4,294,967,295), this object is set to this value and ifHighSpeed is set to the interface's speed.
ifPhysAddress	The physical layer address of this interface. See the interface's media-specific MIB for the specific format of the value of this object. For interfaces that do not reside at the physical layer, this value will contain an octet string of zero length (i.e., an empty octet string with the length byte set to zero).
ifAdminStatus	The desired state of the interface. An operator specifies the **testing (3)** state if it is possible to set the interface into a testing mode. Some interfaces may not have such a mode, and in these cases, the agent will not allow the value to be set to this state (i.e., it will return an error). When a device initializes, all interfaces should have their administrative status set to the corresponding value that they have been configured for in nonvolatile storage. If no such value exists, then they should be initialized to the **down (2)** state. The interface will also be set to this state when the operator desires that the interface not pass traffic. Note that it is possible that a significant delay in time will exist between when the manager sets this value and when the device actually takes on this value (if it can). Thus, the interface may continue to pass traffic during this time if it should transition from the **up (1)** to the **down (2)** state.

continued

Table 6.2 continued

Object name	Object description
ifOperStatus	This value reflects the operational or *actual* state of the interface as determined by the device's hardware or operating system. The manager cannot directly set this value; instead, the state of this variable should eventually contain the same value as the ifAdminStatus variable if and only if that new state is possible. For example, if the ifAdminStatus is set to **testing (3),** but the interface has no testing model, the agent may choose to reject the request (i.e., return an error), or it may ignore the request. The **testing (3)** state indicates that no operational packets can be passed and that some diagnostic function is being performed. The transmission of traffic on this interface is disallowed during this time. The device will set this value to **dormant (5)** if the interface is waiting for external actions to happen. For example, a corresponding entry in a media-specific MIB may be in the process of being created. The **lowerLayerDown (7)** value is set if a lower-layer interface exists, but its state is down. It should remain in the **down (2)** state if and only if there is a fault that prevents it from going to the **up (1)** state; it should remain in the **notPresent (6)** state if the interface has missing (typically, hardware) components.
ifLastChange	This value is set to the value of sysUpTime when the interface changes state. This value is useful for managers who do not wish to rely on notifications emitted from this interface when its state changes. If notifications are lost, this value also makes it possible to know that some change in state has occurred.
ifInOctets*	The total number of *all* octets (bytes) received on this interface. This value includes framing data.
ifInUcastPkts*	The number of packets delivered by this sublayer to a higher (sub-)layer that was not addressed to a multicast or broadcast address at this sublayer. This includes any packets that were discarded or not transmitted for any reason.
ifInNUcastPkts	Deprecated.†
ifInDiscards*	The number of correctly received packets that were discarded on this interface. This might occur in cases where buffer space had been exhausted or exceeded or when packets did not fit into the interface's queuing discipline.
ifInErrors*	The number of inbound packets or transmission units (for nonpacket interfaces) that contained errors and were discarded as a result.

Table 6.2 continued

Object name	Object description
ifInUnknownProtos*	The number of inbound packets or transmission units (for nonpacket interfaces) that could not be delivered to the upper layer because that upper layer did not exist. Interfaces not supporting protocol multiplexing will always set this value to zero.
ifOutOctets*	The total number of octets (bytes) transmitted out of the interface. This value includes framing data.
ifOutUcastPkts*	The number of packets delivered by this sublayer to a higher (sub-)layer that was not addressed to a multicast or broadcast address at this sublayer. This includes any packets that were discarded or not transmitted for any reason.
ifOutNUcastPkts	Deprecated.†
ifOutDiscards*	The number of packets that were discarded on this interface. This might occur in cases where buffer space had been exhausted or exceeded, or when packets did not fit into the interface's queuing discipline.
ifOutErrors*	The number of outbound packets or transmission units (for nonpacket interfaces) that contained errors and were discarded as a result.
ifOutQLen	Deprecated.†
ifSpecific	Deprecated.†

* The time of any discontinuities encountered by the agent for this counter is indicated by setting ifCounter-DiscontinuityTime to the value of sysUpTime at the time of the discontinuity. For example, the agent should set this value to sysUpTime after a reinitialization of the device.

† This object has been deprecated from the table. This means that it has been viewed as no longer being useful and is in the process of being deleted from the table. This also means that subsequent revisions of this MIB will likely not contain this variable. Do not expect agents to implement it, as this classification denotes that agents may (and probably will) begin to phase out this variable in the near future. Consult the agent capabilities statement for information on whether or not a specific implementation currently implements this variable, or if it too is phasing it out.

The ifStackTable contains objects that were not originally part of the IfTable, but were deemed necessary and were added in later revisions of the MIB. In an effort to preserve the original table structure, these objects were added into their own table that augments the IfTable. Most implementations will find that the code generated for the IfTable contains all of the objects for the IfXTable; thus additional coding for a separate table is unnecessary. The objects in the IfXTable are defined in Table 6.3.

Table 6.3 IfXTable objects.

Object name	Object description
ifName	The textual name of the interface as it was assigned by the local device or by the operator at the CLI.
ifInMulticastPkts	The number of packets addressed to a multicast address at this sublayer and delivered by this sublayer to a higher (sub-)layer.
ifInBroadcastPkts	The number of packets addressed to a broadcast address at this sublayer and delivered by this sublayer to a higher (sub-)layer.
ifOutMulticastPkts	The total number of packets addressed to a multicast address that higher-level interfaces requested be transmitted by this interface. This includes those that were discarded or not sent.
ifOutBroadcastPkts	The total number of packets addressed to a broadcast address that higher-level interfaces requested be transmitted by this interface. This includes those that were discarded or not sent.
ifHCInOctets	High Capacity Counter object representing the 64-bit version of the "basic" ifTable counter. This object has the same basic semantics as its 32-bit counterpart except its syntax has been extended to 64 bits.
ifHCInUcastPkts	High Capacity Counter object representing the 64-bit version of the "basic" ifTable counter. This object has the same basic semantics as its 32-bit counterpart except its syntax has been extended to 64 bits.
ifHCInMulticastPkts	High Capacity Counter object representing the 64-bit version of the "basic" ifTable counter. This object has the same basic semantics as its 32-bit counterpart except its syntax has been extended to 64 bits.
ifHCInBroadcastPkts	High Capacity Counter object representing the 64-bit version of the "basic" ifTable counter. This object has the same basic semantics as its 32-bit counterpart except its syntax has been extended to 64 bits.
ifHCOutOctets	High Capacity Counter object representing the 64-bit version of the "basic" ifTable counter. This object has the same basic semantics as its 32-bit counterpart except its syntax has been extended to 64 bits.
ifHCOutUcastPkts	High Capacity Counter object representing the 64-bit version of the "basic" ifTable counter. This object has the same basic

Table 6.3 continued

Object name	Object description
	semantics as its 32-bit counterpart except its syntax has been extended to 64 bits.
ifHCOutMulticastPkts	High Capacity Counter object representing the 64-bit version of the "basic" ifTable counter. This object has the same basic semantics as its 32-bit counterpart except its syntax has been extended to 64 bits.
ifHCOutBroadcastPkts	High Capacity Counter object representing the 64-bit version of the "basic" ifTable counter. This object has the same basic semantics as its 32-bit counterpart except its syntax has been extended to 64 bits.
ifLinkUpDownTrapEnable	This boolean value indicates whether ifLinkUp/ifLinkDown notifications should be generated for this interface.
ifHighSpeed	An estimate of the interface's current bandwidth in units of 1 million bits per second. If this object reports a value of n, then the speed of the interface is somewhere in the range of $n - 500{,}000$ to $n + 499{,}999$. For interfaces that do not vary in bandwidth or for those where no accurate estimation can be made, this object should contain the nominal bandwidth. For a sublayer that has no concept of bandwidth, this object should be zero.
ifPromiscuousMode	Takes on the value of **true (1)** only when the agent accepts all packets/frames transmitted on the media attached to this interface. Note that this does not apply to some types of media.
ifConnectorPresent	This value is set to **true (1)** if the interface sublayer has a physical connector attached to it.
ifAlias	This object is an alias or alternative name for the interface as specified by a network manager. This value typically differs from ifName, which is assigned by the agent/CLI. This value provides a nonvolatile "handle" for the interface. An example of the value is the circuit number/identifier of the interface.
ifCounterDiscontinuityTime	The value of sysUpTime during the most recent discontinuity experienced by one or more of the counters belonging to this interface. The relevant counters are any Counter32 or Counter64 objects contained in the ifTable or ifXTable that are associated with this interface. This value is set to zero if no discontinuities have occurred or after a reboot of the agent.

The ifStackTable contains objects that define the stacking relationships among the sublayers of interfaces in a device. These objects are defined and enumerated in Table 6.4.

Table 6.4 IF-MIB ifStackTable objects.

Object name	Object description
ifStackHigherLayer	The ifIndex value corresponding to the interface that represents the higher sublayer, that is, the interface that runs on top of this one. If no corresponding higher sublayer (not including the internetwork layer) exists, then this object is set to zero.
ifStackLowerLayer	The ifIndex value corresponding to the interface that represents the lower sublayer, that is, the interface that runs below this one. If no corresponding lower sublayer exists, then this object is set to zero.
ifStackStatus	The status of the relationship between two sublayers. If this object is set to **active,** then the interface stacking is operational and functioning. Any other state indicates that traffic is not flowing between the layers described therein. Note that changing the value of this object from **active** to **notInService** or **destroy** will result in a disruption of traffic between the layers. The destruction of an intermediate layer may result in "dangling" layers that will not be able to operate until a new interface is assigned to replace the deleted one. The aforementioned are all potentially catastrophic consequences of changing the state of a single variable; therefore, operators and managers will typically find that many implementations deem write access to this object inappropriate for all or some types of interfaces.

The ifRcvAddressTable contains objects that are used to define the media-level addresses that this interface will receive. Because this table is a generic table, media-specific MIBs should define exactly how this table applies to the specific media type MIB. The objects that can be found in this table are enumerated and defined in Table 6.5.

Table 6.5 IF-MIB ifRvcAddressTable objects.

Object name	Object description
ifRcvAddressAddress	The specific address that the system will accept packets/frames addressed to on behalf of this entry's interface.

Table 6.5 continued

Object name	Object description
ifRcvAddressStatus	This object is provided so that an operator can create and delete rows in this table. This value takes on the common values of the RowStatus type as defined in the SMIv2.
ifRcvAddressType	The value **Nonvolatile (3)** is used for entries that are valid and will *not* be deleted after the next reboot of the device. The value **volatile (2)** is reserved for those entries that have been created, but that have not yet been saved to nonvolatile storage. Any entry with this value will not exist after the next restart of the managed system. Any entry that has been created and has this value set to **other (1)** may or may not exist after the next restart. The default value and behavior for entries created in this table is **volatile (2).**

6.3 Evolution of the IF-MIB

If you are attempting to locate this MIB by searching for "interfaces MIB" at the IETF Web site's search engine, you will probably be surprised to find that several entries are returned from your search. The reason why you will find more than one reference for the IF-MIB is simply because the MIB has gone through several revisions since its inception as part of the original "MIB II" MIB. The IF-MIB was originally part of the Interfaces Group in RFC 1213, "Management Information Base for Network Management of TCP/IP-Based Internets: MIB-II." This was later updated and published as RFC 1213 with the same name. It was later removed from MIB II and placed into its own MIB module called "Evolution of the Interfaces Group of MIB-II." This document appeared as RFC 1573. This MIB further evolved and was updated again and, as of the time of the publication of this book, appears as RFC 2863. This document still carries the same name as RFC 2863. In addition, a companion extension entitled "The Inverted Stack Table Extension to the Interfaces Group MIB" was produced as RFC 2864. This document introduces an inverted stack table that can be used to traverse the interface entry stacking in reverse order from that which is provided in RFC 2863.

6.4 Applying the IF-MIB to Classic MPLS Networks

The IF-MIB defines a mechanism by which generic managed objects representing interfaces can be implemented by devices. However, as was noted earlier, additional MIBs (and objects) must be defined to properly extend these general interfaces to support media-specific extensions and features. Some such extensions have been made in the name of MPLS-related media types. Specifically, extensions have been created in order to support MPLS traffic-engineered tunnels and MPLS-enabled interfaces. Let us discuss the extensions for general MPLS networks.

Figure 6.11 illustrates the IF-MIB stacking relationship between MPLS-enabled interfaces and other ifTable entries. In particular, when using MPLS-enabled interfaces, a subinterface should be created for each MPLS-enabled interface. Note that an MPLS-enabled interface is sometimes referred to as a "link" on some systems. An MPLS interface is an interface over which labeled and unlabeled packets are transmitted and received. These packets include the MPLS control packets. When this interface is in the **down (2)** state, no data is transmitted or received on this interface. The MPLS ifEntry is assigned an Interface Type (ifType) equal to mpls (166) and should be stacked above whatever underlying transport layer supports the MPLS interface. On some systems this may be an Internet Protocol subinterface, and on others it may be a direct physical interface such as an Ethernet-type, ATM VC–type or Frame Relay–type interface, to name only a few. This is demonstrated in Figure 6.11. Due to the nature of MPLS, it is possible—and sometimes necessary—to facilitate different underlying layers depending on the implementation. This is why it was mentioned that the underlying layer could, on some devices, actually be the IP routing layer's subinterfaces or a physical layer interface. The definition of an MPLS interface is formally defined in the MPLS-LSR MIB.

With the exception of the different stacking relationship, an MPLS-type subinterface should behave similarly to any other packet-based interface. That is, it should support all of the byte-related and packet-related counters including the high-capacity versions, operational status, administrative status, and of course, notifications for transitions into and out of the up and down states. MPLS-type interfaces should also behave as any other packet-oriented interface entry in terms of how its stacking relationship behaves. That is, if the underlying layer is down for

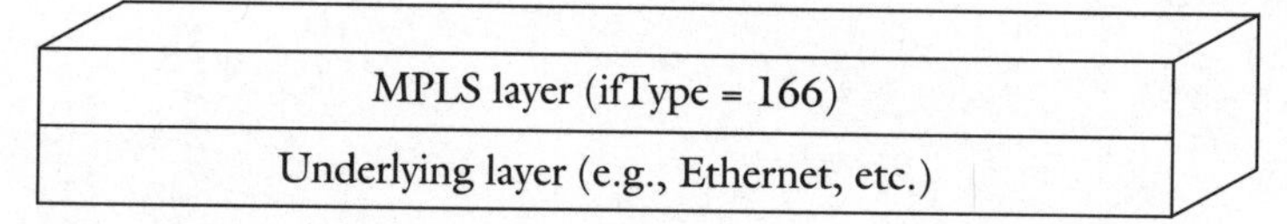

Figure 6.11 Conceptual MPLS interface stacking.

some reason, it too should no longer forward traffic and enter the lowerLayerDown state. It should also be treated as its own sublayer and notifications suppressed appropriately when necessary.

6.5 Applying the IF-MIB to MPLS TE Networks

Some extensions have been made to allow MPLS traffic-engineered tunnels to be represented as individual ifTable entries. Figure 6.12 illustrates the IF-MIB stacking relationship between MPLS-enabled interfaces, MPLS traffic-engineered tunnel interfaces, and other ifTable entries. In particular, when using MPLS TE tunnel interfaces, a subinterface should be created for each TE tunnel. This tunnel interface should be stacked above its corresponding outgoing MPLS-enabled interface (i.e., an interface with ifType = 166). Traffic engineering tunnel interfaces should have an ifType equal to mplsTunnel (150). The definition of an MPLS TE tunnel interface is formally defined in the MPLS-TE MIB. The benefit of this approach is that traffic accounting can be filtered in terms of the total amount of MPLS traffic by viewing the counters on the MPLS interface, while the fraction of traffic used by the TE tunnels can be accounted for by subtracting their counts from the MPLS interface's counters. Furthermore, the accounting for each tunnel is neatly kept with each ifEntry representing the TE tunnel.

It is important to note that TE tunnels may not be implemented as ifTable entries by some vendor implementations. Some vendors chose not to do so for a variety of reasons. For this reason, this approach is allowed in the standard. We, however, recommend that TE tunnels be implemented as interfaces. A manager can determine whether a TE tunnel has been implemented as an ifTable entry by simply checking the value of mplsTunnelIsIf in the MPLS-TE MIB's mplsTunnelTable entry. If this value is set to **true (1),** then the value of the mplsTunnelIfIndex will contain a valid ifIndex value. If the relationship has not yet been configured, but the mplsTunnelIsIf is true, then this value must be set to zero. The value will also be set to zero if the device does not support the TE tunnel as an ifTable entry. One

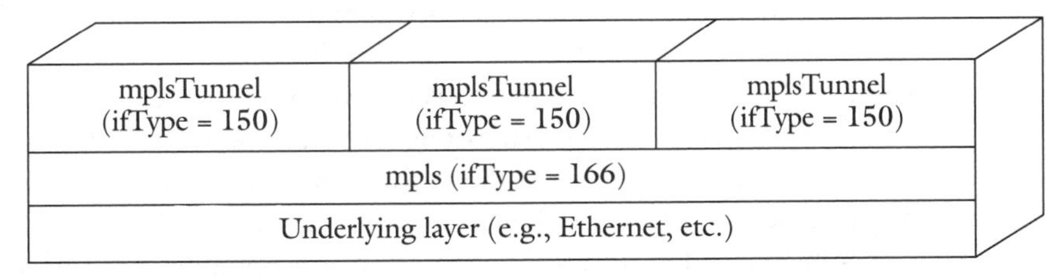

Figure 6.12 Conceptual MPLS traffic engineering interface stacking.

additional caveat regarding MPLS TE tunnels is that if the entry is implemented as an ifTable entry, there are certain variables that should be found in the IF-MIB instead of on the mplsTunnelEntry itself. This includes any counter that is duplicated from the IF-MIB. These counters were duplicated on the mplsTunnelEntry to facilitate interfaces that were not implemented as ifTable entries only. When an mplsTunnelEntry is implemented as an ifEntry, the mplsTunnelEntry counters should be set to zero and not be used.

A traffic engineering tunnel is represented in the mplsTunnelTable as a unique entry therein. The tunnel's "head" (i.e., configured) interface can be represented in this table, as well as the tunnel's LSP instances. The LSP instances are called *tunnel instances* in the MPLS-TE MIB and are typically assigned entries in the mplsTunnelTable with a tunnelInstance greater than zero. It is recommended that the tunnel head interface be configured as assigned tunnelInstance 0. Using this convention will allow operators to quickly identify a tunnel head from its signaled LSPs. LSP instances typically do not have corresponding mplsTunnelIfIndex values assigned to them, so their mplsTunnelIsIf should be set to false. It is possible, however, to assign ifIndexes to these entries in order to, for example, load-share across these virtual interfaces.

When attempting to expose the outgoing MPLS-enabled interface for the tunnel, the vendor has two options depending on the implementation. If the tunnel interface supports a single LSP at any given time, the device should stack the tunnel interface above the ifIndex of the outgoing MPLS interface that carries that tunnel's active LSP. If the active LSP changes, so should the stacking relationship. In addition, the tunnel instance LSP should point at the appropriate cross-connect entry in the MPLS-LSR MIB to indicate which labels are in use for that LSP. This LSP will be stacked on top of the appropriate outgoing MPLS interface. If an implementation supports multiple signaled LSPs per tunnel interface simultaneously, then the tunnel interface should be stacked upon one of the actively signaled LSP's underlying MPLS interfaces. The operator can find all outgoing interfaces by interrogating all of the signaled LSPs' MPLS-LSR MIB cross-connect entries, as just described for the single LSP case.

MPLS TE-type subinterfaces should behave similarly to any other packet-based interface. These interfaces should support all of the byte-related and packet-related counters including the high-capacity versions, operational status, administrative status, and of course, notifications for transitions into and out of the up and down states. MPLS TE tunnel-type interfaces should also behave as any other packet-oriented interface entry in terms of how its stacking relationship behaves. That is, if the underlying MPLS layer is down for some reason, it too should no longer forward traffic and enters the lowerLayerDown state. It should also be treated as its own sublayer and notifications suppressed appropriately when necessary.

6.6 Summary

This chapter first went into the details of the IF-MIB. Then, we investigated the components of the ifTable entry. We discussed the purpose of each entry, when it should be created, and how it should be used. We investigated how virtual subinterfaces could be created in order to represent conceptual software component interfaces such as those for routing protocols. In addition to virtual subinterfaces, we discussed how those interfaces that have a one-to-one correspondence with physical connectors on a device could also be modeled using the ifTable.

Next, we discussed how ifTable entries could be created and destroyed dynamically, as well as the issues related to these operations. We explored how each component in the ifTable could be conceptually stacked into sublayers. This conceptual stacking reveals how a layering of subinterfaces both above and below each interface can depict the internal flow of traffic through these interfaces within a device. The discussion of interface stacking provided guidance for when it was appropriate to create new subinterface layers and when it was not. The discussion then moved to a definition of the various states that an ifTable entry interface could take on. The distinction between operational and desired or administrative status was made and explained in depth. We then looked at what types of notifications are available for interfaces, when it is appropriate (or desirable) to emit them, and how to control their emission. The section on the IF-MIB concluded with brief overviews of each table, as well as an enumeration and a short definition of each object. This was provided for quick reference. We explained the history behind the MIB and how to find each progressive version.

Finally, we discussed how the IF-MIB has been extended to support MPLS. We introduced the concept of an MPLS-type interface and what its behaviors are expected to be. These interfaces can, for example, count MPLS labeled and unlabeled traffic that is transmitted or received by an LSR. We also discussed how MPLS traffic engineering tunnels could be stacked above the MPLS interfaces to represent ifTable entries for TE tunnels. Among other things, we explained that this approach facilitates consistent accounting and autodiscovery for all MPLS-related interfaces.

Further Reading

RFC 1213. McCloghrie, K., and M. T. Rose. "Management Information Base for Network Management of TCP/IP-Based Internets: MIB-II." March 1991. *ftp://ftp.isi.edu/in-notes/rfc1213.txt.*

IETF Standard 0017. McCloghrie, K., and M. T. Rose. "Management Information Base for Network Management of TCP/IP-Based Internets: MIB-II." March 1991. *ftp://ftp.isi.edu/in-notes/std/std17.txt.*

RFC 1573. McCloghrie, K., and F. Kastenholz. "Evolution of the Interfaces Group of MIB-II." January 1994. *ftp://ftp.isi.edu/in-notes/rfc1573.txt.*

RFC 2864. McCloghrie, K., and F. Kastenholz. "The Inverted Stack Table Extension to the Interfaces Group MIB." June 2000. *ftp://ftp.isi.edu/in-notes/rfc2864.txt.*

RFC 2233. McCloghrie, K., and F. Kastenholz. "The Inverted Stack Table Extension to the Interfaces Group MIB." November 1997. *ftp://ftp.isi.edu/in-notes/rfc22233.txt.*

RFC 2494. Fowler, D. "Definitions of Managed Objects for the DS0 and DS0 Bundle Interface Type." January 1999. *ftp://ftp.isi.edu/in-notes/rfc2494.txt.*

RFC 2496. Fowler, D. "Definitions of Managed Objects for the DS3/E3 Interface Type." January 1999. *ftp://ftp.isi.edu/in-notes/rfc2496.txt.*

Internet Assigned Numbers Authority. "IANAifType-MIB." Requests for new ifTypes should be emailed to iana@iana.org. *www.iana.org/assignments/ianaiftype-mib.*

Srinivasan, C., A. Viswanathan, and T. Nadeau. "Multiprotocol Label Switching (MPLS) Traffic Engineering Management Information Base." IETF Internet Draft. August 2001. *www.ietf.org/internet-drafts/draft-ietf-mpls-te-mib-08.txt.*

Srinivasan, C., A. Viswanathan, and T. Nadeau. "MPLS Label Switching Router Management Information Base Using SMIv2." January 2001. *www.ietf.org/internet-drafts/draft-ietf-mpls-lsr-mib-07.txt.*

To find out more about the IETF, visit their Web page at *www.ietf.org/.*

For more information about IANA, check out their Web site at *www.iana.org/.*

Cisco's Web site also provides a great deal of information regarding MPLS: *www.cisco.com.*

Interview with the Expert

Adrian Farrel

is a director of protocol development with Movaz Networks, Inc., which provides next-generation all-optical solutions for delivery of cost-effective wavelength services. He gained over 15 years' experience developing portable, scalable, and fault-tolerant protocol implementations at Data Connection Ltd., where he was MPLS architect and development manager, leading a team that produced a carrier-class MPLS solution for customers in the optical space. Adrian is very active in the IETF's MPLS and CCAMP Working Groups, where he has co-authored and contributed to numerous Internet drafts on MPLS-related technologies. He is also the editor of the GMPLS MIB drafts. He was a founding board member of the MPLS Forum and is the author of several white papers on GMPLS.

Adrian, you have been involved with the standards process of the IETF as well as other standards bodies. How does the standards process of the IETF differ from others? In your opinion, is the IETF standards process better, worse, or the same in the end as other standards bodies such as the ITU or ATM Forum?

The standards process within the IETF is certainly different from that of other bodies in the same space. Membership is free to any individual regardless of (and in fact disregarding) company affiliations, and the work of developing standards is devolved to volunteers. Agreement to proceed with standards is arrived at through consensus of the people that have registered an interest in the area by joining a Working Group under the guidance of elected chairs.

This can lead to extremely rapid standards development in new areas where there is a healthy interest. That's good because technological innovation will not wait for a slow standards process, and too many different and proprietary protocols might be invented if the authors of the standards cannot keep up. On the other hand, the IETF has to be constantly alert to the risk of rushed work being shoddy—fortunately the large number of interested parties and the high level of experience helps to keep things on track.

You have recently become involved in the standards related to MPLS and GMPLS network management. What do you see in your crystal ball of the future for these technologies with regard to management?

The next steps will be to consolidate the existing MIBs so that they can be used to model and manage systems running the many extensions to MPLS, from DiffServ and

VPNs to optical networking and GMPLS. In doing this, it is important that the authors consider the migration path from older MPLS systems running earlier MIBs, and also that they look to the future to make sure that the MIBs are extensible.

They must also make sure that the new MIBs are usable! There are a very large number of small features that have already been added to MPLS, and there are bound to be many more in the near future. The authors must ensure that support for these additions within the MIBs does not obscure or overcomplicate the management of the average MPLS system.

The longer-term goal is to allow a single network management station to seamlessly and fully manage a network of MPLS switches from disparate vendors. For this to be achieved, vendors must obviously also cooperate on interworking at the network level. More importantly, they must implement the standard MIBs and offer full SNMP access so that objects can be controlled as well as inspected.

7

Offline Traffic Engineering

> "Nature seems . . . to reach many of her ends by long circuitous routes."
>
> —Rudolph Lotze

Introduction

Traffic engineering (TE) is a process whereby a network operator can engineer the paths used to carry traffic flows that vary from those chosen automatically by the routing protocol(s) in use in that same network. This is done in an effort to steer traffic through the network, which may result in more efficient use of network resources, protect against network node or link failures, as well as provide certain customers with custom services such as guaranteed bandwidth connections. It is possible to create a continuously operating traffic engineering system within MPLS environments that can be used to enhance the quality, availability, and robustness of network infrastructure and services, as well as reducing the costs of operating that same network without traffic engineering. Traffic engineering systems are sometimes referred to as *offline* traffic engineering systems because they are often run on systems that are separate from the network nodes they control. Offline systems can be used for a variety of tasks including precalculation (prediction) of head-end per LSP usage as well as other dynamic TE within the network, or how TE-based modification of the network might look when implemented. Offline systems can also be implemented centrally or in a distributed manner.

Traffic engineering is not a new phenomenon, nor is the offline traffic engineering application. Offline traffic engineering tools have often been used to maintain the original switched telecommunications networks. Traffic engineering was later adapted to routed IP networks to tackle some of the same problems they were used to solve in earlier networks. Recently, traffic engineering and offline traffic engineering applications have been employed by operators for use within MPLS networks. This chapter provides an overview of traffic engineering and then introduces offline traffic engineering with a particular focus on how this tool can be applied to MPLS.

7.1 Traffic Engineering

Traffic engineering is a process whereby a network operator (or an application or applications that the operator uses) can craft a specific course taken by traffic flows through the network. It is desirable in some networks to achieve more efficient use of network resources, protect against network node or link failures, as well as to provide certain customers with specialized services such as guaranteed bandwidth connections. The paths that are chosen to achieve these goals typically vary, at least in part, from those calculated by the routing protocol(s) used in the network. The reason is that the paths selected by the routing protocols typically apply the same metrics to all paths in an effort to produce the "best" paths for all traffic flowing throughout the network. However, a frequent result is that large amounts of traffic migrate to links that are not heavily used. This leads to oversubscribed or heavily subscribed links. As more and more traffic migrates to preferred links, other less-preferred links begin to become lightly used. This is unfortunate, as many alternative paths to the same destinations using these alternate links remain unused. One problem encountered is that as more and more traffic is forwarded onto the same set of links, eventually other lightly used links will become preferred by the routing protocol, and thus traffic patterns will start to shift from the congested links onto these more lightly used links. Ultimately this can result in traffic oscillating between heavily used links and lightly used or unused ones. The problem is that, at any point in time, some link resources go underutilized while others may be overutilized.

Traffic engineering can be used to alleviate the problems just described by allowing an operator to override the paths chosen by the routing protocol for some fraction of network traffic. This allows traffic to be spread more evenly across the network's available resources (i.e., links and/or nodes). This also allows an operator to direct specific types of traffic down more appropriate paths, thereby enabling several other possible features, including alternative backup paths that can be used to protect traffic in the event of a network failure.

Another added benefit of traffic engineering is that once a traffic-engineered path has been deployed, it will typically persist within the network and will continue to carry traffic along the same fixed path. The path will only change in the event of a network failure or disruption. However, when an alternative path is chosen by network nodes maintaining the traffic-engineered path, the nodes may be instructed to choose this path using the same traffic parameter constraints as the original path. This can result in the new path using the same constraints as the original path with the exception of the route it takes through the network. In some cases, this can be used to forward traffic for that tunnel without interruption. Use of this mechanism also allows operators to assign traffic to the traffic-engineered path based on a traffic policy without worrying about which physical path the traffic takes, since this policy will be used not only to compute the primary path, but alternative paths as well.

7.2 Traffic Engineering in MPLS Networks

When used to engineer MPLS networks, traffic engineering refers to functionality provided by an MPLS LSR that allows it to route a subset of overall network traffic through a *traffic-engineered tunnel*. A traffic-engineered tunnel represents a series of network flows that all share the same path across a network, as well as the same class of service (or path constraints). A TE tunnel does not constitute an end-to-end path for any single microflow; rather, microflows are *aggregated* together into the tunnel, thereby sharing the resources and attributes of the path taken by the tunnel. New microflows can be added to the tunnel after the tunnel has been established without imparting any additional state within the network. Traffic flows may constitute traffic from several IP prefixes, thus aggregating traffic flows together across a single traffic-engineered path.

The path a tunnel takes can be chosen explicitly by the operator or can be left to a constraint-based shortest-path first (CSPF) calculation to compute. Thus, the path this tunnel takes may or may not be the same as would be calculated by the routing software. The CSPF can be left to compute the whole path, or in some cases, just parts of the path with some other parts explicitly specified. The paths taken by a tunnel can be chosen using a variety of constraints including desired minimum, maximum, or average bandwidth; minimum or maximum delay; or minimum number of hops. MPLS TE tunnels can also be chosen according to a link *affinity* mask. This bit mask is assigned to all of the TE-enabled links within a network. An analogy sometimes used is that the mask can *color* each of the links a specific color. A specific tunnel can then be instructed to prefer or avoid (include or exclude) links with a particular mask (or color) when choosing a path.

You may find it interesting to read further about the requirements for the traffic engineering process as defined in the first document produced by the IETF MPLS Working Group (RFC 2702). This document describes how each traffic-engineered tunnel may allow additional traffic parameters that include such things as QoS parameters, which may allow them to carry traffic flows from many different sources, but using a common traffic characterization (e.g., low delay, high bandwidth).

MPLS traffic engineering systems may also provide backup paths of various types for any given tunnel. These include backup paths for single tunnels, groups of tunnels, or even any tunnel traversing a particular link. These backup paths can be programmed to avoid or include certain links, nodes, or complete paths in order to provide sufficient redundancy. Backup tunnels are usually configurable by the operator or by some traffic engineering application that can be used to automate the process.

A point of presence (POP) is simply a place where a service provider has a local presence. POPs are often a place where a service provider allows customers to join their network. Figure 7.1 shows how traffic from POPs A and B destined for Net A would normally be assigned to LSPs that traverse the MPLS network. These LSPs would follow a shortest path as calculated by the routing protocol. However, this sometimes can result in congestion as traffic from many links can be assigned to LSPs that take the same path or portion of a path. More specifically, paths can merge onto the same links without necessarily considering this. This can lead to network congestion. This scenario is shown in the figure as traffic from POPs A and B converging onto the same links (see the top left of the figure). The additional traffic on some of these links results in congestion.

Figure 7.2 shows how the proper placement of several traffic-engineered tunnels can alleviate the situation. First, a tunnel is placed from POP A to Net A using the constraints that it cannot include any of the nodes included in the original routing protocol path except for the penultimate hop at the top right of the figure. A second TE tunnel is set up to explicitly route traffic from POP B to Net A over a path that largely does not overlap with the one chosen by the routing protocol. This can be done by specifying that it avoid the rightmost nodes except for the penultimate node connecting the core network cloud to Net A. The result is a series of paths across the network that largely resembles the original ones in terms of traffic parameters, but that result in less network congestion and better overall utilization of network resources.

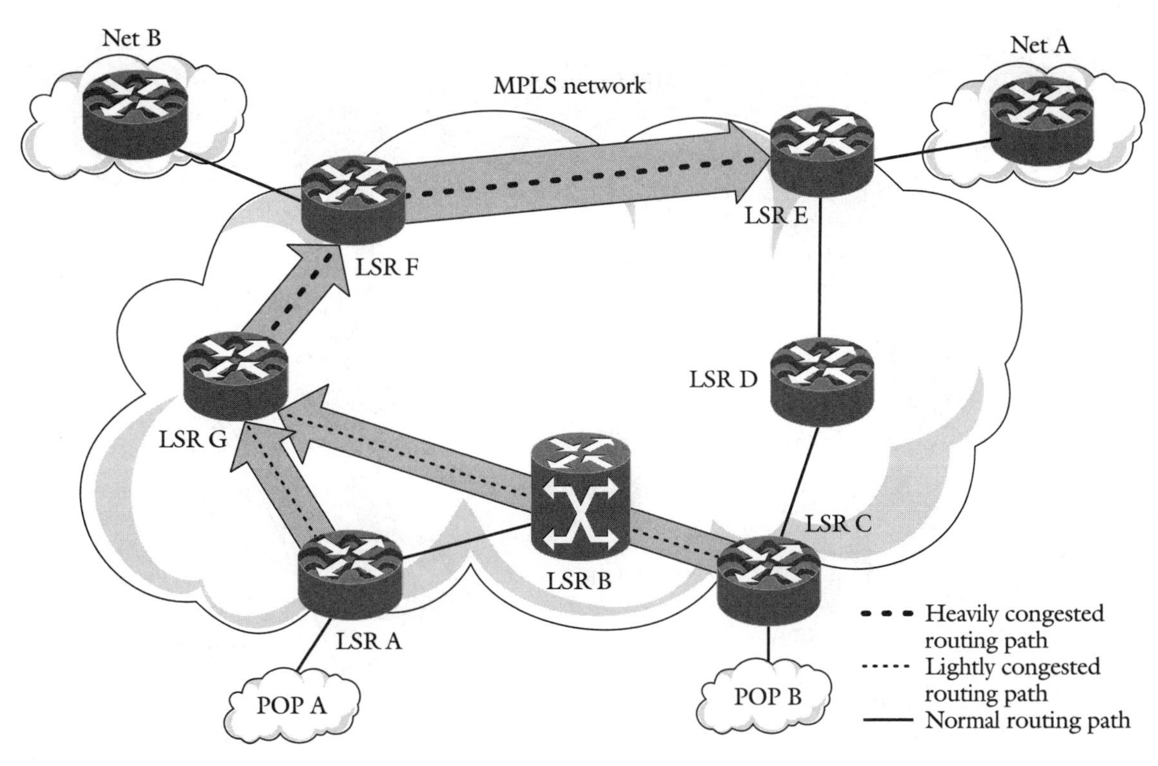

Figure 7.1 Normally, traffic from POPs A and B destined for Net A is assigned to an LSP that takes that traffic along a shortest path as calculated by the routing protocol. However, this sometimes can result in congestion as traffic from many links can be assigned to LSPs that take the same path. Here traffic from POPs A and B converges on links that become congested with this traffic. Use of a traffic engineering tunnel to steer some of the traffic away from the overused links may abate the problem.

A traffic-engineered tunnel typically resembles a subinterface on the LSR where the tunnel begins, and resembles a virtual interface that forwards traffic from its ingress to egress points. Some implementations provide another interface on the LSR where the tunnel terminates. In cases where the tunnels are represented by subinterfaces, the device also typically models this interface as an entry in the IF-MIB (see Chapter 6). This provides a commonly accepted means for gathering link counter statistics from these interfaces. In addition, traffic-engineered tunnels are typically represented as MPLS Tunnel Table entries in the MPLS-TE MIB. We will discuss this in further detail in Chapter 8.

Once the tunnel has been configured as an interface, one or more label switched paths must be associated with that tunnel. To achieve this, one of two signaling protocols can be used to allocate the label switched path across the MPLS network. These are the ReSerVation Protocol (RSVP) (RFC 3209) and the Constraint-based Routing Label Distribution Protocol (CR-LDP) (RFC 3212). The

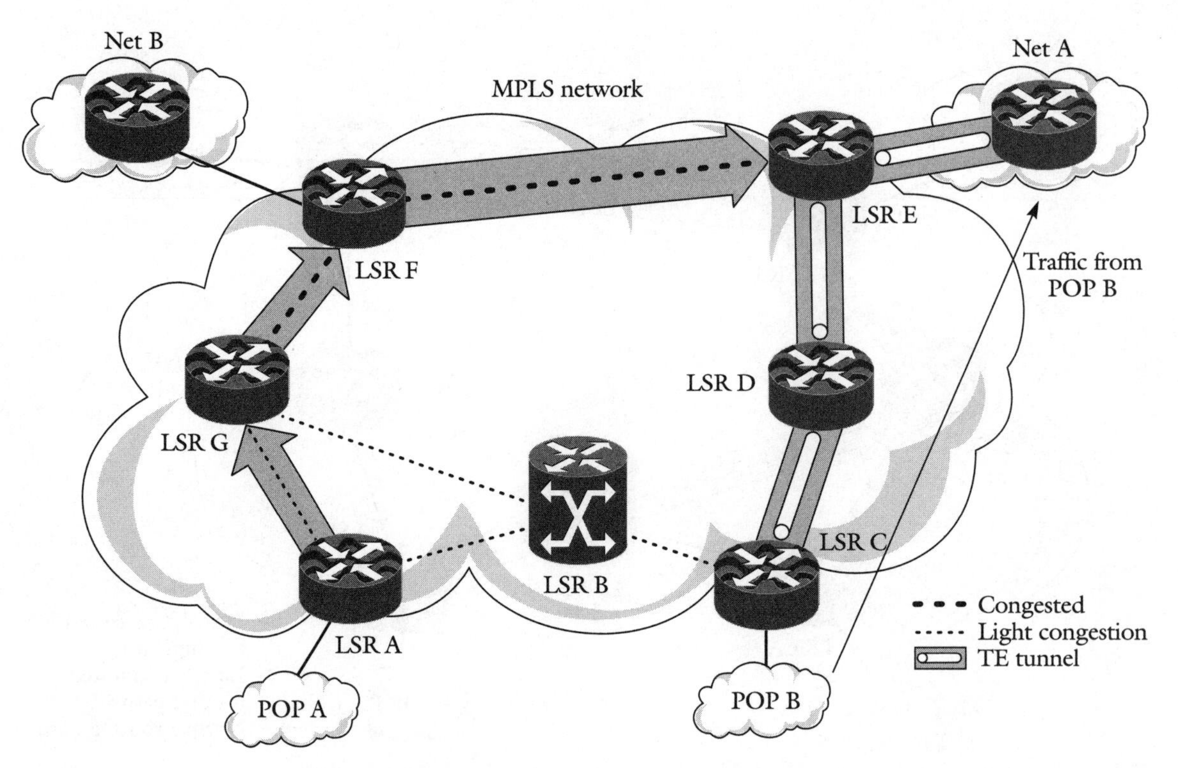

Figure 7.2 Traffic from POPs A and B destined for Net A would normally use the shortest path as calculated by the routing protocol, which would result in traffic from POPs A and B using many of the same links in the network. In the network shown, this results in congestion and oversubscription on some links. To alleviate this, a TE tunnel is set up to explicitly route traffic from POP B to Net A over a path that largely does not overlap with the one chosen by the routing protocol.

tunnel's interface persists regardless of whether there is an underlying LSP signaled for the tunnel. Using this approach, an operator can simply program traffic to enter that tunnel at the ingress point and assume that it will exit at the specified egress point, all the while with the confidence that the path used to convey this traffic will obey the traffic parameter constraints outlined by the operator.

7.3 Deliberate MPLS TE Models

In general, MPLS traffic engineering can be used to benefit most deployments of MPLS networks at least to some degree. Figures 7.1 and 7.2 showed an example of how MPLS TE can be used to overcome a specific network overcapacity problem. In particular, the problem described is one that sometimes surfaces in both MPLS and non-MPLS networks that rely solely on a routing protocol to compute the paths that network flows take throughout a network. These examples showed how traffic engineering could be deployed to correct network "hot spots" by routing some traffic around them. This is sometimes referred to as "hot spot" or *deliberate* deployment of MPLS traffic engineering. This method is typically used on an ad hoc basis to correct specific situations. Once the situation is alleviated, the "patches" that avoid the hot spots are usually removed. Hot spot analysis is one function of a traffic engineering system that is directed specifically at redistributing utilization from a particular link or set of links to other less-used links. This can effectively result in better network resource utilization.

7.3.1 Full-Mesh Traffic Engineering

Another more proactive deployment method of MPLS TE is available. This method is usually described as *full-mesh traffic engineering*. This model is employed by network operators to enable measurement that results in a sufficient level of control over link utilization, service levels, and backup tunnel placement. Full-mesh MPLS TE involves the establishment and active management of a mesh of traffic engineering tunnels between all LSRs within a particular MPLS domain. These LSRs then typically use some combination of CSPF and explicit routing to manage traffic loads within operational parameters. Offline TE systems can be used to automate and fully achieve the potentials of this model by managing the placement, removal, and periodic reoptimization of the tunnels in the mesh. This poses some practical challenges over the ad hoc mechanism described earlier, in that the amount of data that must be collected and managed in order to perform the computations necessary to fully realize the potential of this model is significant. On the other hand, the results of such an approach can be quite beneficial when compared to the results of not using them.

Two common models are used to realize a fully meshed network of LSRs within an MPLS network. Both models are shown in Figure 7.3. Figure 7.3(a) shows how a full mesh of traffic engineering tunnels and their corresponding LSP can be established among core routers, while Figure 7.3(b) depicts how a more hierarchical deployment can be deployed. When comparing these different models, notice the number of tunnels that are deployed. In the case of Figure 7.3(a), roughly N^2 tunnels are deployed; in the hierarchical model in Figure 7.3(b), far fewer tunnels are required while achieving similar results.

7.4 Tunnel Sizing

Traffic engineering systems need to be careful when computing or recomputing the sizes of TE tunnels within a network. In particular, the resources such as bandwidth

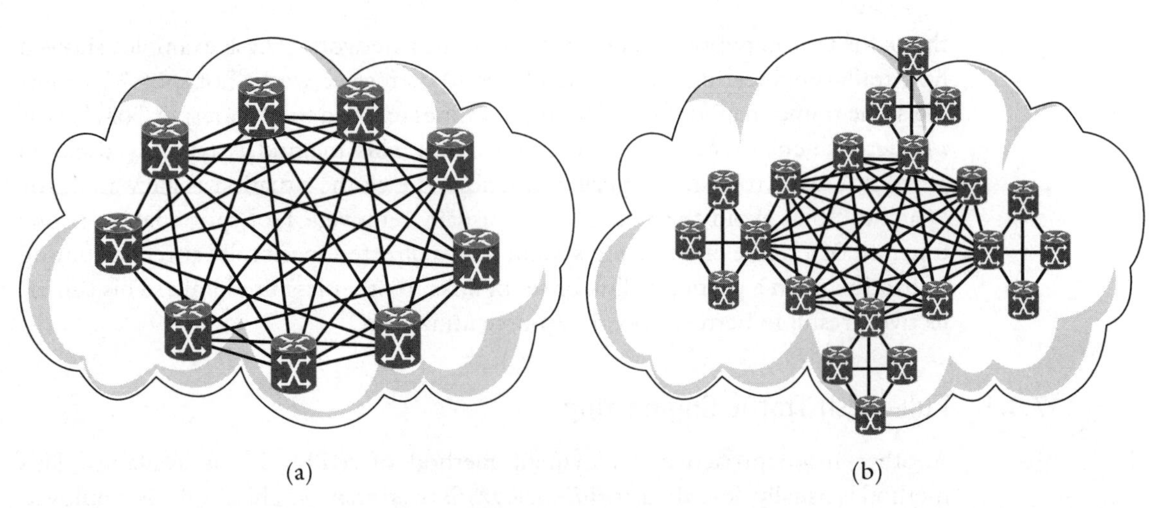

Figure 7.3 Two models of fully meshed MPLS TE deployments: (a) a full mesh and (b) a hierarchical model.

reserved for each TE tunnel should be chosen appropriately to carry just enough of the predicted aggregated traffic flows over that path given the constraints associated with the tunnel. Failure to do so will result in an overallocation of network resources, thus making the operator's network potentially less profitable. In particular, inappropriate sizing predictions can result in unused bandwidth that can be allocated to new customer flows or to customers wishing to pay for additional bandwidth. In the worst cases, incorrect tunnel sizing can result in needlessly delayed or dropped traffic through nodes where overallocation has occurred, thus degrading a normally functioning network.

As part of the traffic engineering cycle, TE tunnels can be resized periodically if measured network usage deviates from the assumptions made when the tunnel was first created. A number of factors may result in the tunnel being resized. First, if the tunnel has been underused for a long period, the system may choose to reduce the allocated bandwidth slightly. It may continue to do this until some lower value is reached. At this point, if the tunnel continues to be underutilized, the traffic engineering system may choose to reallocate the traffic entering that tunnel by moving it into another tunnel and removing the current one from the network. On the other extreme, if the system measures that a tunnel has been overused for some period of time (i.e., lots of dropped packets are measured that should have entered the tunnel), the operator may wish either to increase the bandwidth allocated to the tunnel or to reroute some traffic destined for the tunnel onto another path or tunnel.

One twist on the theme of tunnel resizing is to let the network node that controls the tunnel resize it on its own. This can be achieved if the node is able to monitor average flow statistics for each TE tunnel. If the traffic patterns associated with any particular tunnel vary from those configured as described earlier, the node may resize the TE tunnel automatically.

Another twist on TE tunnel resizing is an approach that does not require them to be resized. This approach is sometimes referred to as "zero-bandwidth" tunnels. In this mode, TE tunnels are allocated zero bandwidth, allowing them to, in effect, function as best-effort interfaces. However, these tunnels still allow an operator to configure the explicit path taken by traffic using the tunnel. In this sense, an operator or TE system might first deploy zero-bandwidth tunnels to simply monitor traffic that traverses the network, meanwhile gathering performance statistics. Some TE systems will then use this information as the initial input to their traffic prediction and analysis software, after which they may choose to assign actual bandwidth values to the tunnels.

7.5 Tunnel Path Selection

Traffic-engineered tunnels can have their paths specified in a variety of ways. The most obvious mode is for an operator to *explicitly* specify each hop that the tunnel should take. This requires that the operator know about the network topology a priori, thus making deployment of such an approach somewhat difficult in large networks. However, this approach has the advantage of allowing the operator to deliberately construct the paths taken by tunnels as opposed to relying on the CSPF to compute the path. Because of the difficulties related to explicit path specification, one option might be to use this mode only in cases where the CSPF algorithm cannot find any path given the additional traffic parameter constraints. In cases where the complete path is specified explicitly, the CSPF algorithm is not executed to compute the tunnel's path; instead, the specified path is used as is. Note that if the operator does not take care in specifying the explicit path, the network nodes may fail to set up the path. This can happen, for example, if the operator has stale network topology information and includes a node that no longer exists or is out of service.

An example of an explicitly specified TE tunnel is given in Figure 7.4. The TE tunnel between nodes A and K was specified by specifying the path {A, B, G, I, K}.

Another mode of operation allows the operator to simply specify the end points of the tunnel, and then instruct the CSPF software to *dynamically* calculate the path taken to connect those points with a tunnel. The advantage with this approach is that CSPF algorithms typically accept many different metrics including bandwidth, delay, and link characteristics. This approach also does not require the operator to possess a topological map of the network; only the ends of the tunnel are important. The drawback to this approach is that the CSPF algorithm may, in some cases, pick a path that, although abiding by the constraints that were specified, does

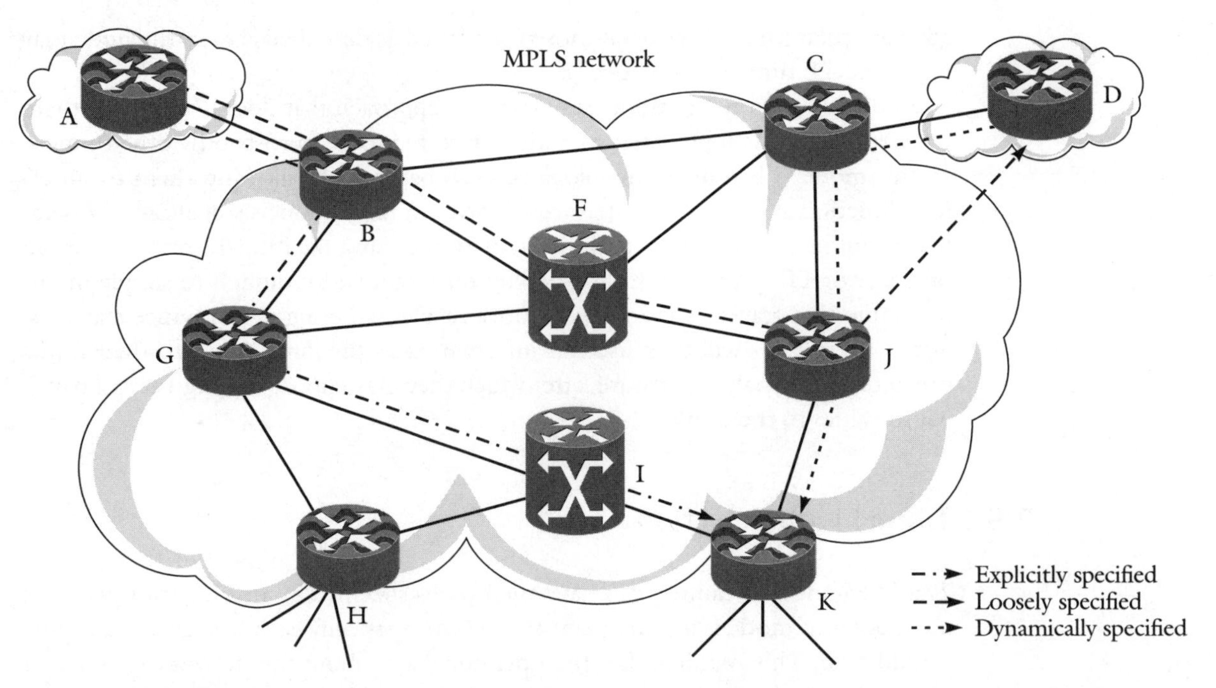

Figure 7.4 An MPLS network containing traffic-engineered tunnels specified in a variety of ways. Note that to emphasize the functions of TE, the TE paths do not take the same path that the routing protocol would have computed. Assuming equal link metrics between the LSRs above, the routing protocol would probably pick the shortest path based on the number of hops.

not achieve all of the desired characteristics that an operator may wish to achieve with the tunnel. For example, the operator may wish that the tunnel avoid certain hops and instead these were included (see the later discussion of the hybrid approach). It is also possible that the operator fails to specify all of the constraints, thereby affording the opportunity that an undesirable path is computed by the CSPF algorithm.

An example of a dynamically specified TE tunnel is given in Figure 7.4. The TE tunnel between nodes D and K was specified by only specifying nodes D and K, but with bandwidth constraints that precluded any other paths. This was then fed to the CSPF algorithm, which calculated the best path between those nodes as {D, J, K}. We should also note that, given this fact, the path taken is not equivalent to the one that the routing protocol would have computed. It is, however, sometimes desirable under certain circumstances (e.g., zero-bandwidth tunnels used for measurement of traffic) for the dynamic path to be equivalent to the one that the routing protocol would have chosen.

Lastly, a hybrid approach combines features of the two previous methods. This method allows the operator to specify the end points of the tunnel, as well as any number of hops that are to be taken or avoided by the tunnel. These are all input into the CSPF, which then attempts to "fill in" the missing hops. This is called a *loosely routed* TE tunnel. A loosely routed tunnel is depicted in Figure 7.4 by the TE tunnel between nodes A and D. The path was specified as {A, B, D} as well as a list of hops to avoid {C, G}. An operator might have wished to avoid hops C and G because the operator knew ahead of time that these nodes where already near capacity or were to be decommissioned soon. In this case, the CSPF algorithm determined that the best path given the constraints was to take {A, B, F, J, D}. Notice that a loosely routed tunnel can be equivalent to a dynamic tunnel if the only two hops specified are the source and destination. A loosely routed tunnel can also be considered a fully explicit tunnel as well if all of the hops are specified in the path, including some to avoid. However, the hops to avoid may be redundant given that the entire path is specified, and thus those hops will be avoided anyhow. One last important point to make is that it is invalid to specify the same hop to be both included and excluded from the tunnel's path, since this is clearly a contradiction.

7.6 Use of Offline TE for Backup Tunnels

Offline traffic engineering applications can be quite useful for redundant or complex TE operations that might require large amounts of time to perform calculations or a large number of steps to apply configuration or reconfiguration operations. Due to the computationally intensive nature of the calculations that determine the best placement of backup TE tunnels, it is often preferable to use a more automated TE approach. Offline traffic engineering tools can be used first to analyze the current network topology, and then to apply this information and other heuristics to calculate backup tunnels for the tunnels that have been placed in the network. This can be used to provide the operator with an effective and time-saving form of insurance against node or link failures. Ease in deploying this feature can potentially improve the profitability of an operational network in that an operator can market backup services as a value-added service.

7.7 The Traffic Engineering System

A *traffic engineering system* typically refers to a two-part approach that includes both network management and traffic engineering control on network nodes. Both facets are incorporated into a periodic cycle. The first step in this cycle entails gathering statistical data from the existing network. This information can then be used to drive predictions of future network traffic growth or network failure scenarios. Once complete, these predictions and calculations may then lead to network redesign in an effort to reoptimize the network for better performance, more efficient use of network resources (i.e., links or nodes), or more fault tolerance under failure conditions. Typical actions resulting from the analysis phase of traffic engineering are the acquisition of new links, the reprovisioning of bandwidth allocated to certain links, the deployment of new networking devices due to a lack of capacity, or a rerouting of some capacity. This sequence of events is typically repeated at some periodic interval that can vary from network to network and can depend on the type of traffic engineering being performed.

This approach is sometimes referred to as *offline traffic engineering,* as sometimes much of the processing occurs on an external network device such as a PC or workstation instead of on the actual network nodes. This term also refers to the fact that this processing does not happen in real time; instead, it happens *after the fact.* Even systems that rely on network nodes performing much of the configuration and path calculation are sometimes referred to as offline TE systems, as they generally still require some amount of offline processing.

The life cycle of a typical traffic engineering system involves periods of analysis and acquisition of statistical data. These periods are long in relation to the relatively short periods of time required by dynamic routing algorithms. However, short-term situations can arise during the analysis period, representing an important part of the traffic engineering system. This area includes predictions and analysis of various types of network failures, unexpected growth, or usage patterns that are unaccounted for during typical network analysis.

The traffic engineering system represents an interactive feedback control system that is typically used to complement a network's supporting of explicit traffic engineering control. It achieves this by providing an integrated set of tools that can be used to perform reporting, statistical measurement, predictive synthesis, and result analysis. These systems may even provide suggestions of how to reengineer the network in order to improve its function. Furthermore, this set of integrated tools typically provides the operator with a modeling function, thereby allowing them to visualize and study the performance impact of any suggested or attempted changes, as well as allowing them to observe the behavior of the existing network. Some systems go as far as even providing the operator with the option of automatically reconfiguring the network once the operator has approved the suggested configuration changes designed by the traffic engineering tool.

7.8 TE System Components

The traffic engineering system consists of two main components: network nodes and traffic engineering tools. Network nodes must be capable of establishing and maintaining traffic engineering tunnels. This includes traffic engineering system software that runs on the network node and allows it to control traffic engineering functions such as establishing, modifying, or deleting traffic-engineered tunnels. Traffic engineering network management tools help network operators determine where and when to place TE tunnels, as well as how and when to modify or delete existing ones. The traffic engineering system may also specify which traffic should be routed into a new or existing tunnel by programming the network device accordingly.

Two informational exchanges must exist between the network nodes and the management tools in a network. First, network nodes must supply *statistics* to the management tools using one or more management interfaces. Figure 7.5 illustrates how information flows between the various components of the TE system. The figure also shows how TE tools gather data from the network nodes using one or more of the CLI, SNMP, or XML management interfaces. Finally, the figure shows how traffic engineering management tools send configuration information to the network nodes after analysis and modeling of the statistics have been completed.

The network nodes create and maintain the traffic engineering tunnels based on a private configuration. Configuration of a TE tunnel only occurs at the origination

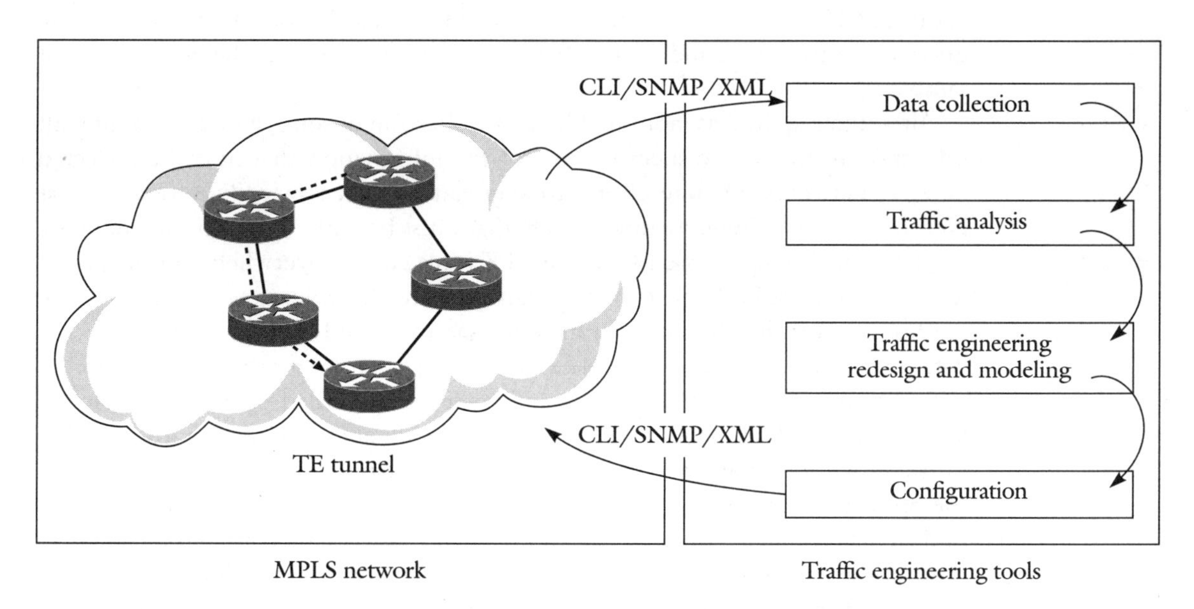

Figure 7.5 Traffic engineering life cycle: first data is collected using one or more management interfaces, then traffic analysis is performed on the data, then traffic engineering modeling of fault contentions or future capacity is fed into a redesign exercise. The redesign effort may then result in configuration changes, shown as the last element of the cycle.

point or "head end" of the tunnel, and thus this is where configuration information is sent by a TE system. Once this information is received, the tunnel's head end can use it to signal it along the path of the tunnel. This act typically must reserve resources and state for the tunnel. Configuration information consists of both the primary path over which the tunnel will carry its traffic, any secondary backup paths that can be used to reroute the tunnel's traffic in the event of a network failure, as well as the traffic parameter constraints that will be used to establish the tunnel (or reroute in the event of a failure).

7.9 Input to Traffic Engineering Tools

Traffic engineering tools require the input of certain pieces of information in order to operate effectively. Input to these tools must be gathered from each node either directly or indirectly using an offline data collection mechanism. Examples of such a mechanism include NetFlow (see Chapter 9); Traffic Matrix Statistics (TMS), which is described in more detail in Chapter 10, as well as the OSPF-TE routing protocol. Regardless of the mechanism used to obtain the information, the information itself typically includes the node's TE configuration, what the node knows about the TE topology (i.e., nodes and/or links), and historical network loading information for the period of time between when the data is gathered and the last sample.

Although required as input to TE tools, collection of some statistics is a difficult undertaking due to the sheer volume of the information that must be collected in some networks. Despite efforts to aggregate statistics by offline tools such as NetFlow, the remaining volume of data that must be gathered from each network element and later processed by offline TE tools can be overwhelming to say the least. Certain methods do exist that can alleviate the volume of data, but these methods unfortunately tend to introduce sampling and accuracy errors into the data because they simply discard data by filtering it out rather than doing more intelligent correlation. This problem is further compounded by the fact that the large amounts of data collected from a single node are multiplied by the number of nodes within a typical MPLS network. This introduces another dimension to the problem of gathering TE data, which can be characterized broadly as an issue of data management. In particular, the vast amounts of data collected for TE purposes must be aggregated, transported, and then stored for periods of time that may vary from a few days to years. Fortunately, many effective offline TE applications have been produced despite these hurdles.

7.10 TE Cycle Components

Offline traffic engineering systems consist of several basic and fundamental components. When used in concert, these tools can provide a continuously or periodically executing traffic engineering system for MPLS-enabled networks. The steps of this cycle include the ability to collect data from network nodes. This information is then analyzed and perhaps fed into models predicting future growth of network traffic. These models allow the operator to more easily approve these changes by providing them with a visualization of the network. Next, configurations are generated that are intended to improve the network. These configurations must then be validated and approved. Finally, if approved, these new configurations are deployed into the network. This process is repeated as often as is designed by the system operator. Typical periods used vary from days to weeks depending on the ultimate goal of the traffic engineering exercise. The following sections describe the steps taken by the cycle of traffic engineering systems. Figure 7.6 illustrates this process pictorially by indicating the cycle of steps and the order in which they are taken in succession.

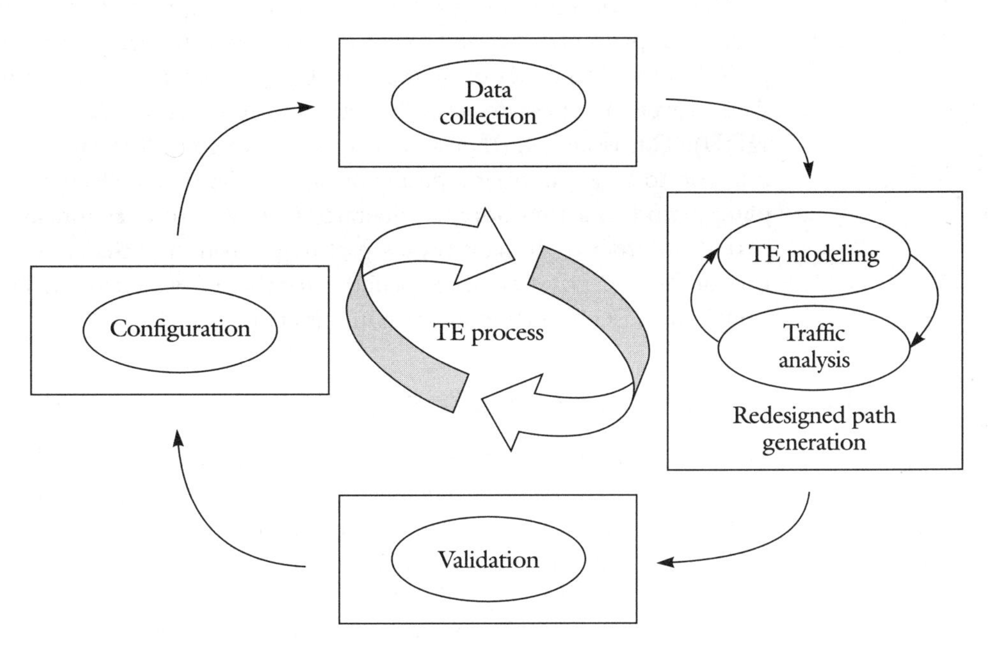

Figure 7.6 Continuous traffic engineering process.

7.10.1 Data Collection

The first step in the TE cycle entails the periodic collection of data required as input to the TE process (see Section 7.9). The length between successive periods of data collection can vary depending on the use of the data, the operational philosophy, and perhaps other factors. In general, data collection can occur in one of two places. The most obvious location is at network nodes. Network nodes that provide one or more management interfaces can provide statistical data in the form of per-flow traffic statistics or interface counters. Some examples are "show" screens on CLIs that show interface accounting information. Another example is the IF-MIB that provides similar interface accounting information. The IF-MIB can be accessed using SNMP. The network transport protocols used to transmit and access the data provided by network nodes can vary depending on the management interface (see Chapter 2) and the needs and requirements of the network.

The second method commonly used to deliver traffic statistics is to use a bulk file transfer mechanism. A file is packed with traffic accounting information and is then "pushed" by the node to a networked server, or is "pulled" from the node by a server device. Examples of this approach include NetFlow (see Chapter 9), TMS (see Chapter 10), and simple bulk file transfer using FTP (see Section 2.5).

The data collection function in itself raises the question of resource utilization scaling. In particular, the volume of the data collected from just a single sampling period can lead to issues of offline storage, both in terms of speed of access to this data as well as where to actually store the information (e.g., tape, CD, DVD, HDD). This issue is further amplified when we consider that in order for the TE function to be accurate, it typically must consider data collected over several sampling periods. Furthermore, an operator may wish to occasionally recall sampling periods in order to perhaps review past usage from months or years ago. Restoration of this data poses another challenge for these days. Fortunately, most TE solutions provide a reasonable solution in this area.

7.10.2 TE Model Building and Traffic Analysis

Once TE data has been collected, this data is then fed into model building software. This software can construct a detailed model of the network using the input data. It is important that this model reflect that actual state of the network given the data it is fed. This model should also remain accurate given the time delay experienced between capturing the collected data and transporting that data to it. Modeling software then uses the collected information to create models of several aspects of the network. These might include representations of the nodes both in terms of physical and logical configurations, or connected peers (logical or physical). Additionally, pertinent configuration information such as the network protocols running on a device might be somehow indicated.

In addition to this general configuration information, the state of each link on each node, as well as the traffic characterizations of per-flow network traffic over these links, is usually taken into account. When modeled, it is this information that provides the operator with visualizations or other representations of overall network resource utilization. It is also important for modeling software to understand the differences between the logical and physical connections because this information will later be used as a constraint when calculating failure scenarios, as well as backup strategies for avoiding them. Finally, the raw traffic demand data collected must include a few important types of sampled data. This includes the start and stop times of each network flow, per-flow bandwidth utilization, and service level criteria (delay, jitter, loss) and how it is being adhered to for each flow. It may also be important to aggregate this information together in order to model gross network behavior.

When complete, the modeling information can be used to model and depict end-to-end network traffic and loading characterizations on a macro level. In large networks, this characterization may be an important tool for visualizing actual network behavior rather than having to aggregate this mentally or by hand by viewing individual flow characteristics. The modeling tool can also point out or highlight areas of overcapacity or congestion that might be difficult to see by simply viewing interface or per-flow statistics at any isolated point within the network.

The modeling and analysis phase also includes the results of failure analysis. This information can be used to predict network node, link, or other network component failures. Once a network has been modeled, an operator can simulate changes to the network by pathologically failing links or nodes, as well as oversubscribing certain links or nodes. It is important that these things can be used to determine how the network might react to such situations without actually having to modify the network. This saves time and effort and, of course, does not disrupt the operational network. This analysis can also be used to compute sufficient backup tunnels that can be placed during the reconfiguration phase.

Tunnel Sizing

At this point, the operator may be presented with several options for tunnel bandwidth allocation and placement for primary tunnels. Many systems are also capable of providing secondary tunnels and understand the intricacies of provisioning them. The decision of how much bandwidth to allocate for each tunnel can be based on existing or predicted link or node utilization. Regardless of these factors, the operator is ultimately constrained to the maximum allowable tunnel size over the links that the tunnel will traverse. This assures that the user has control over the maximum fraction of link bandwidth that an individual tunnel can require. This also assures that the user configuration will be valid and reasonable once pushed into the network nodes. Most TE tools will provide the operator with a reasonable topology, but it should be double-checked nonetheless.

7.10.3 Redesigned Path Generation

Path generation is the step that follows data collection, modeling, and analysis. This step typically produces new configurations for existing TE tunnels, deletes existing tunnels, or perhaps produces configurations that create new tunnels. This step may also include a network optimization phase. The new configuration is based on the outputs of the modeling and analysis phase, as well as the input of any new constraints by the operator. The process of analysis, modeling, and path generation is an iterative process that continues until the operator is happy with the outcome. The outcome of the modeling and path generation phase will typically be tweaked by the operator several times before accepted, especially when the tool is being used initially.

Of particular concern during this phase is the correct generation of new configurations for network nodes. Failure to generate correct configurations for a network node could result in an operational network falling to its knees. Special attention should be paid to any configuration changes that alter the existing network. For example, if either a tunnel is deleted, or explicit routes that steer traffic into a tunnel are deleted or modified, then it is imperative for the TE tool (and the operator) to scrutinize these changes to make sure that no customer traffic will be disabled or "black holed." It is better to uncover this shortcoming now rather than having a customer uncover it later. Another type of configuration change that should involve a high degree of scrutiny might be the addition of new tunnels. More specifically, attention should be paid to what order the tunnels will be added to the network during the configuration phase. New tunnels can and will take bandwidth away from other tunnels that run over those same links unless care is taken to provision them appropriately. The configuration of one tunnel can also preempt the configuration of another depending on its preemption priority and several other factors. Preemption of existing tunnels can also happen if a TE tool erases the existing configuration of network nodes and starts with a blank one. In this case, new tunnels may be placed before existing tunnels and allocated remaining bandwidth before they can be signaled. Tunnel signaling times should also be taken into account during this process for these same reasons.

Route Computation

Three basic route computation types may be used to generate newly computed TE tunnel paths and their associated configuration. The first entails the computation of *incremental* changes to existing tunnels rather than their wholesale replacement and/or reconfiguration. In general, these changes are only used to improve portions of an existing tunnel's path or configuration. For this reason, these changes typically result in less drastic modifications to the existing operational network and its associated configuration. For example, if a tunnel is deemed to be underutilizing the bandwidth it has been allocated, then a smaller amount might be reserved without having to tear down and resignal the tunnel. Another example might be to correct a network "hot spot" where several tunnels may collide onto the same link, thus overutilizing its resources. An operator may wish to move one of the tunnels such that it avoids that link while leaving the others alone, as well as most of the moved tunnel's configuration. Such changes are useful from the perspective that the changes are usually minimal and therefore typically do not have the potential of resulting in overall chaos within the network. This is because they leave most of the network to operate as it did before the changes (presuming it worked before).

Another method used when generating new or modified network configuration entails changes to the network that are more drastic in nature than those in the incremental approach just described. This method is referred to as *global optimization*. This approach involves a more holistic approach to tunnel placement and network reoptimization by considering the global implications of tunnel, interface, and node loading that can result from the placement of a new tunnel or the modification of an existing one. This includes the issue of the order in which tunnels are placed that was described earlier. This approach also may involve the computation of backup tunnels for each primary tunnel that is placed. Backup tunnels can be used to create secondary (or any number) of backup paths that, while maintaining constraints specified for the tunnel traffic, maintain a path that may be link, resource group, or node disjoint from the primary path. It should be noted that the incremental method can also be used to compute backup paths, but these paths may be of lower quality given that a global view of their placement is not necessarily taken.

The advantage of the global approach over the incremental method is that since this calculation takes into account the entire network, it may be required to run less often (i.e., presumably it would produce the same or similar result if it were executed too often). This results in its being potentially far less disruptive than the incremental approach, which often runs much more frequently. The disadvantage to the global approach is that it typically takes much longer to compute and realize within the network than an incremental approach.

7.10.4 Validating the Generated Configuration

Once the resulting configuration has been generated, it is advisable that it be validated using both the operator's eyes and other automated mechanisms (because an operator's eyes can sometimes become tired). Validation may also be performed before and after the optimization phase is executed when the new configuration is generated. Validation can be used to point out invalid configurations such as topologies that do not reflect the current routing protocol topology, incorrect resource utilizations (e.g., tunnels traversing a link oversubscribe the maximum bandwidth available), or incorrect QoS parameters given the constraints required by the TE tunnels. In addition to failure conditions that have not been accounted for, other performance-related situations can be uncovered using this analysis. Once pointed out, the operator is typically given the option of returning to the modeling/path generation steps if a fault or warning is encountered, or ignoring it and continuing.

Once validated, the final configuration can then be analyzed one last time to provide the operator with information and statistics that may be interesting. This might include summaries of link utilization; percentage of link bandwidth that was used versus what was left unused; average, maximum, and minimum delay across any given path; or insufficient backup resources for any given TE tunnel. This information can provide the operator with an additional high-level view of the new network configuration. It is fair to say that such additional information may also inundate the operator with too many things to process and decide on.

7.10.5 Network Reconfiguration

Once the final configuration has been computed and validated, the last step is to push the configuration back down into the network nodes so that the configuration can be realized. This is not as easy as it may first seem. Simply using a random or brute-force approach to visit each node and reconfigure it may result in network behavior that is undesirable, unpredictable, or even disastrous. Several challenges therefore must be overcome by the TE system at this point, starting with the order in which to configure each node.

Therefore, one important piece of information that the resulting configuration should include at this point is a listing of the order in which to configure the nodes in the network. This list should also include the specific order in which to configure each node's tunnels and other configurable items. Inclusion of this ordering information is necessary as it may result in network behavior that differs from those predicted and assumed during the modeling and analysis phases. Some TE systems add an additional layer of confidence for operators by allowing them to confirm each action performed on the network. This is typically coupled with a function that allows any changes to be reverted to their original state. Ultimately an operator may completely trust a TE system outright and simply allow it to download all network configurations automatically, but that generally only happens after a long period of use, if ever.

In reality, reconfiguration takes place in the form of using one or more management interfaces that the devices in the network support. These may or may not be the same set of interfaces that were used to gather the network data from the nodes during the data collection phase. In fact, in many cases it is not, since the management interfaces that are suitable for high-volume data collection are not used for configuration. The most prevalent management interface used for configuration is the CLI because of the reluctance of many operators to use any other management interface for configuration. This unfortunately has resulted in TE systems having to generate differing sets of CLI configuration strings depending on the node being reconfigured, and sometimes depending on the version of software running on that node. Although nearly 100 percent of devices support CLI for configuration, other effective mechanisms can be used for configuration such as SNMP, XML, and CORBA.

7.11 Offline versus Online Calculations

Now that we have gone over the basic functions and operation of a TE system, and what its benefits and drawbacks are, it is appropriate to think about the overall advantages and disadvantages of using such a system on a more or less frequent basis. TE systems are typically referred to as operating in an "offline" or delayed capacity. That is, the time taken to collect data, process and model it, and then generate a new configuration that is pushed down to the network is typically significant—on the order of hours, days, or weeks—especially in terms of how time is typically measured for computational devices, which is on the order of microseconds. Some traffic engineering systems that are available today are approaching near real-time analysis and reconfiguration times. These systems are referred to as "online" or "near online" systems. Some of these systems may even be built into the network nodes themselves. The advantage to these systems is that they can provide the benefits of both global and incremental optimizations due to the quick turnaround time they offer from data collection to network reconfiguration, while still providing the quick turnaround that might result from a quicker, more targeted TE strategy.

In general, whatever system is being used to do traffic engineering, the operator is faced with several challenges related to the timing of the network reconfiguration. If the steps following data gathering have taken too long, and the view of the network that was modeled and then used to compute the modified configuration has become stale, these steps must be repeated or the resulting reconfiguration of the network may produce undesired results. Therefore, care must be taken in how long the period of the TE cycle is specified. In general, systems that are "online" cannot completely obviate these problems, but these systems are sometimes optimized in order to produce configurations that are produced in less time, so the problems may be mitigated. This often comes with monetary cost of high-powered computation platforms or storage systems. Worse yet, if the TE cycle is too short and not enough analysis is done or not enough data is collected, a less accurate global optimization may be employed that can result in less optimal traffic engineering. The choice of which cycle interval to ultimately utilize is up to the network operators and their operational needs. Since this interval may change from time to time, and may not be set on any fixed period of time such as one hour or one week, TE software that is most flexible will probably best suit the largest audience of operators.

As with any tool available for MPLS network management, there are pros and cons to their use. Wide variations of tools exist and are available in any given category. Thus, the operator's mileage with any one may vary, and so we recommend thorough investigation and comparison of the available offerings.

7.12 Summary

This chapter introduced traffic engineering—a process whereby a network operator can engineer the paths used to carry traffic flows that vary from those computed by the routing protocol(s) in use in that network. Operators may desire to engineer their traffic in an effort to steer it through custom-chosen paths to achieve more efficient use of network resources, to protect against network node or link failures, or to provide certain customers with specially designed services such as guaranteed bandwidth paths.

We then introduced how TE is applied in MPLS networks. MPLS intrinsically provides the operator with traffic engineering functions that allow for the control and deployment of traffic engineering. The basis of this control is a traffic engineering tunnel. This concept is typically implemented as an interface at the point where the tunnel begins. This tunnel is signaled and a path reserved (if possible) that meets the traffic parameter constraints specified by the operator. If successful, a label switched path is created that will be used to carry labeled traffic for that tunnel. One advantage of nodes implementing TE tunnels as interfaces is that they can provide statistics using commonly understood measurement means such as the SNMP IF-MIB and NetFlow statistics.

We then investigated the three general ways in which traffic engineering can be deployed. These included deliberate models, which involve ad hoc placement of new TE tunnels as needed. An operator might place tunnels in this manner to repair ongoing network "hot spots" and may later remove them once the situation is abated. Additionally, two more proactive deployments of MPLS TE are available. These are usually described as *full-mesh traffic engineering,* involving the creation of TE tunnels between all nodes within an MPLS domain. When properly deployed, it was shown that these models might result in effective control over all or most link utilization, service levels, and backup tunnel placement within an MPLS network. Two variations of this model were illustrated.

Next, we discussed the traffic engineering cycle. Traffic engineering was described as a cyclic process that involved several steps. These steps included data collection using a variety of mechanisms. Once the data was collected, we investigated how it was modeled and analyzed. Once processed, a new network configuration could be generated in order to make the network more efficient using optimization parameters specified by the operator.

We discussed various mechanisms that are used to generate and analyze the new network topology. The configuration was then refined using a closed-loop iterative process that only ended when the operator was satisfied that the new configuration was sufficient to satisfy their goals. We described how the configuration was then validated using a series of different mechanisms that could further ensure the correctness of the configuration. Finally, we discussed how the new configuration could be pushed down into the network devices in order to realize the optimized configuration.

The last section discussed the various issues related to offline versus online traffic engineering, and the benefits of either approach. We talked about how some TE systems offered both approaches to suit different operational environments and the operational consequences of such systems.

With the traffic engineering capabilities described, we learned that it is possible to create a continuously operating traffic engineering system within MPLS environments that can be used to enhance the quality, availability, and robustness of network infrastructure and services, as well as reducing the costs of operating that same network without traffic engineering enabled.

Further Reading

RFC 2702. Awduche, D., J. Malcolm, J. Agogbua, M. O'Dell, and J. McManus. "Requirements for Traffic Engineering over MPLS." September 1999. *www.ietf.org/rfc/rfc2702.txt.*

Multi-Protocol Label Switching (MPLS). Cisco Systems documentation. *www.cisco.com/univercd/cc/td/doc/product/software/ios120/120newft/120limit/120s/120s5/mpls_te.htm.*

RFC 3209. "RSVP-TE: Extensions to RSVP for LSP Tunnels." *www.ietf.org/rfc/rfc3209.txt.*

"Constraint-Based LSP Setup using LDP." *www.ietf.org/internet-drafts/draft-ietf-mpls-cr-ldp-06.txt.*

"Applicability Statement for CR-LDP." *www.ietf.org/internet-drafts/draft-ietf-mpls-crldp-applic-01.txt.*

Kodialam, M., and T. Lakshman. "Dynamic Routing of Bandwidth Guaranteed Tunnels with Restoration." Bell Laboratories.

Kodialam, M., and T. Lakshman. "Minimum Interference Routing with Applications to MPLS Traffic Engineering." Bell Laboratories.

Srinivasan, C., A. Viswanathan, and T. Nadeau. "Multiprotocol Label Switching (MPLS) Traffic Engineering Management Information Base." August 2001. *www.ietf.org/internet-drafts/draft-ietf-mpls-te-mib-08.txt.*

Davie, B. S., and Y. Rekhter. *MPLS: Technology and Applications.* First edition. San Francisco: Morgan Kaufmann Publishers. 2000.

Gray, E. W. *MPLS: Implementing the Technology.* Reading, Mass.: Addison-Wesley Professional. 2001.

The IETF Traffic Engineering Working Group is a good source of information about current topics in traffic engineering: *www.ietf.org/html.charters/tewg-charter.html.*

Data Connection, Ltd., provides an informative discussion of the different signaling mechanisms for MPLS traffic engineering at *www.dataconnection.com/download/crldprsvp.pdf.*

Xiao, X., A. Hannan, B. Bailey, and L. Ni. "Traffic Engineering with MPLS in the Internet." *www.juniper.net/techcenter/app_note/350000.pdf.*

Ross Callon

is a distinguished engineer in the protocols group at Juniper Networks. He is co-author of the Layer 3 VPN Framework document and is co-chair of Network Reliability and Interoperability Council 6, Focus Group 2, advising the Federal Communications Commission (FCC) on network reliability. He is a long-standing participant in multiple IETF Working Groups and has previous experience in the ATM Forum, Internet Engineering Steering Group, IEEE, ANSI, and ISO. He has authored or contributed toward VPN, MPLS, IPv6, IS-IS, and CLNP networking standards. He was a major contributor to the ATM Forum PNNI standard, which provided the first standard supporting dynamic distributed traffic engineering, and is a former co-chair of the IETF IP Next Generation (IPv6) Working Group. He has published numerous articles and been awarded 12 patents. He has an S.B. in mathematics from the Massachusetts Institute of Technology and an M.S. in operations research from Stanford University.

Some people feel that offline traffic engineering is the only way to correctly engineer and maintain an MPLS network. What is your take and advice in this area? Where do you see offline TE as fitting into a complete strategy for managing an MPLS network today, and in the future for a GMPLS network?

Offline traffic engineering can consider the entire capabilities of and demands on the overall network, and can use relatively computationally intense algorithms. These characteristics allow for computation of very good routes to balance network traffic effectively. However, offline computation is inherently based on a priori computation of routes. This implies that offline traffic engineering does not by itself respond rapidly to network failures. I see offline traffic engineering therefore as *part* of the overall solution, but in many cases it will not be the entire solution.

In contrast, constraint-based routing can be used to allow dynamic adjustment of traffic in the network, for example, to respond quickly to changes in the network, such as link or equipment failures, and changes in the traffic demand. The distributed nature of constraint-based routing also eliminates concern about a single point of failure.

In many cases, therefore, constraint-based routing will be deployed in network devices, with offline traffic engineering used to seed the network with very good routes, or to verify the performance of the overall traffic engineering solution. Offline computation may also be used to place large and/or long-term flows.

This combination of offline computation plus dynamic distributed routing is likely to continue to be used in GMPLS networks. However, the greater range of devices supported by GMPLS implies that there are a wider range of ways that offline computation and constraint-based routing may be combined.

8

The MPLS Traffic Engineering MIB (MPLS-TE MIB)

"I shall be telling this with a sigh
Somewhere ages and ages hence:
Two roads diverged in a wood, and I—
I took the one less traveled by,
And that has made all the difference."

–Robert Frost, "The Road Not Taken"

Introduction

In previous chapters of this book, we discussed how an operator can exploit the process of traffic engineering to custom-engineer the paths used to carry traffic flows within a network. We also investigated how the first step in this process was to gather traffic statistics from network nodes. One mechanism by which this data can be gathered for MPLS traffic engineering tunnels is the MPLS Traffic Engineering MIB (MPLS-TE MIB). This MIB can also be used by a traffic engineering system to reprovision and modify existing tunnels as well as to deploy new tunnels.

In this chapter, we focus on the study of the MPLS-TE MIB. We first present an introduction to MPLS traffic engineering in terms of how the protocols have been defined or extended to support MPLS traffic engineering. This will form a foundation for why the MIB was structured as it is, as well as for many of the objects found therein. We will then discuss in detail how the MPLS-TE MIB can be used by operators who deploy traffic engineering in their networks. MPLS traffic-engineered tunnels can be deployed either manually or by a more automated offline traffic engineering system. In either case, however, the MPLS-TE MIB can be used to manage these tunnels.

Device vendors implementing this MIB will benefit from reading this chapter as we provide a few pointers on how to make the implementation easier and more useful for operators using their devices. Operators reading this chapter will benefit from the details provided for how they can expect the MIB to be presented by a device.

This chapter will focus on the draft version 08 of the MPLS-TE MIB, which is typically referred to as draft-ietf-mpls-te-mib-08.txt. This draft version may have been updated or replaced with an IETF RFC document after this book was published. Please keep this in mind when searching for the document on which this chapter is based.

Chapter 7 described how a network could be traffic-engineered using offline traffic engineering techniques in order to make its operation more efficient. The techniques described in that chapter showed a means by which an operator could override paths selected by the routing protocols for LSPs within the MPLS network. An operator could then use these techniques to engineer some or even all of the paths in a network depending on the needs of that network. The results of these efforts would be the potential gain in network utilization and resiliency.

The data that feeds into the traffic engineering tool must first be gathered from network nodes. As was described earlier, mechanisms for doing so include NetFlow (see Chapter 9), Traffic Matrix Statistics (see Chapter 10), or interface counters including those on tunnel interfaces (see Chapter 6, as well as later in this chapter). Once the traffic engineering tool has collected the traffic statistics data for input, it will then process it in a variety of ways. Once the data is processed, the TE system is then ready to propagate a solution back onto the network either by modifying existing tunnels or by placing new ones. This is typically achieved by the traffic engineering tool performing a variety of configuration changes at one or more network nodes. These configuration changes can be realized either by making explicit modifications to configuration files on the devices in question, or by making modifications to the MPLS-TE MIB on those devices.

This chapter will focus on how the MPLS-TE MIB can be utilized as an integral part of the traffic engineering process, both in terms of gathering input statistics (and monitoring the health of tunnels in general) as well as for configuration of the traffic engineering tunnels.

8.1 Constraint-Based Routing

Traditional intradomain routing systems or interior gateway protocols (routing protocols) used in IP networks compute routes based on algorithms that optimize paths based on a single scalar metric. For example, the minimum number of hops might constitute such a metric that could be used to compute a shortest path through a network. This is in fact how the Routing Information Protocol (RIP) computes what it considers the shortest path. RIP uses the Bellman-Ford algorithm to compute the shortest path and minimizes its path computation based on a *distance vector algorithm*. RIP can be used in smaller networks, but when used in moderate-sized ones reveals intrinsic difficulties related to allowing for routing loops and the time required to converge on the solution. It was for this reason that OSPF and IS-IS were chosen for use in moderate- to large-sized intranets (but not the Internet).

Both IS-IS and OSPF are *link state algorithms*. These algorithms are based on a single scalar metric that is used to minimize the shortest path. This metric is chosen and assigned to each link by the network operator. Both algorithms then use this metric to minimize the path from one node to another. This is accomplished using the Dijkstra Shortest Path First (SPF) algorithm. This algorithm is known as a *greedy* algorithm because it does local minimization as it iterates over the network graph based on the metric. Another fundamental difference between these algorithms and RIP is that they rely on each node distributing the state of its adjacent neighbors to the rest of the network. This information is then gathered at each node so that each can have an up-to-date graph of the network. On the other hand, RIP passes around distance vectors that, at least in part, lead to its slow convergence.

Although IS-IS and OSPF are successful when using a single scalar metric, a need arose to specify additional constraints in addition to the normal scalar metric. For example, in addition to the bandwidth of a path, an operator might wish to avoid or prefer certain links. This, of course, is not possible (at least easily) using a single scalar metric. Therefore, the traditional IS-IS and OSPF algorithms were further enhanced by adding additional constraint checks during the minimization phase of the SPF algorithm. This enhancement then resulted in the algorithms choosing paths based on enhancements that now minimize the shortest path based on a set of constraints *and* a single scalar metric.

For example, Dijkstra's SPF algorithm has a goal of choosing the next-hop along a path based on that next-hop being optimal with respect to the scalar metric. Now if constraint-based routing is used, the SPF algorithm's optimality goal is extended to choose the next-hop not only based on the most optimal scalar metric, but also such that it does not violate any of the constraints specified for that path. For example, a set of constraints could constitute a minimum amount of bandwidth that should be available to the data flow on that path. Another example might be some administrative constraints such as the coloration of links such that the path calculation explicitly avoids those paths because they might be, for example, part of a private network. This new version of the SPF algorithm that adds constraint-based functionality is referred to as constraint-based SPF (CSPF).

Constraint-based routing algorithms are useful because they can then be used to calculate paths based on constraints that may result in different paths from those chosen by the typical metric-based routing algorithms. As was explained, the paths that are chosen will still be based on the scalar metric, but will also be chosen based on the set of constraints. These constraints can then be used to specify paths that are engineered by an operator for traffic engineering purposes.

Once calculated, the path must be signaled. This is accomplished using one of the two signaling protocols defined by the IETF MPLS Working Group. A discussion of the two mechanisms standardized by the IETF's MPLS Working Group is the topic of the next section.

8.2 Signaling Constraint-Based Paths

The IETF MPLS Working Group first defined the requirements for traffic engineering in an MPLS network in RFC 2702. This document specifies that constraint-based routing be used for path calculation. Since SPF-based algorithms require link state information about all of the links in the network to function correctly, only link state–based routing algorithms such as OSPF and IS-IS can be used within networks where CSPF is in use. For example, RIP cannot be used because it is a distance vector algorithm. To facilitate the correct operation of CSPF, additional extensions were made to OSPF and IS-IS to support traffic engineering and constraint-based routing. These are specified in Katz et al. (2001) and Li and Smit (2001).

Two other documents defined two mechanisms that can be used to establish and later modify traffic engineering tunnels. These involved extending the RSVP and LDP protocols to allow them to signal traffic-engineered tunnels based on constraint-based routing calculations. Both mechanisms signal tunnels by reserving resources along a path that matches the constraints specified for that path, and then assigning MPLS labels along this path to form a label switched path to carry the traffic along this path across the MPLS network. Both protocols piggyback MPLS labels within protocol messages as well as other traffic engineering parameters to achieve these goals. Both mechanisms are described in the following.

8.2.1 RSVP-TE

Several years ago, the IETF's Integrated Services (int-serv) Working Group developed an architecture that described how overall end-to-end QoS guarantees could be achieved in the Internet. Such guarantees would ensure that an application that required a minimum amount of bandwidth and/or a limit on the end-to-end delay of its connection could have these constraints met. One of the documents produced by this Working Group was the Resource Reservation Protocol (RSVP). This protocol was designed to allow such requests for QoS parameters.

RSVP is a protocol that allows applications to signal QoS requirements to the network. To achieve this goal, RSVP defines several protocol messages that are processed by each of the nodes along the requested connection path. Each node must agree on the resource reservations. Once processed, a special message is returned that indicates success or failure of the request.

RSVP defines two basic message types: PATH (path) and RESV (reservation) messages (for a more detailed examination of RSVP, please see the Further Reading section at the end of this chapter). When a host wishes to signal a reserved network path, it will issue a PATH message that is addressed to the *session* or destination address of where the session should terminate. This address typically indicates the single destination of the connection, but may also designate a multicast address if more than one destination is desired. The PATH message may also contain an explicit route object (ERO) that can contain the entire explicit path (i.e., a source route) that the message should take from start to finish. Every PATH message contains TSpecs (sender traffic specifications) and classification information that specify the QoS parameters for the connection. When received, a PATH message is processed by the node to determine if it can meet the reservation requirements (i.e., does it have enough buffer space left for the connection). If the request is successfully processed, a network node or host will temporarily reserve some resources for that connection and then pass the message onto the next-hop for successive processing. This must be done so that interceding reservations do not use the reserved resources while the remainder of the connection path is verified. If the message contained an ERO, then the next host in the path is examined and popped from the message just before it is delivered to the next node. If no explicit paths are specified, the path as specified by the routing software is chosen.

When the node addressed by the *session* address is reached, it will issue a new RESV message with an RSpec and send this back to the sender. This message indicates that the reservation is successful. Each hop along the path receiving the message should then replace the tentative resource reservation with an active designation. If the record route option is specified in the original PATH message, the session node that issued the RESV message will indicate this. Each node along the path will then indicate that it received the RESV message by inserting its address into the RRO object. This message can thus be used to indicate the actual path taken by the reservation. This is useful if a dynamic path is specified (i.e., just the source and destination) instead of an explicit one.

An RSVP flow reservation is illustrated in Figure 8.1. The leftmost host issues a PATH message that travels along a path through the network. Each node that receives the message makes a tentative reservation of resources. Once the request reaches the host on the far right of the diagram, a RESV message is issued by that host. This is used to confirm the tentative reservations along the path taken by the PATH message. Note that an explicit path could have been specified by the host originating the request; thus a different path could have been taken by the reservation. The figure also illustrates the unidirectional nature of RSVP. That is, reservations form a unidirectional connection. If a similar reservation needs to be made to

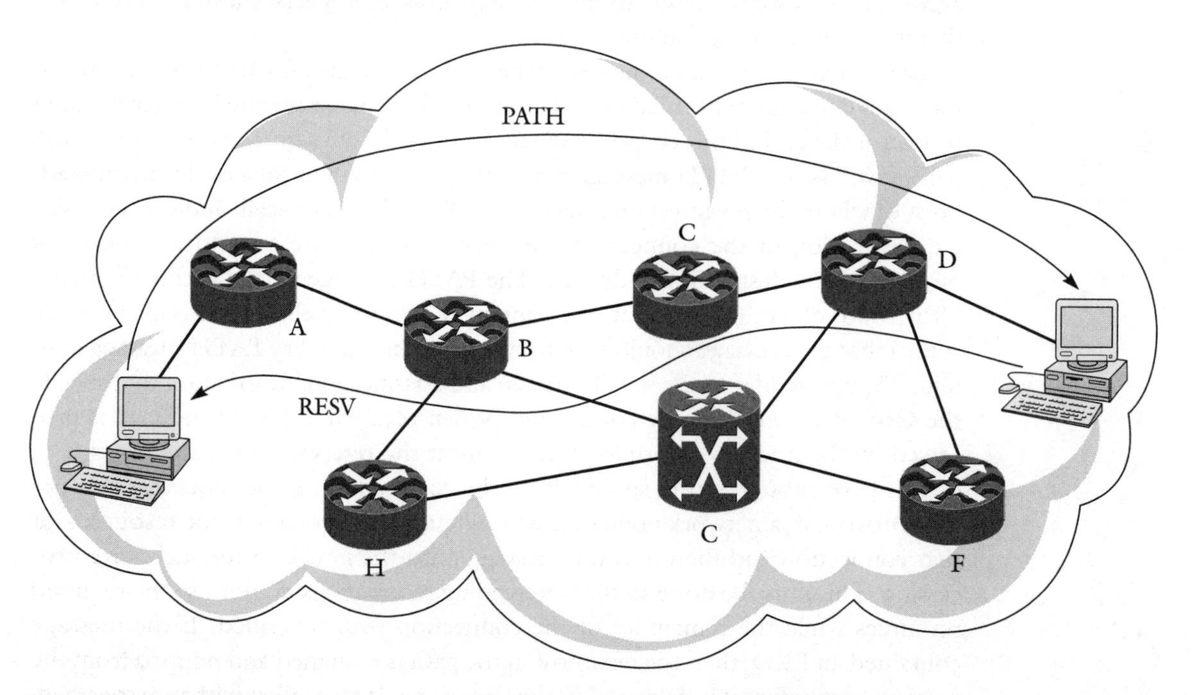

Figure 8.1 RSVP PATH messages start at the leftmost host and traverse a path to reach the host on the far right. Along the way resources are tentatively reserved by each node the request reaches. The reservation message (RESV) is emitted from the host on the far right once the PATH message reaches it. The RESV message flows along the same path to return to the host on the far left to confirm the resource reservations.

form a bidirectional connection, then a reservation from the opposite end of the connection needs to be made as well.

When it came time to decide on a mechanism for signaling constraint-based paths through the MPLS network, many chose to extend RSVP (RFC 3209). One of the primary reasons was that the RSVP protocol was proven in networks and was known to work. RSVP was extended to bind labels to reserved flows (for unicast) in order to enable MPLS LSRs to recognize packets that belong to flows for which reservations had been made. At the ingress LSR this amounts to a FEC that contains an RSVP reservation instead of a set of IP prefixes. A new RSVP object, called the LABEL object, was defined to carry the MPLS label. This object is contained in an RSVP RESV message. Instead of sending back the standard RESVmessage that contains an IP address and confirmation of a reservation, the LSR allocates a label, creates an entry in its LFIB, and includes the label it just allocated for that flow as the incoming label in that entry. The allocated label is then returned for the flow within the LABEL object in the RESV message. Each LSR along the flow's path that receives this message makes an entry in its LFIB with this label as the outgoing label and allocates a new label to use as the incoming label. It then inserts this label into the LABEL object and forwards the RESV message along to the session sender.

Figure 8.2 illustrates how an RSVP-TE tunnel is established. Node A attempts to establish a tunnel between itself and node D. The tunnel is initiated by node A transmitting an RSVP-TE PATH message toward B. The message traverses the explicit path as specified in the ERO to reach node D. Along the way resources are tentatively reserved by each node the request reaches. When the message is received at D, D allocates label 32 and inserts it into the RSVP-TE LABEL object. Each hop back to the session sender allocates a different label and binds it to this flow's resources. When the message returns to the session sender, the traffic engineering tunnel is ready to carry MPLS traffic. As you can see, the majority of the basic RSVP protocol is left intact and reused.

8.2.2 CR-LDP

The IETF MPLS Working Group has defined one alternative to using RSVP-TE to signal traffic-engineered tunnels. This is referred to as the Constraint-Based Routed Label Distribution Protocol, or CR-LDP (Jamoussi 2001).

In Chapter 4 we discussed how the Label Distribution Protocol (LDP) was defined to distribute labels and label forwarding state between adjacent neighbors. In an effort to satisfy the traffic engineering requirements, CR-LDP was defined to extend this basic functionality to support the additional signaling of state along an explicit path as well as to reserve resources along this path. Both of these extended

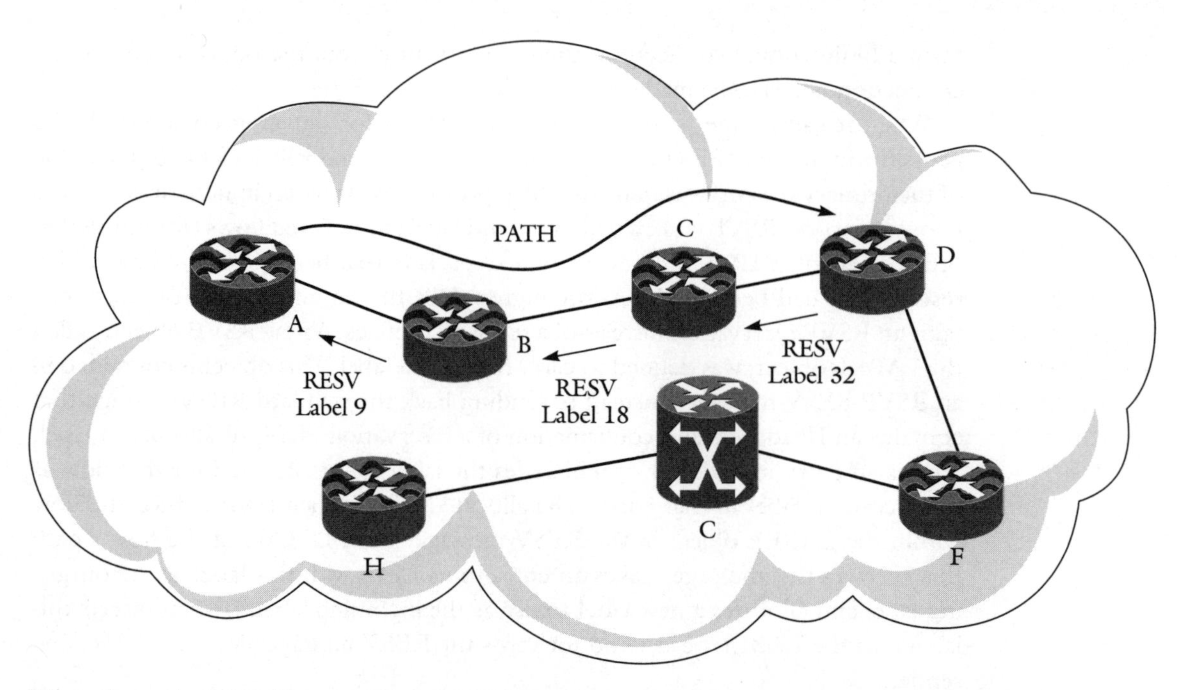

Figure 8.2 An RSVP-TE tunnel is established between nodes A and D. This is initiated by node A first sending an RSVP-TE PATH message that traverses the explicit path specified in the ERO to reach node D. Along the way resources are tentatively reserved by each node the request reaches. When the message is received at D, D allocates label 32 and inserts it into the RSVP-TE LABEL object. Each hop back to the session sender allocates a different label and binds it to this flow's resources. When the message returns to the session sender, the traffic engineering tunnel is ready to carry MPLS traffic.

functions are analogous to what is achieved by using the aforementioned extensions to RSVP-TE, but they are done within the context of the Label Distribution Protocol; thus they have the advantage of not requiring LSRs within an MPLS domain to support the RSVP protocol.

To achieve these goals, CR-LDP introduces new Type-Length-Values (TLVs), called the explicit route (ER) object, into LDP. This object contains many of the same things as the explicit route object defined for RSVP-TE that was introduced earlier. In particular, it is defined as a new TLV within the base LDP LABEL_REQUEST object. For this to work, downstream-on-demand label advertisement must be used so that the LDP LABEL_REQUEST message can be used. This message will include the ER object to support explicit routing. When an LSR wishes to establish a tunnel along a path, it will first use its CSPF algorithm to construct an ER object that contains a sequence of abstract nodes. The abstract nodes are really just IP addresses of either a link or the loopback address of the LSRs along the path in question. The LSR would then create a LABEL_REQUEST message and inserts the ER object within it. Once created, the LSR then inspects the ER object, determines the first hop along the path, determines which next-hop link to use to reach this node, and forwards the message to this node. At the next-hop along the path, the LSR will receive the message. It will examine the message and find one of its IP addresses in the message. It will then remove this entry from the list of hops, create some state for the LSP (i.e., tentatively reserve some resources), examine the next-hop, and forward the message to this LSR. This continues until we reach the last hop in the ER object. When the message is received by the last node in the ER list, the node will create a LABEL_MAPPING message and allocate a label, and will include this label in the message and send it back to the node adjacent to it.

Figure 8.3 illustrates how a CR-LDP tunnel is established between nodes A and D. The traffic-engineered tunnel signaling is initiated by node A by first sending a CD-LDP LABEL_REQUEST message with an ER object set to abstract nodes (A, B, C, D). This object contains information specifying what resources need to be reserved along the path such as peak data rate and committed data rate, but we will ignore the specifics of these values for the sake of keeping this example straightforward. The message then traverses the explicit path specified in the ER object to reach node D. At each node along the path, resources are tentatively reserved by each node the request reaches, and each node reached is removed from the ER

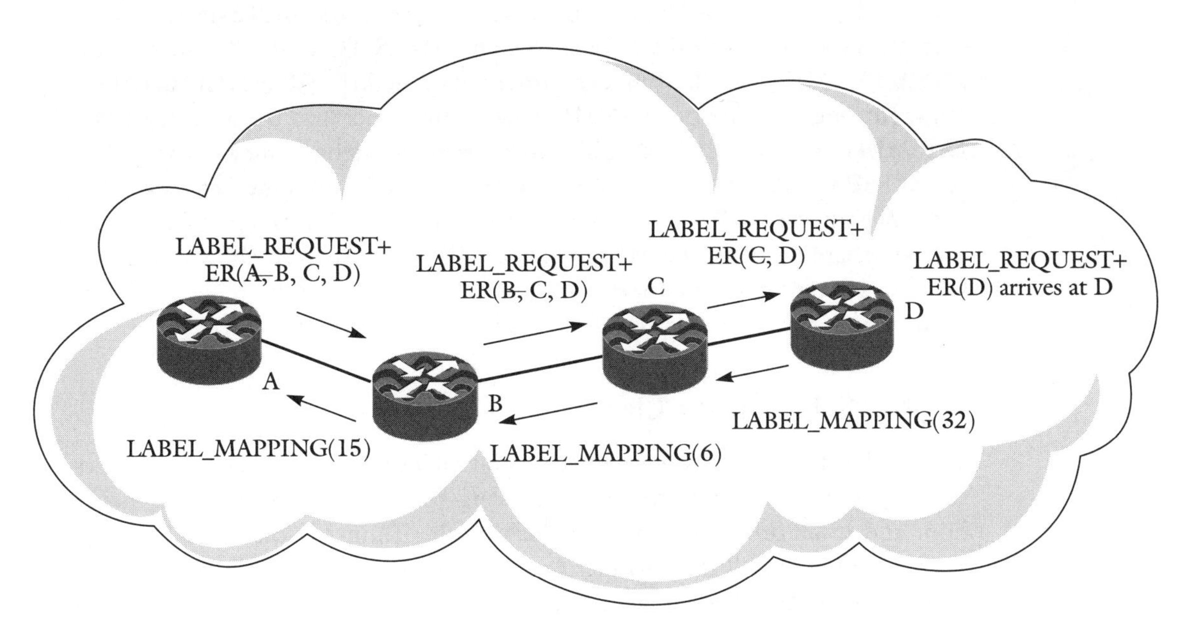

Figure 8.3 A CR-LDP tunnel is established between nodes A and D. ER() indicates the hops contained by the ER object.

object. When the message is received at D, D allocates label 32 and inserts it into the LABEL_MAPPING message that it then returns to abstract node C. Each hop back to the originator allocates a new label, binds it to this flow's FEC, and actually reserves the resources requested. When the message returns to the originating LSR, the CR-LDP signaled tunnel is ready to carry MPLS traffic.

8.3 MPLS-TE MIB Overview

Let us now focus on the MPLS-TE MIB and how it can be used to manage the traffic-engineered tunnels. The MIB allows a user to configure MPLS tunnels. These tunnels may be static (configured manually or through SNMP at each node) or signaled by any signaling protocol. The MIB is structured to contain enough information for signaling using RSVP-TE or CR-LDP. Additionally, the MIB allows a user to configure a subset of the full route and apply a traffic engineering algorithm (such as CSPF), either at the head end or within the network, to compute the explicit route.

We begin our discussion of the MPLS-TE MIB with an introductory overview of its layout, explain at a high level what the MIB allows an operator to manage, and introduce terminology that might be unique to the MPLS-TE MIB. We will then dive into the details of the MIB.

The MPLS-TE MIB fits into the larger picture of the MPLS-related MIBs in that it depends on both the MPLS-LSR and MPLS-TC MIBs. In the case of the MPLS-LSR MIB, each TE tunnel represented in the MPLS-TE MIB can optionally point (through the use of an SNMP RowPointer object) to the associated MPLS-LSR MIB cross-connect entry. This can be useful for debugging a tunnel, since the actual LSP used to route the tunnel can be followed across the network. The dependency on the MPLS-TC MIB is for a few common SNMP textual conventions that are shared among the MPLS-related MIBs. Figure 1.10 shown originally in Chapter 1 is repeated here for the reader's convenience and to highlight where this MIB fits into the larger picture of the MPLS-related MIBs.

8.3.1 The MPLS-TE MIB at a Glance

The MPLS-TE MIB is composed of several tables that are all designed to coordinate together to facilitate traffic engineering tunnel creation, deletion, modification, and monitoring. These tables include the Tunnel Table, Tunnel Resource Table, Tunnel Hop Table, Tunnel Actual Routed Hop Table, Tunnel Computed Hop Table, Tunnel Performance Table, and the Tunnel CR-LDP Resource Table. Figure 8.4 presents a high-level view of how the MIB is organized and how its

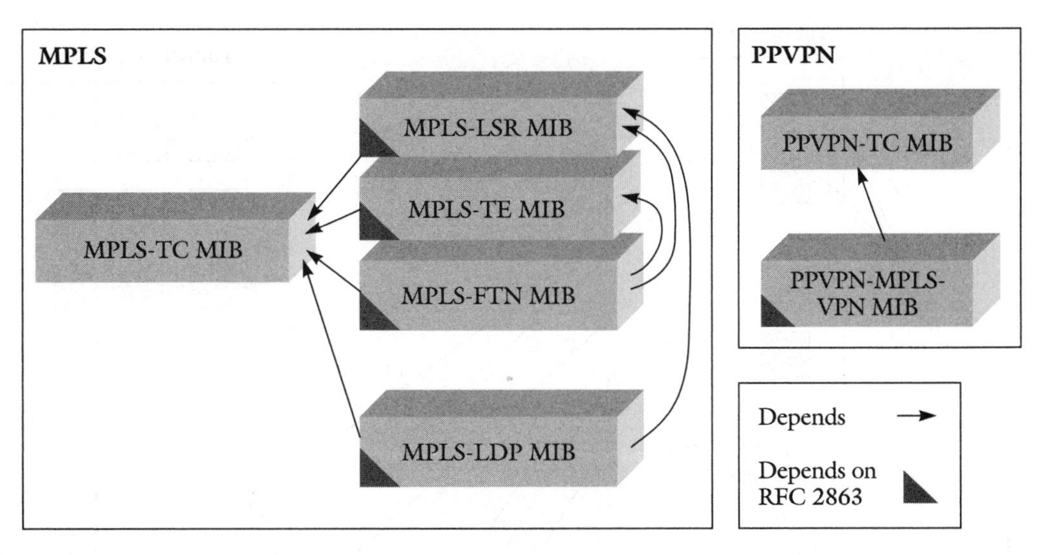

Figure 1.10 (*repeated*) MIBs for MPLS network management discussed in this text.

tables interact. The high-level explanation of this interaction is covered in the following subsections.

8.3.2 Tunnel Table

The Tunnel Table allows new MPLS tunnels to be created between the MPLS LSR supporting the MIB and a remote LSR. Existing tunnels can be deleted, reconfigured, or monitored using the mplsTunnelTable in coordination with the other tables in the MIB. Each MPLS tunnel that is active and operational will have one active LSP that is signaled in order to carry the traffic for that tunnel. This LSP should be associated with one MPLS-LSR MIB cross-connect entry (mplsXCEntry). This cross-connect entry is associated with one outsegment (mplsOutSegmentEntry) that originates at the LSR supporting the MIB and shows the next-hop of the LSP (and the tunnel). Each MPLS Tunnel Table entry (mplsTunnelEntry) should be associated with the LSP by "pointing" at its associated cross-connect entry through the use of the SNMP RowPointer defined in the mplsTunnelTable's mplsTunnelXCpointer. This is achieved by assigning the OID of the first accessible row of the associated instance of the MPLS-LSR MIB's cross-connect entry (mplsXCEntry) to this value. Note that this value will include the cross-connect index.

The Tunnel Table is closely related to the Tunnel Resource Table (mplsTunnelResourceTable), Tunnel Hop Table (mplsTunnelHopTable), Tunnel Actual Route

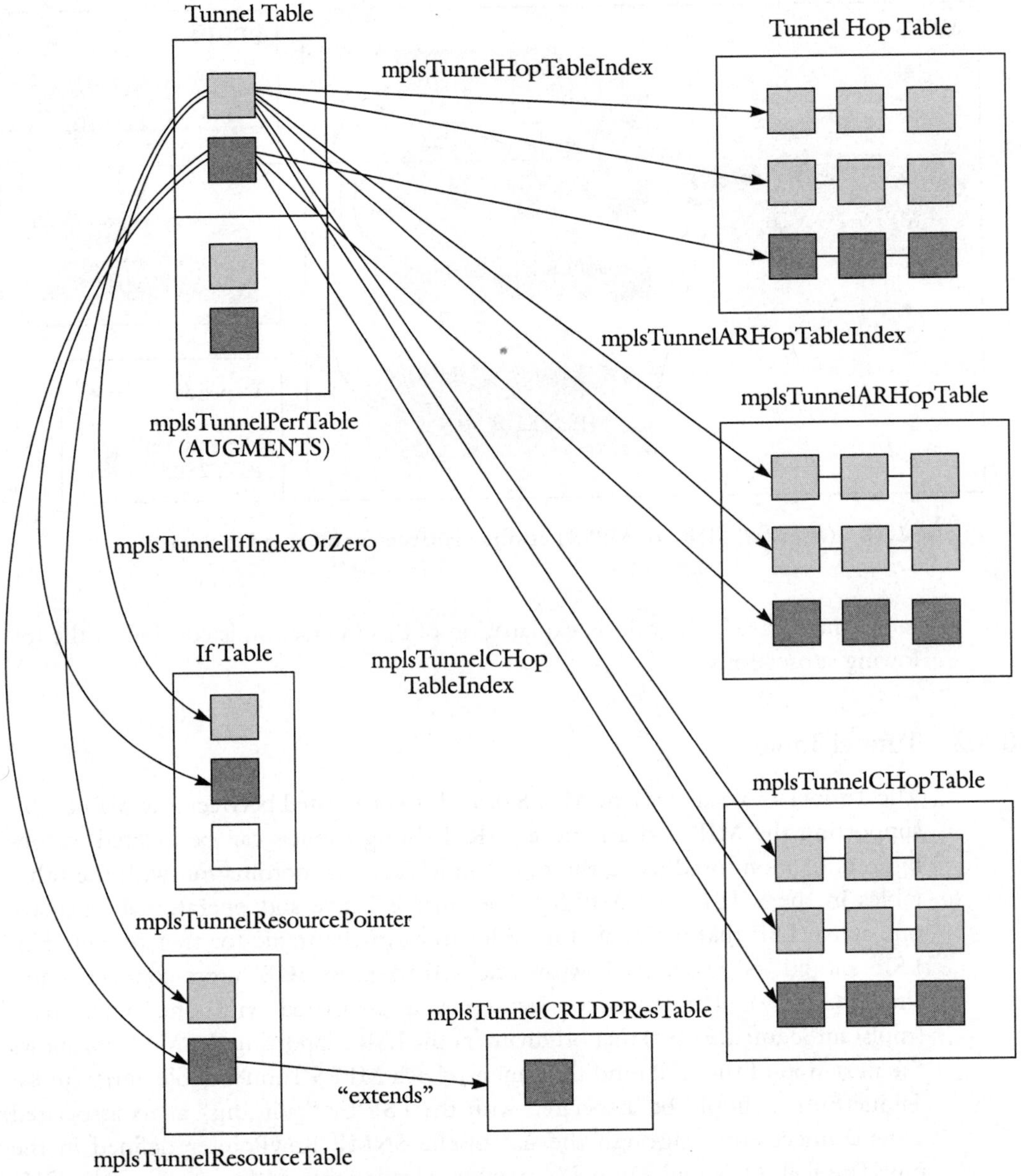

Figure 8.4 MPLS-TE MIB table relationships.

Hop Table (mplsTunnelARHopTable), Tunnel Computed Hop Table (mplsTunnelCHopTable), Tunnel Performance Table (mplsTunnelPerfTable), Resource Table (mplsTunnelResourceTable), and optionally the CR-LDP Resource Table (mplsCRLDPReseTable) either through direct reference, the SNMP AUGMENTS, or the SNMP *extends* relationships. In implementations that choose to model TE tunnels as interfaces, Tunnel Table entries are also closely related to entries in the IF-MIB's IfTable. If they are implemented as interfaces, the mplsTunnelIfIndex will refer to a valid ifIndex present in the IfTable. All of these relationships are detailed at a high level in Figure 8.4.

8.3.3 Tunnel Resource Table

The Tunnel Resource Table (mplsTunnelResourceTable) is used to indicate the resources required for a tunnel. The fact that a pointer is used allows an operator to specify that more than one tunnel entry shares the same resources by pointing each tunnel entry at the same resource table entry. Tunnels that do not share resources have to point at separate entries in this table. Note that this table shows the configured resource sharing. The actual resource sharing must be noted by examining the LSPs that are present in the MPLS-LSR MIB.

8.3.4 Tunnel Hop Table

The Tunnel Hop Table (mplsTunnelHopTable) contains entries that indicate the path hops (i.e., LSRs) that an MPLS tunnel is desired to take when it is established by the selected signaling protocol. Each hop may indicate a strict or loose preference for routing. Since this table can specify loose hops, the path specified here can therefore differ from both the Computed Hop and Actual Route Hop Tables. This table contains three indexes that, when used together, indicate a group of paths associated with a tunnel, each path, and then the hops that belong to that specific path.

8.3.5 Tunnel Actual Route Hop Table

The Tunnel Actual Route Hop Table (mplsTunnelARHopTable) contains entries that indicate the actual path hops (i.e., LSRs) that an MPLS tunnel is traversing. This table is populated only after the selected signaling protocol has established the actual path through the network. Each hop in this table indicates an actual LSR traversed by the tunnel. This table contains two indexes that, when used together, indicate the actual path associated with a tunnel, and the hops that belong to the actual path. Note that support of this table is optional since not all MPLS signaling protocols or implementations may support this feature.

8.3.6 MPLS Tunnel Computed Hop Table

The MPLS Tunnel Computed Hop Table (mplsTunnelCHopTable) contains entries that indicate the path hops (i.e., LSRs) that the selected tunnel signaling protocol has computed and will use (or is using) to signal the tunnel. This table is populated only after the tunnel is enabled and constraint-based path computation is instructed to compute and then signal the tunnel path, or if a completely explicit path is selected, when the tunnel is instructed to signal the explicit path. Because this table contains the path(s) that may have been computed by the CSPF, it may vary from that found in the hop table. However, the Actual Route Hop Table paths should not differ from the ones specified here if they are signaled properly unless a tunnel reroute has happened. As with the other tunnel hop tables, this table contains two indexes that, when used together, indicate a computed path associated with a tunnel, and each computed hop that belongs to that specific path. Note that support of this table is optional since not all MPLS signaling protocols or implementations may support this feature.

8.3.7 MPLS Tunnel Resource Table

The MPLS Tunnel Resource Table (mplsTunnelResourceTable) contains the desired resource association and allocation for a tunnel entry. Multiple tunnel entries can point at the same resource entry. Note that this table indicates the desired and configured tunnel resource constraints. The actual reserved resources should be checked by interrogating the associated MPLS-LSR MIB LSP entry or entries.

8.3.8 MPLS Tunnel Performance Table

The MPLS Tunnel Performance Table (mplsTunnelPerfTable) provides several performance-related counters that can be used to measure the performance of the MPLS tunnels. In particular, many RSVP-TE tunnel resource parameters can be specified here. Note that this table augments the MPLS Tunnel Table, so an entry in this table exists for every entry in the Tunnel Table. Additional performance information can be gathered from the IF-MIB entry corresponding to the tunnel (if such an entry exists).

8.3.9 MPLS Tunnel CR-LDP Resource Table

The MPLS Tunnel CR-LDP Table (mplsTunnelCRLDPResTable) contains tunnel resource information for those tunnels that signaled using CR-LDP (Jamoussi 2001). This table is related to the resource table using an *extends* relationship due to the potentially sparse relationship between its entries and those in the Tunnel Resource Table.

8.4 Definition of Terms Used in the MIB

The MIB uses some terminology either that is not included in the standard MPLS traffic engineering specifications for RSVP-TE or CR-LDP, or that is difficult to understand—even for those who have used the MIB for some time. In particular, we are often asked by those first reading the MIB about the terms *tunnel head, tunnel instance,* and *tunnel LSP.* The terms *cross-connect, insegment,* and *outsegment* were defined earlier in Chapter 3, which discussed the MPLS-LSR MIB.

Table 8.1 enumerates and defines these terms in a manner that is hopefully clear and concise. Note that for the purposes of this chapter, we will use the term *this LSR* to denote the LSR on which the MIB is being run (or queried).

Table 8.1 MPLS-TE MIB terminology.

Term	Definition
Tunnel head or tunnel head interface	This refers to the ingress point of a traffic engineering tunnel. In some implementations, this refers to the subinterface and corresponding IF-MIB entry that are created to represent the tunnel. An entry in the Tunnel Table is created for every tunnel head.
Tunnel instance	A tunnel instance refers to one or more label switched paths that are currently signaled for the tunnel head. An LSP represents an actively signaled path for the tunnel that is used to carry its traffic.
Tunnel LSP	A tunnel LSP refers to the label switched path that is created by the TE signaling protocol (RSVP-TE/IS-IS) or statically configured by the operator. This path then carries

continued

Table 8.1 continued

Term	Definition
	the traffic for the tunnel across the network between the ingress to egress points.
Tunnel hop	A tunnel hop refers to a node along the tunnel's path. A node currently is identified by an LSR ID (IPv4 address).
Tunnel path option	Each tunnel can have one or more groups of hops specified. Each group of hops indicates a different path from the source to the destination, and thus represents an optional path for the tunnel to take. Each group of hops is referred to as one of a tunnel's path options.

8.5 RowPointer Usage in the MPLS-TE MIB

RowPointer is a textual convention used to identify a conceptual row in an SNMP table by pointing to one of its objects. The MPLS Tunnel Table contains two RowPointer objects, called the mplsTunnelXCPointer and mplsTunnelResource-Pointer. The first object, mplsTunnelXCPointer, is used to associate the tunnel entry with the signaled LSP for that entry. Since the signaled LSP is represented as an XC (and associated in-/outsegments) in the MPLS-LSR MIB (Srinivasan et al. 2000b), this RowPointer is set to the corresponding row plus the first accessible column in that row of the mplsXCTable. The second RowPointer object, mpls-TunnelResourcePointer, points at a specific entry in the MPLS Tunnel Resource-Table. There are two reasons why a full index was not used. First, it allows a tunnel entry to refer to the same resource entry as one or more tunnel entries to indicate resource sharing. Second, this approach was taken to allow implementations that wished to expand upon the MPLS-TE MIB's resource table by pointing at that table. As with the previous RowPointer, this one should also point at a valid row in the selected resource table and should refer to the first accessible column in that table. Implementations should take care when assigning the OID value to either of the above RowPointers because an NMS assumes that the OID that is read will contain a valid value and will follow it to the appropriate row. If the row is either invalid or does not exist, the NMS could easily run into problems.

8.6 Scalars

The MPLS-TE MIB contains several scalar objects (i.e., globally accessible, single instances) that can be of interest. These include the number of tunnels that are configured, the number of active tunnels, the maximum number of hops that can be specified for a tunnel's explicit hop table, and the traffic engineering distribution protocol or protocols that are currently available for traffic engineering. The scalars are enumerated in Table 8.2.

Table 8.2 MPLS-TE MIB scalar objects.

Object	Description
mplsTunnelConfigured	This variable indicates the number of traffic engineering tunnels that exist in the Tunnel Table. The definition states that this includes the numbers that have been configured on the device where this variable is queried. The wording of this variable is a little misleading, as it goes on to indicate that a tunnel is considered configured if the mplsTunnel-RowStatus is **active (1).** This subtly means that this variable also counts entries that a device may have made in the Tunnel Table to represent active LSPs, since their RowStatus would also be **active (1).**
mplsTunnelActive	This variable indicates the actual number of tunnels that are up and ready to carry traffic. A tunnel entry is considered to be in this state if the value of mplsTunnelOperStatus for the corresponding row is set to **up (1).**
mplsTunnelTEDistProto	This variable indicates the global setting for which traffic engineering protocol(s) are enabled for use. This variable is specified as a BITS object because more than one distribution protocol may be enabled simultaneously. Valid values for this variable include bit position **other (0)** for a proprietary TE protocol, **ospf (1)** for the IETF OSPF protocol, and **isis (2)** for the OSI IS-IS protocol.
mplsTunnelMaxHops	This variable indicates the maximum number of tunnel hops that can be specified for a tunnel. Take careful note that this variable can be different depending on the implementation and software version of that implementation. Furthermore, this is not an indication of the maximum number of hops supported through the network. That is, some nodes along a path may not support the number of hops included by another node.

8.7 The Tunnel Table

The Tunnel Table lies at the heart of the MPLS-TE MIB. Entries in the other tables are generally created to correspond to tabular rows first created in this table. The Tunnel Table allows new MPLS tunnels to be created between an MPLS LSR where the MIB is running and a remote LSR. In fact, the structure of the MIB does not preclude a proxy-like tunnel management device or application from creating tunnels on behalf of a user and then creating the tunnel entries on the appropriate LSRs.

8.7.1 Tunnel Table Indexing

The MPLS-TE MIB's Tunnel Table is indexed by four objects. Due to the numerous questions we have been asked regarding the nature of these indexes, we will now spend a few paragraphs explaining why they are as they are and what each index corresponds to in terms of the protocol.

The Tunnel Table is indexed by a 4-tuple that includes mplsTunnelIndex, mplsTunnelInstance, mplsTunnelIngressLSRId, and mplsTunnelEgressLSRId. The indexing for this table was chosen this way to facilitate easy referral back to both the RSVP-TE and CR-LDP signaling system software that is used to signal the LSPs for the configured tunnels. Most implementations that were used as references for the MIB contained a table of configured tunnels (i.e., tunnel head interfaces) that is stored separately from the signaled LSPs. Each of these entries then referred into the TE signaling software if one or more LSPs were signaled for that entry. Head entries could quickly refer to tunnel instances because the indexes represented the parameters that were passed to the signaling software to signal the LSP. For example, the first two indexes, Tunnel Index and Tunnel Instance pairs, are meant to correspond to the destination and source ports used in RSVP. In the case of CR-LDP, the Tunnel Instance is always set to zero and the Tunnel Index always corresponds to a unique index that refers to the tunnel, as there is no such concept of a port in that protocol.

In most cases, the signaling table also contained transit LSPs, that is, those LSPs that belong to tunnels that originated at LSRs other than the one in question, and either traversed that LSR or terminated there. Since most LSRs stored data structures for this information, it was thus thought that it would be useful for the Tunnel Table to represent these entries as well. Unfortunately, some vendors were unable to support this, so support for tunnel midpoints and tunnel tails is optional.

Although not mandated in the standard, we recommend that the following convention be used by devices supporting the MPLS-TE MIB when representing tunnel entries in the Tunnel Table. This convention defines two simple cases. First, the tunnel head interface is represented as an entry in the Tunnel Table with tunnel instance set to zero. Second, all other entries indexed by this same primary tunnel index should use a tunnel instances value that is greater than zero. Use of this convention facilitates easy lookup and indexing of tunnel heads and their corresponding instances by network management systems. This is especially important at transit nodes that may have many thousands of heads and instances, as it allows an NMS or operator to quickly discern tunnel heads from transit LSPs by filtering out either the heads or the LSPs. This convention also makes other indexing and lookup easier when you examine the other tables in the MIB.

When using RSVP-TE, the third index, mplsTunnelIngressLSRId, is interpreted as the Extended Tunnel ID field from the RSVP-TE SESSION object. When the MPLS signaling protocol is set to **crldp (3),** this value is gathered from the Ingress LSR Router ID field in the LSPID TLV object. The last index is defined as containing the egress LSR ID, regardless of signaling protocol. A summary of the interpretation of each of the indexes for each signaling protocol is shown in Tables 8.3 and 8.4.

Table 8.3 MPLS Tunnel Table indexing when RSVP-TE signaling is used.

Object	Description
mplsTunnelIndex	RSVP destination port.
MplsTunnelInstance	RSVP source port.
MplsTunnelIngressLSRId	Extended Tunnel ID field in the RSVP-TE SESSION object.
MplsTunnelEgressLSRId	Represents the egress LSR ID of where the tunnel terminates.

Table 8.4 MPLS Tunnel Table indexing when CR-LDP signaling is used.

Object	Description
mplsTunnelIndex	RSVP source port.
MplsTunnelInstance	RSVP destination port.
MplsTunnelIngressLSRId	Extended Tunnel ID field in the RSVP-TE SESSION object.
MplsTunnelEgressLSRId	Represents the egress LSR ID of where the tunnel terminates.

One interesting question was raised recently with regard to those implementations that might implement both CR-LDP and RSVP-TE signaled tunnels: How could they coexist within the same tunnel table? The answer was simple: if an implementation wishes to support such configurations, then the implementation must guarantee that the instances be indexed in a nonoverlapping manner. The same applies to those implementations that allow for hand-routed tunnels. The standard MIB implies this subtle point, but never explicitly mentions it. However, it is important that implementations do this or disaster could ensue.

8.7.2 Tunnel Table Contents

This table supports the reconfiguration, modification, interrogation, or deletion of existing tunnel entries. This table is a repository for tunnel head interfaces existing on the LSR where the table is queried. The MIB does not care which management interface was used to create the tunnels (or modify existing ones); it will just show what is loaded into the device at the time the MIB is queried. The Tunnel Table also contains preconfigured backup tunnels that are used to provide redundancy for the primary tunnels. Tunnel head interfaces may or may not be represented as IF-MIB ifIndexes, but the table functions largely the same for either flavor of implementation: tunnel entries will exist in the table; the only difference is that the InterfaceIndexOrZero value that normally contains the ifIndex will be set to zero to indicate that no association exists.

The Tunnel Table may optionally contain entries for tunnel LSPs that are signaled on behalf of a configured tunnel interface. These LSP entries will always exist on the head end LSR, but may also exist optionally at midpoints along the tunnel's path if an implementation supports this. Many implementations in fact do create entries for tunnel midpoints in this table since the RSVP and CR-LDP "soft" state information for each tunnel is stored at all nodes along the tunnel's path. Some implementations also support the concept of an egress tunnel interface, which the MIB is also capable of supporting.

Table 8.5 enumerates and defines each of the objects found in the mplsTunnelTable.

Table 8.5 MPLS Tunnel Table objects.

Object	Description
mplsTunnelIndex	Primary index into the tunnel table. This index frequently corresponds with the same name for the tunnel found on a device's CLI.

Table 8.5 continued

Object	Description
mplsTunnelInstance	This value further differentiates a tunnel head from its LSPs. We suggest that tunnel heads be indexed using instance 0 and the LSPs for that tunnel use instances greater than 0.
mplsTunnelIngressLSRId	When using RSVP-TE, the mplsTunnelIngressLSRId is interpreted as the Extended Tunnel ID field from the RSVP-TE SESSION object. When the MPLS signaling protocol is **crldp(3),** this value is gathered from the Ingress LSR Router ID field in the LSPID TLV object. The last index is defined as containing the egress LSR ID, regardless of signaling protocol.
mplsTunnelEgressLSRId	Represents the egress LSR ID of where the tunnel terminates. This same value is used for both signaling protocols.
mplsTunnelName	This variable represents the human-readable name assigned to the tunnel. This name should be consistent with the one used on the LSR's CLI. One important caveat for implementations with this variable is that if mplsTunnelIsIf is set to true (i.e., the tunnel is represented by a corresponding IF-MIB ifEntry), then the ifName of the interface corresponding to this one should be set to the same display string as this variable.
mplsTunnelDescr	This variable contains a human-readable string that should contain descriptive information about the tunnel that can also be found through other management interfaces (e.g., CLI). It is possible that no description is set for the tunnel, in which case this object will be set to a zero-length string. However, we strongly discourage this option, as this information can be quite useful for an NMS or operator.
mplsTunnelIsIf	This boolean object indicates whether an IF-MIB ifEntry has been created to correspond to the tunnel entry in this row. It is important to point out that if this value is set to true, the standard requires that the mplsTunnelName be set to the same string as the ifEntry is set to.
mplsTunnelIfIndex	This value indicates the ifEntry's ifIndex that corresponds to this tunnel interface represented by this row. This variable will be set to zero if the interface does not represent an ifEntry. This can

continued

Table 8.5 continued

Object	Description
	occur either if tunnel interfaces do not correspond with ifEntries and/or if the Tunnel Table entry represents a signaled instance for that tunnel.
mplsTunnelXCPointer	This object is set to the OID for the row representing the XC for the LSP. This object is set to 0.0 if no XC entry corresponds to the tunnel entry or this is unsupported by the implementation. Typically, tunnel instances will not set the XC pointer since they are used to represent the signaled session for the tunnel that does have a corresponding XC entry. However, if tunnel instances are represented by an ifIndex, then they may have this value set.
mplsTunnelSignallingProto	This value indicates the protocol(s) supported to signal traffic-engineered tunnels on this LSR. Valid values include **none (1), rsvp (2), crldp (3),** and **other(4).** The value of **none (1)** can be used to indicate that only static tunnel routing is supported. The last option, **other (4),** is used to indicate that a proprietary signaling protocol is in use.
mplsTunnelSetupPrio	This value is used to specify the tunnel setup priority as specified in the RSVP-TE (RFC3209) and CR-LDP (Jamoussi 2001) specifications. Valid values are 0 to 7 (inclusive).
mplsTunnelHoldingPrio	This value is used to specify the tunnel holding priority as specified in the RSVP-TE (RSVPTE) and CR-LDP (Jamoussi 2001) specifications. Valid values are 0 to 7 (inclusive).
mplsTunnelSessionAttributes	This object contains a bitmask that allows an operator to indicate optional session attributes for this tunnel. Since this field is a bitmask, keep in mind that more than one option can be specified simultaneously. We will refer to bits in the bit field by their positions from left to right, if the rightmost bit is the most significant. Actual implementation order may vary, but must represent the bits in this order to the external SNMP management interface. The first bit indicates that Fast Reroute is allowed at any tunnel hop along the path of the tunnel. More precisely, any LSR along this path may choose to reroute this tunnel without first tearing it down. Setting this bit also permits LSRs to use local repair mechanisms that may result in violation of the explicit routing of this tunnel, but will protect the tunnel in the case of a node or

Table 8.5 continued

Object	Description
	link failure. In summary, this bit indicates that any LSR along the path of this tunnel can reroute traffic for fast service restoration in the event that a failure is detected. Note that the specific mechanisms for how the restoration is achieved are not specified. The second bit indicates that merging of tunnel LSPs is permitted. Specifically, when RSVP-TE is used to signal the tunnel, LSRs along the tunnel path are allowed to merge this session with other RSVP sessions for the purpose of reducing resource overhead on downstream transit routers, thereby providing better network scalability. This assumes that RSVP-TE is used for tunnel signaling. The third bit in the bit field indicates if this tunnel should be restored automatically after a failure occurs—specifically, if the tunnel should be restored after a node failure. The fourth bit indicates if the loosely routed hops of this tunnel are to be strictly enforced or if they can be replaced in order to find a route for the tunnel. The last bit field indicates whether the signaling protocol should record and advertise the tunnel path after it has been signaled. This is particularly useful when dynamic or loosely routed paths are specified, as it will give the operator the exact path the tunnel is taking. Note that this assumes that the record route function is available in the LSRs along the tunnel path.
mplsTunnelOwner	This object is used to indicate which protocol or management interface created the entry and is responsible for it. Acceptable values are **admin (1),** which is used to indicate that all management entities should manage this entry; **rsvp (2),** for the RSVP signaling protocol; **crldp (3),** for the CR-LDP signaling protocol; **policyAgent (4),** indicating that some policy-based agent created the tunnel (e.g., COPS). A catchall value of **other (5)** is also available to indicate that some other agent created the entry. The principal idea behind this object is that it can be used to indicate to system software or an NMS who "owns" an entry in the table. For example, if a policy agent creates an entry in this table, other management interfaces or signaling protocols should

continued

Table 8.5 continued

Object	Description
	not delete the interface without first checking with the policy agent.
mplsTunnelLocalProtectInUse	This object is used to indicate whether or not a local repair mechanism is in use on this entry in order to protect the tunnel's path. Some implementations may only implement this indication on tunnel instances, as they are truly bound to the signaling protocol, whereas the tunnel head interface represents the configuration and thus may not have insight into the signaling.
mplsTunnelResourcePointer	This variable is used to indicate which Tunnel Resource Table entry is associated with this tunnel. Some implementations may choose to implement a proprietary tunnel resource table, in which case this object is pointed at an entry in that table. An agent or manager will set this value to **zeroDotZero (0.0)** to indicate best-effort treatment for the tunnel. It is also possible to indicate resource sharing between one or more tunnels by pointing them at the same Tunnel Resource Table entry.
mplsTunnelInstancePriority	This object is used to indicate which priority a tunnel instance is considered to have. The tunnel instance priority is specified in descending order, with 0 indicating the lowest priority. The instance priority applies to tunnel instances within a group of tunnels indexed by the same primary tunnelIndex, but with a different mplsTunnelInstance. We recommend that the tunnel head interface be specified with a tunnelInstance of 0 to indicate that it is the head interface. All other instances of the tunnel should be numbered starting at 1. Tunnel group priorities are used to denote the priority at which a particular tunnel instance will supersede another in the event of a reroute or reoptimization. Tunnel instances can be specified with the same instance priority to indicate load sharing, but it is up to the LSR to understand that it must load-share traffic across the tunnels. A more common method of load sharing is to create multiple tunnel head entries and let the routing protocol load-share into each interface based on destination. This can be indicated using the MPLS-FTN MIB (see Chapter 5) to route traffic into these tunnel interfaces. This mechanism also provides a cleaner mapping to the MPLS-LSR MIB's XC.

Table 8.5 continued

Object	Description
mplsTunnelHopTableIndex	This object contains an index into the mplsTunnelHopTable. The Hop Table is used to specify the explicit route hops and path options for this entry.
mplsTunnelARHopTableIndex	This object contains an index into the mplsTunnelARHopTable. The Actual Route Hop Table is used to specify the actual route hops for this entry.
mplsTunnelCHopTableIndex	This object contains an index into the mplsTunnelCHopTable. The Computed Hop Table is used to specify the route hops for this tunnel as computed by the CSPF algorithm.
mplsTunnelPrimaryInstance	This object specifies the tunnel instance index of the primary instance (i.e., head interface) of this tunnel. This allows tunnel instances to quickly "point" back to the head interface. Note that this object is unnecessary if implementations follow our guidance of indexing tunnel heads with instance index 0. Also, note that if the entry being queried is a tunnel head, this index must include its instance index.
mplsTunnelPrimaryTimeUp	This object indicates the total time that the primary instance of this tunnel has been active. In particular, if a tunnel head has many instances that have each been active individually over the course of time, then the total amount of time that the tunnel has been forwarding traffic is indicated. Thus, this value may not be very useful for tunnel entry instances that are not head interfaces, and in some implementations, it may not be possible to retrieve so it may be set to zero.
mplsTunnelPathChanges	This variable contains the number of times the actual path for this tunnel has changed.
mplsTunnelLastPathChange	This indicates the number of SNMP timeTicks that have elapsed since a path option was chosen for a tunnel. This value gives an indication of how long the current tunnel path has been in use.
mplsTunnelCreationTime	This value contains the sysUpTime when the tunnel instance was first signaled. The time when a tunnel head interface is configured is not reflected here, just when it was first signaled and capable of carrying traffic. This time should coincide closely with an mplsTunnelUp notification, if this notification is

continued

Table 8.5 continued

Object	Description
	supported and was enabled before the tunnel instance was signaled.
mplsTunnelStateTransitions	This contains a count of the number of times this tunnel has changed states. For example, if a tunnel changes from the up to down state, this is counted as one state transition.
mplsTunnelIncludeAnyAffinity	This value contains the configured tunnel "include any" affinity mask (RSVPTE) for this tunnel. TE links are configured with a similar mask, so that when the CSPF algorithm is calculating acceptable paths for this tunnel, one of the constraints any given segment in the path must meet is a successful matching of these masks.
mplsTunnelIncludeAllAffinity	This value contains the configured tunnel "include all" affinity mask (RSVPTE) for this tunnel. TE links are configured with a similar mask, so that when the CSPF algorithm is calculating acceptable paths for this tunnel, one of the constraints any given segment in the path must meet is a successful matching of these masks.
mplsTunnelExcludeAllAffinity	This value contains the configured tunnel "exclude all" affinity mask (RSVPTE) for this tunnel. TE links are configured with a similar mask, so that when the CSPF algorithm is calculating acceptable paths for this tunnel, one of the constraints any given segment in the path must meet is a successful matching of these masks.
mplsTunnelPathInUse	This variable is used to indicate which configured path option is in use for this tunnel. This variable, when nonzero, contains the secondary index into the mplsTunnelHopTable. The primary index into the table is the tunnel index. It should be noted that when examining this path, do not be surprised if it is different from the one found in the corresponding entry in the mplsTunnel-ARHopTable. Remember that the Hop Table contains *configured* path options, and the mplsTunnelARHopTable contains the *actual* signaled path as returned by the signaling protocol. This discrepancy can result because some CSPF modification to the configured path may have taken place if allowed. For example, if some loosely routed hops were specified and they were not

Table 8.5 continued

Object	Description
	"pinned," the CSPF could have replaced them with other, more optimal ones. Also note that this object is set to zero to indicate that no path is currently in use or available for this tunnel entry. This could happen, for example, if a tunnel has not been fully configured before an NMS queries it.
mplsTunnelRole	This value is used to indicate the segment of the tunnel that is being queried. Valid values are a head interface (1), transit or intermediate hop along the path of the tunnel (2), or the tunnel's tail (3). Note that some implementations may indicate a tunnel's tail at the penultimate hop of the tunnel if this is enabled.
mplsTunnelTotalUpTime	This indicates the total aggregate time that this tunnel head has been able to carry traffic. That is, if a tunnel has four path options, and has switched many times between each, but has never gone into the down state (i.e., it was able to continue carrying traffic), then this value will indicate the overall time that each LSP has been up and running. Note that this value is of no use on tunnel entries that represent tunnel instances other than the head, as these entries may only be able to return the total time that specific instance was active.
mplsTunnelInstanceUpTime	If the entry being queried is a nonhead tunnel instance, then this object is used to indicate to the operator how long this instance has been carrying traffic. Note that this value is not very useful for tunnel head entries, and thus may be returned as zero.
mplsTunnelAdminStatus	This object contains the SNMP administrative status of this tunnel entry. Only three states are valid: **up (1), down (2),** and **testing (3).** The last is used to indicate that the tunnel is up, but is undergoing some local testing function. If the value is set to **down (2),** this indicates that the operator desires that the tunnel discontinue forwarding traffic.
mplsTunnelOperStatus	This object indicates the actual operational status of this entry. A value of **up (1)** indicates that the tunnel is ready to send and receive data. A value of **down (2)** is the opposite of **up (1).** A value of **testing (3)** indicates that the tunnel is currently in some local test mode and cannot pass data. A value of **unknown (4)**

continued

Table 8.5 continued

Object	Description
	denotes that the current state cannot be determined. A value of **dormant (5)** indicates that some component is missing and is preventing the tunnel from operating correctly. For example, if the signaling protocol that is specified for this entry is not enabled on the system, the tunnel will not be able to signal its corresponding LSP(s). A value of **notPresent (6)** indicates that the tunnel has been taken out of service. A value of **lower-LayerDown (7)** specifies that the tunnel is down because the state of the corresponding MPLS layer is down.
mplsTunnelRowStatus	This object contains the standard SNMP RowStatus that is used for creating, modifying, and deleting rows in this table.
mplsFTNMapStorageType	The SNMP storage type for this entry. Valid values are **other (1),** meaning that a storage type other than the ones defined below is available; **volatile (2),** meaning that the row will be stored in RAM and will disappear after the device reboots; **Nonvolatile (3),** meaning that the value is stored in some sort of nonvolatile RAM and will be preserved across reboots of the system; **permanent (4),** meaning that the row is stored partially in ROM; or **readOnly (5),** meaning that the row is stored completely in ROM. Operators will typically find this value to be set to either **Nonvolatile (3)** or **volatile (2).**

8.8 MPLS Tunnel Resource Table

The MPLS Tunnel Resource Table (mplsTunnelResourceTable) is used to indicate the resources required for a tunnel. Multiple tunnels may share the same resources by pointing to the same entry in this table. Tunnels that do not share resources must point to separate entries in this table. The objects found in the mplsTunnel-ResourceTable are enumerated and defined in Table 8.6.

Table 8.6 MPLS Tunnel Resource Table objects.

Object	Description
mplsTunnelResourceIndex	This object contains an index that uniquely identifies a resource table entry. It is important to note that the index does not necessarily correspond to the tunnel index or any other index in the MPLS-TE MIB as this table may have entries that are used by more than one tunnel entry. However, for those implementations that wish to have a one-to-one correspondence between tunnel entries and tunnel resource entries, it is certainly acceptable to reuse the tunnel entry index here to simplify the implementation. However, NMS systems should not make this assumption.
mplsTunnelResourceMaxRate	This object is set to contain the desired maximum rate in bits/second for the associated tunnel entry. Operators and implementers of management systems should note that setting this object to zero indicates best-effort treatment for the associated tunnel entry. If the tunnel is signaled successfully, the value contained in this object is copied into an instance of mplsTrafficParamMaxRate in the MPLS-LSR MIB for the corresponding mplsInSegmentTraffic-ParamPtr.
mplsTunnelResourceMeanRate	This object is set to contain the desired mean rate in bits/second for the associated tunnel entry. Operators and implementers of management systems should note that setting this object to zero indicates best-effort treatment for the associated tunnel entry. If the tunnel is signaled successfully, the value contained in this object is copied into an instance of mplsTrafficParamMeanRate in the MPLS-LSR MIB for the corresponding mplsInSegmentTraffic-ParamPtr.
mplsTunnelResourceMaxBurstSize	This object is set to contain the desired maxium burst rate in bits/second for the associated tunnel entry. Operators and implementers of management systems should note that setting this object to zero indicates best-effort treatment for the associated tunnel entry. If the tunnel is signaled successfully, the value contained in this object is copied into

continued

Table 8.6 continued

Object	Description
	an instance of mplsTrafficParamMaxBurstRate in the MPLS-LSR MIB for the corresponding mplsInSegment-TrafficParamPtr.
mplsTunnelResourceMeanBurstSize	This object is set to contain the desired maximum burst rate in bytes for the associated tunnel entry. Operators and implementers of management systems should note that setting this object to zero indicates best-effort treatment for the associated tunnel entry. NMS systems should be aware that implementations that do not support this variable will indicate this by setting it to zero. If this value is read as zero, an NMS should not allow a user to set this value to zero.
mplsTunnelResourceExcessBurstSize	This object is set to contain the desired excess burst size in bytes for the associated tunnel entry. Operators and implementers of management systems should note that setting this object to zero indicates best-effort treatment for the associated tunnel entry. NMS systems should be aware that implementations that do not support this variable will indicate this by setting it to zero. If this value is read as zero, an NMS should not allow a user to set this value to zero.
mplsTunnelResourceFrequency	This variable is set to indicate the granularity of the availability of committed rate. The standard directs implementations that do not support this variable to return **unspecified (1)** when this object is queried. Implementations that do not support this value should not allow a user or NMS to set this value (i.e., they should return an error if any SETs are received). Values for this object are **unspecified (1), frequent (2),** and **veryFrequent (3).**
mplsTunnelResourceWeight	This object contains the desired relative weight for using excess bandwidth above its committed rate. Implementations that do not support this value should not allow a user or NMS to set this value (i.e., they should return an error if any SETs are received). Implementations

Table 8.6 continued

Object	Description
	that do not support this variable are required to return a value of 0 to indicate this fact.
mplsTunnelResourceRowStatus	This object contains the standard SNMP RowStatus that is used for creating, modifying, and deleting rows in this table.
mplsTunnelResourceStorageType	The SNMP storage type for this entry. Valid values are **other (1),** meaning that a storage type other than the ones defined below is available; **volatile (2),** meaning that the row will be stored in RAM and will disappear after the device reboots; **Nonvolatile (3),** meaning that the value is stored in some sort of nonvolatile RAM and will be preserved across reboots of the system; **permanent (4),** meaning that the row is stored partially in ROM; or **readOnly (5),** meaning that the row is stored completely in ROM. Operators will typically find this value to be set to either **Nonvolatile (3)** or **volatile (2).**

8.9 The CR-LDP Resource Table

The MPLS Tunnel CR-LDP Resource Table (mplsTunnelCRLDPResTable) contains resource information that is specific for tunnels that are signaled using CR-LDP (Jamoussi 2001). The variables contained in this table are meant to augment those in the base Tunnel Resource Table. Due to the possibility of a sparse relationship between the entries in this table and the base Tunnel Resource table, this table must *extend* rather than AUGMENT the base table. Therefore, this table is indexed by the mplsTunnelResourceIndex. As with the mplsTunnelResourceTable, multiple tunnels may share the same resources by pointing to the same entry in this table. Tunnels that do not share resources must point to separate entries in this table.

It is important to note that the version of the MPLS-TE MIB that is used for this chapter contains four variables that are redundant with the base Tunnel Resource Table: mplsTunnelCRLDPResMeanBurstSize, mplsTunnelCRLDPResExcess-BurstSize, mplsTunnelCRLDPResFrequency, and mplsTunnelCRLDPResWeight. These variables were mistakenly repeated from the base table and were subsequently removed from future versions of the MIB.

The objects that can be found in the mplsTunnelCRLDPResTable are enumerated and defined in Table 8.7.

Table 8.7 MPLS Tunnel CR-LDP Resource Table objects.

Object	Description
mplsTunnelCRLDPResFlags	This variable contains the desired CR-LDP traffic parameter flags. This byte is included as part of the traffic parameters included in the CR-LDP tunnel establishment message when setting up a tunnel. Each bit in the flag bitmask is set or not set to indicate that it is either a negotiable or a nonnegotiable traffic parameter. If a bit is set, this indicates that the corresponding traffic parameter is negotiable. Setting a value to zero indicates that it is not negotiable. Note that the low-order 8 bits of the 32-bit word are valid in this object, and the remaining 25 (6–31) bits are reserved. An agent must return the reserved bits as zero. An agent cannot allow these bits to be set and should ignore them if they are set. An agent is required to return these as zero even if an NMS attempts to set them. The bits in this object are interpreted as follows: Bits 6–31 5 4 3 2 1 0 Reserved F6 F5 F4 F3 F2 F1 F1 corresponds to the PDR. F2 corresponds to the PBS. F3 corresponds to the CDR. F4 corresponds to the CBS. F5 corresponds to the EBS. F6 corresponds to the weight.
mplsTunnelCRLDPResRowStatus	This object contains the standard SNMP RowStatus that is used for creating, modifying, and deleting rows in this table.
mplsTunnelCRLDPResStorageType	The SNMP storage type for this entry. Valid values are **other (1)**, meaning that a storage type other than the ones defined below is available; **volatile (2)**, meaning that the row will be stored in RAM and will disappear after the device reboots; **Nonvolatile (3)**, meaning that the value is stored in some sort of nonvolatile RAM and will be preserved across

Table 8.7 continued

Object	Description
	reboots of the system; **permanent (4),** meaning that the row is stored partially in ROM; or **readOnly (5),** meaning that the row is stored completely in ROM. Operators will typically find this value to be set to either **Nonvolatile (3)** or **volatile (2).**

8.10 MPLS Tunnel Hop Table

The MPLS Tunnel Hop Table (mplsTunnelHopTable) is used to indicate the hops, strict or loose, that are associated with one or more tunnel entries in the Tunnel Table. The entries found in this table specify those network nodes that must be either included or excluded from a tunnel's path. To realize this, a list of hops to include and exclude is passed to the constraint-based routing software as part of the constraints associated with that tunnel. The result of this computation will then be signaled by the signaling software.

Conceptually, the mplsTunnelHopTable represents one or more lists of tunnel hops that are associated together with the primary index of the table. This allows an operator to define what are sometimes referred to as *tunnel path options*—a list of paths that can be signaled in succession until one is realized. This list of paths can also be consulted in the future if a tunnel needs to be rerouted due to a link or node failure.

The Tunnel Hop Table is indexed by three indexes: mplsTunnelHopListIndex, mplsTunnelHopPathOptionIndex, and mplsTunnelHopIndex. The primary tunnel index refers to a group of hops that can then be associated with one or more Tunnel Table entries. The secondary index refers to a subgroup of hops that are then chained together using the tertiary index as a means of specifying an order to the hops. Each hop defined with the same primary and secondary indexes is thus associated with the same path option.

Hops can be marked as being included or avoided. Each hop in this list then instructs the SPF on how to treat the constraint when encountered. A note about excluded hops: When hops are marked as being excluded, the order where they appear in a list of loosely routed hops associated with a path option is generally considered not important. However, when a strict explicit path is specified, excluded hops must not be specified in the list. This is because this list is not passed to the

SPF for calculation; instead, the list of hops now represents an explicit path that forms the actual path that is used to route the tunnel signaling messages. Inclusion of hops to be avoided does not make sense and, in fact, will probably cause the tunnel not to be signaled correctly.

Multiple tunnels may share the same hops by pointing to the same entry in this table, but it is generally not advised that they do so, since it makes the implementation more difficult. The reason is that most implementations that have been consulted create a configuration data structure for each tunnel head. Each structure has embedded within it a list of path options associated with the particular tunnel head. This approach does require more memory to store path options that may be redundant, but it does seem easier on the bookkeeping algorithms to take this approach. In either case, we recommend that if you are implementing this MIB, it is best to keep the MIB implementation close to the actual implementation, since inconsistencies between the two, however harmless, seem to often lead to bugs.

The Tunnel Hop Table is composed of the objects in Table 8.8.

Table 8.8 MPLS Tunnel Hop Table objects.

Object	Description
mplsTunnelHopListIndex	This object represents the primary index into the Tunnel Hop Table. This is used to denote a group of path options that are typically configured together for a tunnel entry. More than one tunnel entry can be set to point at a group of path options if desired.
mplsTunnelHopPathOptionIndex	This object is used as a secondary index into this table. It is intended to identify a specific group of hops, which when viewed together, represent a single path option for a tunnel entry.
mplsTunnelHopIndex	This object represents the tertiary index into this table. This index represents the actual hop object. Note that the relative positioning of hops is denoted using this index. It is recommended that hops that are to be excluded/avoided be listed after those that should be included.
mplsTunnelHopAddrType	This object is used to specify the type of address used in this hop entry. Valid values are **ipV4 (1)** and **ipV6 (2)** for IP addresses, **asNumber (3)** is used when an autonomous system is used to identify a hop, and finally, **lspid (4)** is used when an LSP ID is specified. It is important to note that **lspid**

Table 8.8 continued

Object	Description
	(4) should only be used when the signaling protocol for the tunnel is specified as CR-LDP since RSVP-TE does not support this option.
mplsTunnelHopIpv4Addr	This object will contain the IPv4 address of the hop entry if the object mplsTunnelHopAddrType is set to **ipV4 (1).** However, if the object mplsTunnelHopAddrType is set to **lspid (4),** then this object will be set to the Ingress LSR Router ID of the tunnel. In all other cases, implementations must set this value to 0, and operators should expect this value to be set to 0.
mplsTunnelHopIpv4PrefixLen	This object will contain the IPv4 address prefix length of the hop entry if the object mplsTunnelHopAddrType is set to **ipV4 (1).** In all other cases, implementations must set this value to 0, and operators should expect this value to be set to 0.
mplsTunnelHopIpv6Addr	This object will contain the IPv6 address of the hop entry if the object mplsTunnelHopAddrType is set to **ipV6 (2).** However, if the object mplsTunnelHopAddrType is set to **lspid(4),** then this object will be set to the Ingress LSR Router ID of the tunnel. In all other cases, implementations must set this value to 0, and operators should expect this value to be set to 0.
mplsTunnelHopIpv6PrefixLen	This object will contain the IPv6 address prefix length of the hop entry if the object mplsTunnelHopAddrType is set to **ipV6(2).** In all other cases, implementations must set this value to 0, and operators should expect this value to be set to 0.
mplsTunnelHopAsNumber	This object will contain the autonomous system number if the object mplsTunnelHopAddrType is set to **asNumber (3).** In all other cases, implementations must set this value to 0, and operators should expect this value to be set to 0.
mplsTunnelHopLspId	This object is set to the LSP identifier of the tunnel hop if the object mplsTunnelHopAddrType is set to **lspid (4).** In all

continued

Table 8.8 continued

Object	Description
	other cases, implementations must set this value to 0, and operators should expect this value to be set to 0.
mplsTunnelHopType	This object indicates the desired routing style of this tunnel hop. Valid values denote strict, **strict (1),** or loosely routed, **loose (2).**
mplsTunnelHopIncludeExclude	This object indicates the operator's desire to either have this hop included **(include (1))** or excluded **(exclude (2))** during the CSPF calculation. Note that it does not make sense to mark hops as **exclude (2)** if mplsTunnelHopEntryPathComp is not set to **explicit (2)** because an explicit path by nature contains a source-routed path. If the hop is noted as being excluded **(exclude (2)),** then an operator can expect that the hop will be avoided when calculating the path for this tunnel. This can be verified by examining the corresponding hop list in the mplsTunnelCHopTable if the tunnel is in the process of being signaled, or the mplsTunnelARHopTable if the tunnel has been signaled. The default action of hop objects is to be included **(include (1));** thus care should be taken when specifying tunnel hops to make sure that they are actually set to the desired status.
mplsTunnelHopPathOptionName	This object contains the human-readable description of the series of hops. This value should be identical, at least in prefix, to the other corresponding hops to facilitate easy identification by an operator or NMS.
mplsTunnelHopEntryPathComp	This object is used to instruct the LSR as to whether or not to run the CSPF algorithm over this path option if a tunnel selects it to be signaled. If this value is set to **dynamic (1),** then only the source, destination, and optionally loosely routed hops to either include or exclude should be specified. If an explicit path is desired, then this object has to be set to **explicit (2)** and the CSPF will not be consulted on the path computation; instead, the LSR will attempt to signal the path directly. It is for this reason that care should be taken to ensure that the complete path that is specified is valid to the best of the knowledge of the NMS or the signaling will fail.

Table 8.8 continued

Object	Description
	The fact that each hop along a specific path can potentially have this object set to a different value may cause some confusion. Implementations should disallow mixed values for hop entries belonging to the same path option if one of the values in the path option is set to **explicit (2).** However, if a user has already specified one or more hops as **dynamic (1),** then an implementation has one of two options. First, only allow additional objects to have the same value for each hop in a path option set to the same type and cause SET operations for new hops that are different to fail. Second, an implementation may allow a user to commingle hop types, which should result in the signaling to fail. We recommend the first option as it appears to be more intuitive from an operational perspective.
mplsTunnelHopRowStatus	This object contains the standard SNMP RowStatus that is used for creating, modifying, and deleting rows in this table.
mplsTunnelHopStorageType	The SNMP storage type for this entry. Valid values are **other (1),** meaning that a storage type other than the ones defined below is available; **volatile (2),** meaning that the row will be stored in RAM and will disappear after the device reboots; **Nonvolatile (3),** meaning that the value is stored in some sort of nonvolatile RAM and will be preserved across reboots of the system; **permanent (4),** meaning that the row is stored partially in ROM; or **readOnly (5),** meaning that the row is stored completely in ROM. Operators will typically find this value to be set to either **Nonvolatile (3)** or **volatile (2).**

Figure 8.5 demonstrates a sample Tunnel Hop Table configuration. In the example, a collection of hops is represented by each box in the figure. Each hop is grouped together as a list of hops. Specifically, hops are grouped into four subsets of hops, and each group of hops is associated together by arrows. Each horizontal row of hops represents a path option that is defined for a single tunnel entry. The indexing of the conceptual table is represented by the parenthesized 3-tuple shown at the bottom of each box. This 3-tuple is defined as (mplsTunnelHopListIndex,

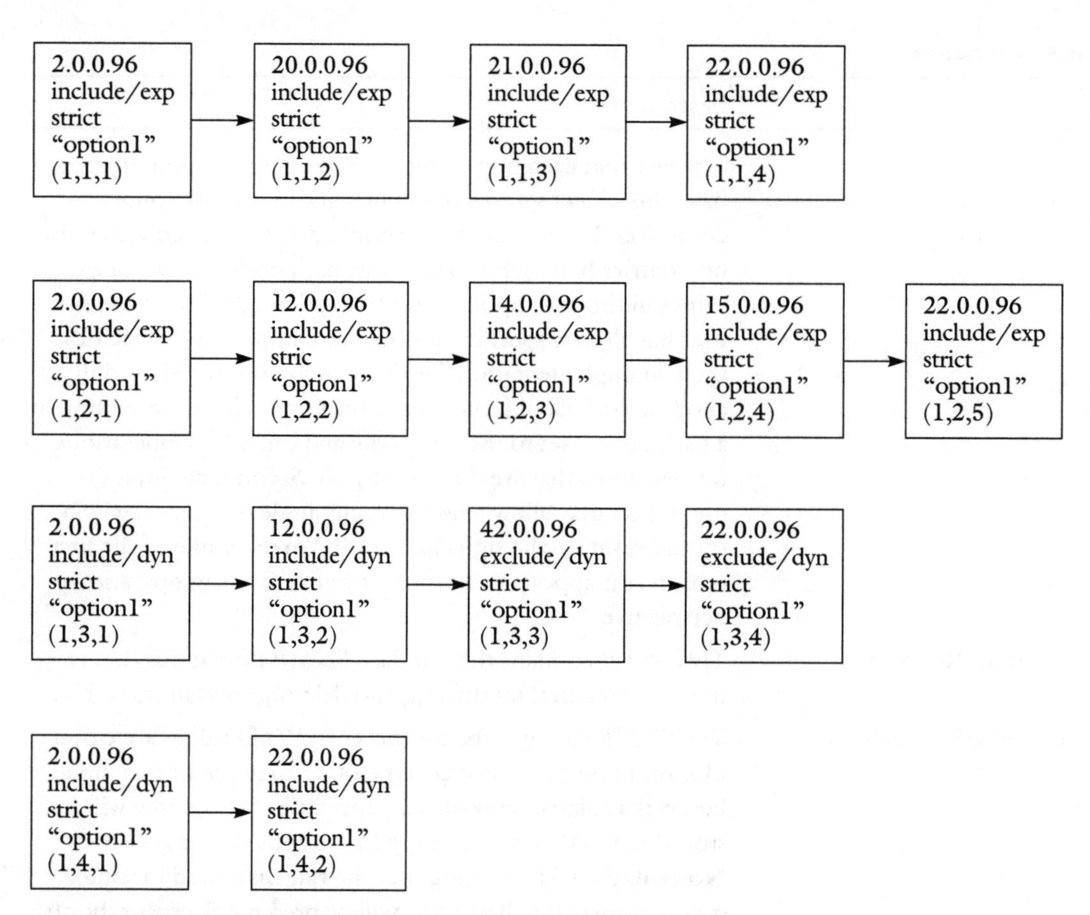

Figure 8.5 Example Tunnel Hop Table configuration containing four path options.

mplsTunnelHopPathOptionIndex, mplsTunnelHopIndex). In addition to the indexing, each hop contains a reasonable value for the following variables (in this order): mplsTunnelHopIPv4Addr, mplsTunnelHopIncludeExclude/mplsTunnelHopEntryPathComp, mplsTunnelHopType, and mplsTunnelHopPathOptionName. Note that the variables mplsTunnelHopIncludeExclude and mplsTunnelHopEntryPathComp are separated by a slash for illustrative purposes only, and rather than spelling out the entire types "dynamic" and "explicit," we instead use "dyn" and "exp" in the example. We have set mplsTunnelHopAddrType implicitly to **ipV4(1)** for each entry since each hop entry in this example contains an IPv4 address. Each entry is assumed to have reasonable values for the remaining columnar variables.

If we first examine the hop entry at the top left of the figure, we see that the hop first belongs to group 1 of tunnel hops. It also belongs to the first path option defined within this group and is the first hop within this path option. Thus the remaining hop entries with the same mplsTunnelHopListIndex, mplsTunnelHopPathOptionIndex values (i.e., 1) will belong to the same path option—in this case path option 1. Also, note that they are named accordingly, to allow for easy recognition by an operator. This path option represents an explicit set of hops, as is noted by the value of mplsTunnelHopEntryPathComp being set to "exp."

The second path option is defined in much the same way as the first path option; it is defined to be interpreted as an explicit path and used verbatim. The third path option is defined as what is referred to as a dynamic path option. This type of path option typically specifies the beginning and end of the path and lets the CSPF choose the hops in between these two points. In this case, we also show how a few nodes must be excluded from the path calculation by specifying their MplsTunnelHopIncludeExclude values as exclude. In the last case, we only specify the start and finish of the tunnel as another dynamic path option, but in this case, we place no other constraints on the path option. This is done as a means of specifying a last resort path option. This option is generally only attempted if none of the others has succeeded. Some implementations will cycle through the path options serially until one succeeds, while others may go back and start at the top if they are partway through the list and a failure occurs.

8.11 The Actual Route Hop Table

The Tunnel Actual Route Hop Table (mplsTunnelARHopTable) is used to indicate the actual hops traversed by a tunnel as reported by the MPLS signaling protocol after the tunnel is set up. In the case of RSVP-TE, entries in this table reflect those returned by the RSVP-TE Record Route Object (RRO). When CR-LDP is used as the signaling protocol, entries will not appear in this table to reflect the actual path, as CR-LDP does not support a record route function. In general, the functionality of this table is identical to that of the previously discussed MPLS Hop Table (see Section 8.3.4). The major difference between the Hop Table and this table is the indexing and the fact that mplsTunnelARHopTable entries are not specified as loose, strict, dynamic, or explicit. Simply put, this table contains a list of hops that have been successfully signaled for a specific tunnel.

This table utilizes two indexing objects—mplsTunnelARHopListIndex and mplsTunnelARHopIndex. Only two are required (as opposed to the three used for the Tunnel Hop Table) because tunnels can only signal one actual path through the network at a time. Therefore, only one group of hops can ever be represented. The remaining objects in the table entries behave the same way as has been defined in the Tunnel Hop Table.

Two critical differences between this table and the mplsTunnelHopTable are first that all of the objects in this table are specified with a MAX-ACCESS clause of read-only, and second, no row status or storage type objects exist. These differences exist for the simple reason that only the signaling software would ever create entries in this table on behalf of a tunnel entry. Therefore, no operator configuration of this table is allowed. Prior to signaling a tunnel's LSP, this table should contain no entries for the tunnel in question. After the first tunnel LSP has been signaled, it will contain the last known tunnel path (if available). It is strongly recommended that implementations remove the entry in this table after a tunnel begins to signal a new LSP for a tunnel. Not doing so could potentially confuse the operator into thinking that the current path option was successful with an old path.

It is important to note that support of this table is specified as optional in the standard. The reasoning is that, since not all implementations of RSVP-TE support the Record Route Object, these implementations may wish not to support this table as well. Furthermore, support for any one RRO for a tunnel is optional since the support for the RSVP-TE RRO is optional along the path of a tunnel. This can mean that although the RRO is *requested* by the operator for a particular tunnel, depending on which LSRs that path traverses, it may or may not be able to realize an RRO. Therefore, it should be expected that given a particular path option, a tunnel's LSP might not be able to return an RRO that can be displayed in this table. This should not be mistakenly thought to indicate that the tunnel was not able to have an LSP signaled. To verify this, the operator should always consult the operational status of the tunnel head interface. If the operational status is up, then an active tunnel instance must be present in the table regardless of whether an mplsTunnelARHopTable entry can be found.

The objects and their corresponding definitions that are found in the Tunnel Actual Route Hop Table are enumerated in Table 8.9.

Table 8.9 MPLS Tunnel Actual Route Hop Table objects.

Object	Description
mplsTunnelARHopListIndex	This object represents the primary index into the Actual Route Hop Tunnel table. This primary index denotes a group of path hop objects that together comprise the actual path taken by a tunnel's signaled LSP. Although more than one tunnel entry can be set to point at an actual path, we recommend strongly that only one tunnel point at any one group of hops in this

Table 8.9 continued

Object	Description
	table to avoid confusion on the part of both the implementation and the operator.
mplsTunnelARHopIndex	This object contains the secondary index into this table and represents an index that specifies the particular hop object.
mplsTunnelARHopAddrType	This object is used to specify the type of address used in this hop entry. Valid values are **ipV4 (1)** and **ipV6 (2)** for IP addresses, **asNumber (3)** is used when an autonomous system is used to identify a hop, and finally, **lspid (4)** is used when an LSP ID is specified. It is important to note that **lspid (4)** should only be used when the signaling protocol for the tunnel is specified as CR-LDP since RSVP-TE does not support this option.
mplsTunnelARHopIpv4Addr	This object will contain the IPv4 address of the hop entry if the object mplsTunnelHopAddrType is set to **ipV4 (1).** However, if the object mplsTunnelHopAddrType is set to **lspid(4),** then this object will be set to the Ingress LSR Router ID of the tunnel. In all other cases, implementations must set this value to 0, and operators should expect this value to be set to 0.
mplsTunnelARHopIpv4PrefixLen	This object will contain the IPv4 address prefix length of the hop entry if the object mplsTunnelHopAddrType is set to **ipV4(1).** In all other cases, implementations must set this value to 0, and operators should expect this value to be set to 0.
mplsTunnelARHopIpv6Addr	This object will contain the IPv6 address of the hop entry if the object mplsTunnelHopAddrType is set to **IpV6 (2).** However, if the object mplsTunnelHopAddrType is set to **lspid (4),** then this object will be set to the Ingress LSR Router ID of the tunnel. In all other cases, implementations must set this value to 0, and operators should expect this value to be set to 0.
mplsTunnelARHopIpv6PrefixLen	This object will contain the IPv6 address prefix length of the hop entry if the object mplsTunnelHopAddrType is set to **ipV6 (2).** In all other cases, implementations must set this

continued

Table 8.9 continued

Object	Description
	value to 0, and operators should expect this value to be set to 0.
mplsTunnelARHopAsNumber	This object will contain the autonomous system number if the object mplsTunnelHopAddrType is set to **asNumber (3).** In all other cases, implementations must set this value to 0, and operators should expect this value to be set to 0.
mplsTunnelARHopLspId	This object is set to the LSP identifier of the tunnel hop if the object mplsTunnelHopAddrType is set to **lspid (4).** In all other cases, implementations must set this value to 0, and operators should expect this value to be set to 0.

8.12 The Computed Hop Table

The Tunnel Computed Hop Table is used to indicate the hops computed by the CSPF algorithm for nonexplicit path options. The CSPF algorithm takes as input a path option that is specified in the Tunnel Hop Table (see Section 8.3.4). The actual algorithm that is used to choose the path option that is fed into the CSPF algorithm is implementation-dependent. The purpose of this table is to show the operator what paths are being calculated by the CSPF for signaling into the network when dynamic path options are specified. This is important for two reasons. It is sometimes important to double-check the output of the CSPF for debugging or sanity checking. For example, if an operator thinks that a path has been correctly specified in the Tunnel Hop Table, but that path cannot be signaled, a first step in debugging the situation might be to determine the exact path that is specified to the signaling protocol.

In general, the functionality of this table is identical to that of the previously discussed Tunnel Hop Table. The major difference between the Hop Table and this table is the indexing and the fact that computed hop entries only specify a single path that is to be signaled. This table utilizes two indexing objects—mplsTunnelCHopListIndex and mplsTunnelCHopIndex. Only two are required (as opposed to the three used for the Tunnel Hop Table) because tunnels can only signal one actual path through the network at a time. For these same reasons this table has all its objects specified with a MAX-ACCESS clause of read-only and does not possess a row status or storage type. Thus, system software can only create entries in this table. Therefore, no operator configuration of this table is allowed.

Similarly to the Actual Route Hop Table, entries should not exist in this table prior to signaling a tunnel's LSP. Entries in this table should correspond to the currently signaled or in-progress LSP. Implementations must make sure to preserve this consistency; otherwise, this could potentially confuse the operator into thinking that the current in-progress path option is incorrect.

This table is specified as optional in the standard. The reason for this is that not all implementations may be able to report the output of the CSPF algorithm.

The Tunnel Computed Route Hop Table is composed of the objects defined in Table 8.10. Aside from the exceptions noted earlier, the remaining objects in the table entries behave the same as has been defined in the mplsTunnelARHopTable.

Table 8.10 MPLS Tunnel Computed Route Hop Table objects.

Object	Description
MplsTunnelCHopListIndex	Primary index into this table identifying a particular computed hop list.
MplsTunnelCHopIndex	Secondary index into this table identifying the particular hop.
MplsTunnelCHopAddrType	This object is used to specify the type of address used in this hop entry. Valid values are **ipV4 (1)** and **ipV6 (2)** for IP addresses, **asNumber(3)** is used when an autonomous system is used to identify a hop, and finally, **lspid(4)** is used when an LSP ID is specified. It is important to note that **lspid (4)** should only be used when the signaling protocol for the tunnel is specified as CR-LDP since RSVP-TE does not support this option.
mplsTunnelCHopIpv4Addr	This object will contain the IPv4 address of the hop entry if the object mplsTunnelHopAddrType is set to **ipV4 (1).** However, if the object mplsTunnelHopAddrType is set to **lspid(4),** then this object will be set to the Ingress LSR Router ID of the tunnel. In all other cases, implementations must set this value to 0, and operators should expect this value to be set to 0.
mplsTunnelCHopIpv4PrefixLen	This object will contain the IPv4 address prefix length of the hop entry if the object mplsTunnelHopAddrType is set to **ipV4**

continued

Table 8.10 continued

Object	Description
	(1). In all other cases, implementations must set this value to 0, and operators should expect this value to be set to 0.
mplsTunnelCHopIpv6Addr	This object will contain the IP v6 address of the hop entry if the object mplsTunnelHopAddrType is set to **IpV6 (2).** However, if the object mplsTunnelHopAddrType is set to **lspid (4),** then this object will be set to the Ingress LSR Router ID of the tunnel. In all other cases, implementations must set this value to 0, and operators should expect this value to be set to 0.
mplsTunnelCHopIpv6PrefixLen	This object will contain the IPv6 address prefix length of the hop entry if the object mplsTunnelHopAddrType is set to **ipV6 (2).** In all other cases, implementations must set this value to 0, and operators should expect this value to be set to 0.
mplsTunnelCHopAsNumber	This object will contain the autonomous system number if the object mplsTunnelHopAddrType is set to **asNumber (3).** In all other cases, implementations must set this value to 0, and operators should expect this value to be set to 0.
mplsTunnelCHopLspId	This object is set to the LSP identifier of the tunnel hop if the object mplsTunnelHopAddrType is set to **lspid (4).** In all other cases, implementations must set this value to 0, and operators should expect this value to be set to 0.
mplsTunnelCHopType	This object indicates if this tunnel hop is routed as a strict or a loose hop. Specifically, if this hop is specified as **loose (2),** then it was either specified by the operator as such, or introduced by the signaling protocol to fill in a portion of the path that was specified as loose. If a tunnel hop is specified as **strict (1),** then it was specified by the operator and preserved by the CSPF calculation.

8.13 The Tunnel Performance Table

The Tunnel Performance Table (mplsTunnelPerfTable) provides the operator with several counters to measure the performance of the MPLS tunnels. These counters closely resemble those found in the IF-MIB (see Chapter 6). Entries in this table should only return values when the tunnel entry they correspond to (i.e., augment) does not have a corresponding IF-MIB entry; otherwise they should return zero. This can be determined if mplsTunnelIsIf is set to the value false.

As was mentioned, this table augments the mplsTunnelTable; therefore entries exist for each tunnel entry. However, counters should only be available for entries that are not implemented as interface entries. For those implementations that do not implement mplsTunnelEntries as interfaces, implementation of this table is mandatory for implementations that wish to comply with the minimum requirements in the MPLS-TE MIB.

Table 8.11 lists and defines each of the objects found in the Tunnel Performance Table.

Table 8.11 MPLS Tunnel Performance Table objects.

Object	Description
mplsTunnelPerfPackets	This object contains the number of packets forwarded by the tunnel.
mplsTunnelPerfHCPackets	This object contains the high-capacity (i.e., 64-bit) version of mplsTunnelPerfPackets.
mplsTunnelPerfErrors	This object contains the number of packets received on this tunnel in error.
mplsTunnelPerfBytes	This object contains the number of bytes forwarded by the tunnel.
mplsTunnelPerfHCBytes	This object contains the high-capacity (i.e., 64-bit) version of mplsTunnelPerfBytes.

8.14 IF-MIB Applicability

In Chapter 6 we discussed the structure and composition of the IF-MIB. The MPLS-TE MIB defines a Tunnel Table that stores tunnel entries as well as signaled LSPs for those tunnels. In most implementations, the tunnel head entries are also reflected as subinterfaces in the IF-MIB (RFC 2863). The advantage to taking this approach was that standard management practices could be employed to manage tunnels once they were signaled.

To realize tunnel entries existing with the IF-MIB, the MPLS-TE MIB had to define a new interface-stacking model for these new subinterfaces. This model specifies that the MPLS traffic engineering interfaces are stacked above an underlying MPLS interface. A specific ifType value of 150 has been assigned by IANA for

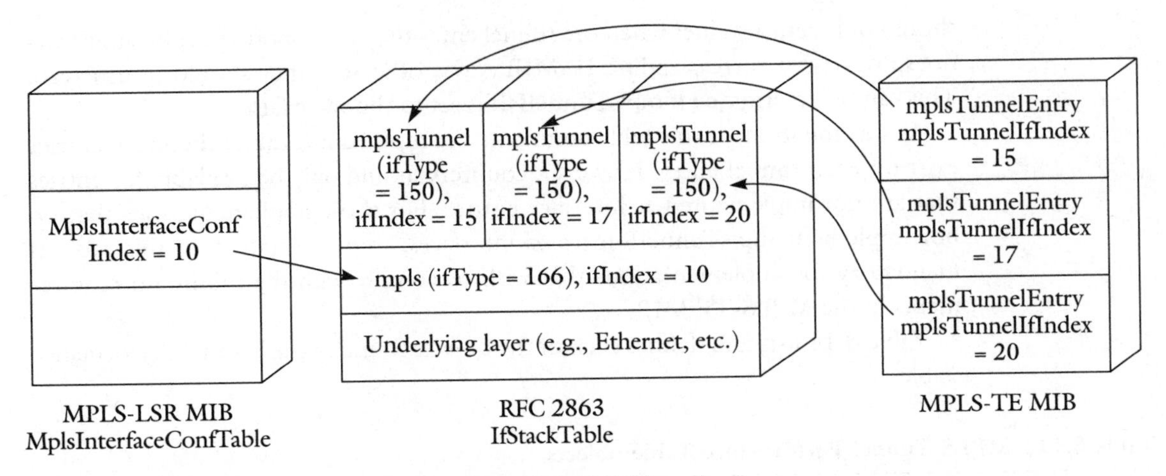

Figure 8.6 Interface stacking for traffic engineering tunnel interfaces.

the MPLS TE interfaces. All MPLS TE tunnel interfaces that are also implemented as IF-MIB entries must use this value for their ifType.

The model for TE tunnel interfaces is shown in Figure 8.6. There are three MPLS tunnel entries represented in the MPLS-TE MIB that refer to ifIndexes in the IF-MIB. These are noted with the arrows from the rightmost MPLS-TE MIB tunnel entries to the IF-MIB's IfStackTable entries in the center. The tunnel interfaces are then in turn stacked upon the MPLS interface that is also cross-referenced with the MPLS-LSR MIB's InterfaceConfTable entry (see Table 3.6 on page 93). This MPLS interface is in turn stacked above some underlying layer-2 media interface such as Ethernet, ATM, or Frame Relay.

When MPLS TE tunnels are represented as IF-MIB entries, some specific interpretations of ifTable attributes are specified to help implementations realize these new interfaces as well as to let management systems understand what to expect from these new interfaces. These variations and interpretations are enumerated in Table 8.12.

Table 8.12 IF-MIB interactions with MPLS Tunnel Table entries.

Object	Use for the MPLS tunnel
IfIndex	One MPLS tunnel is represented by a single ifEntry.
IfDescr	This contains the description of the MPLS tunnel as it appears on the device management console.

Table 8.12 continued

Object	Use for the MPLS tunnel
IfType	IANA has allocated the value 150 for MPLS TE tunnel interfaces.
IfSpeed	This object contains the total bandwidth in bits per second that is reserved for the tunnel.
ifPhysAddress	This value is unused, since it does not apply to tunnels.
ifAdminStatus	This value must behave as is defined in RFC 2863.
ifOperStatus	If the MPLS tunnel cannot pass traffic, it is set to **down (2).** It is set to **up (1)** if it is capable of passing traffic.
IfLastChange	This value is defined the same as in RFC 2863.
ifInOctets	This contains the number of bytes (octets) that have been received on the tunnel.
ifOutOctets	This contains the number of bytes (octets) that have been transmitted on this tunnel.
ifInErrors	This contains the number of labeled packets that were not transmitted on this tunnel due to some error.
ifInUnknownProtos	This is defined to be the count of the number of packets that were received on the tunnel, but were then discarded during packet header validation. This includes packets that are received and then discarded because they contain unrecognized label values.
ifOutErrors	This object is defined using the definition found in RFC 2863.
ifName	This object must be set to the textual or human-readable name of the tunnel. This name must be unique among all tunnels configured on this system. If no such name is available, then an octet string of zero length must be returned.
ifLinkUpDownTrapEnable	This object indicates whether Link Up/Down notifications should be issued for the tunnel in question. The default value for this object is **disabled (2).** This default value prevents unexpected notifications from being emitted by devices. Operators should take care to set this value appropriately since the number of tunnel interfaces can be large and therefore the potential for issuing large numbers of notifications may exist.
ifConnectorPresent	This value is always set to **false (2).**

continued

Table 8.12 continued

Object	Use for the MPLS tunnel
ifHighSpeed	Please see RFC 2863 for additional details.
IfHCInOctets	This object contains the 64-bit version of ifInOctets. Implementations must support this value if required by the compliance statements in RFC 2863.
ifHCOutOctets	This object contains the 64-bit version of ifOutOctets. This value must be supported if required by the compliance statements in RFC 2863.
ifAlias	This object is set to the nonvolatile name for the tunnel interface as specified by the operator. This value frequently contains a convenient "alias" for the tunnel.

8.15 Tunnel Table and MPLS-LSR MIB Interaction

The MPLS-TE MIB interacts with the MPLS-LSR MIB (Srinivasan et al. 2000b) in that once a tunnel has been signaled, the LSP used to transport the tunnel's traffic will be reflected in the MPLS-LSR MIB. More precisely, the labels used to carry that LSP's traffic will be reflected in the MPLS-LSR MIB by entries in the mplsInSegmentTable, the mplsOutSegmentTable, or the mplsXCTable.

The MPLS-TE MIB supports point-to-point tunnels. It does not support multipoint tunnels. Therefore, the MPLS-LSR MIB where a specific MPLS TE tunnel originates will contain a single mplsOutSegmentEntry and mplsXCEntry corresponding to the LSP, but no mplsInSegmentEntry. Intermediate LSRs will have mplsOutSegment, mplsInSegment, and mplsXC entries populated for the LSP. The tail end of the tunnel will have a single mplsInSegmentEntry and mplsXCEntry for the terminating LSP. At each LSR where entries are created in the MPLS-LSR MIB's tables, each entry is related to the MPLS-TE MIB by the Tunnel Table mplsTunnelXCPointer object. This object contains the OID of the corresponding MPLS-LSR MIB's mplsXCTable entry's first accessible conceptual row. It is possible that no association has yet been made, in which case the pointer is set to 0.0.

Figure 8.7 illustrates the interaction between the MPLS-LSR, MPLS-TE, and IF-MIBs by showing a sample configuration of three originating TE tunnels and two transient LSPs. The transient LSPs are denoted in the MPLS-LSR MIB tables

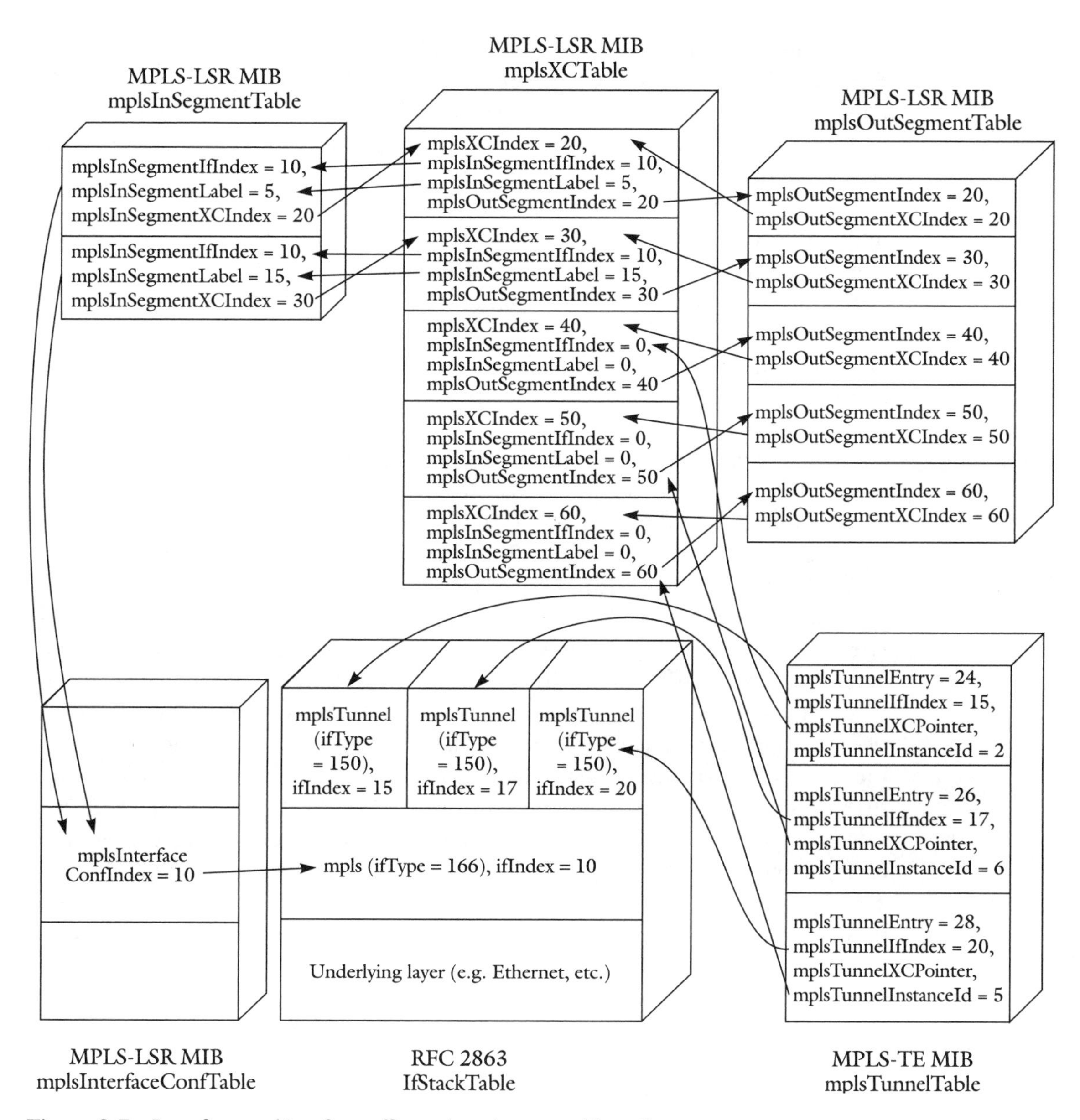

Figure 8.7 Interface stacking for traffic engineering tunnel interfaces.

by the indexes (mplsXCIndex = 30, mplsInSegmentIfIndex = 10, mplsInSegmentLabel = 15, mplsOutSegmentIndex = 30) and (mplsXCIndex = 20, mplsInSegmentIfIndex = 10, mplsInSegmentLabel = 5, mplsOutSegmentIndex = 20), while the TE tunnels are denoted by (mplsXCIndex = 40, mplsInSegmentIfIndex = 0, mplsInSegmentLabel = 0, mplsOutSegmentIndex = 40), (mplsXCIndex = 50, mplsInSegmentIfIndex = 0, mplsInSegmentLabel = 0, mplsOutSegmentIndex = 50), and (mplsXCIndex = 60, mplsInSegmentIfIndex = 0, mplsInSegmentLabel = 0, mplsOutSegmentIndex = 60).

Starting at the bottom of the ifStackTable you will notice that an "underlying layer" interface is stacked. We did not choose a specific ifType entry for this example simply because this entry can be of any type that is designed to carry MPLS traffic. For example, this could be Ethernet, POS, or ATM. We are not concerned with the specific type, just that it can support MPLS traffic. Stacked above this interface is a single MPLS interface whose ifIndex is set to 10. A corresponding entry in the MPLS-LSR MIB's mplsInterfaceConfEntry is created for this ifEntry. This allows an NMS to quickly peruse the MPLS-enabled interfaces without having to examine the entire ifStackEntry or ifEntry tables. The mplsInterfaceConfTable entry is related back to the arrows between it and the mplsInSegmentEntry.

Next, notice that the MPLS interface is stacked below three TE tunnel interfaces. The figure shows three mplsTunnelTable entries configured as entries 24, 26, and 27. These are associated with the IF-MIB via ifIndexes 15, 17, and 20, respectively, and all have an appropriate ifType of mplsTunnel (150). The arrows between the mplsTunnelTable and the ifStack table illustrate the relationship between the two. Note that the ifEntry values also exist in the ifTable. These entries are then associated with the originating LSPs in the MPLS-LSR MIB's mplsXCTable via the mplsTunnelXCPointer that "points" at the corresponding entries. For example, mplsTunnelEntry 24 points at the cross-connect entry referenced as the 4-tuple mplsXCIndex = 50, mplsInSegmentIfIndex = 0, mplsInSegmentLabel = 0, mplsOutSegmentIndex = 50. The remaining entries are cross-referenced similarly.

It is important to note that the mplsTunnelEntries shown above do not represent the head tunnel interfaces; instead, they represent signaled instances of those tunnels (i.e., their LSPs). As we suggested in Section 8.7.1, this is indicated by the mplsTunnelInstanceId being greater than zero. Note that the values chosen are inserted just for the example and do not correspond to any special values other than being nonzero. Notice that these cross-connect entries represent originating LSPs for these tunnel entries as well. There should not be any corresponding MPLS-LSR table entries for tunnel heads as they generally represent the configured tunnel head interface and not the underlying LSP.

8.16 Multiple Tunnels across MPLS Network Example

Figure 8.8 shows a sample MPLS-enabled network containing 10 interconnected MPLS LSRs. POPs connected to the edges of the network direct traffic to and from those POPs into the MPLS-enabled core. Let us assume that all LSRs are running OSPF, but note that the configuration of the MIB objects would be identical if IS-IS were in use. Normally, traffic from the POP connected to node A (LSR ID = 2.0.0.96) that is destined for networks that are reachable via D traverse the shortest path. In the example network this path takes traffic along the path A, B (LSR ID = 2.0.0.98), C (LSR ID = 2.0.0.97), and D (2.0.0.99). However, this may cause congestion because traffic from node H may also take this same path to reach D. To alleviate the congestion, a network operator can create a traffic-engineered tunnel and reroute some of the congested traffic using this alternative path.

In the example in Figure 8.8, an operator has realized two TE tunnels, one from A to D, and another from F to C, in an effort to explicitly forward traffic through

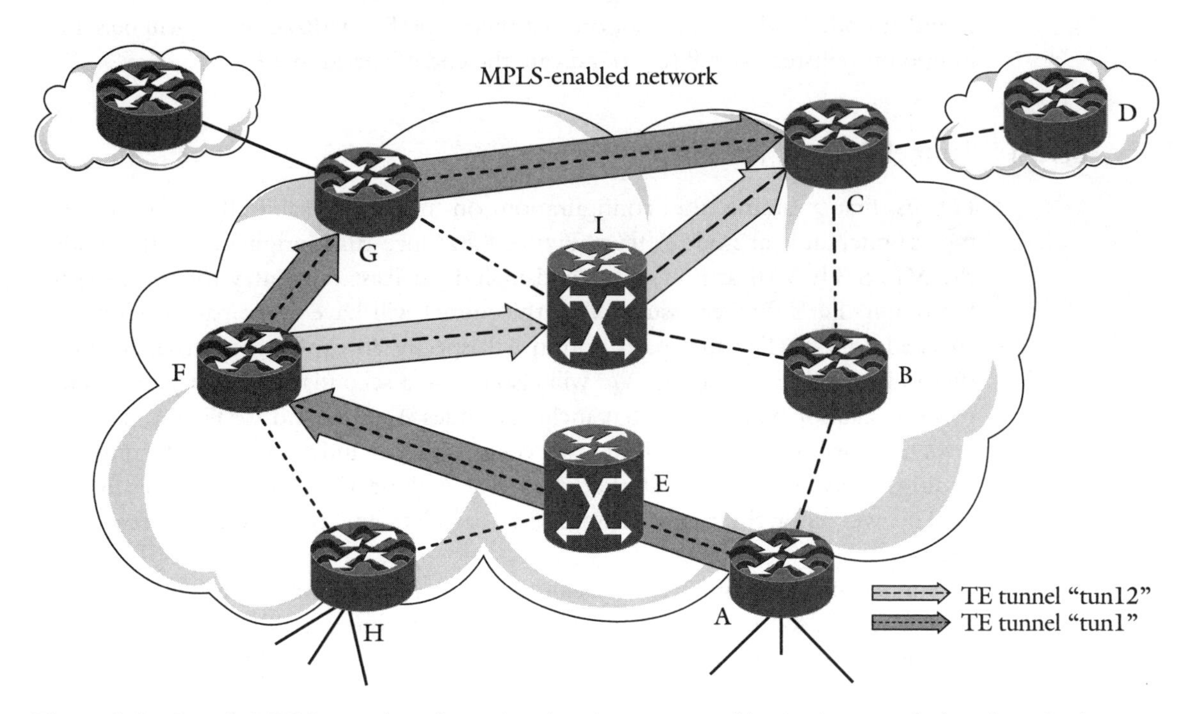

Figure 8.8 Sample MPLS tunnel configuration showing one tunnel beginning at node A and continuing to node C, as well as a tunnel starting at node F and terminating at node C.

his network using paths that differ from those computed by the routing protocols. On node A, a tunnel interface has been created and assigned the identifier "tun1," while node F has had an interface created with the identifier "tun12." Note that the names have been made different for pedagogical purposes only; they can be the same on all nodes within the network because the source and destination ports and addresses are what disambiguate the RSVP-TE sessions. In both cases, each tunnel interface will have a series of traffic engineering constraints specified with it. These include a minimum bandwidth of 1 Mbps and default tunnel affinity bit settings. In addition, each tunnel will have one or more path options associated with it.

We will now examine what the MPLS-TE MIB's tables will look like at nodes A and F, as well as node C for each configured tunnel. Nodes A and F will demonstrate how a tunnel head will appear with its signaled LSPs. Node F will demonstrate how a tunnel head will appear with a midpoint of another tunnel. Finally, node C will show what the MIB will look like at a node where no tunnel heads are present, and only midpoints/tails. We will not examine each node along the tunnel path since the entries for each specific tunnel instance will be identical to the ones found at node C with the exception that their mplsTunnelRole object will be set to midpoint(s) instead of tail (3) to indicate the end of the tunnel.[1]

8.16.1 Tunnel Table at Node A

Let us first examine the configuration on node A. Table 8.13 shows the mplsTunnelTable at node A from Figure 8.8. Since tun1 originates at this node, the MPLS-TE MIB at node A should include at least one entry for tun1's configuration. Let's further assume that the tunnel will have three path options associated with it. The first path option will specify an explicit path that includes the nodes A, E, F, G, and C. We will also create a secondary explicit path in case this one cannot be signaled that includes nodes A, E, I, and C. Finally, as a last resort, we will configure a third path option that includes only the starting and ending nodes A and C, but that does not include (i.e., avoids) nodes B and C. This will allow the tunnel head end to calculate any available path between A and C.

Tunnel tun1 is configured to begin on node A. First, notice that the head interface can be found in the table as an entry with indexes of TunnelIndex = 5, TunnelInstance = 0, TunnelIngressLsrId = A, TunnelEgressLsrID = C. Notice

1 Note that when penultimate hop popping is used, the node indicating that it is the tunnel tail point is not truly the tunnel's tail; rather, it is the next-to-last point.

again that the tunnel instance index appears as 0 to distinguish this entry from any of the others related to this tunnel. The first path option has been signaled successfully, and so it appears in the figure with the same indexes as the head interface except that its tunnelInstance = 4.

Table 8.13 Tunnel Table at node A in Figure 8.8.

TunnelIndex = 5, TunnelInstance = 0, TunnelIngressLSRId = A (2.0.0.96), TunnelEgressLsrId = C (2.0.0.97)	TunnelIndex = 5, TunnelInstance = 4, TunnelIngressLSRId = A (2.0.0.96), TunnelEgressLsrId = C (2.0.0.97)
mplsTunnelName = "tun1"	MplsTunnelName = "tun1 inst4"
mplsTunnelDescr = "tunnel from A to C"	MplsTunnelDescr = "tunnel instance 4 from A to C"
mplsTunnelIsIf = true (1)	MplsTunnelIsIf = false (0)
mplsTunnelIfIndex = 14	MplsTunnelIfIndex = 0
mplsTunnelXCPointer = 0.0	MplsTunnelXCPointer = mplsXCIndex.2.1.16.2
mplsTunnelSignalingProto = rsvp (2)	MplsTunnelSignalingProto = rsvp (2)
mplsTunnelSetupPrio = 1	MplsTunnelSetupPrio = 1
mplsTunnelHoldingPrio = 1	MplsTunnelHoldingPrio = 1
mplsTunnelSessionAttributes = 0x0000	MplsTunnelSessionAttributes = 0x0000
mplsTunnelOwner = admin (1)	MplsTunnelOwner = rsvp (3)
mplsTunnelLocalProtectInUse = false	MplsTunnelLocalProtectInUse = false
mplsTunnelResourcePointer = mplsMIB.3.2.6.1.1	MplsTunnelResourcePointer = mplsMIB.3.2.6.1.1
mplsTunnelInstancePriority = 0	MplsTunnelInstancePriority = 1
mplsTunnelHopTableIndex = 1	MplsTunnelHopTableIndex = 1
mplsTunnelARHopTableIndex = 1	mplsTunnelARHopTableIndex =1
mplsTunnelCHopTableIndex = 1	mplsTunnelCHopTableIndex =1
mplsTunnelPrimaryInstance = 4	mplsTunnelPrimaryInstance = 4
mplsTunnelPrimaryTimeUp = 100	mplsTunnelPrimaryTimeUp = 100
mplsTunnelPathChanges = 1	mplsTunnelPathChanges = 1
mplsTunnelLastPathChange = 100	mplsTunnelLastPathChange = 100
mplsTunnelCreationTime = 1234	mplsTunnelCreationTime = 1238

continued

Table 8.13 continued

TunnelIndex = 5, TunnelInstance = 0, TunnelIngressLSRId = A (2.0.0.96), TunnelEgressLsrId = C (2.0.0.97)	TunnelIndex = 5, TunnelInstance = 4, TunnelIngressLSRId = A (2.0.0.96), TunnelEgressLsrId = C (2.0.0.97)
mplsTunnelStateTransitions = 1	mplsTunnelStateTransitions = 1
mplsTunnelIncludeAnyAffinity = 0x0000	mplsTunnelIncludeAnyAffinity =0x0000
mplsTunnelIncludeAllAffinity =	mplsTunnelIncludeAllAffinity = 0x0000
mplsTunnelExcludeAllAffinity =	mplsTunnelExcludeAllAffinity = 0x0000
MplsTunnelPathInUse = 1	mplsTunnelPathInUse = 1
MplsTunnelRole = head (1)	mplsTunnelRole = head (1)
MplsTunnelTotalUpTime = 100	mplsTunnelTotalUpTime = 100
MplsTunnelInstanceUpTime = 100	mplsTunnelInstanceUpTime = 100
MplsTunnelAdminStatus = up (1)	mplsTunnelAdminStatus = up (1)
MplsTunnelRowStatus = active (5)	mplsTunnelRowStatus = active (5)
MplsFTNMapStorageType = volatile (2)	mplsFTNMapStorageType = volatile (2)

8.16.2 Tunnel Hop Table at Node A

Table 8.14 contains the representative path options as configured for "tun1" on node A. This table specifically includes three path options. The first is configured as an explicit path option that includes the nodes. The second is also an explicit path option that instead includes nodes {E, I, C}. Finally, as a last resort, a dynamic path option between nodes A and C is specified.

8.16.3 Tunnel Table at Node F

Now let's examine what the tunnel table would look like on node F. The tunnel interface configured at node F possesses a simpler configuration than the previous tunnel that emanated at node A. The tunnel is configured to require the same bandwidth and affinity constraints as the previous tunnel. The tunnel is also configured with two path options that are specified as {F, I, C} and {F, C, but avoid G}.

Table 8.15 shows the mplsTunnelTable objects at node F in Figure 8.8. The table shown at this node is interesting because it will contain three entries: one for tun12's head interface, one for tun12's signaled LSP, and one for the transit node portion of the LSP for tun1.

Table 8.14 Tunnel Hop Table at node A in Figure 8.8.

Nonindex objects	(mplsTunnelHopListIndex, mplsTunnelHopPathOption-Index,mplsTunnelHopIndex)	"	"	"	"	"
	5 (chosen to correspond to mplsTunnelIndex 5), 1, 1	5, 1, 2	5, 1, 3	5, 1, 4	5, 1, 5	
mplsTunnelHop-AddrType	ipV4 (1)	ipV4 (1)	ipV4 (1)	ipV4 (1)	ipV4 (1)	
mplsTunnelHop-Ipv4Addr	2.0.0.96 (A)	2.0.0.102(E)	2.0.0.101 (F)	2.0.0.100(G)	2.0.0.97 (C)	
mplsTunnelHop-Ipv4PrefixLen	16	16	16	16	16	
mplsTunnelHop-Ipv6Addr	0.0.0.0	0.0.0.0	0.0.0.0	0.0.0.0	0.0.0.0	
mplsTunnelHop-Ipv6PrefixLen	0	0	0	0	0	
mplsTunnelHop-AsNumber	0	0	0	0	0	
mplsTunnelHopLspId	0	0	0	0	0	
mplsTunnelHopType	strict (1)	strict (1)	strict (1)	strict (1)	strict (1)	
mplsTunnelHop-IncludeExclude	include (1)	include (1)	include (1)	include (1)	include (1)	
mplsTunnelHopPath-OptionName	"path option 1"	"path option 1"	"path option 1"	"path option 1"	"path option 1"	

continued

Table 8.14 continued

Nonindex objects	(mplsTunnelHopListIndex, mplsTunnelHopPathOption-Index,mplsTunnelHopIndex)	"	"	"	"	"
mplsTunnelHopEntry-PathComp	explicit (2)	explicit (2)	explicit (2)	explicit (2)	explicit (2)	
mplsTunnelHop-RowStatus	active (1)	active (1)	active (1)	active (1)	active (1)	
mplsTunnelHop-StorageType	Nonvolatile (3)	Nonvolatile (3)	Nonvolatile (3)	Nonvolatile (3)	Nonvolatile (3)	
	5, 2, 1	5, 2, 2	5, 2 ,3	5, 2, 4	5, 3, 1	5, 3, 2
mplsTunnelHop-AddrType	ipV4 (1)	ipV4 (1)	ipV4 (1)	ipV4 (1)	ipV4 (1)	ipV4 (1)
mplsTunnelHop-Ipv4Addr	2.0.0.96 (A)	2.0.0.102 (E)	2.0.0.104 (I)	2.0.0.97 (C)	2.0.0.96 (A)	2.0.0.112 (D)
mplsTunnelHop-Ipv4PrefixLen	16	16	16	16	16	16
mplsTunnelHop-Ipv6Addr	0.0.0.0	0.0.0.0	0.0.0.0	0.0.0.0	0.0.0.0	0.0.0.0
mplsTunnelHop-Ipv6PrefixLen	0	0	0	0	0	0
mplsTunnelHop-AsNumber	0	0	0	0	0	0

Table 8.14 continued

Nonindex objects	(mplsTunnelHopListIndex, mplsTunnelHopPathOption-Index,mplsTunnelHopIndex)	"	"	"	"	"
mplsTunnelHopLspId	0	0	0	0	0	0
mplsTunnelHopType	strict (1)	strict (1)	strict (1)	strict (1)	strict (1)	strict (1)
mplsTunnelHopInclude-Exclude	include (1)	include (1)	include (1)	include (1)	include (1)	include (1)
mplsTunnelHopPath-OptionName	"path option 1"	"path option 2"	"path option 2"	"path option 2"	"path option 3"	"path option 3"
mplsTunnelHopEntry-PathComp	explicit (2)	explicit (2)	explicit (2)	explicit (2)	dynamic(1)	dynamic(1)
mplsTunnelHop-RowStatus	active (1)	active (1)	active (1)	active (1)	active (1)	active (1)
mplsTunnelHop-StorageType	Nonvolatile (3)	Nonvolatile (3)	Nonvolatile (3)	Nonvolatile (3)	Nonvolatile (3)	Nonvolatile (3)

Table 8.15 MPLS Tunnel Table objects at node F in Figure 8.8.

TunnelIndex = 2, TunnelInstance = 0, TunnelIngressLSRId = A (2.0.0.96), TunnelEgressLsrId = C (2.0.0.97)	TunnelIndex = 2, TunnelInstance = 2, TunnelIngressLSRId = A (2.0.0.96), TunnelEgressLsrId = C (2.0.0.97)	TunnelIndex = 5, TunnelInstance = 4, TunnelIngressLSRId = A (2.0.0.96), TunnelEgressLsrId = C (2.0.0.97)
mplsTunnelName = "tun12"	MplsTunnelName = "tun2 inst2"	MplsTunnelName = "tun1 inst4"
mplsTunnelDescr = "tunnel from A to C"	MplsTunnelDescr = "tunnel instance 4 from A to C"	MplsTunnelDescr = "tunnel instance 4 from A to C"
mplsTunnelIsIf = true (1)	MplsTunnelIsIf = false (0)	MplsTunnelIsIf = false (0)
mplsTunnelIfIndex = 14	MplsTunnelIfIndex = 0	MplsTunnelIfIndex = 0
mplsTunnelXCPointer = 0.0	MplsTunnelXCPointer = TBD	MplsTunnelXCPointer = TBD
mplsTunnelSignalingProto = rsvp (2)	MplsTunnelSignalingProto = rsvp (2)	MplsTunnelSignalingProto = rsvp (2)
mplsTunnelSetupPrio = 1	MplsTunnelSetupPrio = 1	MplsTunnelSetupPrio = 1
mplsTunnelHoldingPrio = 1	MplsTunnelHoldingPrio = 1	MplsTunnelHoldingPrio = 1
mplsTunnelSessionAttributes = 0x0000	MplsTunnelSessionAttributes = 0x0000	MplsTunnelSessionAttributes = 0x0000
mplsTunnelOwner = admin (1)	MplsTunnelOwner = rsvp (3)	MplsTunnelOwner = rsvp (3)
mplsTunnelLocalProtectInUse = false	MplsTunnelLocalProtectIn-Use = false	MplsTunnelLocalProtectIn-Use = false
mplsTunnelResourcePointer = mplsMIB.3.2.6.1.1	MplsTunnelResourcePointer = mplsMIB.3.2.6.1.1	MplsTunnelResourcePointer = mplsMIB.3.2.6.1.1
mplsTunnelInstancePriority = 0	MplsTunnelInstancePriority = 1	MplsTunnelInstancePriority = 1
mplsTunnelHopTableIndex = 1	MplsTunnelHopTableIndex = 1	MplsTunnelHopTableIndex = 1
mplsTunnelARHopTable-Index = 1	mplsTunnelARHopTable-Index = 1	mplsTunnelARHopTable-Index = 1
mplsTunnelCHopTableIndex = 1	mplsTunnelCHopTableIndex = 1	mplsTunnelCHopTableIndex = 1

Table 8.15 continued

TunnelIndex = 2, TunnelInstance = 0, TunnelIngressLSRId = A (2.0.0.96), TunnelEgressLsrId = C (2.0.0.97)	TunnelIndex = 2, TunnelInstance = 2, TunnelIngressLSRId = A (2.0.0.96), TunnelEgressLsrId = C (2.0.0.97)	TunnelIndex = 5, TunnelInstance = 4, TunnelIngressLSRId = A (2.0.0.96), TunnelEgressLsrId = C (2.0.0.97)
mplsTunnelPrimaryInstance = 4	mplsTunnelPrimaryInstance = 4	mplsTunnelPrimaryInstance = 4
mplsTunnelPrimaryTimeUp = 100	mplsTunnelPrimaryTimeUp = 100	mplsTunnelPrimaryTimeUp = 100
mplsTunnelPathChanges = 1	mplsTunnelPathChanges = 1	mplsTunnelPathChanges = 1
mplsTunnelLastPathChange = 100	mplsTunnelLastPathChange = 100	mplsTunnelLastPathChange = 100
mplsTunnelCreationTime = 1234	mplsTunnelCreationTime = 1238	mplsTunnelCreationTime = 1238
mplsTunnelStateTransitions = 1	mplsTunnelStateTransitions = 1	mplsTunnelStateTransitions = 1
mplsTunnelIncludeAnyAffinity = 0x0000	mplsTunnelIncludeAnyAffinity = 0x0000	mplsTunnelIncludeAnyAffinity = 0x0000
mplsTunnelIncludeAllAffinity = 0x00	mplsTunnelIncludeAllAffinity = 0x00	mplsTunnelIncludeAllAffinity = 0x00
MplsTunnelExcludeAllAffinity = 0x00	mplsTunnelExcludeAllAffinity = 0x00	mplsTunnelExcludeAllAffinity = 0x00
MplsTunnelPathInUse = 1	mplsTunnelPathInUse = 1	mplsTunnelPathInUse = 1
MplsTunnelRole = head (1)	mplsTunnelRole = head (1)	mplsTunnelRole = head (1)
MplsTunnelTotalUpTime = 100	mplsTunnelTotalUpTime = 100	mplsTunnelTotalUpTime = 100
MplsTunnelInstanceUpTime = 100	mplsTunnelInstanceUpTime = 100	mplsTunnelInstanceUpTime = 100
MplsTunnelAdminStatus = up (1)	mplsTunnelAdminStatus = up (1)	mplsTunnelAdminStatus = up (1)
MplsTunnelRowStatus = active (5)	mplsTunnelRowStatus = active (5)	mplsTunnelRowStatus = active (5)

continued

Table 8.15 continued

TunnelIndex = 2, TunnelInstance = 0, TunnelIngressLSRId = A (2.0.0.96), TunnelEgressLsrId = C (2.0.0.97)	TunnelIndex = 2, TunnelInstance = 2, TunnelIngressLSRId = A (2.0.0.96), TunnelEgressLsrId = C (2.0.0.97)	TunnelIndex = 5, TunnelInstance = 4, TunnelIngressLSRId = A (2.0.0.96), TunnelEgressLsrId = C (2.0.0.97)
MplsFTNMapStorageType = volatile (2)	mplsFTNMapStorageType = volatile (2)	mplsFTNMapStorageType = volatile (2)

8.16.4 Tunnel Table at Node C

Finally, let us examine what the tunnel table would look like on node C in Figure 8.8. Table 8.16 shows the mplsTunnelTable objects that exist at node A in Figure 8.8. In this case, both tunnels tun1 and tun12 terminate at this node. Therefore, no head interfaces will be present at this node in this network configuration. Upon examining the table, only two entries should be viewed—one for each signaled LSP of each tunnel. Both should indicate that they are the tails of the tunnel by setting tunnelRole = tail (3).

Table 8.16 MPLS Tunnel Table objects at node A in Figure 8.8.

TunnelIndex = 5, TunnelInstance = 4, TunnelIngressLSRId = A (2.0.0.96), TunnelEgressLsrId = C (2.0.0.97)	TunnelIndex = 5, TunnelInstance = 4, TunnelIngressLSRId = A (2.0.0.96), TunnelEgressLsrId = C (2.0.0.97)
mplsTunnelName = "tun1"	MplsTunnelName = "tun1 inst4"
mplsTunnelDescr = "tunnel from A to C"	MplsTunnelDescr = "tunnel instance 4 from A to C"
mplsTunnelIsIf = true (1)	MplsTunnelIsIf = false (0)
mplsTunnelIfIndex = 14	MplsTunnelIfIndex = 0
mplsTunnelXCPointer = 0.0	MplsTunnelXCPointer = TBD
mplsTunnelSignalingProto = rsvp (2)	MplsTunnelSignalingProto = rsvp (2)
mplsTunnelSetupPrio = 1	MplsTunnelSetupPrio = 1
mplsTunnelHoldingPrio = 1	MplsTunnelHoldingPrio = 1
mplsTunnelSessionAttributes = 0x0000	MplsTunnelSessionAttributes = 0x0000

Table 8.16 continued

TunnelIndex = 5, TunnelInstance = 4, TunnelIngressLSRId = A (2.0.0.96), TunnelEgressLsrId = C (2.0.0.97)	TunnelIndex = 5, TunnelInstance = 4, TunnelIngressLSRId = A (2.0.0.96), TunnelEgressLsrId = C (2.0.0.97)
mplsTunnelOwner = admin (1)	MplsTunnelOwner = rsvp (3)
mplsTunnelLocalProtectInUse = false	MplsTunnelLocalProtectInUse = false
mplsTunnelResourcePointer = mplsMIB.3.2.6.1.1	MplsTunnelResourcePointer = mplsMIB.3.2.6.1.1
mplsTunnelInstancePriority = 0	MplsTunnelInstancePriority = 1
mplsTunnelHopTableIndex = 1	MplsTunnelHopTableIndex = 1
mplsTunnelARHopTableIndex = 1	mplsTunnelARHopTableIndex =1
mplsTunnelCHopTableIndex = 1	mplsTunnelCHopTableIndex =1
mplsTunnelPrimaryInstance = 4	mplsTunnelPrimaryInstance = 4
mplsTunnelPrimaryTimeUp = 100	mplsTunnelPrimaryTimeUp = 100
mplsTunnelPathChanges = 1	mplsTunnelPathChanges = 1
mplsTunnelLastPathChange = 100	mplsTunnelLastPathChange = 100
mplsTunnelCreationTime = 1234	mplsTunnelCreationTime = 1238
mplsTunnelStateTransitions = 1	mplsTunnelStateTransitions = 1
mplsTunnelIncludeAnyAffinity = 0x0000	mplsTunnelIncludeAnyAffinity = 0x0000
mplsTunnelIncludeAllAffinity = 0x00	mplsTunnelIncludeAllAffinity = 0x00
MplsTunnelExcludeAllAffinity = 0x00	mplsTunnelExcludeAllAffinity = 0x00
MplsTunnelPathInUse = 1	mplsTunnelPathInUse = 1
MplsTunnelRole = head (1)	mplsTunnelRole = head (1)
MplsTunnelTotalUpTime = 100	mplsTunnelTotalUpTime = 100
MplsTunnelInstanceUpTime = 100	mplsTunnelInstanceUpTime = 100
MplsTunnelAdminStatus = up (1)	mplsTunnelAdminStatus = up (1)
MplsTunnelRowStatus = active (5)	mplsTunnelRowStatus = active (5)
MplsFTNMapStorageType = volatile (2)	mplsFTNMapStorageType = volatile (2)

8.17 Notifications

The MPLS-TE MIB defines several notifications that can be useful for networks that deploy MPLS traffic-engineered tunnels. Each notification is useful on its own, but when combined with the others they can be used to provide a more complete operational picture of a device to a manager.

8.17.1 MPLS Tunnel Trap Enable

The purpose of this object is to control the emission of all MPLS-TE MIB-related notifications from a device. If this object is set to **true (1),** then it enables the generation of mplsTunnelUp, mplsTunnelDown, mplsTunnelRerouted, and mplsTunnelReoptimized notifications whenever appropriate. Otherwise, this value is set to **false (0),** which results in no notifications being emitted by the agent.

As with all notification enable/disable objects, the default value for this scalar object is false to prevent the emission of notifications from a device unless explicitly programmed to do so. Device vendors are encouraged to follow this guideline and only enable this value when an operator does so via configuration, as enabling notifications on TE tunnels can result in potentially large amounts of notification traffic if the number of configured tunnels is large and the number of state fluctuations is significant. If an operator is not expecting this traffic, they may be quite surprised (or upset) by it.

Devices and management stations should be programmed to know and understand the interface stacking relationship as shown in Figure 8.6. Doing so makes it possible to suppress higher-layer notifications in cases when a lower layer or layers are no longer operational. This can prevent notification storms from occurring. For example, in Figure 8.7, if the MPLS type subinterface is disabled, it is not necessarily important or desirable to issue mplsTunnelDown notifications for each one of the TE tunnels stacked above the MPLS type subinterface. If an NMS understands that these tunnels are already configured on these interfaces, then it is possible to only receive the ifDown notification from the MPLS interface and deduce that the tunnels stacked above it are also no longer operational.

8.17.2 MPLS Tunnel Down

The mplsTunnelDown notification is generated when the mplsTunnelOperStatus object associated with one of the configured tunnels is about to enter the **down (2)** state from any state other than the **notPresent (4)** state. When issued, this notification contains the following objects: mplsTunnelAdminStatus and

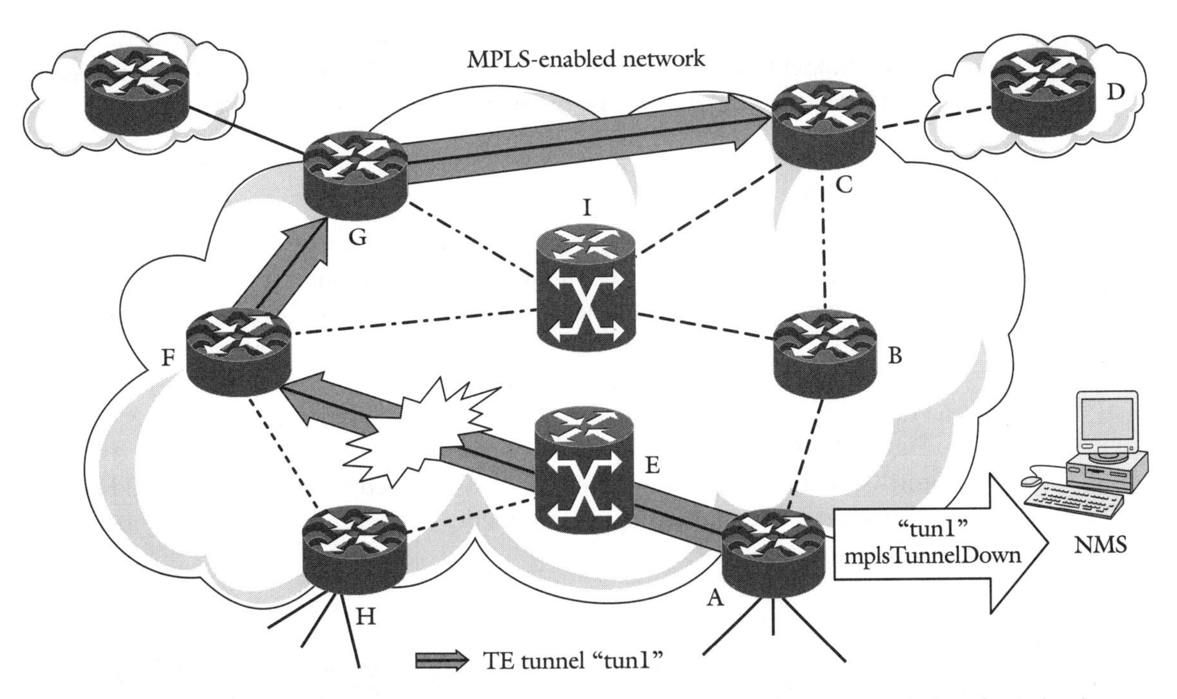

Figure 8.9 Sample MPLS tunnel configuration showing one tunnel beginning at node A and terminating at node C. This tunnel is configured with only one path option, which is an explicit route including A, E, F, G, and C. The tunnel will enter the **down (2)** operational state and issue an mplsTunnelDown notification (if enabled) as soon as the path breakage message is received at A.

mplsTunnelOperStatus. These objects are used to inform the NMS operator of the existing state of the tunnel, its configured state, as well as the indexes of that tunnel.[2]

Figure 8.9 shows an example of a tunnel named "tun1" that just had its mplsTunnelOperStatus object set to **down (1)** because the link that carried that tunnel's traffic between nodes F and E failed. This resulted in the head end for that tunnel being notified of the path breakage at node A. When node A was notified of this, it immediately issued an mplsTunnelDown notification containing mplsTunnelAdminStatus = **up (1)** and mplsTunnelOperStatus = **down (2)** for "tun1." Notice that node E did not issue the notification. This is because the MPLS-TE

2 Indexing of the tunnel object emitting the notification is implicitly indicated by the fact that the instance ID of the tunnel must be included in the OID for both the mplsTunnelAdminStatus and mplsTunnelOperStatus objects. In general, most notifications defined today are done this way to obviate the need to include an explicit index object within the notification varbind.

MIB states that configured tunnels should only issue the notification, implying that only the tunnel head interface (which is configured) should issue the notification once it knows that the state of the tunnel has transitioned to **down (2).**

8.17.3 MPLS Tunnel Up

The mplsTunnelUp notification is generated when the mplsTunnelOperStatus object associated with one of the configured tunnels is about to enter the **up (1)** state from any state other than the **notPresent (4)** state. When issued, this notification contains the following objects: mplsTunnelAdminStatus and mplsTunnelOperStatus. These objects are used to inform the NMS operator of the existing state of the tunnel, its configured state, as well as the indexes of that tunnel.[3]

Figure 8.10 shows an example of a tunnel named "tun1" that just had its mplsTunnelOperStatus object set to **up (1).** This occurred because an LSP could be signaled to carry that tunnel interface's traffic. The example shows how once the

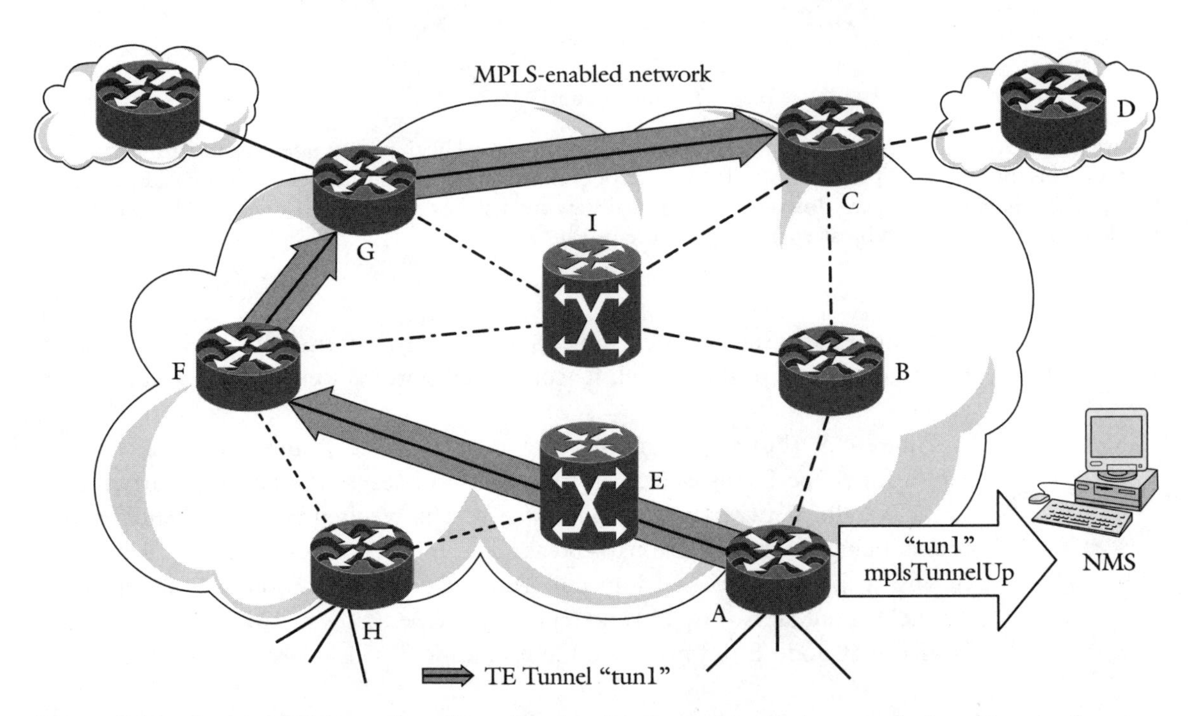

Figure 8.10 Sample MPLS tunnel configuration showing a tunnel beginning at node A and terminating at node C. This tunnel has been configured as an explicit route including A, E, F, G, and C. Since this tunnel has been successfully signaled, the tunnel head interface at A can now issue an mplsTunnelUp notification.

3 See footnote 2.

tunnel has been signaled, the agent issues the mplsTunnelUp notification for this tunnel. Note that although not shown, the notification issued carries the current values of mplsOperStatus = **up (1)** and mplsAdminStatus = **up (1)** for the instance corresponding to "tun1." This message is delivered to the NMS that is identified as a notification target for that LSR.

8.17.4 MPLS Tunnel Rerouted

This notification is generated when a tunnel is rerouted. Specifically, a tunnel is considered to have rerouted itself if a local repair operation has occurred at some midpoint along the tunnel path. This notification is typically issued from midpoints along the tunnel's path to indicate that a local repair is currently in place. A path break message is typically sent back to the tunnel interface when this event occurs. This typically results in the tunnel choosing a different path option and resignaling itself. If a tunnel can choose a new path and reroute using the "make before break" algorithm, then an mplsTunnelReoptimized notification will be emitted shortly thereafter. However, if not, then it is possible that either or both of the mplsTunnelDown and mplsTunnelUp notifications will follow such an event. During the time of path fluctuations, it is possible that the tunnel's mplsTunnelARHopTable does not reflect the actual path that the tunnel has taken, since this table reflects the actual path as seen by the signaling software. If a local repair has taken place, this will not be signaled back to the head interface; instead, only the fact that a local repair has taken place is made known to the tunnel head interface.

Figure 8.11 depicts a sample MPLS tunnel configuration with one tunnel identified as "tun1" beginning at node A and continuing to node C and another identified as "tun12" from F to C. When the link between G and C breaks, node G performs a local repair of "tun1," which traverses this link by rerouting it over the link between G and I. It is at this time that node G may issue an mplsTunnelRerouted notification for "tun1." If an NMS has been configured to receive this notification, as is the case in the figure, it will receive this notification and inform the operator. After the local repair happens, node G signals to the head end node A that a local repair has taken place. It is at this time that the tunnel's head end may choose to reroute the tunnel over some other path option if one exists.

8.17.5 MPLS Tunnel Reoptimized

This notification is generated when a tunnel is reoptimized. A tunnel is defined as being reoptimized when some change occurs to the tunnel to make it perform more optimally, and this change does not disrupt the tunnel's traffic flow. This implies that any path changes are performed using the "make before break" algorithm, which specifies that the new path should be signaled *before* the old path is

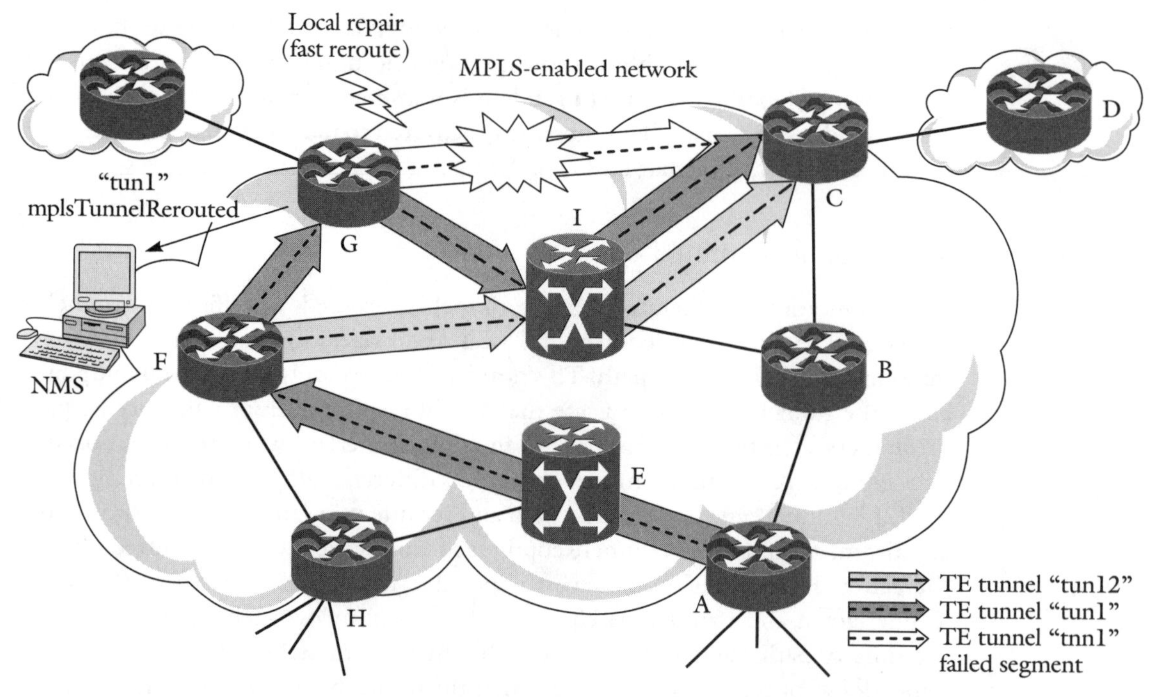

Figure 8.11 Sample MPLS tunnel configuration showing one tunnel beginning at node A and continuing to node C that traverses node G. Another tunnel starts at node F and terminates at node C. When the link between G and C breaks, node G performs a local repair of the tunnel traversing that link by rerouting it over the link between G and I. After the local repair happens, node G issues an mplsTunnelRerouted notification for "tun1" that arrives at the NMS, and signals to the head end node A that a local repair has taken place.

decommissioned and traffic moved onto the new one. Tunnel reoptimizations include re-reserving the bandwidth to better utilize system or network resources, or to change to a more optimal path if one should become available. If a tunnel is reoptimized the mplsTunnelARHopTable should contain the actual path that is in use shortly after this notification is issued by the agent. The mplsTunnel-Reoptimized notification contains two mandatory objects: mplsTunnelAdmin-Status and mplsTunnelOperStatus.

Figure 8.12 depicts a sample MPLS tunnel configuration with one tunnel beginning at node A and continuing to node C along the path {A, E, F, G, C}. At some point the tunnel interface at node A decides that a more optimal path is available along the path {A, E, F, G, I, C} and thus resignals this path. However, this is signaled using the "make before break" algorithm, which signals the new path before tearing down the old one, thus preserving the flow of traffic over that tunnel. Node A would issue an mplsTunnelRerouted notification as soon as the new path was engaged. The mplsTunnelARHopTable would reflect this change as well.

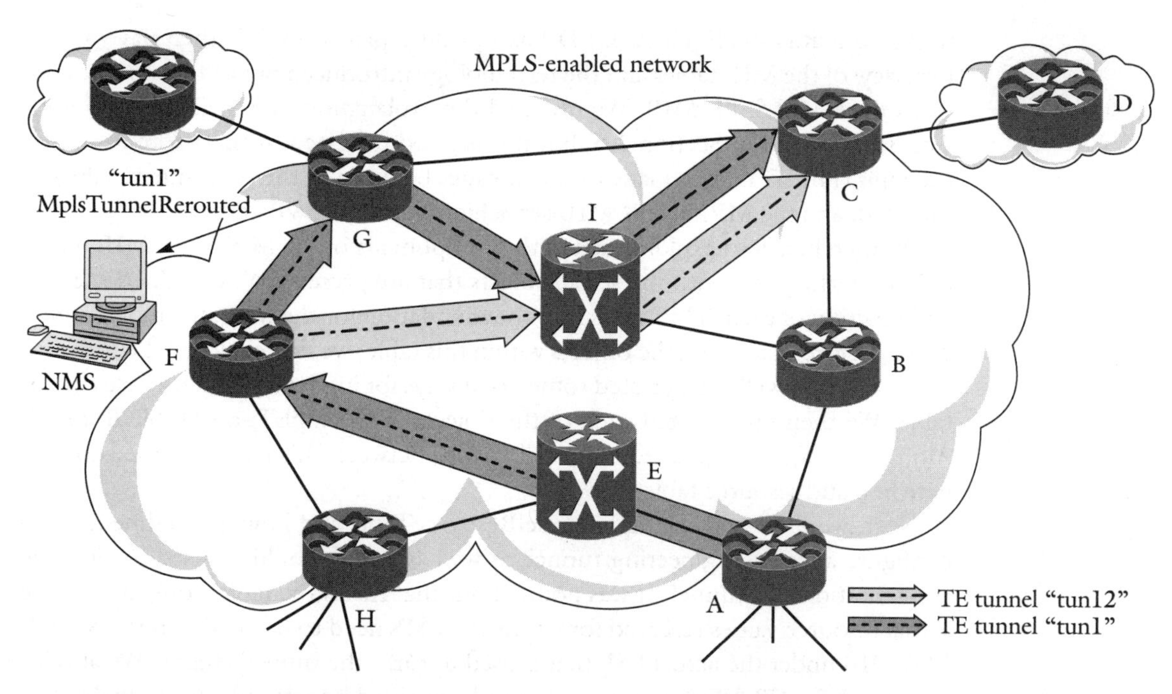

Figure 8.12 Sample MPLS tunnel configuration showing one tunnel beginning at node A and continuing to node C along the path {A, E, F, G, C}. At some point the tunnel interface at node A decides that a more optimal path is available along the path {A, E, F, G, I, C} and thus resignals this path. However, this is signaled using the "make before break" algorithm, which signals the new path before tearing down the old one, thus preserving traffic. Node A would issue an mplsTunnelRerouted notification as soon as the new path was engaged. The mplsTunnelARHopTable would reflect this change as well.

Another version of this example occurs when an additional backup tunnel is signaled ahead of time. In this case, a second LSP appears in the tunnel table as a second tunnel instance. The only difference between this instance and the one being used to actively forward traffic for the tunnel is that its operational status will be set to **down (2).** It will also have a tunnel instance priority that is higher than the one being actively used. This tunnel LSP can then be switched to transparently by the head end without interrupting traffic. Upon switching to a different path, the head end interface should issue an mplsTunnelReoptimized notification.

8.18 Summary

This chapter introduced and described the MPLS Traffic Engineering MIB (MPLS-TE MIB). The chapter began with an introduction to basic constraint-based routing, and then explained how constraint-based paths could be signaled using both RSVP-TE and CR-LDP as signaling protocols. We then gave a brief overview of the MIB tables and the terminology introduced or used within the context of the MPLS-TE MIB. We revisited the explanation of SNMP RowPointers. This was covered in Section 2.6, but it was good to review again before exploring the mplsTunnelTable, as many of the linkages between it, the subordinate tables in the MIB, and the MPLS-LSR MIB are achieved using RowPointers.

We then began to explore each of the components of the MPLS-TE MIB one at a time, starting first with the scalar objects that are present. We explained the role and function of each. This led us into the foundation for the MIB: the mplsTunnelTable. Before examining the objects within this table, we first explored the indexing of this table, as well as suggested some good ways for implementing instances in this table. We then investigated each of the objects in the mplsTunnelTable in detail. Along the way, we explained the relationships between this table and the subordinate hop and resource tables.

Next, we discussed the mplsTunnelResourceTable and how it could be used to configure a traffic engineering tunnel, as well as how it could be used to view the configuration of a tunnel. It was pointed out that this is the table's only use, as the actual resource values reserved for a tunnel's LSPs need to be viewed in the MPLS-LSR MIB under the actual LSP that is used to carry the tunnel's traffic. We also investigated the CR-LDP resource table and explained how these objects could augment (literally) those found in the basic resource table that was used primarily for RSVP-TE–based implementations. We then examined the tunnel hop tables. These included the mplsHopTable, mplsARHopTable, and the mplsCHopTable. The differences and purposes of each table were explained, as well as how each fit together with and related to the base mplsTunnelTable.

Next, we examined the interaction and relationship the MPLS-TE MIB has with the IF-MIB. We explained how mplsTunnelEntries were implemented as interfaces and specifics of how each object (i.e., counters) should be interpreted when implementing tunnels as interfaces. We then examined a complex example of a tunnel configuration and how the tables in the MIB would be configured to support this example.

Finally, we defined and investigated the notifications as they are defined in the MPLS-TE MIB. We went through each notification individually by first explaining the definition and purpose of the notification. We then examined a brief example that illustrated when each notification should be issued by an agent. This also gave managers an idea of when to expect to receive such notifications.

Further Reading

Traffic engineering extensions to OSPF: *www.ietf.org/internet-drafts/draft-katz-yeung-ospf-traffic-06.txt.*

IS-IS extensions for traffic engineering: *www.ietf.org/internet-drafts/draft-ietf-isis-traffic-04.txt.*

Srinivasan, C., A. Viswanathan, and T. Nadeau. "Multiprotocol Label Switching (MPLS) Traffic Engineering Management Information Base." IETF Internet Draft. August 2001. *www.ietf.org/internet-drafts/draft-ietf-mpls-te-mib-08.txt.*

Durham, D., and R. Yavatkar. *Inside the Internet's Resource Reservation Protocol.* New York: John Wiley & Sons Publishers. 1999.

Davie, B. S., and Y. Rekhter. *MPLS: Technology and Applications.* First edition. San Francisco: Morgan Kaufmann Publishers. 2000.

Gray, E. W. *MPLS: Implementing the Technology.* Reading, Mass.: Addison-Wesley Professional. 2001.

To find out more about the IETF, visit their Web page at *www.ietf.org/.*

For more information about IANA, check out their Web site at *www.iana.org/.*

To locate information about the ITU: *www.itu.int/.*

To locate information about the MPLS Forum: *www.mplsforum.org/.*

To locate additional information about the OIF: *www.oiforum.com/.*

Cisco's Web site also provides a great deal of information regarding MPLS: *www.cisco.com.*

Harmen van der Linde

is a principal member of the technical staff at the AT&T Frame Relay Service department in Middletown, New Jersey. Harmen is currently the network management architect for AT&T's IP-enabled Frame Relay (IPeFR) service, one of the leading IP VPN service offerings in the United States. In this role, he is leading the design and implementation of a broad range of management solutions for IPeFR, which are focused on MPLS VPN provisioning, performance, and fault management. Harmen is also actively involved in the Internet Engineering Task Force, where he participates in the standardization efforts for MPLS VPN management.

As an operator who is deploying MPLS technology, what do you see as the most important challenges to the complete management and thus deployment of MPLS in large service provider networks?

The problem space for management of MPLS technology can be divided into two areas.

The first one is service provisioning: activation of new services in MPLS-based networks. Increasing demand for IP VPN services and growing expectations of service quality require fast and efficient activation capabilities for new services. Due to limited support of transactional provisioning interfaces, such as SNMP, and limited IP VPN management software solutions, provisioning of IP VPN services in large-scale deployment environments still poses many challenges for operators.

The second problem area for managing MPLS technologies is service maintenance or life cycle management. Still open areas for research are MPLS fault, configuration, and performance management, for which currently only limited management interfaces and tools exist.

Do you think that MPLS will ever be completely manageable, or is this simply something that is the Holy Grail of networking?

Despite the limited management interfaces and management applications for MPLS technologies, progress is being made to address these problems, which is reflected by the MPLS standardization efforts in the IETF and the growing number of vendors offering MPLS management solutions. Although there is still a long way to go, effective MPLS management solutions seem to be feasible and will probably become a reality with the increasing market demand for IP VPN services.

9

NetFlow Accounting

> "Not everything that can be counted counts, and not everything that counts can be counted."
>
> —Albert Einstein

Introduction

NetFlow technology provides an efficient and scalable means for metering flow-based network traffic. The information captured by NetFlow can later be aggregated by a NetFlow collector and then fed into offline applications including network traffic accounting, usage-based network billing, network capacity planning, network monitoring, and data mining capabilities. Vendors such as Cisco Systems and Juniper Networks offer a variety of devices that produce NetFlow data records that can be exported for use by offline NetFlow-aware management stations and applications, or aggregated by NetFlow-based collection applications. Both types of applications are available from many third-party vendors, as well as from some of those that produce NetFlow-capable network devices. This chapter introduces NetFlow and how it can be used within the context of an effective MPLS network management strategy.

9.1 NetFlow Overview

NetFlow technology enables a set of applications including network traffic accounting, usage-based network billing, network monitoring, outbound marketing, network planning, and data mining capabilities for network operators. Devices from vendors such as Cisco Systems and Juniper Networks, among others, export NetFlow information that is captured during the actual flow of traffic. This information is then made available in a bulk file–oriented format for NetFlow applications that can further process the information or display it for operators. The reason why a special bulk file format is used is due to the potentially large volume of per-flow accounting information that can be stored in the file. In fact, on some systems where continuous monitoring is performed, the file may grow for the life of the system. NetFlow collector applications retrieve data from devices and process it into a form that is more suitable for end user applications. NetFlow data mined from network devices can be used for network monitoring, user billing, and capacity planning or verification purposes. NetFlow data can also be used to verify Quality of Service (QoS) functions within a network by showing that each flow is actually receiving the bandwidth it was promised.

There are several benefits of using NetFlow in an operational network. NetFlow can be used to monitor a network while simultaneously gathering data. Later, an offline tool can process this data to determine what specific network traffic patterns exist. This information may later feed into a capacity planning exercise that can determine if new capacity such as additional physical links, traffic-engineered tunnels, or application servers need to be deployed or repositioned to compensate for the increased or changed capacity. NetFlow allows an operator to observe what traffic patterns exist within a network after such network modifications are made. NetFlow can be used as a means of monitoring network usage on a per-customer basis in near real time. This usage can then be used for billing purposes. NetFlow information can also be used to detect and resolve potential security and policy violations. NetFlow data, or the data that is derived through correlation or aggregation, can then be warehoused for later retrieval and analysis. For example, this information might be used later by a marketing organization at some Internet Service Provider (ISP) to determine which users are using what applications. This correlation can then be used to market specific services or applications more effectively to improve sales.

9.2 Flow-Based Accounting

Flow-based accounting was originally developed to enable service providers and network operators to collect accounting information from individual user flows that consumed network services. This information could then be used as a basis for the provider to bill customers for a specific service or level of service. This information could also be used as part of their periodic billing statement to substantiate charges.

An individual network flow can be defined as a unidirectional sequence of packets traveling between a source and destination address. However, a finer granularity of flows may also be possible if identified with a few additional fields from the IP packet. These fields include the IP protocol, the IP precedence bits, and the source and destination layer-4 ports (TCP, UDP, etc.). Because of their unidirectional nature, flows from a client to a server are differentiated from flows from the server to the client. Flows may also be differentiated by protocol. For example, Hypertext Transfer Protocol (HTTP) packets from a particular source host to a particular destination host constitute a separate flow that is different from a File Transfer Protocol (FTP) file transfer flow between the same pair of hosts. Note that the IF-MIB Interface Index associated with a flow is sometimes useful as an additional parameter, for example, to differentiate all traffic flows that traverse a particular interface, or simply to bill a customer for all flows traveling over their access interface.

Figure 9.1 illustrates how designated ports of each router can be configured to track NetFlow accounting data. Specific ports are typically configured with NetFlow traffic flow descriptor filters that are used to match against traffic. When traffic matches a particular filter, one or more counters may be incremented in the NetFlow record. Notice that ports of "edge" and "core" nodes can be configured with this feature. It may be beneficial to configure NetFlow on outward-facing interfaces at network cloud boundaries for accounting purposes. However, general traffic monitoring may require other NetFlow collection points. Later we will discuss under which conditions it is desirable to configure which type of port for the most effective NetFlow collection.

9.3 NetFlow Architecture

The NetFlow architecture is composed of several network devices and applications including the network node capturing the accounting information. When used in concert, these components can provide the operator with a comprehensive, flow-based network accounting solution that can be used to deliver the many benefits described earlier.

Figure 9.1 NetFlow can be used to measure flows at particular device ports. The triangles denote where NetFlow might be measured.

The NetFlow architecture is illustrated in Figure 9.2. The architecture describes three general categories of components that must interact to successfully implement the architecture. From left to right, the first general category contains the actual network node or nodes responsible for capturing the raw accounting statistics. Network nodes must be configured with specific rules that are used to match against traffic flows. The network node must make these accounting statistics available sometime later, via a file formatted using the NetFlow record format. The NetFlow record typically contains one entry per flow. In addition to rote data collection, network devices implementing NetFlow can be configured to aggregate the accounting information to some degree, as well as filtering using routing filters. Both methods, when applied in an appropriate manner, may achieve a reduction in the size of the NetFlow record, and thus the file transferred to the NetFlow collector. NetFlow functionality is typically configured by an operator explicitly because some NetFlow operations can be performance intensive and, if configured incorrectly, can result in degraded forwarding performance of the network node capturing the accounting information. As with all measurements, they have a varying effect on the information being measured.[1] Therefore, it may be desirable for an

1 Werner Heisenberg's Uncertainty Principle states: "The more precisely the position is determined, the less precisely the momentum is known." For more information about the Heisenberg Uncertainty Principle, see *www.aip.org/history/heisenberg/*.

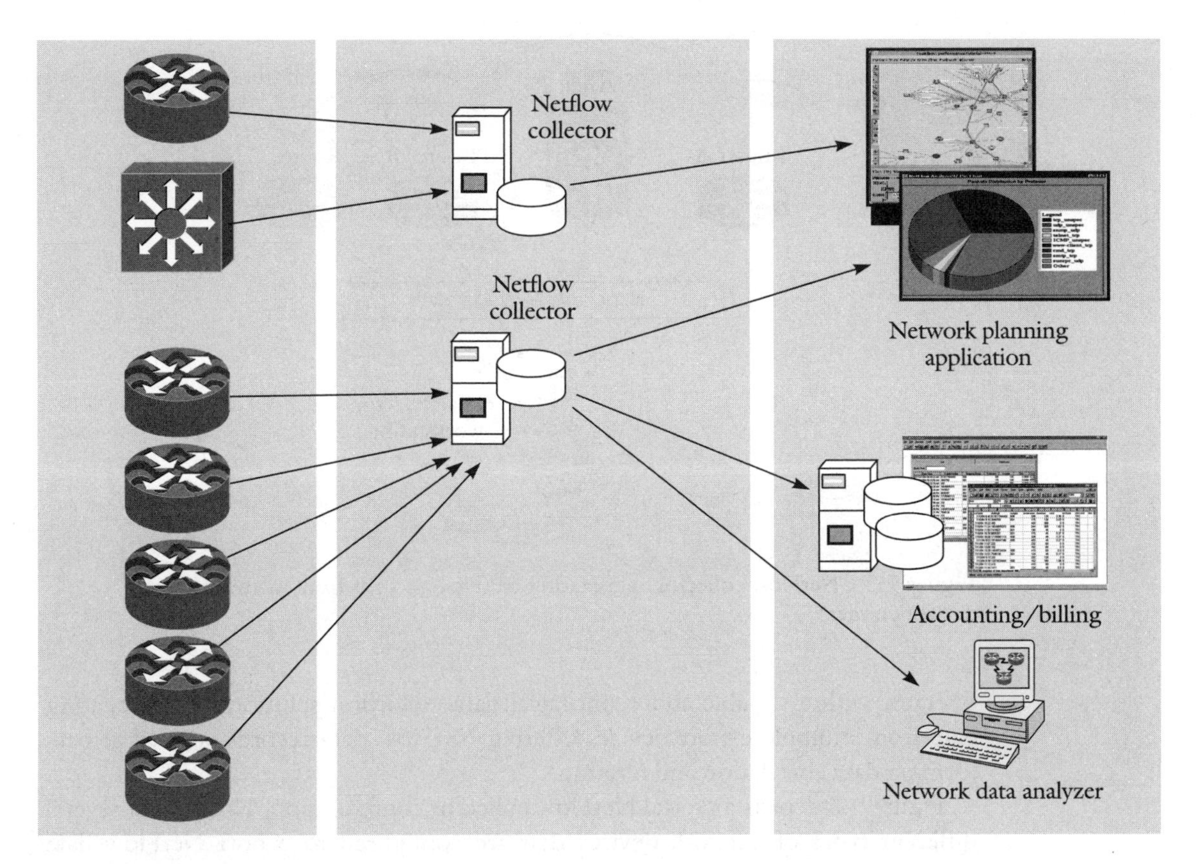

Figure 9.2 NetFlow achitecture.

operator to enable NetFlow accounting only during certain intervals, or during certain times of the day. The following sections detail the remaining portions of the NetFlow architecture.

9.3.1 NetFlow Collector

Central to the NetFlow architecture are devices used to centralize the collection of NetFlow records from network nodes. These devices are called *NetFlow collectors.* These devices typically consolidate and aggregate data from many different network nodes before it is given to offline processing applications. Vendors that provide NetFlow data for offline consumption by NetFlow-aware applications typically offer a midlevel manager application called a NetFlow collector. This application is used in conjunction with the NetFlow data export feature on the network device to gather and aggregate NetFlow records. The NetFlow collector can provide the

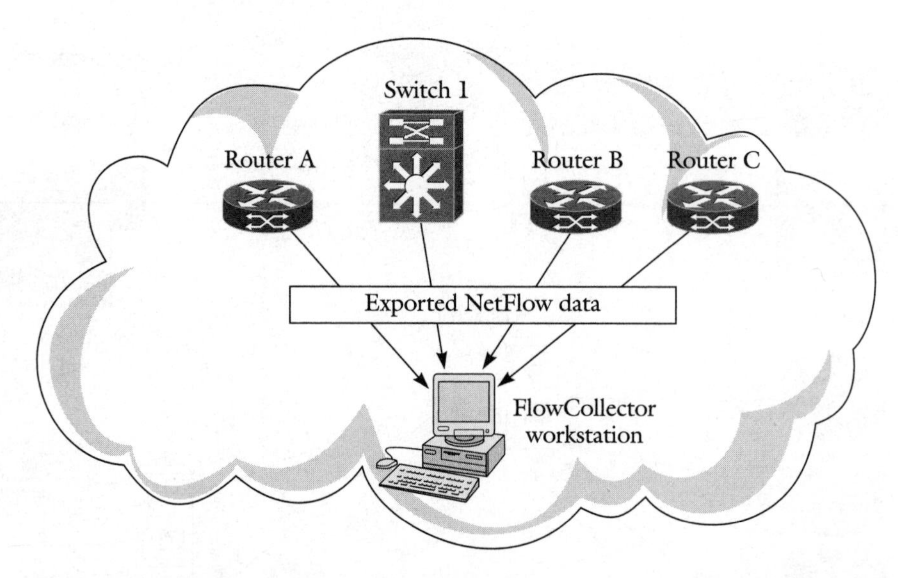

Figure 9.3 NetFlow collector aggregating NetFlow records from many different network nodes.

operator with a scalable and economical data collection solution by aggregating data from multiple export devices offering NetFlow data records, as well as performing data correlation and filtering.

Figure 9.3 depicts a typical NetFlow collector configuration. Notice that several different types of network devices that are configured to export NetFlow data transfer that data to a specific NetFlow collector device. Devices are typically capable of being configured to export their data to one or more collectors. This configuration is sometimes necessary to facilitate redundant data collection that can overcome the failure of one or more NetFlow collectors, or the network between the nodes and the collector device.

Some operators choose to deploy NetFlow collectors in a distributed manner, whereby one NetFlow collector gathers data from a subset of the network's nodes. This can facilitate the use of less expensive machines that run the collection software, but also adds a layer of fault tolerance to the network in that no one single point of NetFlow collection failure will result in all NetFlow data being lost. This approach also allows multiple collectors to collect data from the same network node, which might afford a dimension of redundancy. Although it is possible to deploy NetFlow collectors in a distributed manner, it is also perfectly acceptable to deploy a single collection device if that configuration suits your network requirements.

Once NetFlow data is collected by the NetFlow collector, several postprocessing steps can be taken to compress and concentrate the data so that NetFlow applications can better use it. Most NetFlow collector applications can be programmed to concentrate data volume through filtering, correlation, and user-programmed heuristics. NetFlow collectors offer many different aggregation schemes that can be configured. The specifics of these schemes are not discussed here, since they are too numerous and are beyond the scope of this section. Please see the related items listed in the Further Reading section at the end of this chapter.

NetFlow collectors also typically employ some sort of hierarchical data storage that can be used to make the aggregated NetFlow data more easily accessible to NetFlow client applications wishing to retrieve that data.

9.3.2 NetFlow Applications

The final component of the NetFlow architecture is the applications that process the data collected by the NetFlow collectors. Numerous NetFlow applications exist. These applications can be as simple as a UNIX command-line application that displays the raw postprocessed data. Applications can also be as complex as an integrated set of very sophisticated billing, network planning, and correlation programs. When these applications are used for offline network engineering, the results of their processing may take the form of network reconfigurations or redeployment of existing network nodes or links. These applications can determine if a particular user is overusing their subscribed service, for example, as well as bill that same user for the services it consumes. In short, many possibilities exist for providing useful tools for operators in this area.

9.4 NetFlow Data Export

The export of NetFlow data by network nodes makes NetFlow traffic statistics available for purposes of collection, and perhaps postprocessing by network planning or billing applications. A network device configured for NetFlow data maintains a flow cache used to capture flow-based traffic statistics. Traffic statistics for each active flow are maintained in the cache and are incremented when packets within each flow are switched. Periodically, a summary of traffic statistics is exported using the NetFlow record file format and some form of file transfer.

NetFlow data is exported and sent to a user-specified destination, such as the workstation running a NetFlow collector application. This can occur either when the number of recently expired flows reaches a predetermined maximum or after some configured interval of time has elapsed. Flows expire after traffic from the flow has not been seen by a network node for some predetermined period of time. Limits on the number of flows that can be exported are imposed by the version of NetFlow supported by the device as well as the processing and switching horsepower that the device is capable of. Therefore, the volume of NetFlow data that can be captured and processed by any single device may vary. In addition, since different versions of the NetFlow record format exist (see Section 9.4.1), different devices within a network may write records in slightly different formats. Therefore, NetFlow collectors that wish to be robust and function with different types of network devices should understand all versions of the format.

9.4.1 NetFlow Export Data Format

There are several NetFlow record formats. Currently, eight different NetFlow record formats exist, not including a new ninth one that supports MPLS. The newer formats are backward compatible with the older ones; the new formats only build upon prior ones by adding new fields to the record. This allows newer collectors to work with older data formats. Each NetFlow record format contains a set of basic information as part of the detailed traffic statistics portion of the record:

- Source and destination IP addresses
- Next-hop address
- Input and output interface numbers
- Number of packets in the flow
- Total bytes (octets) in the flow
- First and last timestamps of packets that were switched as part of a particular flow
- Source and destination port numbers
- Type of service (ToS)
- Layer-4 protocol

Additional information such as an MPLS label or VPN identifier can also be included in the NetFlow record. The newest formats of the NetFlow record have specifications for such fields to accommodate new network technologies that accounting information is required for.

The NetFlow record format used is a de facto standard. Cisco Systems introduced the format several years ago, and many companies including Cisco continue to use it today. Despite its age, the format continues to be supported by their devices as customers find NetFlow accounting information to be quite useful. The NetFlow record format is freely available via the Cisco Web site. For these reasons, numerous network device vendors now produce devices that can export NetFlow records in one of the specified formats. Vendors have also produced interoperable NetFlow collectors, as well as NetFlow applications. Due to its popularity, there is currently a move afoot to standardize the NetFlow record format with the IETF. This work is new, so we will not investigate it further here; however, if you are interested in more information about this effort, the IETF's Web site is the best place to look.

9.5 Deploying NetFlow

As was noted in the introduction, NetFlow can be deployed at several points within a network. The specific points where NetFlow is enabled depends on whether flow-based statistics need to be captured for billing and accounting purposes or for capacity-planning purposes. Furthermore, it is important to understand that NetFlow accounting can be deployed in *both* places depending on the accounting requirements for a specific network. Figure 9.4 shows the measurement points that are most effective for either type of application.

9.6 NetFlow Accounting for MPLS

MPLS provides network operators with several advantages including many useful applications such as traffic engineering, virtual private networking, and pseudo-wire

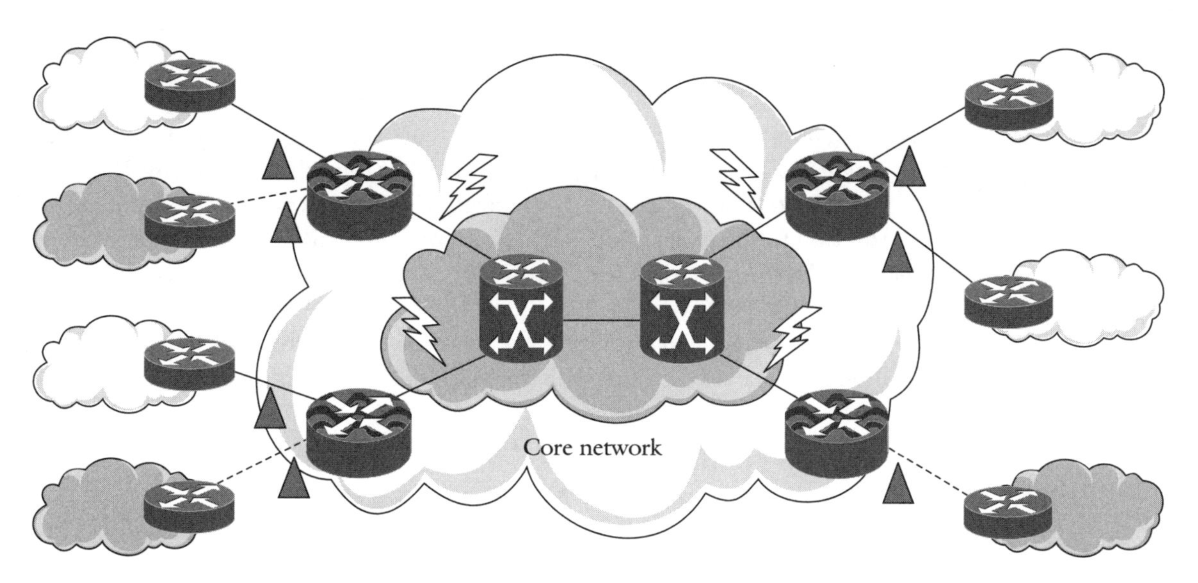

Figure 9.4 Where to deploy NetFlow when using it for billing and accounting purposes. The triangles show where NetFlow should be deployed to capture billing and accounting statistics, while the lightning bolt icons show where to capture NetFlow statistics.

emulation services. However, operators still need to monitor the traffic flows that utilize the MPLS network. NetFlow can be used to collect per-flow statistics at the ingress LERs by monitoring flows that enter a device. Flows that enter the core will not be visible using NetFlow accounting until they have traversed the core MPLS network and are ready to have their labels deposed. Once at the egress edge LER, the MPLS header is stripped and the packets are ready for transmission as native IP frames. At this point it is possible to collect NetFlow billing and accounting information using a feature that some MPLS device vendors call *egress NetFlow accounting*.

MPLS NetFlow accounting allows an operator to capture IP flow information for packets that traverse an MPLS network by accounting for it as the flows enter and exit the MPLS network. IP packets arriving at an ingress edge LSR are typically matched against a Forward Equivalency Class (FEC) that is stored in the LER's forwarding database. A FEC typically contains one or more IP destination prefixes and perhaps other IP packet information to match against the packet's header. When a match is made, this information is used to match IP packets with a next-hop MPLS LSR. Once an incoming packet has been assigned to a FEC, an MPLS header is imposed using the label stack corresponding to that FEC's MPLS next-hop. There are, however, two problems with accounting for the packet at this point or anywhere within the MPLS network. First, per-flow statistics can no longer be kept for an IP packet once the MPLS header and corresponding label stack have been assigned. This is because intervening LSRs are technically not allowed to look beneath the MPLS header, and therefore cannot discern any IP-related information, much less which flow the packet belongs to. Second, per-flow accounting is no longer possible since a FEC can contain more than one IP prefix that it aggregates onto a single MPLS label. This can result in the commingling of many IP flows onto a single label switched path, which effectively prevents per-flow accounting.

The solution to the problem of NetFlow accounting within an MPLS network is to collect the NetFlow statistics at the *edges* of the MPLS network. Specifically, this is possible at the ingress edge before the IP packets have MPLS headers imposed, and at the egress edge after the MPLS headers have been stripped. In both cases, the IP header is available to the LER so that per-flow IP information can be accessed easily.

9.6.1 Using NetFlow Accounting for MPLS VPN

The use of NetFlow at the egress edges of an MPLS network is essential to monitoring flows as they traverse the MPLS VPN network. In particular, within the context of an MPLS VPN configuration, it is typical to capture IP packet flow information as packets arrive from customer edge (CE) devices. These packets arrive as

IP traffic from customer sites. Since the traffic arrives as native IP packets, it can be readily classified on a per-flow basis. However, as was described earlier, when this traffic is subsequently forwarded into the MPLS core, per-flow information is no longer visible until it reaches the point where the MPLS header is again removed. Within the MPLS VPN environment, this is not possible until the packet has traversed the MPLS network and arrives at the corresponding customer edge device.

MPLS egress NetFlow accounting can be used to capture the site-to-site MPLS VPN IP flows that traverse the service provider MPLS backbone. Figure 9.5 shows a sample MPLS VPN topology. The Xs mark where NetFlow accounting can be enabled to allow the LERs to capture per-flow statistics for the intersite VPN traffic. Notice that the leftmost PEs can monitor traffic on a per-VPN basis for IP flows entering from both VPN A and VPN B, as the traffic travels to its respective VPN site located on the other side of the MPLS core. The same can be done by the PEs that reside on the right-hand side of the figure. The flows that are captured are made available to the respective NetFlow collectors. Dual NetFlow collectors are deployed in the figure and are responsible for gathering NetFlow records from a subset of PE devices. This information is then transferred to some set of NetFlow-aware applications that can then represent this data for an operator.

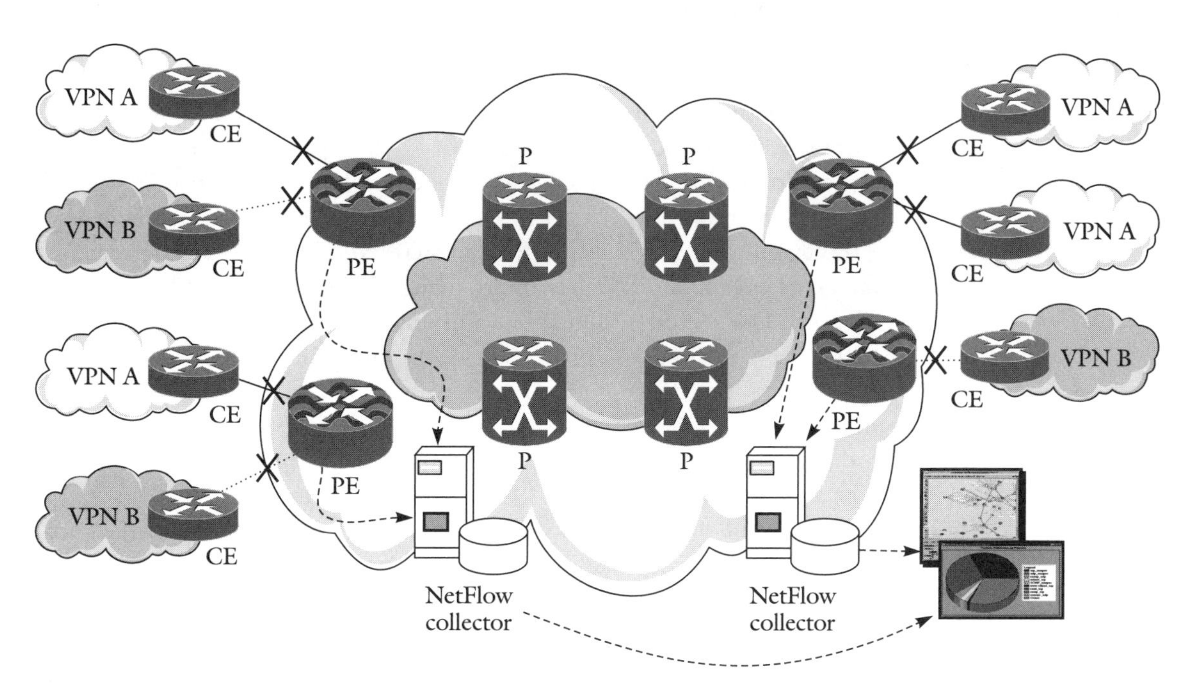

Figure 9.5 NetFlow can be used to measure flows on a per-VPN basis in an MPLS VPN environment. The Xs mark where NetFlow can be measured.

9.6.2 NetFlow Impact on Packet Forwarding Performance

NetFlow routers must maintain a record of each packet flow in order to keep track of each one. However, some flow-based routers in the past have had difficulty maintaining large volumes of per-flow information for two reasons. First, much of the traffic at the edges of the network tends to be quite short-lived. Therefore, a node may be called upon to maintain a database of flow history that becomes stale quite quickly. Second, this database can become very, very large in a short period—especially for an edge device. This can impact the performance of the device in that its flow cache may not be big enough to be effective.

The key factor that is required for NetFlow-enabled routing and switching devices to scale desirably is a highly intelligent flow cache management system that can overcome both of these shortcomings. This is especially important for highly used edge routers that may handle large numbers of concurrent and "bursty" flows. Devices running NetFlow must typically employ a highly sophisticated set of algorithms that can be used to perform several operations in an efficient manner. These include the ability to dynamically update per-flow accounting statistics maintained within the device's NetFlow cache; the ability to quickly and efficiently decide if an incoming packet either requires a new NetFlow cache entry to be created or if one exists and should be updated; and the ability to correctly age cache entries when flows expire.

Even with efficient NetFlow cache implementations, NetFlow accounting may still have a noticeable impact on the packet forwarding performance of a device. This can depend on many factors such as the overall packet processing ability of the device, as well as the number of NetFlow rules that must be used to match against incoming traffic. In general, those platforms where NetFlow does impact performance may experience a performance improvement if the number of flows maintained in the NetFlow record is minimized. This may be particularly useful on edge devices, which tend to have less processing horsepower and may lack hardware-based or hardware-assisted switching and/or routing table lookups. Another improvement gain may be achieved in carefully selecting NetFlow collection strategies that are granular enough to capture all of the necessary flow information, while being broad enough to keep the volume of this information manageable. This set of information should be just adequate enough so that the necessary accounting information can be produced after further (offline) processing. This approach can save on network node processor usage and may reduce the amount of network traffic attributed to NetFlow record file transfers.

While many network operators and service providers are comfortable using flow-based accounting because it can fit into their traditional billing and accounting methods, it is important for operators to consider the performance and operational implications of using such an approach to accounting. As with any of the accounting mechanisms discussed in this book, we recommend investigating all of the relevant approaches prior to using any particular one. While there are many devices available that offer NetFlow accounting, these same devices also offer a variety of other accounting solutions as well. Some examples are TMS statistics (see Chapter 10) and per-interface accounting (see Chapter 6), among others. While some vendors will make wild claims about the efficacy of one accounting mechanism over another (usually based on some ulterior motives), we recommend that these different mechanisms be weighed together within the context of your specific network before passing judgment on a particular accounting approach. One or more of these tools may be appropriate for a particular network.

9.7 Summary

This chapter introduced NetFlow-based accounting as one form of per-flow-based accounting that could be utilized by operators to capture per-flow IP traffic statistics from network devices. This information could enable many important applications including network traffic accounting, usage-based network billing, network monitoring, outbound marketing, network planning, and data mining capabilities for network operators.

We first defined what it means to gather statistics on a per-flow basis. We explained that devices at the edge of a network could be configured to look for certain types of IP packet configurations and, once seen, could mark this information down. This information could then be used to monitor per-flow traffic. This is important because per-flow traffic usage is a means by which operators can monitor and eventually bill customers of their networks for usage.

We next described the NetFlow architecture. This architecture is composed of three different types of applications. The first are the network nodes that are responsible for forwarding the IP traffic. When these devices support NetFlow accounting, they are able to capture per-flow statistics for IP packets that traverse their interfaces. Many network device vendors produce network nodes that are capable of exporting NetFlow information. We next discussed NetFlow collectors, which comprise the middle section of the architecture. These devices are responsible for collecting NetFlow accounting information from multiple network nodes, and then pass this information on to interested NetFlow-aware applications. We pointed out that NetFlow collectors sometimes do intermediate processing of the aggregated NetFlow information in order to either compress it or otherwise make it more useful for the NetFlow-aware applications. We discussed NetFlow applications as the final element of the NetFlow architecture. The benefits of these applications are numerous, and include billing, usage, and network planning.

We then discussed the NetFlow record format and pointed out that, although this format was invented by Cisco Systems, many other vendors have adopted this format for their devices. This format is also being standardized by the IETF.

Next, we investigated how NetFlow could be deployed, first in a typical IP network, and then within the context of an MPLS network. In the case of MPLS, we specifically looked at an example of how it could be used for MPLS VPN networks.

Finally, we discussed potential performance issues related to deployment of NetFlow. We touched on scalability issues and provided some hints as to how to safeguard deployments from potentially devastating network performance degradation due to not configuring NetFlow appropriately.

Further Reading

Cisco NetFlow Services and Applications White Paper available at *www.cisco.com/warp/public/cc/pd/iosw/ioft/neflct/tech/napps_wp.htm*.

Cisco's Web site provides a great deal of information regarding NetFlow and MPLS: *www.cisco.com*. Some relevant links for NetFlow at Cisco are
www.cisco.com/univercd/cc/td/doc/cisintwk/intsolns/netflsol/nfwhite.htm
www.cisco.com/warp/public/cc/pd/rt/12000/tech/splm_wp.htm
www.cisco.com/univercd/cc/td/doc/product/software/ios121/121newft/121t/121t5/egress.htm

Cflowd Web site provides additional insight into NetFlow: *www.caida.org/tools/measurement/cflowd/*.

Juniper's Web site provides insight and information into their NetFlow offerings, as well as an interesting white paper on per-flow accounting. These can be found at *www.juniper.net* under the "white paper" section of their Web site.

Davie, B. S., and Y. Rekhter. *MPLS: Technology and Applications*. First edition. San Francisco: Morgan Kaufmann Publishers. 2000.

Gray, E. W. *MPLS: Implementing the Technology*. Reading, Mass.: Addison-Wesley Professional. 2001.

To find out more about the IETF, visit their Web page at *www.ietf.org/*.

XiPeng Xiao

is director of technical marketing at Photuris Inc., where he interacts with carriers and service providers and defines product architecture. Prior to Photuris, he was senior manager of advanced technology at Global Crossing Telecom. He designed network architecture, deployed VPNs and QoS, and engaged in operation/management of Global Crossing's IP network. The MPLS network he built from scratch at Global Crossing was the largest in the world as of October 2001.

XiPeng's work on MPLS, traffic engineering, multiservice networking, and QoS is well recognized in the Internet community. He has published RFC 2873 (standard track), several IETF drafts, and many widely referenced journal papers on MPLS, traffic engineering, and QoS.

Xipeng, you have been an active member of the IETF community for some time now. Today MPLS is a mature technology and has been deployed in lots of production networks. What do you see as the issues and hurdles (and possible solutions) for current and future deployments of MPLS? Where do you see MPLS and GMPLS headed in the not-so-distant future?

I see two issues for the current and future deployments of MPLS.

The first one is a philosophical issue. IP has been used as the sole network protocol in the Internet for a long time. Some people feel that IP is simple and good enough, and are philosophically against introducing anything new or more sophisticated. There is little you can do to change these people's opinion. While this opinion slows down deployment of MPLS, it also helps MPLS in the sense that it curbs some "perfectionists" from making MPLS omnipotent and overly complex.

The second one is a practical issue. For MPLS to be widely deployed, people need to have experience working with MPLS. However, people will not get the experience without deploying MPLS. This "chicken and egg" problem implies that the wide adoption of MPLS will be a slow process. The quality of equipment vendors' MPLS implementations plays an important role—if badly implemented, even a good protocol cannot be deployed.

I started deploying MPLS in 1999 at Global Crossing. Over the years, I have seen that the quality of MPLS code from major vendors has improved significantly, and many more operators have gained operational experience with MPLS. Therefore, I think the second issue is being addressed.

While being path-oriented, MPLS uses the same equipment as IP (i.e., IP routers as LSRs) and uses IP's control protocols (e.g., BGP, RSVP). This enables MPLS to be a good complement to IP. However, people should not be too ambitious about MPLS and try to make it too sophisticated. Such "perfectionists," in my opinion, pose a bigger threat to MPLS today than the "IP fundamentalists" do.

10

Traffic Matrix Statistics

"There are three kinds of lies; lies, damned lies and statistics."

—Benjami Disraeli

Introduction

Traffic engineering is a tool used by operators of networks as part of an overall program of network management. Use of this tool includes a periodic cycle of network analysis and/or predictions of network traffic growth. This process is fed and driven by first gathering network statistics from network nodes or intermediate collection devices. This data is then used as input into intermediate traffic engineering analysis processes that may determine if any adjustment to either the physical or the logical configuration of the network is required. One popular mechanism that is used today to collect and view traffic engineering data from network nodes is called Traffic Matrix Statistics (TMS). This chapter will present an overview of TMS, how it may be used, and where it might be most useful.

10.1 The Traffic Engineering Problem

In many cases, networks that function correctly and efficiently do so because the nodes and their corresponding links that comprise the network are not overloaded, and yet have traffic optimally distributed across these links and nodes. It is often an important goal of traffic engineering to solve the problem of how to program the traffic flow distribution across network nodes and their corresponding links such that the traffic dispersal best meets the constraints of the network resources. In essence, traffic engineering can be thought of as a problem that can be solved by constantly reprogramming the network for optimal network resource utilization through proper traffic flow distribution. In addition to optimizing traffic load dispersion throughout the network, traffic engineering can also be used to program backup tunnel paths that can overcome different failure scenarios, as well as a means by which an operator can offer preferred service to selected customers.

The essential ingredient in any traffic engineering system is the persistent characterization of network demand in terms of network node resources such as link bandwidth and buffer resources. Specifically, traffic characterization needs to be performed at regular intervals over the course of a network's lifetime. This information can in turn be used to optimize the current network configuration when fed into a traffic engineering system. Such a system can examine the totality of traffic characterization information for a large number of network nodes. The sum of traffic information can be characterized as a matrix containing traffic information indexed by each network node and then each link operated by that node. The most practical and cost-effective means of forming such a traffic matrix is to employ the actual network nodes to construct a matrix where traffic is being measured, and then make it subsequently available for offline retrieval. This chapter focuses on such a mechanism, called Traffic Matrix Statistics.

10.2 Traffic Matrix Statistics Objectives

Traffic engineering tools require that traffic-engineered tunnels and the flows traversing those tunnels be represented in a form containing control-level information. This information can be collected based on several different types of traffic flow identifications.

Traffic engineering and the performance analysis phase of traffic engineering generally require that the ingress and egress points for any flow be specified based on the network controls used to set up that flow. This may include end user, external interface of a backbone node, host identifier, IP network number, BGP next-

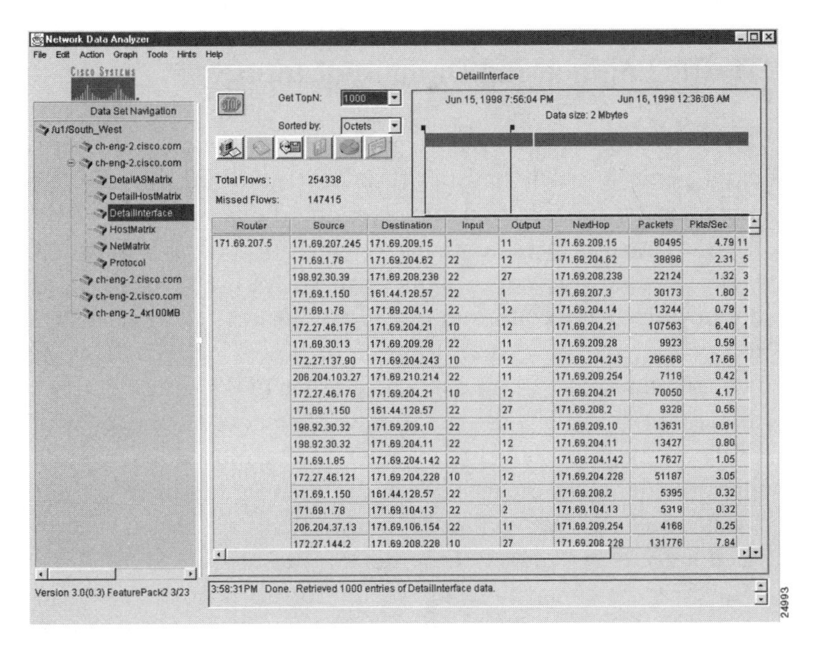

Figure 10.1 Traffic Matrix Statistics represented as a matrix using the Cisco Network Data Analyzer to display the TMS matrix.

hop, administrative domain, backbone service provider, a backbone node, grouping by application, and a single backbone at large or multiple company/agency networks. In addition to those flow parameters just specified, traffic matrix information may be gathered for MPLS flows based on BGP next-hop AS, BGP next-hop address, destination prefix, MPLS precedence bits, and source port or destination port. Figure 10.1 illustrates how Cisco's NetFlow Analyzer can be used to display TMS statistics as a matrix of traffic accounting information. This matrix is indexed by network node and then destination prefix.

When considered together, the ingress and egress points for any given flow constitute an end-to-end traffic matrix. The data used in traffic engineering must necessarily have the same degree of resolution as the traffic to be mapped to tunnels; otherwise, the result of processing the data may return incorrect or inaccurate results. In general, only two requirements exist for any flow represented in the traffic matrix. First, despite the specific type of classification for any flow, it must be uniquely identified so that it can represent a unique "cell" within the traffic matrix. Second, the traffic engineering system must be able to discern the device where the flow entered the traffic-engineered domain. This is in part because the TE system may need to modify the entry point for that flow after subsequent traffic analysis.

10.3 Traffic Engineering Domain of Interest

Traffic Matrix Statistics can be gathered in many different network domains. The exact point at which statistics should be gathered depends largely on the intent and focus of the operators implementing the network. However, regardless of the specific network, most networks contain some points of measurement that can generally be used for traffic engineering analysis. For example, most ISP networks are organized as a collection of *points of presence* (POPs). A POP is typically a collection of geographically co-located (or closely located) network nodes. Each POP is interconnected with one or more other POPs, and it is this collective of POPs that forms the ISP network. Bandwidth between devices within a POP is relatively inexpensive compared to inter-POP connections, typically because the service provider owns the facility and can therefore easily interconnect nodes residing within the POP. Network nodes within POPs are usually wired with very fast connections such as Gigabit Ethernet or Packet over SONET (POS) connections. However, the connections between a service provider's POPs are typically much more expensive than these local ones because of the greater distance these connections must span. These connections are also costly because the service provider usually does not own these connections and therefore must pay another service provider for the use of these lines. It is for these reasons that operators wishing to perform traffic engineering on this type of network may wish to focus on optimizing the use of expensive inter-POP connections to maximize their use. Likewise, measurement of Traffic Matrix Statistics within this network would focus on inter-POP connections.

The configuration illustrated in Figure 10.2 shows how an ISP can be constructed from four disparately located points of presence. Notice that the POPs in the figure contain links that are relatively large compared to any of the inter-POP connections. Link bandwidth is indicated by the relative width of each link. These links represent high bandwidth and relatively low-cost connections such as Gigabit Ethernet. Each POP interconnects the ISP with other ISPs and service providers (SPs), providing its customers with access to other networks including the Internet. Operators performing traffic engineering measurements would capture statistics between the New York, Los Angeles, Warsaw, and Paris POPs as well as the other ISP and SP links to assure that traffic was balanced between these expensive links as much as possible.

It is important to understand that the minimum amount of end-to-end traffic statistics should be gathered that could be used to optimize the specific network configuration of interest according to the goals of the operator. Since the locations

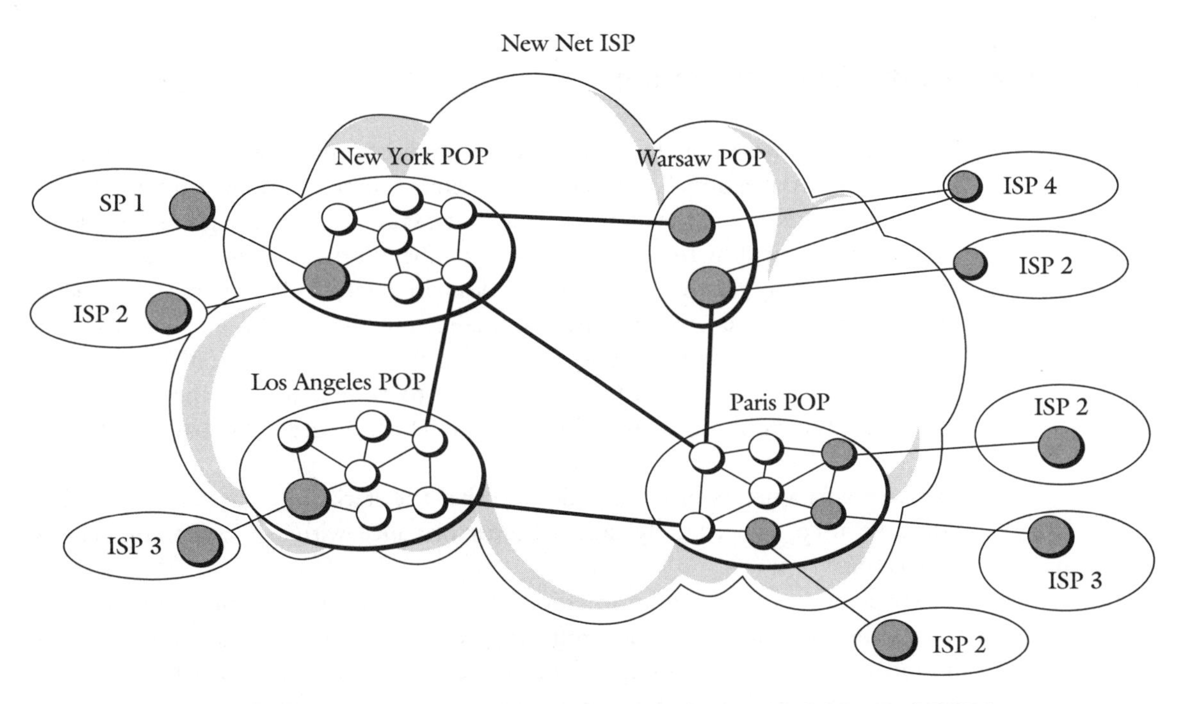

Figure 10.2 A typical ISP network organized as many geographic-specific POPs. Each POP is interconnected to the other POPs as well as to other SPs and ISPs using a variety of link technologies. Thicker lines between nodes represent higher-bandwidth links.

specified for gathering traffic matrix data are user-defined, it therefore behooves the operator to only collect as much traffic engineering information as is necessary. Since the collection of traffic accounting information results in the consumption of network node resources (i.e., processing power, disk space, network bandwidth), operators should take care in collecting only the information they need. Collection of superfluous information will incur additional processing time by the traffic engineering tool, node resources, as well as additional network resources for off-loading that information to the tool. Since it is typical of the ISP deployment shown earlier to connect only a small percentage of the routers in a POP to routers within other POPs, only a few wide-area inter-POP links need be configured to have traffic data gathered. Monitoring other links will not provide data that will additionally enhance the traffic eengineering results. The TMS traffic matrix in this case would be configured to represent traffic entering and exiting the POPs. This traffic information would be periodically off-loaded to the traffic engineering system for processing and analysis.

10.4 Traffic Characterization

Traffic in IP networks can be represented as unidirectional flows that enter and exit the network in question. Traffic flow parameters can be captured at different measurement points within the network. The traffic statistics gathered provide a relative measure of traffic flow through the domain. The reason why these measurements are noted as being relative to the specific domain in which they are measured is because traffic flows may traverse many domains, but the measurements of the flow that is necessary for traffic engineering within that domain are relative to *that domain*. Therefore, traffic statistics do not need to be measured at the actual starting and ending points of each traffic flow, just where it enters and exits the domain of interest. However, the statistics captured about each specific flow must indicate the salient parameters of a traffic flow so that the statistics gathered for that flow could be used to accurately model its performance within a traffic engineering application.

Traffic flow accounting information should be measured at the points where it makes the most sense for the traffic engineering system. This is typically at the boundary of a network domain. The conceptual node in Figure 10.3 lies at the boundary of a network domain. It is connected to some number of nodes that sit at the boundary of another network domain. In addition, the conceptual node is connected to other nodes within its network domain.

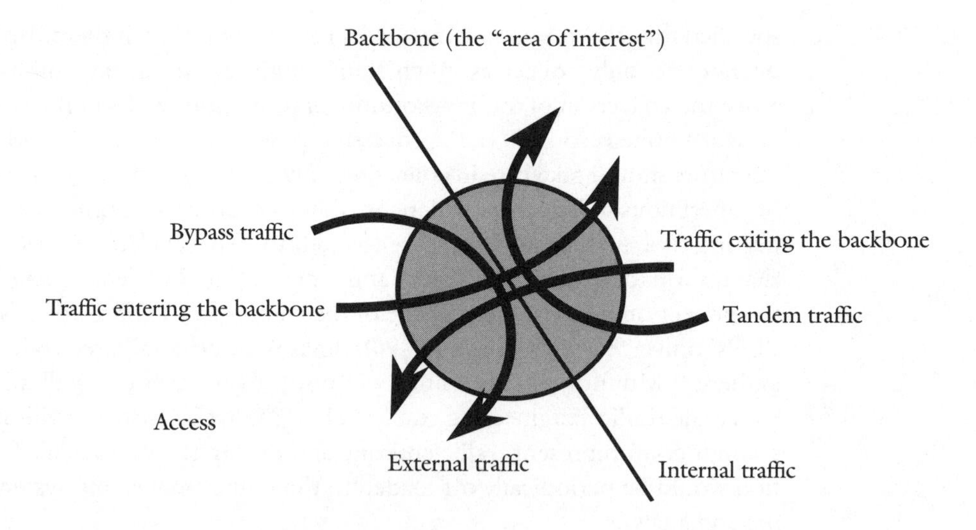

Figure 10.3 TMS traffic characterizations.

Four different types of traffic flows can be measured in general. These types are illustrated in Figure 10.3. The first classification is *bypass* traffic. This traffic merely uses the service provider as a transit node along a path that does not actually utilize the SP's internal network for transit. For example, if three networks A, B, and C are interconnected with a link from A to B and a link from B to C, then traffic from A to C arriving at B would be classified as bypass traffic, since it need not traverse B's network to reach C. The second type of traffic is referred to as *tandem* traffic. This traffic is contained within the network in question and therefore never leaves it. The last two categories of traffic deal with traffic that either is leaving or entering the network in question. Figure 10.3 also classifies traffic as being *internal* or *external* to a particular domain. Internal traffic includes both tandem traffic and traffic entering the domain, while external traffic includes both flows that bypass and exit the domain.

The minimum information typically gathered for a traffic flow includes the number of bits and packets transferred or received along a particular flow over some period. For example, the number of bits and packets transmitted over the course of an hour might be measured for an IP connection between an entry and exit point within a transit network. The reason why bit and packet counters are gathered is that both provide insight into the traffic being carried across that flow. For example, bit counters provide statistics important for link loading, while packet counters are useful in estimating network node loading. Furthermore, packet counters can sometimes be used to model Quality of Service behavior. In addition to these two basic counters, other information may be necessary to accurately measure traffic flow performance depending on what types of things a traffic engineering system will be instructed to model and analyze. Additional information that might be gathered includes the overall percentage of utilization of a particular link (or links), and packets dropped or discarded by interface (or interfaces). All of this data may be gathered for a specific traffic engineering tunnel within the context of an MPLS network.

Additional information may enhance the traffic matrix and may allow the traffic engineering system to more accurately model and predict network behavior. This information may, for example, provide insight into which links may fail in the near future. When gathered accurately, and during appropriate periods of time, some or all of this information can also be used to identify peak utilization or "hot spot" periods. All of the aforementioned counters are provided within standard IETF MIBs such as the IETF IF-MIB (see Chapter 6), the IETF MPLS-TE MIB (see Chapter 8), or the IETF IP-Forwarding MIB (RFC 2096). Packet and byte counters are provided by the IF-MIB for each interface represented therein. This may also include traffic engineering tunnel interfaces if the implementation in question supports such entries. Additional traffic statistics may be provided in other MIBs that may or may not be provided by the IETF.

10.5 Selecting Sampling Periods

The traffic matrix represents the *actual* measured traffic accounting as seen by the network devices that measured the traffic flows. Due to the essentially random nature of the network traffic being accounted for by the network nodes, the traffic matrix used to store traffic information should follow a specific temporal period so that each flow that the operator desires to monitor is captured accurately. The operator should make every effort to allow the traffic matrix to also account for any extremes of traffic traversing those flows. This is usually achieved by utilizing an extended sampling period.

Traffic Matrix Statistics must be gathered over some period of time to accurately reflect the reality of the network in question. Although the specific amount of time and methods used to gather statistics vary by the operational nature of the network being monitored, and can vary from over the period of a week or even a month, the target periods of activity should be the same times during which statistics are captured; otherwise the subsequent analysis will be inaccurate. At a minimum, it is recommended that data be collected during several known or predicted peak periods of network activity so that at least a somewhat accurate assessment of network and traffic flow behavior can be obtained. Sometimes gathering data over much longer periods has the advantage of providing the traffic characterization over an extended period without requiring any normalization of the data to remove normal usage patterns that are exhibited during normal network utilization. This approach also has the advantage of capturing abnormal or unexpected peaks in flow traffic that might not have been captured using shorter intervals.

The intervals during which traffic statistics are sampled play an important factor in determining how accurate the final traffic engineering analysis is. Therefore, we recommend that operators using TMS pay particular attention to the intervals specified. In particular, we recommend experimentation with intervals, starting with the longest intervals that your traffic analysis tools will tolerate.

Once the interval or intervals at which traffic statistics are gathered have been determined, the data can be analyzed and traffic patterns determined. However, the potential for missing data exists due to either loss or outages of network nodes, the links that interconnect them, or the vehicles used to transport the per-flow statistics to the offline traffic engineering tools. Therefore, statistics gathered at each router should be examined prior to analysis in order to identify any missing samples of traffic. If traffic samples are gathered from each node at a specific interval, missing samples are easy to spot. If a missing interval is found at a particular node, all is not yet lost. If a networkwide traffic matrix summary is used to identify periods of overall adequate or inadequate network capacity, then accurate estimates of missing data can be ascertained by simply comparing traffic leaving the network with the same traffic entering the network. If there is a difference in these numbers, some data was lost due to a lack of capacity or a network failure. If it is determined that a portion of data was lost during some period of time, the amount of data lost can be estimated by examining the amounts of traffic passing through the network elements neighboring the devices that the traffic should have traversed while in the network. This approach alone can often identify which flows were missing data with sufficient accuracy. Additionally, the traffic engineering system can use the sum of the TMS data to verify the accuracy of the routing functions in the devices measured. This is possible because actual routing measured from the gathered data can be compared against that of the routing predicted using a simulation model. If the two compare, then the actual routing functions are adequate.

10.6 Traffic Matrix Structure

A traffic matrix is a two-dimensional array comprised of elements that contain traffic accounting information such as bits and packets per second. The rows represent individual routers and the columns represent next-hop information such as BGP next-hop addresses. The packet counts captured are useful in estimating router loading, link loading, sanity-checking data (using an overall analysis), as well as compensating for any packet overhead not reported in the bit counts (i.e., some bit counts do not account for packet headers).

The following tables represent the information stored in the TMS matrix depending on the type of flow. Table 10.1 denotes the information that can be stored based on the Label Forwarding Information Base. These fields include the IP destination prefix and IP destination mask that are associated with an outgoing label binding, incoming label, interior and exterior packet counts, as well as total bytes received and transmitted for this flow.

Table 10.1 LFIB TMS information.

TMS field name	Description
IP destination prefix	Destination prefix
IP destination mask	Destination mask
Outgoing label binding	Outgoing label binding
Incoming label value	Incoming label value
Interior packet count	Number of frames sourced from within the area

continued

Table 10.1 continued

TMS field name	Description
Exterior packet count	Number of frames sourced from the outside of the area
Count of total bytes received*	Count of total bytes received
Count of total bytes transmitted*	Count of total bytes transmitted

* Both 32- and 64-bit counters are represented.

Table 10.2 denotes the information that can be stored based on the global Forwarding Information Base. This information includes the IP destination prefix and mask, BGP next-hop, the associated interior and exterior packets seen, and overall bytes sent and received for a particular traffic flow.

Table 10.2 FIB TMS information.

TMS field name	Description
Destination prefix	Destination prefix
Destination mask	Destination mask
BGP next-hop	May be the next-hop AS or the destination AS
Interior packet count	Number of frames sourced from within the area
Exterior packet count	Number of frames sourced from the outside of the area
Count of total bytes received*	Count of total bytes received
Count of total bytes transmitted*	Count of total bytes transmitted

* Both 32- and 64-bit counters are represented.

Table 10.3 illustrates the data that a node can track for an MPLS traffic engineering tunnel. This information includes the MPLS label used to carry the traffic for that tunnel, its head end IP address, and the total count of bytes received on the tunnel head interface.

Table 10.3 Traffic engineering tunnel TMS information.

TMS field name	Description
Tunnel label	IP destination address
Head IP address	IP destination mask

Table 10.3 continued

TMS field name	Description
Count of total bytes received*	Count of total bytes received
Count of total bytes transmitted*	Count of total bytes transmitted

* Both 32- and 64-bit counters are represented.

10.7 Measurement Architecture Options

Depending on the specific operational environment, several considerations should be made with regard to the manner in which TMS data is collected. In particular, these considerations may place several constraints on the collection process. These various options are outlined below.

10.7.1 Volume of Data

Operators who deploy and run large networks need the ability to manage these networks in an effective and efficient manner despite the large number of links and nodes deployed therein. Traffic engineering promises to empower these operators with just such a tool for managing these networks. Consequently, traffic engineering and the TMS mechanism for gathering traffic accounting information tend to be deployed in very large networks. As a result, the sheer volume of per-flow accounting information that must be gathered tends to be quite large. The size of this information poses some interesting challenges for those deploying TMS.

In an effort to speed the process of gathering the vast amounts of TMS data, tools have been created that can automatically collect this information, thus easing the burden on the operator. Due to the potentially large number of network nodes from which simultaneous statistics must be gathered, most of these tools are distributable. Specifically, more than one collection system is deployed throughout the operator's network and is designated to collect data from a subset of network nodes. Once collected, the information can be aggregated or repackaged before it is transported to the main offline traffic engineering system.

Due to the large volume of data that must be collected by network nodes—sometimes several megabytes per collection period—many systems employ a bulk file transfer mechanism instead of requiring individual SNMP operations to fetch each counter. However, SNMP is still used in some cases and is sometimes used as the mechanism to configure and activate the bulk file transfer.

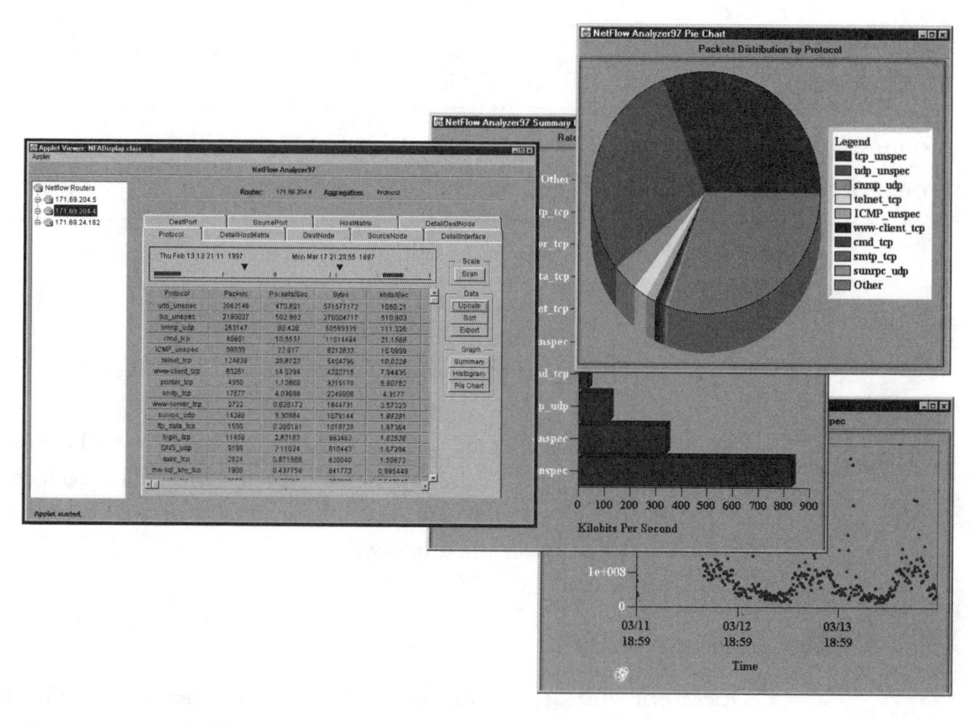

Figure 10.4 NetFlow collector using the Network Data Analyzer.

10.7.2 Aggregation of Collection Prior to Sending

One strategy for reducing the volume of data is to summarize prior to transmitting it to the TMS offline traffic engineering system. One approach is to allow the actual network node to do some aggregation and summarization. However, although this might result in a varying reduction in the volume of TMS data transported to the collection agents, the cost of such an operation may be expensive and may impact actual forwarding performance. Thus, it is recommended that any aggregation operations be performed on the TMS collection device instead of on the actual network node. One such application is the Cisco NetFlow Collector when used with the Network Data Analyzer.[1] A screenshot from this application is shown in Figure 10.4.

Second, an argument against any local summarization and aggregation is that it sometimes results in a reduction in the detail of the data to the point that it is not

1 See *www.cisco.com* for more information on Cisco's NetFlow collector and Network Data Analyzer products.

very useful to the operator using the traffic engineering system. Thus, the operator needs to carefully examine the costs and benefits of local summarization and aggregation prior to activating it. In some cases, it can result in a substantial improvement in network performance without loss of generality, but unfortunately this is not always the case.

There are several options in cases where it is decided to use an intermediate collection agent. One important option exists in how to configure the collector to interact with the network nodes. In particular, a proper interval needs to be selected after which a collector device will "pull" statistics from a network node, or the device will "push" the statistics to the collector. If the time interval is too short, the network node may run out of buffer space and begin to lose traffic statistics because the data will have not been off-loaded. Similarly, if the interval is set too long, this can also occur. In both cases, care should be taken so that the offline traffic engineering tool will have enough data to produce sufficient analysis. In addition, if the collector receives data too often, this may result in unwanted network congestion or overloading of the collector. The actual decision of which interval to select depends largely on the performance of the collector, the intermediate network, the overall collection architecture, and the overall traffic engineering strategy.

10.7.3 Protocol for Transmitting Bulk Files

The primary statistics required to design traffic engineering tunnels are the raw data from which a traffic matrix may be reduced. Much of this consists of a large volume of data collected simultaneously from a large number of routers. In many cases, SNMP cannot be used to transfer this volume of data due to network and node processing overhead. Therefore, TMS data collection is typically performed using some bulk file delivery mechanism. This mechanism can be configured to either "push" a virtual file using some form of network file system protocol to a collector or traffic engineering system or file transfer protocol (FTP), or it can be configured to allow a collector to periodically "pull" a file from it using either of these mechanisms. Vendors implementing a TMS system typically provide some form of bulk file transfer mechanism.

10.8 Cost and Performance Considerations

Traffic Matrix Statistics must be captured at network nodes that have links supporting traffic flows in question. The traffic accounting data captured must include bit- and packet-oriented counters that can be used to later model the behavior and loading of traffic flows across a particular node's interface. Depending on the goals of the traffic engineering being performed on a network, different sampling times and interval lengths can be selected by an operator. Within these confines, it is important to consider that the measurement, collection, and retrieval of TMS data might noticeably degrade the performance of network nodes where the data is being sampled. In these cases, the statistics that are gathered are tainted by the tool measuring them, since their values might otherwise be different if TMS data were not gathered. The Heisenberg Uncertainty Principle rears its ugly head again.

In addition to the basic network resources that can be impacted by using TMS, the cost of deploying a successful TMS system should also be considered. For instance, it may be possible to reduce the overhead of TMS data transfers from network nodes by deploying additional data collection agents, but a monetary cost is associated with each new agent. These costs take on several dimensions beyond the simple purchase cost of the agent. For example, each node requires system administration and software maintenance.

It is for these reasons that it is important to understand the performance and cost trade-offs inherent in a traffic flow measurement approach such as TMS. Tools and mechanisms designed to capture traffic flow statistics usually come with trade-offs in accuracy and usefulness in relation to node resource consumption (i.e., memory and bandwidth).

10.9 Summary

This chapter discussed how Traffic Matrix Statistics could be used as one means of collecting the data used by traffic engineering systems used by operators of networks as part of their overall program of network management. Network operators are motivated to squeeze every bit of profitability out of their networks to maximize profitability. Traffic engineering systems include a periodic cycle of operations including the data collection phase, the network analysis and predictive phase, and the network reoptimization phase. We discussed how this process is driven by gathering specific types of network statistics. In particular, we emphasized how per-flow byte and packet counters were required as input and how this information could provide network loading information that a traffic engineering system could use to gauge network resource utilization.

Next, we discussed how traffic could be characterized at different points within the network. The points at which traffic is characterized depended largely on the type of network and the goals of the traffic engineering. Specifically, we discussed how traffic flows could be classified as internal or external traffic flows. In terms of internal traffic, traffic flows either enter a traffic engineering domain, or emanate and remain there (tandem traffic). External traffic flows were described as those flows beginning within the traffic eengineering domain and exiting it on their way to another TE domain, or as bypass traffic that always remains external to the domain and only uses one of its edge routers as a transit point to another AS.

We discussed the salient components of the traffic matrix and made distinctions between which data needed to be collected for which type of network. We then discussed the importance of selecting an appropriate sampling period for both the data being gathered, as well as the intervals at which the gathered data is off-loaded to an external TMS collection agent. We again discussed the potential performance impacts of certain choices that an operator could make when configuring the system of network nodes, collection agents, and offline traffic engineering system. We also discussed some of the options available for off-loading the TMS data.

Finally, the chapter ended with a discussion of cost and performance considerations inherent when using any traffic flow measurement approach including TMS. Specifically, we discussed how measuring the data could impact the actual data if proper attention were not paid to the overall impact on network node performance and resource utilization. We also discussed how measurement systems could introduce less impact onto the network elements being interrogated, but in exchange for additional monies spent to purchase, manage, and maintain additional network data statistics gathering agents.

Further Reading

Cisco Systems Traffic Matrix Statistics Documentation, *www.cisco.com/univercd/cc/td/doc/product/software/ios121/121newft/121t/121t5/tms.htm.*

AT&T Research Traffic Matrix Information, *www.research.att.com/projects/daytona/flatflproc/raw2netmat/help.html.*

Juniper Networks White Paper, *www.juniper.net/techcenter/techpapers/200003–04.html.*

Davie, B. S., and Y. Rekhter. *MPLS: Technology and Applications.* First edition. San Francisco: Morgan Kaufmann Publishers. 2000.

Gray, E. W. *MPLS: Implementing the Technology.* Reading, Mass.: Addison-Wesley Professional. 2001.

Cisco's Web site also provides a great deal of information regarding MPLS: *www.cisco.com.*

Danny McPherson

Most recently, Danny McPherson served as director of architecture at Amber Networks (acquired by Nokia), an edge routing company focused on service convergence and systems availability. He has held technical leadership positions with four large service providers (Qwest, GTEI, Genuity, and internetMCI), where he's been responsible for everything from network and product architecture to routing design, peering, and other business- and policy-related issues.

Danny is an active contributor to numerous IETF Working Groups, serves on a number of directorates, and currently co-chairs the PWE3 Working Group. He's also actively involved in other technical forums such as NANOG, OIF, and the MPLS Forum. Danny has authored a number of RFCs, standards documents, and other publications and is an acknowledged expert in Internet architecture, routing protocols, and large-scale network design.

Danny, you have been an active participant within the IETF community for some time now, and especially now as co-chair of the Pseudo-Wire Emulation Edge-to-Edge (PWE3) Working Group. What do you see as the issues and potential solutions to these issues, with MPLS and its deployment in current networks, and (near) future ones where new applications such as pseudo-wire emulation services are being deployed over existing MPLS infrastructures?

Before the true potential of services such as pseudo-wire emulation can be realized, the MPLS substrate must first be ubiquitously deployed throughout service provider networks. The initial deployments of MPLS have been fostered by more "traditional" MPLS applications such as traffic engineering and fast restoration techniques, which limited the scope of MPLS-enabled devices primarily to the network core.

Edge-to-edge MPLS deployments are beginning to occur, and as network operators begin to better understand the components (e.g., associated MPLS signaling protocols) and characteristics of their MPLS-enabled networks, converging other "revenue streams" onto their larger and more visible IP/MPLS core networks comes more naturally.

It's obvious that decreased OPEX and quicker ROI will be a direct result of "converged networks." However, until the operator knowledge base and tools that provide ways to effectively and efficiently manage and understand MPLS networks are more common, MPLS-based services and ubiquitous MPLS deployments will occur only gradually.

Timely progression of PWE3 and similar specifications within the responsible standards bodies will hopefully assist in producing more interoperable implementations in the marketplace. This is especially important for service providers that operate multivendor networks.

What do you see as the issues and hurdles (and possible solutions) for current deployments of MPLS?

Tools to better understand and manage MPLS networks will play a key role in its deployment. Training and hands-on experience by network operators will assist as well.

Protocol and software developers in the MPLS space have been working extremely hard over the past several years to keep providing additional MPLS services and feature sets to service providers. Many of these are so complex or convoluted that they can barely be understood by protocol developers, much less network operators. Some subtle attempt to qualify and quantify feature sets or services before clogging the systems with "any and everything MPLS" would likely benefit everyone.

Having worked in the area of network operations in the past, how effective do you see the new crop of commercial MPLS network management tools being in today's networks? What features/functionality do you see as lacking from all (or most) vendor offerings?

Multivendor support is typically the largest problem in a single vendor's network management product. Issues that arise often result in finger pointing rather than problem solving. There's clearly a lot of room in this market, though getting in and being established will likely be a bit of a challenge. This is often the case because either the folks a service provider devoted engineering resources to just ran out of money or they were acquired by some behemoth that'll likely ruin the product (i.e., either by losing focus or making it overly vendor-specific). As such, the set of in-house developed PERL scripts that manage most of the routers in the service provider world will be added to yet again.

11

The MPLS Virtual Private Networking MIB (PPVPN-MPLS-VPN MIB)

"The pure and simple truth is rarely pure and never simple."

—Oscar Wilde

Introduction

This chapter focuses on the management of MPLS VPN networks. In particular, it investigates how the PPVPN-MPLS-VPN MIB can be a useful addition to the existing tools in the network manager's quiver to manage MPLS VPN networks—specifically, the management of provider edge routers as well as the customer edge routers in the carrier's carrier model of MPLS VPN. The PPVPN-MPLS-VPN MIB is but one tool of many available to the operator for managing MPLS VPN networks, but we assert that it is quite a useful one. Some of the other tools available are those that have been mentioned in previous chapters of this book; therefore, we suggest that you become familiar with the existing tools before reading this chapter. In particular, the existing MPLS MIBs described can be useful for provisioning, troubleshooting, or performance management when deployed side by side with the PPVPN-MPLS-VPN MIB.

Device vendors implementing the PPVPN-MPLS-VPN MIB will benefit from reading this chapter as we provide pointers on how to make the implementation easier and more useful. Operators reading this chapter will benefit from the details and analysis provided for how they can expect the MIB to be implemented by a device.

A note to the reader: This chapter will focus on the draft version 04 of the PPVPN-MPLS-VPN MIB, which is typically referred to as draft-ietf-ppvpn-mpls-vpn-mib-04.txt. This draft version may have been updated or replaced with an IETF RFC document after this book was published. Please keep this in mind when searching for the document on which this chapter is based.

11.1 MPLS Virtual Private Networks (VPNs)

In previous chapters, we introduced various tools that could be included in the operator's toolbox for managing basic and traffic-engineered MPLS networks. We now shift our focus onto another application of MPLS called MPLS VPNs as defined in RFC 2547bis. We will also later investigate how to manage this new application. MPLS is leveraged once again to realize a new application that service providers can sell to customers who purchase their network services. Specifically, MPLS VPNs realize segregated and private routing and security for end user sites across an MPLS-enabled core network. Each site can communicate directly with other sites that are within its private network, while others cannot. The benefit of this approach is that all existing MPLS applications may coexist with this new application. Let us now briefly introduce and explore some of the details of MPLS VPNs.

The basic concept behind MPLS VPNs is that one or more customer sites wish to be connected in a manner that is both *virtual* and *private*. In terms of a virtual connection, it is desirable to connect sites logically across one or more service provider core networks to leverage their existing infrastructure. In the past, in general it was only possible to physically connect sites together using various layer-2 technologies such as Frame Relay or ATM. In these deployments, leased lines were essentially run from site to site, which was quite expensive both for the operator to deploy and manage, but also for the customer whom the provider had to pass on the added costs to. Furthermore, providers who wished to serve customers with

IP, ATM, and Frame Relay had to support multiple parallel networks, each with its own unique management and operational requirements, challenges, and costs. MPLS VPNs allow a provider to leverage an existing MPLS-enabled network to offer dedicated connectivity over the same network that is shared with other services. This results in lower management and infrastructure costs for the service provider.

In terms of privacy, it is desired that sites that are virtually connected remain as securely segregated from other sites that are not part of the virtual private network as if physical wires connected them together. In essence, it is not only desirable but a requirement for MPLS-based VPNs to have the same or better security characteristics as ATM or Frame Relay VPNs. Fortunately MPLS VPNs satisfy both the requirement of providing logically connected sites securely, while leveraging one or more existing MPLS-enabled networks.

11.1.1 Virtual Private Networks Defined

When deployed logically, a virtual private network boils down to a set of routing policies that provide site-to-site network connectivity, while simultaneously preventing unwanted access from those that are not part of the site-to-site relationship. The exception to this *might* be a network operator that wishes to gain access to the devices supporting the VPNs in question in order to manage them. For example, exceptions to this rule may be acceptable when operators need to either verify service level agreements or debug the network when routing is broken. We emphasize that this might be the case, since a true VPN is a closed connection between sites that cannot be violated by anyone not authorized for access to the network.

When thinking about a set of VPN sites, we arrive at what is in essence building sets of policies that allow or disallow access to one or more sites. This also applies for management access, since it is typically desired that management traffic also be routed to and from these sites using the normal routing processes. For example, we can view the VPN sites depicted in Figure 11.1 in different ways to form different VPNs. It may be the case that perhaps VPN sites A, B, and C might represent three sites for Corporation X. We combine them together into a single VPN by associating the same routing policies for them while implicitly excluding all other sites (e.g., sites D, E, and F). To add new sites using this model, we simply add an additional routing policy for each new site. This can simply include new routing table entries, but may also include such things as traffic filtering.

It should be noted that although it is typically desirable for management access to a VPN site to occur over the normal routing infrastructure, it is sometimes useful to provide "back door" access to these sites for management purposes. This again may technically violate the rule that a VPN is private to only those with access to the VPN, but this is sometimes necessary. Examples of such exceptions may be what are

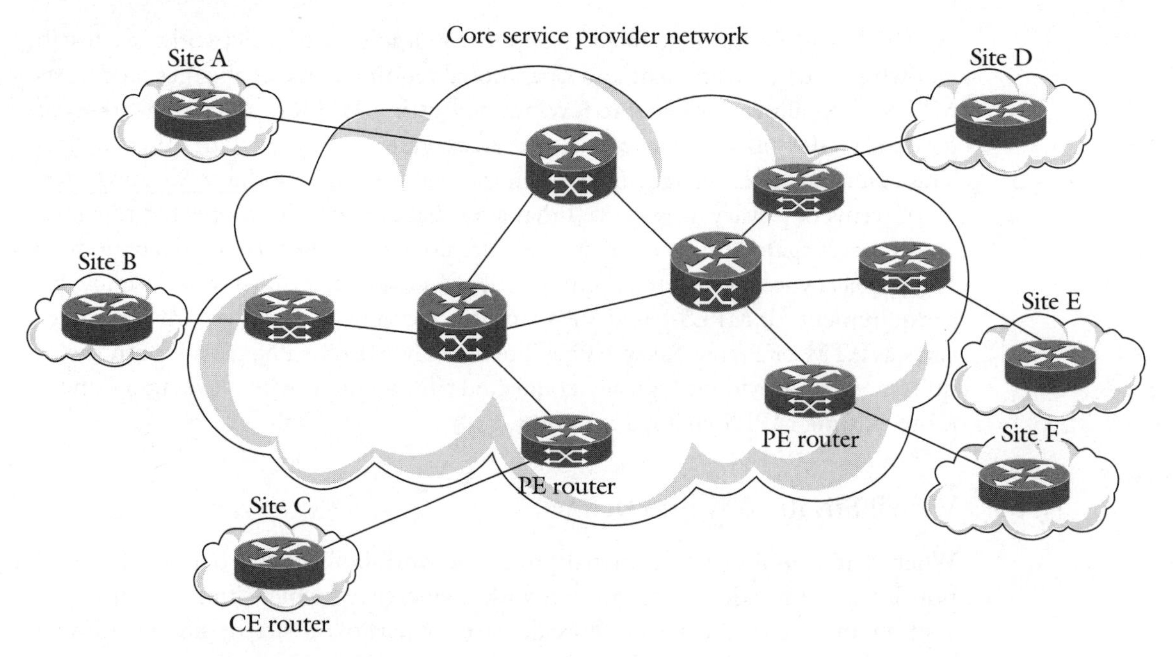

Figure 11.1 Service provider core network with six VPN sites. Different customer sites can be included in different VPNs depending on how they are configured.

sometimes referred to as "shadow" or "meta" VPNs that grant an operator access to more than one VPN—typically all of the VPNs in their management domain. Another possibility is to connect the VPN sites with a different physical or logical network connection that carries strictly management traffic to and from the operator's management station or center.

11.1.2 MPLS VPN Terminology

Let us now introduce some common terminology that will be used whenever discussing MPLS virtual private networks. It is our intent for this discussion to also clarify certain points in the architecture as well that may be potentially confusing. This will also assist us later when we discuss the PPVPN-MPLS-VPN MIB, as the names of objects in the MIB as well as many concepts used are derived directly from the terminology discussed in the standard.

Figure 11.2 shows a simple MPLS-enabled core network that supports MPLS VPNs. Each of the fundamental components of an MPLS VPN network are shown in the figure. The figure includes an MPLS-enabled core network because one is required in order to support MPLS VPNs. Imagine that this MPLS-enabled core also

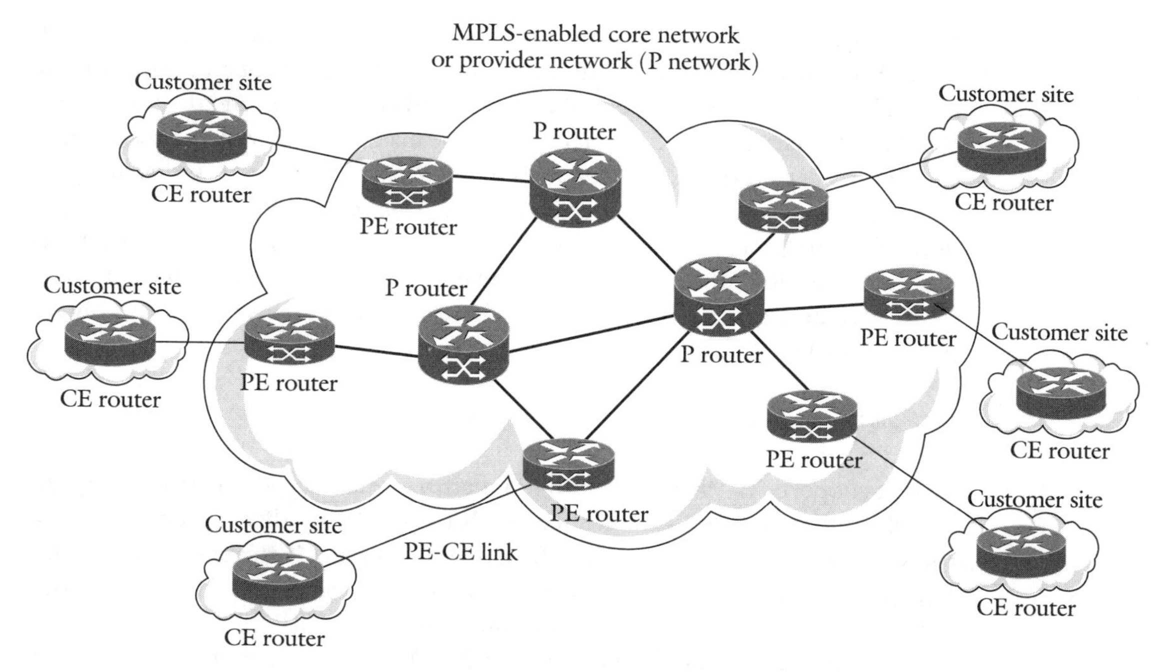

Figure 11.2 Examples of MPLS VPN network terminology.

supports MPLS traffic engineering and LDP. This core network is realized with provider routers, also referred to as "P" routers. P routers typically have no knowledge of the VPNs that traverse their links other than the label swapping information required to forward the site-to-site VPN traffic, as well as perhaps some iBGP or eBGP information. P routers then attach to provider edge routers, also known as "PE" routers. The P-PE connections are typically MPLS-enabled; thus the site-to-site traffic will flow over these links as MPLS-encapsulated packets. The PE routers are then connected to customer edge routers, also known as "CE" routers. CE routers are attached to PE routers via PE-CE links. This link interconnects the customer site to the provider's core network and is the medium over which all of the customer's traffic will enter the network on its way to other VPN sites. Of course, more than one PE-CE link may exist.

PE-CE links must be configured for some sort of routing distribution. The PE-CE links typically run OSPF, RIP, ATM, Frame Relay, Ethernet, PPP, and GRE tunnels, although static routing is also an acceptable configuration. The PE-CE links typically do not run the MPLS protocol unless the VPN is configured as a carrier's carrier MPLS VPN, in which case the Label Distribution Protocol (LDP) may be required to distribute routes over that link. We will explore this scenario in detail later in this chapter.

A *customer site* is one or more (sub)networks that comprise the customer network. Each customer site is connected to the VPN through one or more PE-CE links. Depending on the type of service offered by the provider and the service purchased by the customer, a customer site may be managed either by the customer or by the provider. In the case where the CE device itself belongs to and is managed by the provider, the CE is then referred to as a *managed CE* device. The type of network management approaches used to manage the CE/PE devices sometimes depends on whether the provider owns and manages the CE device. This, for example, may require that the provider have full management access to the VPN and all devices therein.

VPN Routing and Forwarding Table

Each PE device contains a VPN Routing and Forwarding Table (VRF) for each VPN it supports. A VRF is a routing table that is maintained separately from the global routing and forwarding table. Each customer site that is hosted by a PE router is attached to that site via a PE-CE link. This link is then associated with a single VRF. All traffic that travels to and from that site will be routed and forwarded based on information in that customer site's associated VRF. In doing so, this isolates the routing policies associated with that customer site from those of any others hosted by the PE. The advantages to this approach are that any PE link to a customer site has a routing and forwarding table that is only populated with routes that lead to other sites within that VPN. This approach prevents sites that are not in the same VPN from sharing IP connectivity. This works simply because packets from sites that are not in the same VPN cannot get forwarded to and from the VPN site because a route to that site will not exist in their VRF. Another benefit to this approach is that IP address spaces between two or more VPNs can overlap. This is possible because no two VPNs will share a VRF. This is a nice feature for operators, since they can provision the same IP addresses for every VPN if they wish, or use some other scheme that is convenient for them.

The conceptual routing table architecture of a typical PE router supporting MPLS VPN is shown in Figure 11.3. The PE router at the bottom of the figure is connected to two VPN sites via two PE-CE links. PE-CE link 1 connects to VPN "blue," while PE-CE link 2 connects to VPN "red." The figure shows how two separate VRFs are created for each site, and how the specific PE-CE link is associated only with that VRF. The global VRF is associated with the non-VRF links on the PE, which routes the traffic between those links as normal.

11.1.3 Route Distribution and the Route Distinguisher

Now that we have discussed how routes are stored for each VPN, let us now discuss how routes from each customer VRF are distributed to other PEs supporting that

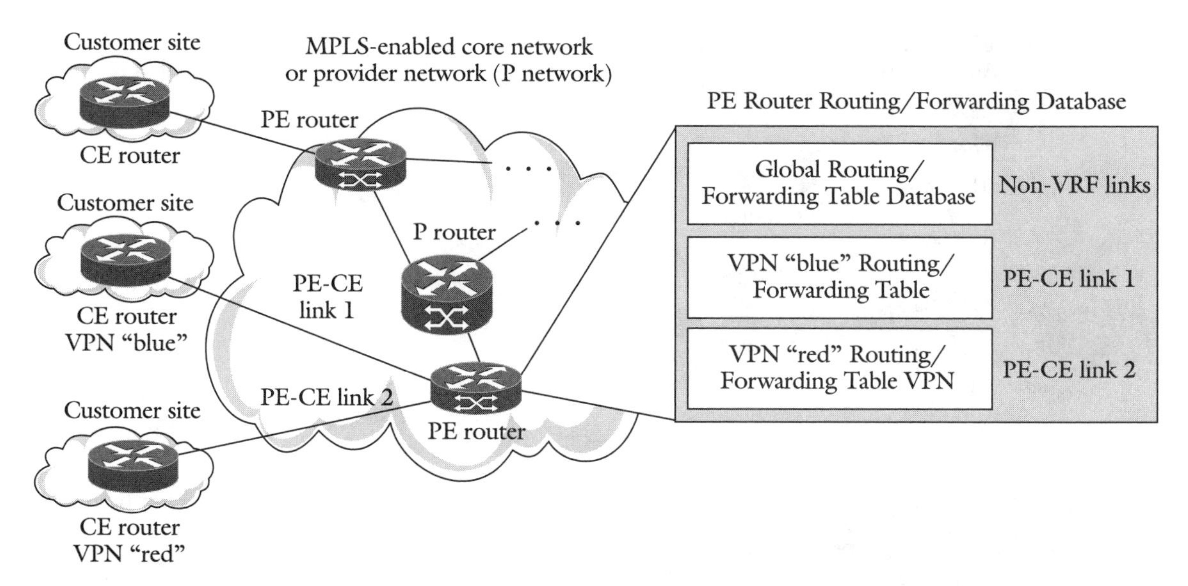

Figure 11.3 PE Routing/Forwarding Database architecture keeps separate routing and forwarding tables for each VPN (VRF) and for the global routing/forwarding.

VPN. Keeping in mind that one of the goals of MPLS VPNs is to allow distinct address spaces for each VPN to coexist with other, potentially overlapping address spaces, it is important then that customer routes are treated in different ways depending on the VPN they belong to. Specifically, traffic is routed according to the VPN it belongs to and does not "leak" into other VPNs or allow traffic from other VPNs to "leak" into it.

The MPLS VPN architecture specifies that BGP be used to distribute VPN routes between PE routers. By itself, BGP cannot support MPLS virtual private networks; therefore, several extensions to the protocol were made to facilitate this additional function. These extensions are referred to collectively as Multi-Protocol BGP (MP-BGP) and can be found in RFC 2858. One of the extensions to BGP defined in MP-BGP is the addition of a route distinguisher (RD) that is combined to make the address unique even if one or more VPNs are using this same address. Together these allow overlapping IPv4 addresses to be used for different VPNs without any difficulties. For example, in the case of IPv4 a new address family was defined called VPN-IPv4 that is specified when a route distinguisher is combined with an IPv4 address.

An example of how overlapping network addressing can be achieved in an MPLS VPN network is given in Figure 11.4. Notice that all six of the sites shown have been assigned the network subnetwork of 198.167.0.0/24. However, also notice that some sites are defined with a route distinguisher (noted as RD in the figure) of

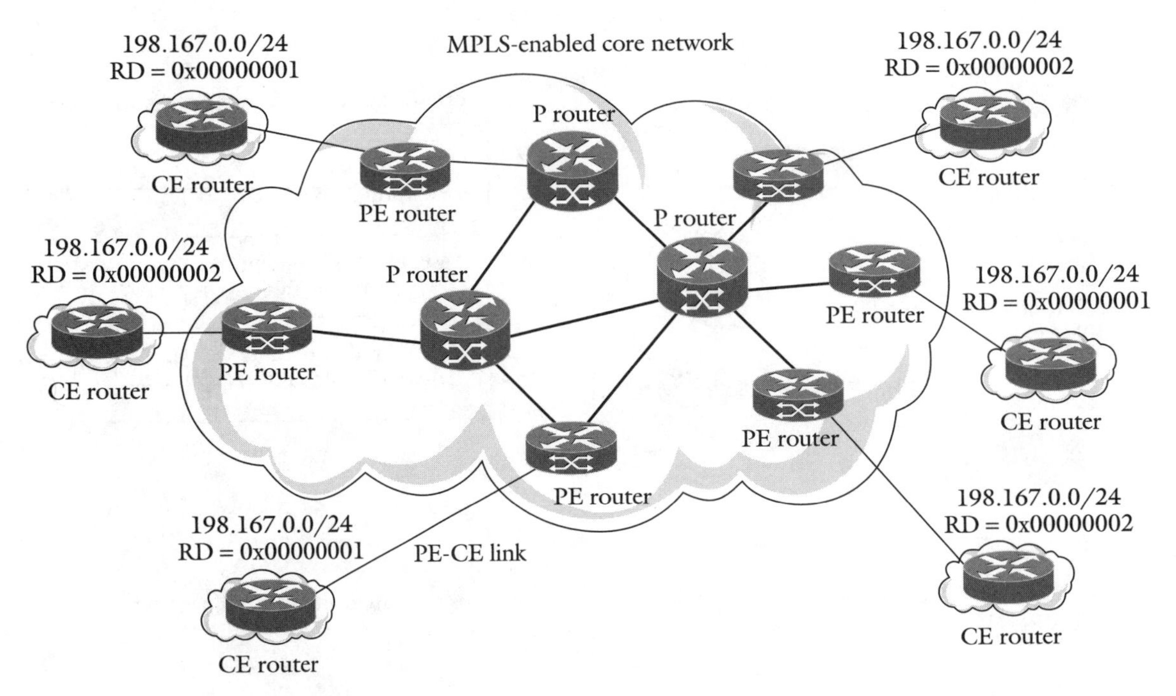

Figure 11.4 Overlapping address assignment in MPLS VPN networks.

0x00000002, while others have an RD of 0x00000001. The sites configured with equivalent RDs are part of the same MPLS VPN. Please note that the RDs used in the example are not necessarily valid VPN-IPv4 route ddistinguishers; their purpose is only to distinguish some sites in one VPN from others. Also, note that we have not shown the specific details of the RD, but it is important to understand that several formats exist for the RD depending on the type of networks used. Please see RFC 2547bis and RFC 2858 for additional details.

Importing or Exporting Route Targets

Each VRF must have at least one route import/export policy configured for it to function properly. When a PE contains a VRF that is programmed with at least one route import/export policy and is either programmed with a new route statically or learns a new route from a CE, this route—called a VPN IPv4 or IPv6 route depending on its contents—is subsequently distributed to other PEs. These routes are distributed so that all PEs can learn about the route. Routing updates between PEs are accomplished by exporting routes from a PE using a routing distribution protocol to other PEs. Exported routes are marked with one or more export route target attributes. Any VPN routes that are received by other PE routers are checked against a list of valid route target attributes (i.e., valid route targets to import from). If a match is made against at least one entry, the new route is deemed eligible and is installed into the remote PE's corresponding VRF. As we will see, the route target mechanism lies at the heart of VPN functionality and allows it the means by which to function correctly.

Earlier we discussed the route target mechanism, but we left out the details of how the route augmented with a route target traverses the core MPLS network on its way to distant PEs. This is accomplished using the multiprotocol extensions to BGP (MP-BGP). When a PE wishes to export a route to one or more PEs, it can distribute this route by associating it with a route target and then distributing it to other interested PE sites via MP-BGP. The information is then distributed during the next BGP update message. Route targets are transported within an Extended Community Path Attribute that exists in the BGP update message. The Extended Community Path Attribute contains a set of one or more extended communities, each coded as an 8-byte value.

The example in Figure 11.5 shows an MPLS-enabled network that supports two VPNs (blue and red). The PE route targets, as well as the import and export configurations to support this configuration, are noted next to each PE and CE. In the case of VPN "blue," both CE2 and CE4 are configured with the network

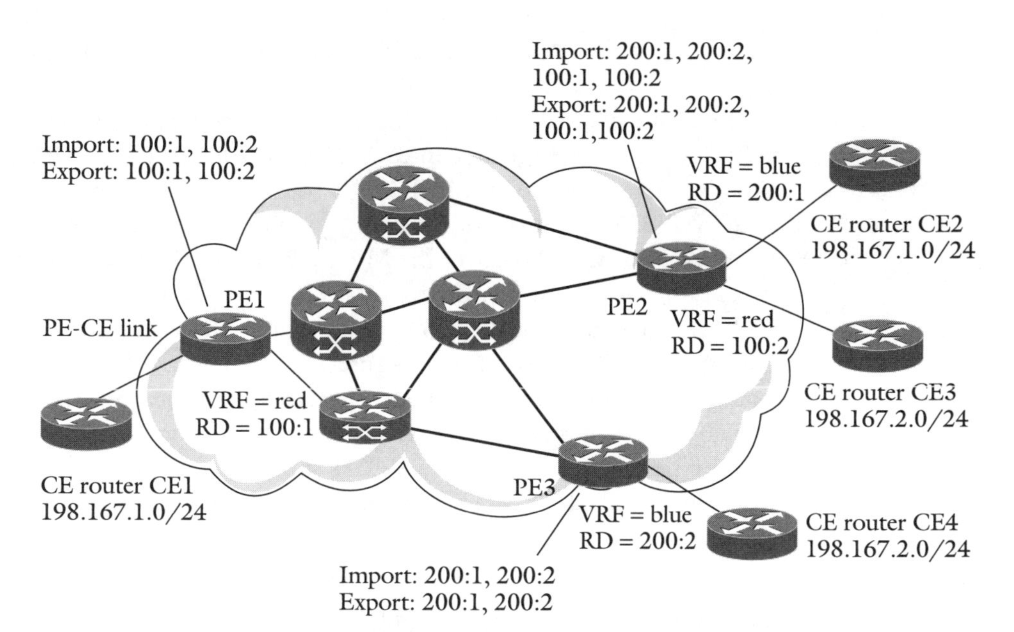

Figure 11.5 An MPLS-enabled network that supports two VPNs (blue and red). The PE route targets, per-VRF route distinguishers, as well as the import and export configurations to support this configuration, are noted next to each PE. The CE network addresses are listed with each CE.

198.167.0.0/24. PE2 has VRF "blue" configured with route distinguisher 200:1, while PE has VRF "blue" configured with RD 200:2. In both cases, the PEs connected to these CEs are configured to import and export the route target for VRF "blue" so that any routes learned from CE1 or CE4 can be distributed to each other. Similarly, VPN "red" requires CE1 and CE3 to be configured for network address 198.167.0.0/24, while PE1 has VRF "red" configured with route distinguisher 100:1, and PE2 has the same VRF configured with RD 100:2. The PEs attached to these CEs are configured to import and export the route targets for VRF "red" so that routes learned on CE1 and CE3 can be distributed accordingly.

Let us investigate how an actual routing update from CE1 is processed, delivered by the MPLS-enabled core network, processed by the PE connected to CE3, and installed into CE3's routing database. First, a new route is learned by CE1 and installed into its local routing table. An IPv4 routing update is sent from CE1 to the PE it is attached to. This route is installed in the VRF associated with the interface on which it was received. The VRF-specific route distinguisher 100:1 is appended to the IPv4 address, which creates a VPN IPv4 address from this route. When it is time to issue a BGP routing update, the export route target "red" is added as an attribute to the Extended Community Path Attribute in the MP-BGP update, and this routing update is sent to all MP-iBGP peers or to a route reflector.

When PE router 2 receives the update, it passes it to its MP-BGP process. The VPN IPv4 routing update is then checked against the configured import route targets that are associated with the VRFs in PE2. The route target "red" that is carried in the routing update matches the configured route target "red," so processing continues. The message is then converted into an IPv4 address and is installed in the VRF table associated with CE3. The routing process then sends the routing update to CE3. The route is subsequently installed into CE2's routing table.

When PE router 3 receives the MP-BGP update from PE router 1, the message is passed to its MP-BGP process just as it was on PE router 2. However, when the VPN IPv4 routing update is compared against those import route targets of the corresponding VRF that are configured, it is found that the routing update does not match any of those in the table. Therefore, PE router 3 will discard this request and not process it further. This results in the request not being installed in any VRF and not passing to any CE router. This simple mechanism prevents route updates from unsupported VPNs from being installed into any VRF on a PE supporting those VRFs.

11.1.4 Virtual Private Network Topologies

Two basic models exist for configuring VPNs: *hub and spoke* and *full-mesh*. Variants of these exist whereby a partial mesh is configured or a partial hub and spoke topology are configured, but the basic premise still exists as with the full configuration. In the case of hub and spoke, all PE routers except one are connected to one other PE router. This central PE router is considered the hub of the conceptual wheel and the remaining PEs the spokes of the conceptual wheel. At first glance, this model may not make sense because it has some obvious disadvantages. First, a single hub presents a single point of failure. Second, the flow of site-to-site traffic in this model is, in general, through the hub, which might result in a bottleneck. However, there are some distinct advantages to this topology over others. First, this topology lends itself well to deployments where it might be desirable to force all traffic to a central site so that security checks and/or address translation are performed. Further, if an operator has a centralized infrastructure, it may be more convenient for configuration and maintenance purposes to have all new PEs configured as spokes and simply have them connect to a single hub. The other advantage is cost. This topology requires, at a maximum, one link between each PE and the PE designated as the hub. We will see that this number of links is quite a few less than the number required for a full-mesh topology.

An example of the hub-and-spoke paradigm is shown in Figure 11.6. Notice that all of the spoke PE sites connect to PE F, which acts as a hub for the topology. Only a single link between each PE and the PE designated as the hub is required. However, additional links may be added for redundancy or to increase site-to-site performance between other spoke PEs. This forms a *partial-mesh* topology.

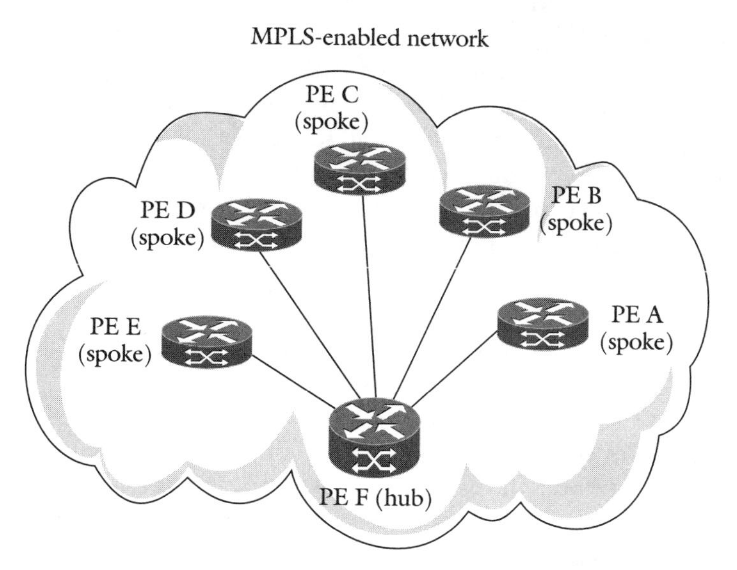

Figure 11.6 VPN hub-and-spoke toplogy.

The second type of topology is called the full-mesh topology. This arrangement requires a connection between each PE in the network and results in approximately N^2 connections, where N is the number of PEs in the network. The full-mesh topology is attractive because it provides each PE with the maximum amount of redundancy over the hub-and-spoke model. It also allows any given PE to route site-to-site traffic using the most optimal route. In the hub-and-spoke model, all traffic had to first arrive at the hub PE and then travel to the nearest spoke PE. This model generally saves at least one hop on the path taken by the data. The costs associated with this model, however, can be significant. As was noted earlier, a connection must exist between any two PEs in this model. If these connections are physical WAN links, for example, they can be quite costly. Second, either network address translation (NAT) functions will have to be centrally managed as was in the case of hub and spoke, or individual NAT functions will have to be programmed into each PE. This can also be costly.

An example of a full-mesh network is given in Figure 11.7. Notice that a connection exists between any two PEs forming a full mesh of connections over the network. These connections provide maximum redundancy and routing options for each PE.

11.1.5 Carrier's Carrier

The MPLS VPN architecture can be extended to allow for a special configuration called carrier's carrier (CsC). This configuration allows a service provider that maintains an MPLS-enabled network (the carrier service provider) to sell VPN services to other service providers (the customer service providers). There are many advantages to this configuration. First, it leverages the existing core provider's network to offer yet another value-added service without the need to deploy a parallel network (and deploy additional hardware). MPLS CsC VPN will run over the same hardware infrastructure as the normal MPLS VPN deployment. Second, the management of this type of network is in essence quite similar to the more traditional deployment of MPLS VPN.

On the surface, this configuration seems like a straightforward and natural extension to MPLS VPN, but it is not necessarily so because the customer service provider will most likely be interested in their VPN carrying the entire Internet routing table. This is an issue for two reasons. First, each PE will have to contain what are in essence duplicates of the Internet routing table (one global copy and one potentially for each customer SP VRF). Second, Interior BGP (iBGP) will have to distribute the routes among PE sites, which is also costly in terms of network bandwidth and processing. Therefore, one of the goals of the CsC architecture is

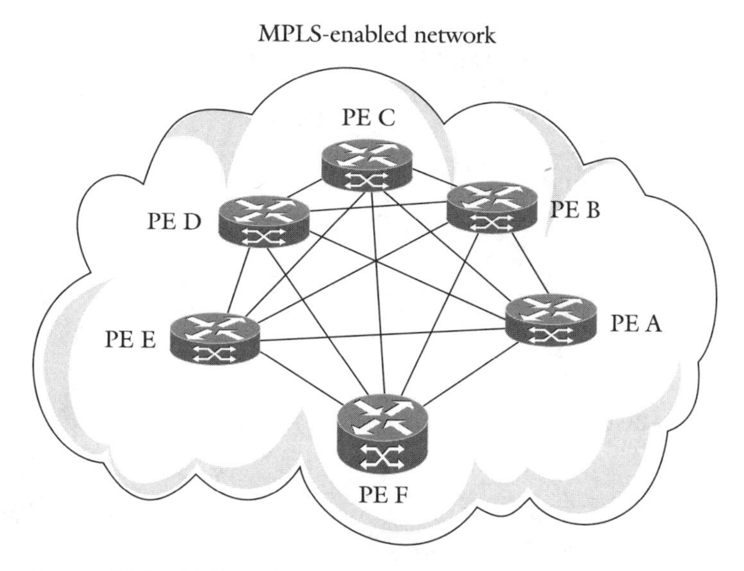

Figure 11.7 Full-mesh VPN topology.

to reduce the number of routes the carrier service provider PEs must exchange. To this end, PEs only include customer service provider routes in the MP-BGP update messages they exchange. Other routes are distributed using LDP between the PE and CE. This requires that LDP be used as the Label Distribution Protocol run over the PE-CE link. iBGP multihop sessions are then used to exchange external routes between client service provider CEs that reside within the same autonomous system (AS), while multihop intraconfederation Exterior BGP (eBGP) is used to distribute client service provider routes that are external. Alternatively, eBGP intraconfederation sessions can be established to exchange this information as well.

Figure 11.8 demonstrates a typical configuration of MPLS carrier's carrier VPN. In the figure, three carrier networks are shown. On the right, the carrier SP's MPLS-enabled network contains MPLS-enabled LSRs. After some time the LSRs connect to provider edge (PE) LSRs that are designated as carrier SP PEs. These PEs then connect to one of two customer SP CE routers. The customer SP CE to carrier SP PE connection is shown to be running LDP. The carrier SP's PEs are shown to be exchanging MP-iBGP information among themselves. Within each client SP MPLS-enabled network, a client SP CE connects the carrier SP PE to the internal client SP network. Traffic is routed from the client SP CE to the appropriate client SP PE, and then onto the client SP customer's CE router. Also notice that

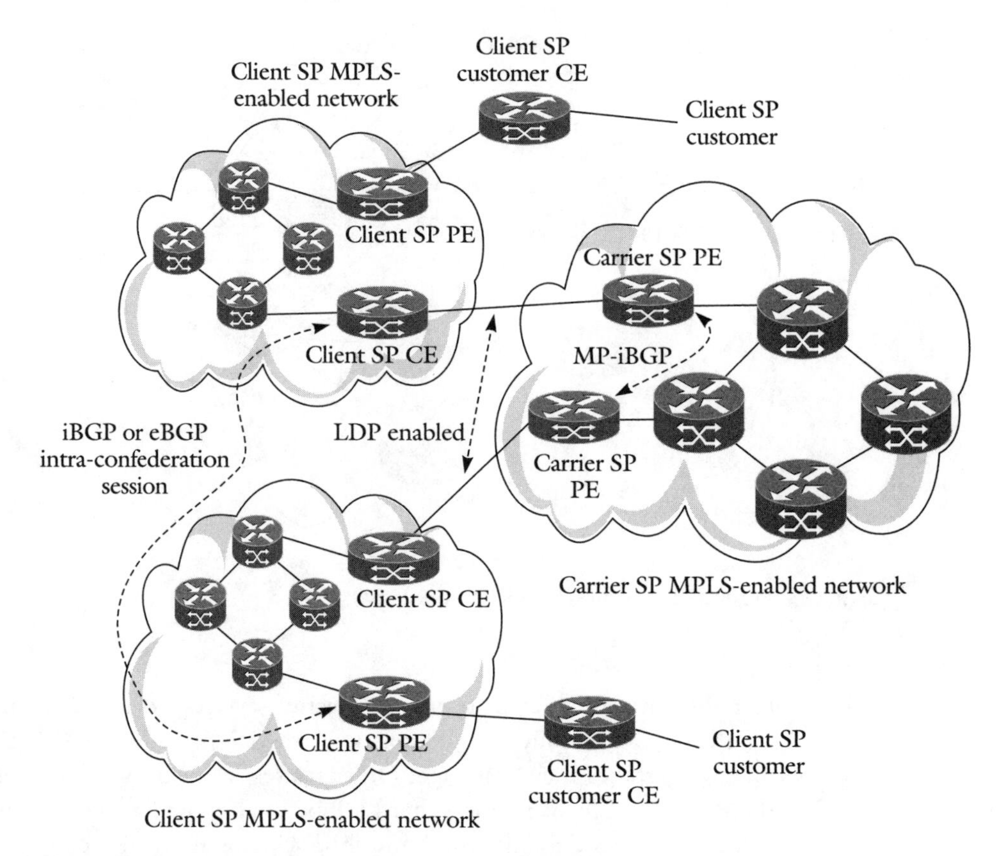

Figure 11.8 Sample MPLS-VPN CsC configuration.

iBGP or eBGP intraconfederation sessions are exchanging routing information between client SP CE nodes, as was described earlier.

As you can see, the configuration of the different protocols that are required to realize MPLS CsC VPN (MPLS, LDP, eBGP, and iBGP) can result in quite an effort on the part of the operator. Complicating things further are the configurations of the non-CsC MPLS VPNs in addition to all of the other things to configure and monitor on the PE devices. Fortunately, many provisioning and monitoring/management systems exist that can assist the operator in provisioning and managing these deployments. In fact, many of these systems can totally automate these processes. Many of these systems are based on using SNMP as the main protocol for monitoring, while using a variety of protocols for provisioning (e.g., SNMP, XML, proprietary CLI). We will now discuss the PPVPN-MPLS-VPN MIB in detail and show how it can be used as one of the basic tools for both the provisioning and monitoring of MPLS VPNs (CsC and otherwise).

11.2 Definition of Terms Used in the MIB

The PPVPN-MPLS-VPN MIB uses some terminology that needs to be explained. Table 11.1 enumerates and defines these terms in a manner that is hopefully clear and concise. Note that for the purposes of this chapter, we will use the term *this LSR* to denote the LSR on which the MIB is being run (or queried).

Table 11.1 PPVPN-MPLS-VPN MIB terminology.

Term	Definition
VRF	Virtual Routing and Forwarding Table. This table contains the forwarding table for the VPN. This table is located at every instance of the VPN and is kept consistent through the distribution of its routes to other VRFs that are in the VPN via the routing distribution protocols such as LDP and BGP.
VPN	Virtual private network. A VPN consists of one or more instances of a VPN, which are really VRFs that exist on a router. The VRF provides the forwarding and security information for the VRF that allows for site-to-site transport of the VPN data, but also disallows non-VPN traffic to enter the VPN sites.
Route target	Route targets provide import and export route policies for a PE and ultimately for the VRFs it supports. All sites that are within a supported VRF must be configured with the same import and export subset so as to allow for the import of routes from other PEs supporting the VPN, and the export of routes learned at a PE to other PEs.
Route distinguisher	The route distinguisher adds an additional 64-bit identifier to the network addresses configured for a VPN. This allows overlapping network addresses between any VPNs supported within a network, as well as overlap with even the global routing within that network. The route distinguisher acts as essentially an internal NAT function. Route distinguishers are appended to VPN network addresses before they are distributed within route distribution protocols such as BGP.
Provider edge (PE) router	Provider edge routers connect to the MPLS-enabled core network, as well as to customer edge (CE) or customer-facing routers. CEs

continued

Table 11.1 continued

Term	Definition
	either can be connected directly to customer sites or can connect to other CEs that have essentially a PE function in a CsC configuration.
Customer edge (CE) router	Customer edge (CE) routers connect to provider edge (PE) routers. CE routers are used to aggregate customer site traffic together and uplink it into the MPLS-enabled core for transport to other VPN sites (or other external sites if NAT is used).
Customer site	Customer network. This network attaches to a CE for uplink into the MPLS-enabled core network.
Network address translation (NAT)	The network address translation (NAT) function provides a mapping from private to public addresses. For example, since VPNs are allowed to use any addresses they choose (given a unique RD appended to the network address), NAT must be used if the VRFs are to connect to sites that are external to the provider core network such as the Internet.
Border Gateway Protocol (BGP)	The IETF Border Gateway Protocol. This protocol is used to distribute routes between VRFs. Internal (iBGP) and external (eBGP) flavors of the protocol exist and have specific uses.
Open Shortest Path First (OSPF)	A link state routing protocol, typically used as a CE-PE link protocol.
OSI Intermediate System to Intermediate System (IS-IS)	The OSI IS-IS protocol is a link state routing protocol used as a CE-PE link protocol.
Router Information Protocol (RIP)	The RIP protocol is a path vector routing protocol used as a CE-PE link protocol.
Label Distribution Protocol (LDP)	The Label Distribution Protocol is a protocol used to distribute labels and route-to-label mappings among LDP peers (see Chapter 4). This protocol is often used as the PE-CE link protocol in the CsC configuration of MPLS VPN.

11.3 The PPVPN-MPLS-VPN MIB at a Glance

The following sections provide a brief overview of the various tables and structure of the PPVPN-MPLS-VPN MIB. Figure 11.9 shows a high-level overview of the PPVPN-MPLS-VPN MIB by showing indexing interdependencies between the

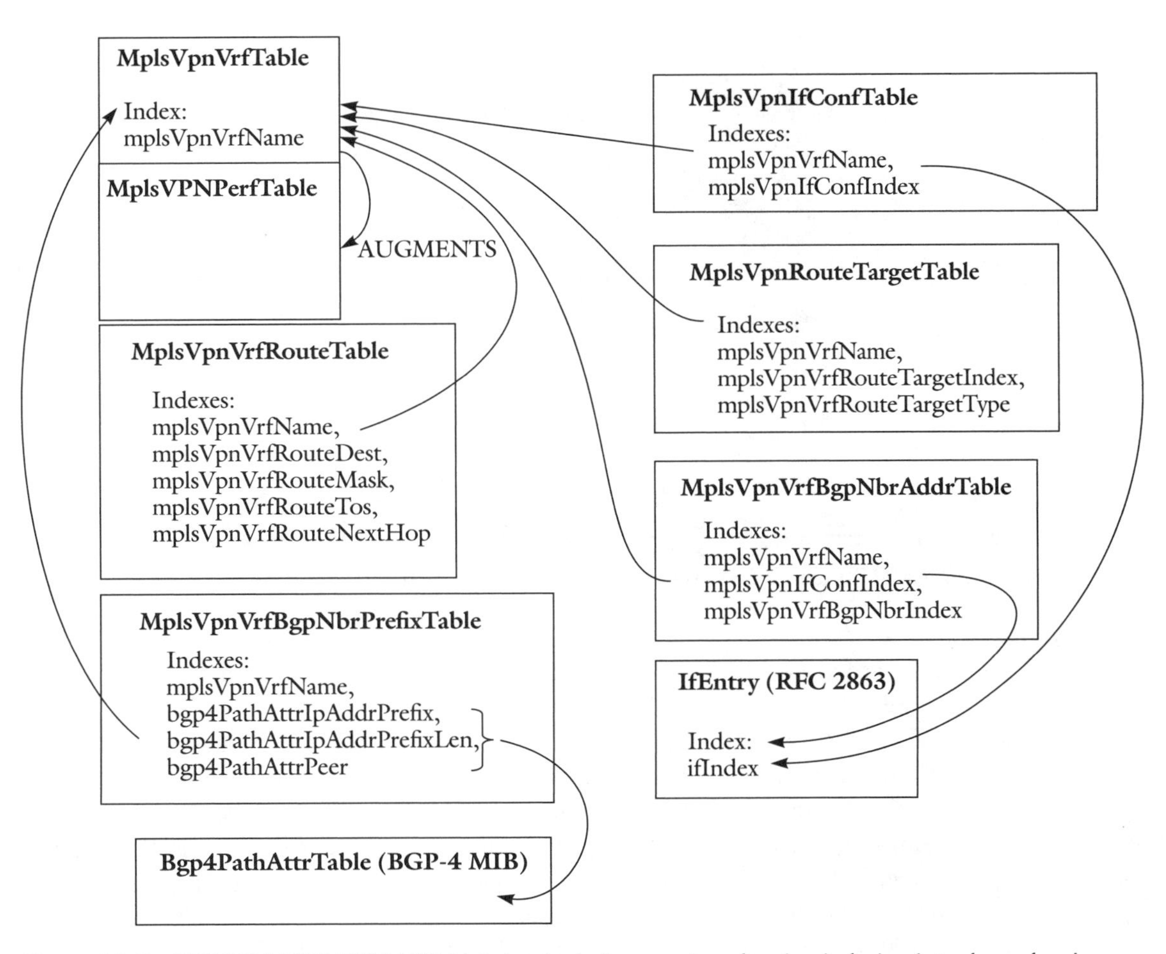

Figure 11.9 PPVPN-MPLS-VPN MIB high-level tabular overview showing indexing interdependencies.

tables in the MIB. Notice that all of the tables in the PPVPN-MPLS-VPN MIB are indexed by the mplsVpnVrfName. This allows easy access to all of the tables in the MIB by cross-referencing each table's entry by the VPN name assigned to the VRF. This also facilitates easy access to the pertinent tabular entries in the event that an operator has received a notification from an agent implementing this MIB. The figure also shows that the performance table AUGMENTS the MplsVpnVrfTable, effectively extending each one of its entries to contain performance-related information. Finally, some of the tables in the MIB extend or cross-reference entries in the BGP-4 and IF-MIB. These relationships are shown as well.

The PPVPN-MPLS-VPN MIB fits into the larger picture of the MPLS-related MIBs in that it depends on the MPLS MIBs discussed thus far: MPLS-LSR MIB, MPLS-FTN MIB, MPLS-LDP MIB, and MPLS-TE MIB. Figure 1.10, repeated

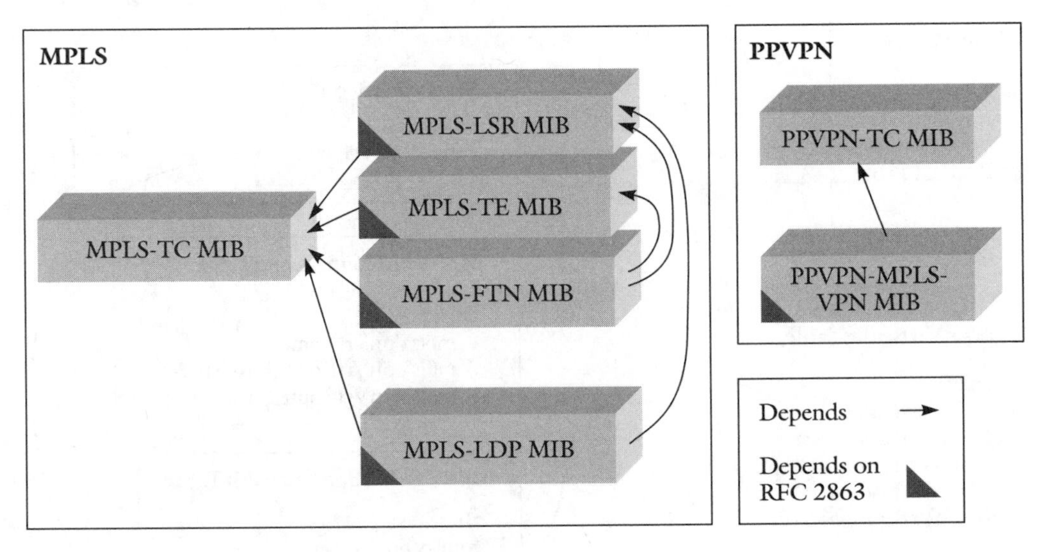

Figure 1.10 (*repeated*) MIBs for MPLS network management discussed in this text.

here once again, shows where the MIB fits into the larger picture of the MPLS-related MIBs.

The PPVPN-MPLS-VPN MIB does not directly depend on any of the MPLS-related MIBs. It does directly depend on the Interfaces and BGP-4 MIBs. It indirectly depends on the MPLS-related MIBs depending on the specific type of situation being managed. For example, if the operator wishes to troubleshoot a problem with a VPN site's connectivity, she may wish to perform a path trace from one of the VPN's core-facing interfaces. This must be accomplished using the MPLS-LSR, MPLS-FTN, and possibly the MPLS-TE MIBs. We will discuss some of these situations in detail later in this chapter.

11.3.1 MplsVpnVrfTable

The MPLS VPN VRF Table (MplsVpnVrfTable) represents the MPLS/BGP VPNs that are configured on the node where the MIB is being queried. Specifically, this table contains the virtual routing and forwarding entries for the VPNs that represent an *instance* of those VPNs. The collection of VRFs as viewed from all network devices comprises the *actual* VPN. This information is typically only known in its entirety at the NMS and not by any individual network node.

An operator or NMS will create an entry in this table for every MPLS/BGP VPN VRF that is desired to run in this MPLS domain. The operator must later associate interfaces with the VRF as well as configure other managed objects mentioned later in the MIB for the VRF to operate correctly. Entries in this table can be created independently using other management interfaces accepted by the network node. In these cases, the agent will represent these entries and may allow an operator to modify them through the SNMP interface.

11.3.2 MplsVpnIfConfTable

The MPLS VPN Interface Configuration Table (MplsVpnIfConfTable) represents those interfaces that are configured with the MPLS/BGP VPN feature enabled. Each entry in this table corresponds to an entry that must exist a priori in the IF-MIB's Interface Entry (ifEntry) Table. The entries found in this table are intended to extend the entries found in the IF-MIB to contain specific BGP/MPLS VPN information. Due to this correspondence, certain objects such as traffic counters are not found in this MIB to avoid overlap, and instead are found in the IF-MIB. Entries can be created or disabled in this table to enable or disable the MPLS VPN feature on an interface on those agents that allow configuration.

11.3.3 MplsVPNPerfTable

The MPLS VPN Performance Table (MplsVPNPerfTable) contains objects that enable an operator to measure the performance of MPLS/BGP VPNs. This table augments the MplsVpnVrfTable; thus an entry in this table will exist for every entry in the MplsVpnVrfTable.

11.3.4 MplsVpnVrfRouteTable

The MPLS VPN VRF Route Table (MplsVpnVrfRouteTable) contains objects representing the routes that are present in a particular VRF. Static entries can be added to this table if an agent allows external configuration; otherwise the entries in the table arrived via one of the routing distribution protocols such as BGP. These entries may come and go depending on the status of the routing distribution protocol, and thus are considered dynamic.

11.3.5 MplsVpnRouteTargetTable

The MPLS VPN Route Target Table (MplsVpnRouteTargetTable) contains the objects necessary to configure and monitor route targets for a particular VRF. As described earlier (see Section 11.1.3), route targets specify the import and export policies for a VRF. This table specifies per-VRF route target association.

11.3.6 MplsVpnVrfBgpNbrAddrTable

The MPLS VPN VRF BGP Neighbor Address Table (MplsVpnVrfBgpNbrAddr-Table) specifies a per-interface MPLS/eBGP neighbor information. Entries are created in this table for every VRF capable of supporting MPLS/BGP VPN that requires eBGP configuration. Agents allowing configuration of the information in this table should coordinate the entries created or modified with the eBGP process on the device as well as the BGP-4 MIB if it is supported.

11.3.7 MplsVpnVrfBgpNbrPrefixTable

The MPLS VPN VRF BGP Neighbor Prefix Table (MplsVpnVrfBgpNbrPrefix-Table) specifies per-VRF VPN-V4 multiprotocol prefixes supported by BGP. Entries are created in this table for every VRF capable of supporting MPLS/BGP VPN that requires BGP configuration. Agents allowing configuration of the information in this table should coordinate the entries created or modified with the BGP process on the device as well as the BGP-4 MIB if it is supported.

11.3.8 MplsVpnVrfSecTable

The MPLS VPN VRF Security Table (MplsVpnVrfSecTable) specifies per-VRF security-related information. This table currently contains two objects that are used to track the number of times illegal labels are received on a VRF interface, as well as to configure the threshold at which the MPLSNumVrfSecIllegalLabelThresh-Exceeded notification will be emitted.

11.4 Scalar Objects

This section describes the global (scalar) objects that are defined in the PPVPN-MPLS-VPN MIB. These objects will be returned by an agent supporting this MIB, regardless of whether or not entries exist in any of the tables described in this section. If an agent does not provide a network manager with an agent capability statement that indicates support for this MIB in part or in whole, finding these scalars returned is a good indication that at least part of the MIB is supported.

11.4.1 mplsVpnConfiguredVrfs

The MPLS VPN Configured VRFs (mplsVpnConfiguredVrfs) scalar contains the number of VRFs that have been configured on this node as seen by the active entries in the MplsVpnVrfTable. This value does not reflect entries that have a row status other than **active (1).** Note that the operational status of the VRF entry is irrelevant to this variable because it only reflects configured VRFs regardless of their operational state (see mplsVpnActiveVrfs for operational VRFs).

The use of this object by an operator can be important in correctly managing a PE router's memory resources. Since any PE is shipped with a finite number of memory resources, this object can be used as a high-water mark for the number of VRFs that can be configured on a particular device. This high-water mark can then be used to prevent a provisioning system from overprovisioning a device. Operators often can determine a useful value for this object using several methods. First, an operator can perform empirical testing on a device to determine its VRF limits. Additionally, this data may be provided by the manufacturer of a device.

It may also be important to reduce the number of VRFs that operate on any given device due to the CPU resources required to update the routes associated with each VRF. It may have been determined that only a certain number of BGP, LDP, or other routing processes can coexist happily on a device before each impacts the others' performance significantly. Finally, this variable might be polled periodically to sanity-test these empirical results given the current operating state of the device. This can be important since the addition of running features on a device will increase the use of memory resources.

11.4.2 mplsVpnActiveVrfs

The MPLS VPN Active VRFs (mplsVpnActiveVrfs) scalar reflects the number of VRFs that are currently active. A VRF is considered "active" if its corresponding mplsVpnVrfOperStatus object value is equal to **operational (1).** If it is set to any other value, it is *not* counted in the set of active VRFs. This distinction is critical, as there may be many pending row status creations for VRFs, many VRFs that have not been assigned interfaces, or even VRFs that are assigned to interfaces that either are not yet or have never been operational. In all of these cases, these VRFs should not be considered active, and thus should not be counted. The point of this scalar is to let the operator know at a glance how many VRFs are *actively capable* of forwarding traffic. For example, imagine a case where one or two interfaces go down and an operator misses the mplsVrfIfDown notifications from these events. In this case, they might also have polling configured that will catch the fact that the number of configured VRFs (mplsVpnConfiguredVrfs) is not equal to the number of active VRFs. This is a very lightweight way to discover this potentially problematic condition.

11.4.3 mplsVpnConnectedInterfaces

The MPLS VPN Connected Interfaces (mplsVpnConnectedInterfaces) scalar object reflects the total number of interfaces connected to all VRFs. By "connected," it is meant that the interfaces are configured in the MplsVpnIfConfTable and associated with a valid primary index (mplsVpnVrfName). For example, if a device has two VRFs, each with two interfaces associated, the mplsVpnConnectedInterfaces object would return four if queried. Entries existing in this table without a valid primary index will not be counted. This scalar is particularly interesting for operators who both perform a distributed configuration through SNMP and wish to validate the consistency of their configurations, or those who allow provisioning through some other mechanism and wish to double-check the provisioning this way. This scalar can also be used by a manager to notice that the number of provisioned interfaces might not equal the number of active interfaces that were provisioned and thought to be active.

11.4.4 mplsVpnNotificationEnable

The MPLS VPN Notification Enable (mplsVpnNotificationEnable) scalar object reflects whether the agent will generate all notifications defined in the PPVPN-MPLS-VPN MIB. The default value of this variable is false. If this object is set to **true (1),** then all notifications defined herein shall be issued.

11.4.5 mplsVpnVrfConfMaxPossibleRoutes

The MPLS VPN VRF Configured Maximum Possible Routes (mplsVpnVrfConfMaxPossibleRoutes) scalar object denotes the maximum number of routes that the device will allow any one VRF to contain. This value does not reflect the overall total of routes available for all VRFs; rather, this number indicates the routes available for each VRF. If this value is set to zero, this indicates that the network node is unable to determine the absolute maximum. In this case, the configured maximum must be determined empirically and configured on each VRF individually by setting the mplsVpnVrfConfMaxRoutes object accordingly.

The judicious use of this variable by an implementation can be critical in managing a PE router's VPN deployment—in particular, because a finite number of resources exist on a PE, especially when the PE must not only support many VRFs in addition to the Internet routing table, but also the many other features supported on the router/switch. Thus, this variable can be used as an absolute high-water mark for VRFs to prevent overuse of the VRF routing table memory resources. This high-water mark can then be used to prevent a provisioning system from overprovisioning a device. Operators often can determine a useful value for this object using several methods. First, an operator can perform empirical testing on a device to determine its limits. Additionally, this data may be provided by the manufacturer of a device.

11.4.6 mplsVpnVrfConfRouteMaxThreshTime

The mplsVpnVrfConfRouteMaxThreshTime scalar object denotes the interval in seconds at which the route max threshold was exceeded. The mplsNumVrfRouteMaxThreshExceeded notification is issued upon the device attempting to cross the configured mplsVpnVrfConfHighRouteThreshold value. It is typically the case that when routes are added, they are not added singularly, but in bunches during routing table updates. Thus, this threshold might be crossed numerous times over the course of a short period of time. This might result in numerous notifications being issued for each threshold crossing. To this end, we defined the mplsVpnVrfConfMaxRoutes notification to not be reissued upon the threshold being crossed until the current sysUpTime is greater than or equal to when the last notification was issued plus the mplsVpnVrfConfRouteMaxThreshTime. This value is specified in seconds since all of the operators we polled thought they would never set this time below 1 second. The default value is actually set to 600 seconds (10 minutes) because this was the typical value operators were interested in using.

A special case exists when this value is set to zero. If it is set to zero, this indicates that the operator desires that the agent should only issue a single notification at the time that the maximum threshold has been reached and should not issue any more notifications until the value of routes has fallen below the configured threshold value. This is useful for deployments that are only concerned with knowing that the VRF has reached its maximum number of routes as viewed by the tabular object mplsVpnVrfPerfCurrNumRoutes, but do not necessarily want to take any corrective action from then on until the value is reduced and then exceeds the amount again. If the mplsVpnVrfConfRouteMaxThreshTime scalar is set to a number of seconds, the operator is actually interested in the number of times the value of mplsVpnVrfConfHighRouteThreshold continues to be exceeded over the specified period of time.

It is important to understand that an implementation should base this time on when the mplsVpnVrfConfHighRouteThreshold tabular object is crossed for a specific VRF and not necessarily when the scalar mplsVpnVrfConfMaxRoutes is reached because the value of mplsVpnVrfConfHighRouteThreshold is guaranteed to be less than or equal to mplsVpnVrfConfMaxRoutes. Furthermore, the value of mplsVpnVrfConfHighRouteThreshold might be set to different values for different VRFs given different customer requirements, so it might be incorrect to base the generation of the mplsNumVrfRouteMaxThreshExceeded notification on this value anyhow.

Figure 11.10 illustrates how the mplsNumVrfRouteMidThreshExceeded and mplsNumVrfRouteMaxThreshExceeded notifications are issued. The figure shows how, as time goes by, the number of routes contained in the VRF fluctuates between zero and mplsVpnVrfConfMaxPossibleRoutes, while sometimes moving between mplsVpnVrfConfMidRouteThreshold and mplsVpnVrfConfMaxRouteThreshold. The figure shows that when the number of routes crosses the threshold mplsVpn-VrfConfMidRouteThreshold first, the mplsNumVrfRouteMidThreshExceeded notification is issued. The number of routes soon falls below the midlevel threshold but then exceeds it once again, and an mplsNumVrfRouteMidThreshExceeded is once again issued. Next, the number of routes rises beyond that of mpls-VpnVrfConfMaxRouteThreshold. At that time, the first mplsNumVrfRoute-MaxThreshExceeded notification is issued. However, notice that as the routing process attempts to add routes to the routing table, additional mpls-

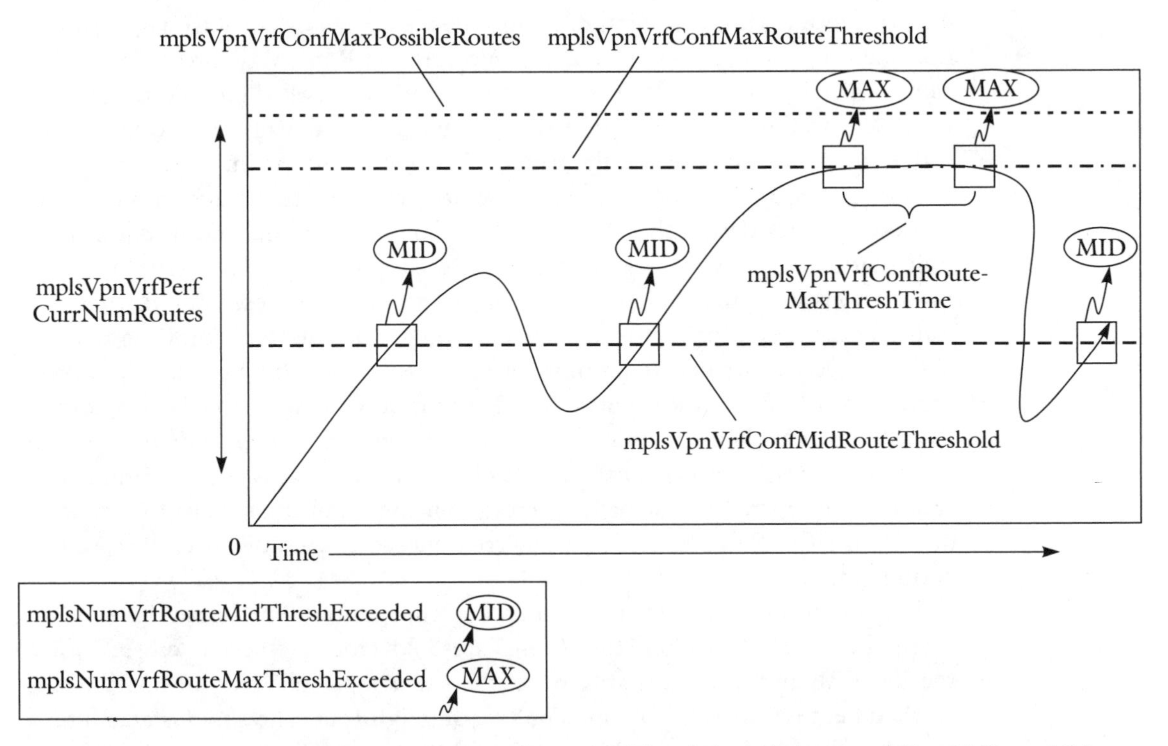

Figure 11.10 Illustration of when mplsNumVrfRouteMidThreshExceeded and mplsNumVrfRouteMid-ThreshExceeded notifications are issued, as well as how they relate to mplsVpnVrfConfMaxPossibleRoutes, mplsVpnVrfConfMaxPossibleRoutes, and mplsVpnVrfConfMaxRouteThreshold.

NumVrfRouteMaxThreshExceeded notifications are squelched until the specified interval of mplsVpnVrfConfRouteMaxThreshTime has passed. When this time has passed, another mplsNumVrfRouteMaxThreshExceeded is issued accordingly.

11.5 MplsVpnVrfTable

The MPLS VPN VRF Table (MplsVpnVrfTable) represents the MPLS/BGP VPNs that are configured on the node where the MIB is being queried. Specifically, this table contains the virtual routing and forwarding entries for the VPNs that represent an *instance* of those VPNs. The collection of VRFs as viewed from all network devices comprises the *actual* VPN. This information is typically only known in its entirety at the NMS and not by any one network node.

An operator or NMS will create an entry in this table for every MPLS/BGP VPN VRF that is desired to run in this MPLS domain. The operator must later associate interfaces with the VRF as well as configure other managed objects mentioned later in the MIB for the VRF to operate correctly. Entries in this table can be created independently using other management interfaces accepted by the network node. In these cases, the agent will represent these entries and may allow an operator to modify them through the SNMP interface.

11.5.1 MplsVpnVrfTable Indexing

The MPLS VPN VRF Table (MplsVpnVrfTable) is indexed by a single object—the mplsVpnVrfName. This object is actually defined in the PPVPN-MPLS-VPN MIB as the MplsVpnName textual convention. The type of this object is in essence an octet string with a maximum length of 32 bytes. The definition of this textual convention states that it represents a unique identifier that is assigned to each VRF. However, since this value is assigned by the system operator or NMS, it can be assigned inconsistently and the VRFs will still be able to forward traffic if the routing and other configuration is correct. For example, if two PEs represent two sites in a VPN, but are configured with the mplsVpnVrfName of "fred" and "barney," this will make identification by an NMS or an operator that may be interrogating the PEs for, say, troubleshooting purposes potentially confusing. However, since there is nothing in the underlying protocols that requires the VRFs to be named consistently, the naming is specified as optional in the MIB. Despite this, we strongly recommended that the VRF naming be consistent and unique throughout the MPLS domain for all VRFs belonging to the same VPN. In fact, we recommend that the name chosen for each VRF be based on the VPN ID, which is globally unique. If this advice is taken, then the indexes in this table will be consistent on all LSRs within the network.

The example shown in Figure 11.11 builds upon the one shown above in Figure 11.5 to demonstrate how the indexing using the consistency method just described can be realized in an MPLS-enabled network. The network shown in the figure supports two VPNs ("BLUE" and "RED"). The PE route targets, as well as the import and export configurations to support this configuration, are noted next to each PE. The mplsVpnVrfTable indexing configuration is shown for each of the PE

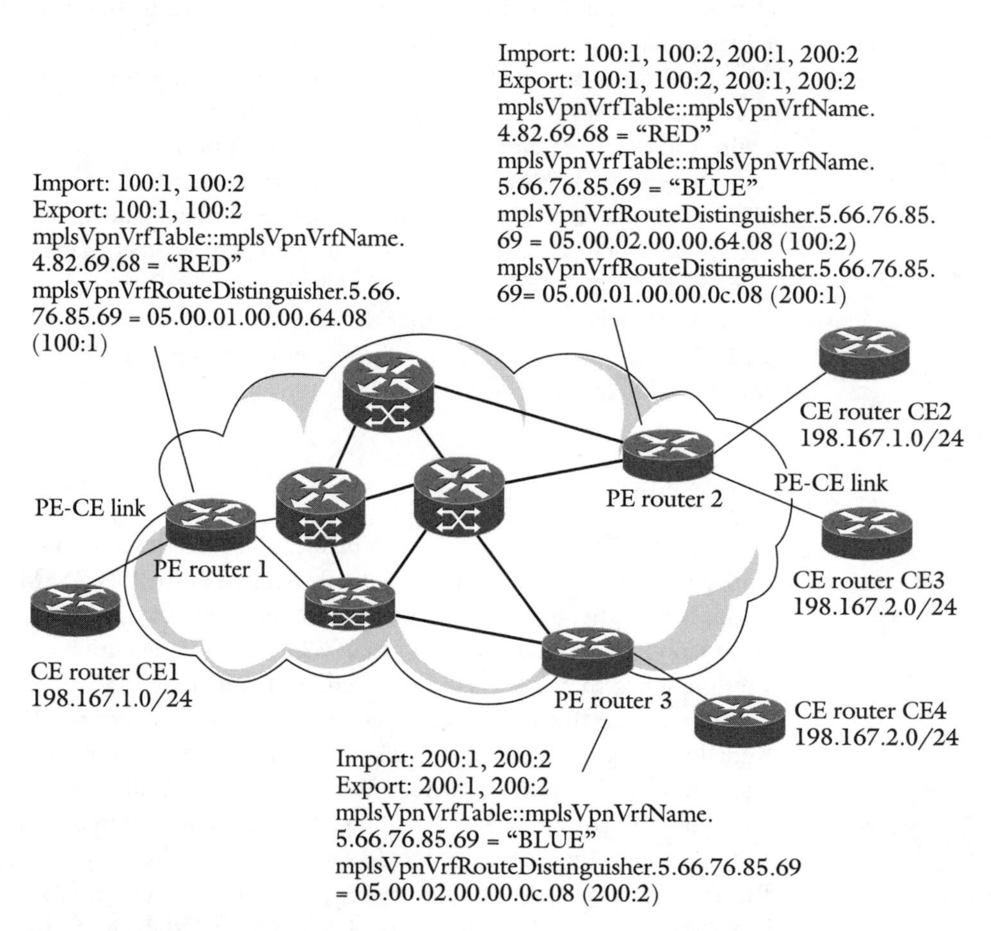

Figure 11.11 An MPLS-enabled network that supports two VPNs ("BLUE" and "RED"). The PE route targets, as well as the import and export configurations to support this configuration, are noted next to each PE. The mplsVpnVrfTable indexing configuration is shown for each of the PE routers. Note that the indexes are represented using hex values.

routers. Notice that PE router 2 contains two VRFs, one for each VPN "RED" and "BLUE," while PE router 1 and PE router 3 contain single VRF entries. Also, notice that the configuration of indexes at each VRF is consistent. If this were not consistent, then imagine how much potential confusion an operator would have if she examined one of the routers that actually had, for example, VPN "BLUE" configured as "PURPLE." Note that the indexes are represented using hex values, and that the first dotted decimal number of the index corresponds to the length of the octet string used to represent the index.

One important facet of this identifier is that some operators may wish to assign the VPN ID as defined in RFC 2685. One advantage of doing so is that this may facilitate the uniqueness of the VPN identifier across MPLS domain boundaries. This may be important in CsC deployments that do in fact cross service provider boundaries. The VPN ID can also be assigned to uniquely identify MPLS/BGP VPNs on a global basis as an IP address can uniquely identify a network.

Table 11.2 enumerates and defines each of the objects found in the MPLS VPN VRF Table (mplsVpnVrfTable). Each object is shown with a corresponding description. In general, the description shown is composed of two parts. First, it explains the definition of the object as defined in the PPVPN-MPLS-VPN MIB standard. Second, it may also contain suggestions on implementing or interpreting the object.

Table 11.2 MplsVpnVrfTable tabular objects.

Object name	Object description
mplsVpnVrfName	This value represents the human-readable name of this VPN and may optionally be set to the RFC 2685 VPN ID. We strongly recommend that this value be kept consistent across all PEs implementing VRFs for this VPN. If it is set to the VPN ID, the standard specifies that it must be set to the same value as mplsVpnVrfVpnId. See the description of the indexing of this table for more information and suggestions for this object.
mplsVpnVrfVpnId	This object contains the VPN ID as specified in RFC 2685 if such a value has been assigned to this VRF. This object should return the value of an empty string if a VPN ID has not been assigned to this VRF. As stated above in the description for mplsVpnVrfName, if the VPN ID is used to index this table, then this value must also match that of mplsVpnVrfName.

continued

Table 11.2 continued

Object name	Object description
mplsVpnVrfDescription	This object contains the human-readable description of this VRF. We recommend that operators should set this to something that is meaningful, especially during debugging situations. For example, if a VRF faces a certain customer and geographic installation, it might be useful to include this information in this field. Implementations should set this value to some reasonable default such as the mplsVpnVrfName and should avoid setting it to something that might be confusing (e.g., first interface name in the VPN, etc.).
mplsVpnVrfRouteDistinguisher	This object is set to the route distinguisher for this VRF. The object type is an octet string. One of the typical encodings of the route distinguisher is the VPN-IPv4 address family. A VPN-IPv4 address is defined as a 12-byte quantity, beginning with an 8-byte route distinguisher and ending with a 4-byte IPv4 address. This allows two or more VPNs to use the same IPv4 address prefix without collisions. This works because the PEs always append each VRF's addresses with unique VPN-IPv4 address prefixes. This approach guarantees that the same address is actually viewed as two different VPN addresses. A route distinguisher is composed of a 2-byte type field, an administrator field, and an assigned number field. The value of the type field determines the lengths of the remaining fields. The value field also implies the semantics of the administrator field that is used to identify which assigned number authority is used to govern the address spaces used in the RD. A VPN-IPv4 address consists of an 8-byte route distinguisher followed by a 4-byte IPv4 address. The RDs are encoded as type field (2 bytes), value field (6 bytes), IPv4 address (4 bytes). The interpretation of the value field depends on the value of the type field. Three values of the type field are defined: 0, 1, and 2. These types are defined as follows: Type 0: *2 bytes* *4 bytes* Administrator Assigned number

Table 11.2 continued

Object name	Object description
	Type 1: *4 bytes* — Administrator (IPv4 address) *2 bytes* — Assigned Number Type 2: *4 bytes* — Administrator (autonomous system number [BGP-AS4]) *2 bytes* — Assigned Number
mplsVpnVrfCreationTime	This object contains the sysUpTime at the time when this VRF entry was created on the device. If this VRF was created via SNMP, it is considered created when its RowStatus value was set to **active (1).** If this VRF was created via a non-SNMP management interface, then this object should contain the value of sysUpTime when the entry was considered created by the management interface.
mplsVpnVrfOperStatus	This object contains the operational status of the VRF. A VRF is considered operational if it is capable of forwarding traffic. A VRF is considered operationally **up (1)** when at least one interface associated with the VRF has an ifOperStatus set to **up (1).** A VRF is considered to be nonoperational or **down (2)** when it cannot forward traffic. A VRF might not be able to forward traffic under two circumstances. In the first case, if at least one interface whose ifOperStatus is **up (1)** does not exist, then the VRF cannot forward traffic and is therefore considered down. In the second case, a VRF may exist without any associated interfaces. In this case, the VRF also cannot forward traffic. A typical example of the second case takes place when a VRF is first created. The row creation operation may not have associated interfaces with the VRF during the operation; thus the second case applies. All of these cases indicate that the VRF is incapable of forwarding traffic; therefore we suggest that implementations

continued

Table 11.2 continued

Object name	Object description
	are careful to properly indicate this condition when the VRF is queried because the operator's NMS may have missed the mplsVrfIfDown notifications associated with the VRF interfaces (assuming they were configured to be issued in the first place). Incorrectly reporting this value may give the operator the false impression that the VRF is operational when, in reality, customer traffic is not flowing.
mplsVpnVrfActiveInterfaces	This variable indicates the total number of interfaces connected to the VRF with ifOperStatus = **up (1).** This object is incremented in two cases: first, when the ifOperStatus of one of the connected interfaces changes from **down (2)** to **up (1),** and second, when an interface with ifOperStatus = **up (1)** is connected to this VRF. Conversely, this object should be decremented when the ifOperStatus of one of the connected interfaces changes from **up (1)** to **down (2),** or when one of the connected interfaces with ifOperStatus = **up (1)** gets disconnected from this VRF. As with the mplsVpnVrfOperStatus object, it is very important that agents correctly reflect the status of attached interfaces for the reasons explained in the description for mplsVpnVrf-OperStatus.
mplsVpnVrfAssociated-Interfaces	This object indicates the total number of interfaces that have been associated with this VRF through configuration action. This value reflects the configured number of interfaces, and is thus independent of the actual value of ifOperStatus for any of those interfaces. This value can be used in collaboration with mplsVpnVrfActiveInterfaces as a sanity check against the number of interfaces that are actually functional versus the number configured. This can be used to quickly ascertain whether all VRF interfaces are operational as have been provisioned by the operator. Implementations that are based on polling the table (versus listening for mplsVrfIfDown notifications) can obtain the status with two simple GET operations on this table.

Table 11.2 continued

Object name	Object description
mplsVpnVrfConfMidRoute-Threshold	This value is used to denote the midlevel-water mark for the number of routes that this VRF has been configured to contain. Once this threshold is crossed, the mplsNumVrfRoute-MidThreshExceeded is issued for this VRF (assuming that notifications are enabled for VRFs). Issuance of this notification indicates to the operator that the midlevel-water mark that was configured has been exceeded by the number of entries in the MplsVpnVrfRouteTable. Please see Section 11.13.3 for more details on the mplsNumVrfRouteMidThreshExceeded notification.
mplsVpnVrfConfHighRoute-Threshold	This value is used to denote the configured maximum level or "high-water" mark for the number of routes that this VRF has been configured to contain. This represents a configured limitation on the implementation in that the device should stop adding routes into the VRF past this limit. It should be noted that this limit may or may not be set to the same as the value of mplsVpnVrfConfMaxPossibleRoutes. However, once this threshold is crossed, the mplsNumVrfRouteMidThresh-Exceeded notification is issued for this VRF (assuming that notifications are enabled for VRFs). Issuance of this notification indicates to the operator that the midlevel-water mark that was configured has been exceeded by the number of entries in the MplsVpnVrfRouteTable. Please see Section 11.13.3 for more details on the mplsNumVrfRouteMidThreshExceeded notification.
mplsVpnVrfConfMaxRoutes	This object indicates the maximum number of routes that this VRF is allowed to hold. Operators are cautioned to configure this value with care, as it has the potential for disrupting VRF traffic. The standard specifies that this value has to be set to a value that is less than or equal to mplsVrfMaxPossibleRoutes unless it is set to zero. Implementations should reject requests to improperly set this value out of allowed bounds.
mplsVpnVrfConfLastChanged	The value of sysUpTime at the time of the last change of this table entry, which includes changes of VRF parameters defined

continued

Table 11.2 continued

Object name	Object description
	in this table or addition or deletion of interfaces associated with this VRF.
mplsVpnVrfConfRowStatus	This variable is used to create, modify, and/or delete a row in this table.
mplsVpnVrfConfStorageType	The storage type for this entry.

11.6 MplsVPNIfConfTable

The MPLS VPN Interface Configuration Table (mplsVpnIfConfTable) contains entries for each interface that is configured with the MPLS/BGP VPN feature enabled. Entries in this table extend those found in the IF-MIB's ifEntry; therefore, entries must exist there prior to existing in this table. Due to this relationship, certain objects such as traffic counters are not found in this table to avoid overlap, and instead are found in the IF-MIB.

Network Management Systems that allow users to provision or configure VPNs must ensure that before an entry can be added to this table, a corresponding one exists in the IF-MIB. Agents implementing these tables must also keep them consistent. For example, if an interface is deleted from the IF-MIB, then the corresponding entry in this table should be destroyed as well. If an IF-MIB entry is disabled, then this entry should reflect that as well. These consistency checks are especially important for logical subinterfaces that may come and go with ease.

11.6.1 MplsVPNIfConfTable Indexing

This table is indexed by two objects: mplsVpnVrfName and mplsVpnIfConfIndex. The second index should be expected from the preceding description of the table. This index is the IF-MIB's ifIndex and is used to extend the corresponding entries in the IF-MIB. The first index provides a sorting of the table by VPN name. Therefore, when the table is walked from top to bottom, the VPN-enabled interfaces are sorted and grouped by the VRF that they are associated with.

The indexing of this table is deliberate. First, it allows an NMS to quickly and efficiently query this table in a targeted manner. When an NMS is managing a VPN, it need only look at the entries related to that VPN (the VRF and the associated interfaces). The indexing of this table is very much in line with that approach.

It should be noted that the secondary index in this table is defined as an object within this table. The next version of this MIB will be released without this as an object within the table because this object is unnecessary, as it is already defined in the IF-MIB. This is one of the few subtle changes between the draft version of the MIB described here, and the next version that the co-authors are working on.

Table 11.3 enumerates and defines each of the objects found in the MPLS VPN Configured Interface Table (mplsVpnIfTable). Each object from the table is shown with a corresponding description. In addition to the basic description, we have sometimes also included suggestions for implementing or interpreting the object, since the standard may have been either vague or incomplete in this regard.

Table 11.3 MplsVpnIfConfTable tabular objects.

Object name	Object description
mplsVpnIfConfIndex	This nonzero index indicates the IF-MIB index for the entry in the mplsVpnIfConfTable. Entries in this table only correspond with IF-MIB entries that are enabled for MPLS VPN; therefore, entries in this table do not necessarily correspond one-to-one with all entries in the IF-MIB, nor do these entries necessarily correspond with all entries with an ifType of MPLS layer. An entry in this table is not required to have an ifType of MPLS layer unless this interface represents the MPLS sublayer. Typically, this only occurs in the carrier's carrier case, where an interface will use LDP as its Label Distribution Protocol, and thereby will have MPLS enabled.
mplsVpnIfLabelEdgeType	This object indicates whether it is considered a customer or provider edge interface. In all non-CsC cases, this value should return **providerEdge (1)** to indicate that this router is a PE and contains that type of interface. However, in a CsC case where this device is a CsC CE device present in this customer SP network, it should return a value of **customerEdge (2).** This value should be consistent with the value of mplsVpnIfVpnClassification (see below).
mplsVpnIfVpnClassification	This object indicates whether this link participates in a carrier's carrier, enterprise, or interprovider scenario. Values are **carrierOfCarrier (1), enterprise (2), interProvider (3).** If this object is set to **carrierOfCarrier (1),** then the value of

continued

Table 11.3 continued

Object name	Object description
	mplsVpnIfLabelEdgeType can be either **providerEdge (1)** or **customerEdge (2).** However, if this object is set to **enterprise (2),** then mplsVpnIfLabelEdgeType must be **providerEdge (1).**
mplsVpnIfVpnRouteDistProtocol	This object indicates the route distribution protocol or protocols that are activated across this interface. Since more than one routing protocol may be enabled simultaneously, this object is defined as a bit field. Valid values are **none (0), bgp (1), ospf (2), rip (3), isis(4),** and **other (5).** The value of **other (5)** is used to indicate that a different, perhaps proprietary, routing distribution protocol is in use. The value of **none (0)** should be set by agents that have no routing protocol, but that are using static routes in their VPN configuration.
mplsVpnVrfConfRowStatus	This variable is used to create, modify, and/or delete a row in this table. The RowStatus of this table cannot go to **active (1)** unless a corresponding ifEntry exists for the entry.
mplsVpnVrfConfStorageType	The storage type for this entry. This object contains the RowStatus for this entry. Agents should not allow entries to become **active (1)** in this table unless corresponding mplsVrfTable and ifTable entries exist for the primary index of mplsVrfName and the ifIndex. Not doing so could result in inconsistent configuration and/or confused management systems.

11.7 MplsVPNPerfTable

The MPLS VPN Performance Table (MplsVPNPerfTable) contains objects that enable an operator to measure certain performance characteristics of a specific MPLS/BGP VRF. Specifically, this table concerns itself with the routing table activity of the VPN. This table AUGMENTS the mplsVpnVrfTable; therefore each entry in this table corresponds to exactly one entry in the MplsVpnVrfTable. This table on its own is not intended to provide performance data related to an entire VPN. It is up to the NMS to combine the statistics for the various VPN instances (i.e., VRFs) together to form a picture of the larger VPN.

The performance information provided in this table is not intended to be the only available information for the VRF or VPN. Other performance characteristics may be interesting as well, but are not kept track of here; instead, they are obtainable from other sources. For instance, the aggregate data throughput of a VRF (i.e., bytes, packets, etc.) is something that operators often ask us for. This information can be obtained by summing the relevant IF-MIB counters for those interfaces that are associated with the VRF. The interfaces can be determined by examining the mplsVrfIfConfTable. From there, each interface's IF-MIB counters can be interrogated and summed. Similarly, the overall throughput of the VPN can be obtained by examining each VRF that comprises the VRF and then summing this information together. Both the aggregate throughput measurements for a VRF and a VPN are functions that many typical network management systems that are available today provide. Many others exist as well.

Figure 11.12 demonstrates how performance statistics might be gathered for the VPN "RED." The figure repeats the basic configuration shown in Figure 11.10, but adds the contents of several performance-related tables. Using this example, if an NMS were interested in gathering route table performance information for VPN "RED" to display to an operator, it would first interrogate the two PEs where the VRFs for VPN "RED" exist. The two PEs containing VRF Table entries for VPN "RED" have an mplsVrfTable with an entry indexed by mplsVpnVrfName = 4.82.69.68 = "RED." Note that since the mplsVpnVrfName object is defined as an octet string, the values we will display will be in the format of an octet string. This format specifies the first dotted decimal as the string's length, and the remaining ones for the values. From this information, an NMS can then interrogate the mplsVpnIfTable using the mplsVpnName.4.82.69.68 = "RED" as the primary index. Each entry returned with this as a primary index will indicate an interface that is associated with the VRF on that PE. A quick glance at both PEs will show that the PE at the bottom left has one entry noted as ifEntry.10, and the top-right PE has an ifEntry.20 that has MPLS VPN enabled and is associated with the VRFs for VPN "RED." From this, the NMS can then gather routing table performance figures such as the number of routes added or deleted, as well as the current number of routes contained in the VRF. For the leftmost PE, these values are shown as mplsVpnVrfPerfEntry.mplsVpnVrfPerfRoutesAdded.4.82.69.68.10 = 10020, mplsVpnVrfPerfEntry.mplsVpnVrfPerfRoutesDeleted.4.82.69.68.10 = 5000, and mplsVpnVrfPerfEntry.mplsVpnVrfPerfCurrNumRoutes.4.82.69.68.10 = 455. Similarly, examining the entry for VPN "RED" and interface 20 on the topmost PE returns mplsVpnVrfPerfEntry.mplsVpnVrfPerfRoutesAdded.4.82.69.68.20 = 120, mplsVpnVrfPerfEntry.mplsVpnVrfPerfRoutesDeleted.4.82.69.68.10 = 5678, and mplsVpnVrfPerfEntry.mplsVpnVrfPerfCurrNumRoutes.4.82.69.68.10 = 455. This information can then be displayed for an operator for each VRF. This information can also be combined (summed) to show a collective VPN statistic. Similarly,

Import: 100:1, 100:2, 200:1, 200:2
Export: 100:1, 100:2, 200:1, 200:2
mplsVpnVrfRouteDistinguisher.5.66.76.85.69 = 05.00.02.00.00.64.08 (100:2)
mplsVpnVrfRouteDistinguisher.5.66.76.85.69 = 05.00.01.00.00.0c.08 (200:1)
IfEntry.ifInOctets.20 = 1250
IfEntry.IfInUcastPkts.20 = 100
mplsVpnVrfName.4.82.69.68 = "RED"
mplsVpnIfConfEntry.mplsVpnIfLabelEdgeType.4.82.69.68.20 = providerEdge(1)
mplsVpnIfConfEntry.mplsVpnIfVpnClassification4.82.69.68.20 = enterprise(2)
mplsVpnIfConfEntry.mplsVpnIfmplsVpnRouteDistProtocol.4.82.69.68.20 = 0x40(ospf)
mplsVpnVrfPerfEntry.mplsVpnVrfPerfRoutesAdded.4.82.69.68.20 = 12000
mplsVpnVrfPerfEntry.mplsVpnVrfPerfRoutesDeleted.4.82.69.68.100 = 5678
mplsVpnVrfPerfEntry.mplsVpnVrfPerfCurrNumRoutes.4.82.69.68.10 = 455

Import: 100:1, 100:2
Export: 100:1, 100:2
mplsVpnVrfRouteDistinguisher.5.66.76.85.69 =
05.00.01.00.00.64.08 (100:1)
IfEntry.ifInOctets.10 = 1250
IfEntry.IfInUcastPkts.10 = 100
mplsVpnVrfTable.mplsVpnVrfName.4.82.69.68 = "RED"
mplsVpnIfConfEntry.mplsVpnIfLabelEdgeType.4.82.69.68.10 = providerEdge(1)
mplsVpnIfConfEntry.4.82.69.68.10.mplsVpnIfVpnClassification = enterprise(2)
mplsVpnIfConfEntry.4.82.69.68.10.mplsVpnIfmplsVpnRouteDistProtocol = 0x40(ospf)
mplsVpnVrfPerfEntry.mplsVpnVrfPerfRoutesAdded.4.82.69.68.10 = 10020
mplsVpnVrfPerfEntry.mplsVpnVrfPerfRoutesDeleted.4.82.69.68.10 = 5000
mplsVpnVrfPerfEntry.mplsVpnVrfPerfCurrNumRoutes.4.82.69.68.10 = 455

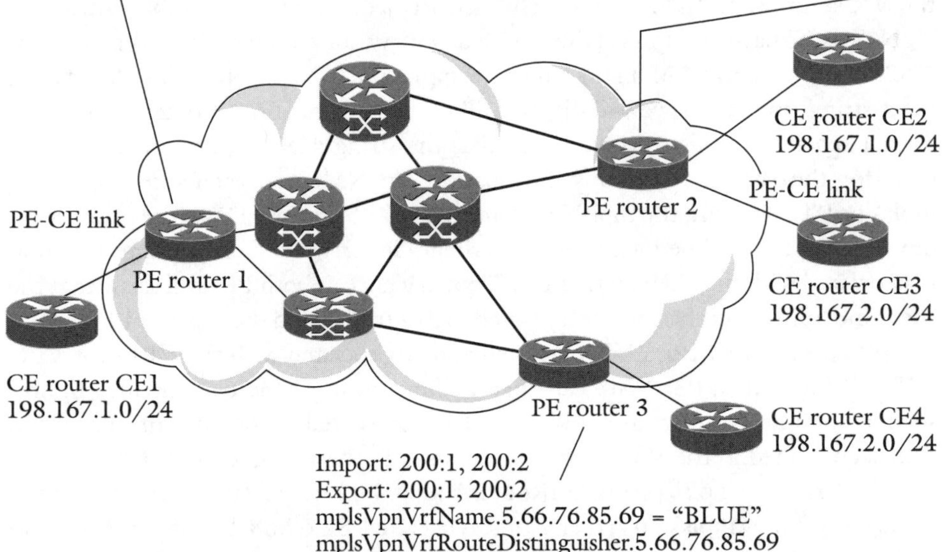

Import: 200:1, 200:2
Export: 200:1, 200:2
mplsVpnVrfName.5.66.76.85.69 = "BLUE"
mplsVpnVrfRouteDistinguisher.5.66.76.85.69
= 05.00.02.00.00.0c.08 (200:2)

Figure 11.12 An MPLS-enabled network that supports two VPNs ("BLUE" and "RED"). The PE route targets, as well as the import and export configurations to support this configuration, are noted next to each PE. The mplsVpnVrfTable configuration is shown for each of the PE routers. Note that the octet string indexes are represented using hex values. The indexing for the mplsVpnVrfPerfTable and IF-MIB's ifTable indexing are shown as well to demonstrate how performance statistics can be gathered from VPN "RED" by gathering statistics from each PE supporting that VPN.

interface-based statistics can be gathered and displayed. This information can then be used, for example, for provisioning, maintenance, or SLA decisions.

One final note regarding the objects found in this table. Since the objects in this table reflect performance-related counters, it is implementation-specific as to how quickly the objects are updated (within reason). For example, it may take a second or two for the routing protocol to actually update the internal counters, and thus this variable may not reflect the exact count at any given time, but should in a very short period of time. It is very typical of high-performance implementations to have their counters lag in being updated by a second or two.

Table 11.4 enumerates each of the objects found in the MPLS VPN VRF Performance Table (mplsVpnVrfPerfTable). Each object from the table is shown with a corresponding description. In addition to the basic description, we have sometimes also included suggestions for implementing or interpreting the object, since the standard may have been either vague or incomplete in this regard.

Table 11.4 MplsVPNPerfTable tabular objects.

Object name	Object description
mplsVpnVrfPerfRoutesAdded	This object indicates the number of routes added to this VPN/VRF over the course of its lifetime. If a VRF is taken out of service (i.e., its RowStatus is changed to something other than **active (1)**), it is implementation-dependent whether or not this variable's value can be preserved when the agent restores the RowStatus to **active (1).**
mplsVpnVrfPerfRoutesDeleted	This object keeps track of the number of routes removed from this VPN/VRF during the course of its RowStatus being set to **active (1).** If a VRF is taken out of service (i.e., its RowStatus is changed to something other than **active (1)),** it is implementation-dependent whether or not this variable's value can be preserved when the agent restores the RowStatus to **active (1).**
mplsVpnVrfPerfCurrNumRoutes	This object indicates the number of routes that are currently in use by this VRF. If a VRF is taken out of service (i.e., its RowStatus is changed to something other than **active (1)),** it is implementation-dependent whether or not this variable's value can be preserved when the agent restores the RowStatus to **active(1).**

11.8 MplsVpnVrfRouteTable

The MPLS VPN VRF Route Table (mplsVpnVrfRouteTable) contains objects representing the routes that are present in a particular VRF. The format of this table is meant to mimic that found in the IP-Forwarding MIB and, in fact, was copied from there. The only difference between the IP-Forwarding MIB and the MplsVpnVrf-RouteTable is the fact that the routing entries are indexed by the addition of an mplsVrfName so that particular route entries can be shown on a per-VRF basis.

The table supports both dynamic and static entries. When routes are added to this table, the creator of the routing entry is specified by the mplsVpnVrfRoute-Proto column for that entry. Static entries can be added to this table if an agent allows external configuration. If a static entry is added, the mplsVpnVrfRouteProto should be set to **netmgmt (3)** to indicate this. Routing entries created by the other supported routing protocols should be amply noted using the mplsVpnVrfRoute-Proto enumeration. Entries added by routing protocols may come and go depending on the status of the routing distribution protocol, and are thus considered dynamic. These entries will typically disappear from the configuration if the device is restarted and will be restored as the routing protocols converge.

Agents allowing configuration of entries in this table should then save configured entries as per their local configuration policy, which should be made clear for NMS operators so that configuration effort is not lost. For example, Cisco devices support two copies of a configuration: running and stored in NVRAM. In the first case, changes are made to the running configuration at will by the operator. However, these changes are only realized in the device's running memory. If the device crashes or is rebooted, these changes are lost. In order to allow a manager to specifically determine when changes to the configuration should be written to permanent memory, Cisco devices provide a "write memory" command for all management interfaces. When this is issued, the current running configuration is copied to the stored one, thus preserving it across reboots of the system.

11.8.1 MplsVpnVrfRouteTable Indexing

The mplsVpnVrfRouteTable is indexed by the same values as the IP-Forwarding MIB's ipCidrRouteTable with the one addition of the primary index that specifies the mplsVpnVrfName to which the route entries belong. This allows operators to filter the routing table entries by VRF rather than having to traverse all of the routing entries. This is especially important if there are a large number of VRFs and/or several of the VRFs contain many routing entries. This may also be useful for those agents that have access to the MIB configured on a per-VRF basis. For security reasons, it may not be desired to show routes from one VRF to someone whose routes belong to another VRF. The indexing of this table makes it easy to hide entries to ensure privacy.

The indexes for this table include mplsVpnVrfName, mplsVpnVrfRouteDest, mplsVpnVrfRouteMask, mplsVpnVrfRouteTos, and mplsVpnVrfRouteNextHop. The routing table entries specify indexing of routing entries on a per-destination-route basis. Routes can be further distinguished by the type-of-service (TOS) bits, as well as the next-hop router address.

Table 11.5 enumerates each of the objects found in the MPLS VPN VRF Routing Table (mplsVpnVrfRouteTable). Each object from the table is shown with a corresponding description. In addition to the basic description, we have sometimes also included suggestions for implementing or interpreting the object, since the standard may have been either vague or incomplete in this regard.

Table 11.5 MplsVpnVrfRouteTable tabular objects.

Object name	Object description
mplsVpnVrfRouteDest	The destination IP address of this route. This object may not take a multicast (Class D) address value. Any assignment (implicit or otherwise) of an instance of this object to a value *x* must be rejected if the bitwise logical-AND of *x* with the value of the corresponding instance of the mplsVpnVrfRouteMask object is not equal to *x*.
mplsVpnVrfRouteDest-AddrType	The address type of the mplsVpnVrfRouteDest entry. This object is of the type InetAddressType.
mplsVpnVrfRouteMask	Indicates the mask to be logical-ANDed with the destination address before being compared to the value in the mplsVpnVrfRouteDest field. For those systems that do not support arbitrary subnet masks, an agent constructs the value of the mplsVpnVrfRouteMask by reference to the IP address class. Any assignment (implicit or otherwise) of an instance of this object to a value *x* must be rejected if the bitwise logical-AND of *x* with the value of the corresponding instance of the mplsVpn-VrfRouteDest object is not equal to mplsVpnVrf-RouteDest.
mplsVpnVrfRouteMask-AddrType	The address type of mplsVpnVrfRouteMask.

continued

Table 11.5 continued

Object name	Object description
mplsVpnVrfRouteTos	This object contains the IP TOS field that is used to specify the policy that is applied to this routing entry. The format of the TOS field is the following: precedence requires 3 bytes, type of service 4, and the last is set to 0. 0 1 2 3 4 5 6 7 8 Precedence Type of Service 0 The encoding of IP TOS is as specified by the following table. Please note that a value of zero indicates the default path unless a more specific policy applies.

Field contents	Policy code	Field contents	Policy contents code
0000	0	0001	2
0010	4	0011	6
0100	8	0101	10
0110	12	0111	14
1000	16	1001	18
1001	20	1001	33
1100	24	1011	26
1110	28	1101	30

Object name	Object description
mplsVpnVrfRouteNextHop	This object indicates the address of the next network node toward the destination indicated by this remote route. It is set to 0.0.0.0. if a next-hop address is not available.
mplsVpnVrfRouteNextHop-AddrType	This object contains the address type of the mplsVpnVrf-RouteNextHopAddrType object. This object uses the InetAddressType textual convention, which is capable of handling IPv4 and IPv6 address families, among others.
mplsVpnVrfRouteIfIndex	This object is set to reflect the local interface through which the next-hop of this route can be reached. This value is set to zero to indicate that no interface is associated with this route; otherwise it contains the IF-MIB's ifIndex value for the interface.
mplsVpnVrfRouteType	This object indicates the type of route this entry in the table represents. Valid values are the following: **other (1).** Not specified.

Table 11.5 continued

Object name	Object description
	reject (2). This type of route refers to one that, if matched, discards any traffic received going to this route as unreachable. This type of entry is also used in some protocols as a means of correctly aggregating routes. **local (3).** Refers to a route for which the next-hop is the final destination and is accessible via a local interface. **remote (4).** This type refers to a route for which the next-hop is not the final destination. Routes that do not result in traffic forwarding or rejection should not be displayed in this table, even if the implementation keeps them stored internally for some reason.
mplsVpnVrfRouteProto	This object indicates the routing protocol that installed the route entry. **other (1)** Not specified **local (2)** Local interface **netmgmt (3)** Static route **icmp (4)** Result of ICMP Redirect **egp (5)** Exterior Gateway Protocol **ggp (6)** Gateway-Gateway Protocol **hello (7)** FuzzBall HelloSpeak **rip (8)** Berkeley RIP or RIP-II **isIs (9)** Dual IS-IS **esIs (10)** ISO 9542 **ciscoIgrp (11)** Cisco IGRP **bbnSpfIgp (12)** BBN SPF IGP **ospf (13)** Open Shortest Path First **bgp (14)** Border Gateway Protocol **idpr (15)** InterDomain Policy Routing **ciscoEigrp (16)** Cisco EIGRP

continued

Table 11.5 continued

Object name	Object description
mplsVpnVrfRouteAge	This object indicates the number of seconds since this route was last updated or otherwise determined to be correct.
mplsVpnVrfRouteInfo	This object optionally includes a reference to MIB definitions specific to the particular routing protocol that is responsible for this route. The specific reference type is implied by the value specified in the route's mplsVpnVrfRouteProto value. Agents not implementing this are strongly advised to set this object to {0.0}.
mplsVpnVrfRouteNext-HopAS	This object contains the autonomous system number of the next-hop. The semantics of this object are determined by the routing protocol specified in the route's mplsVpnVrfRouteProto value. When this object is unknown or irrelevant, the agent should set it to zero.
mplsVpnVrfRouteMetric1	The primary routing metric for this route. The semantics of this metric are determined by the routing protocol specified in the route's mplsVpnVrfRouteProto value. If this metric is not used, its value should be set to −1.
mplsVpnVrfRouteMetric2	An alternative routing metric for this route. The semantics of this metric are determined by the routing protocol specified in the route's mplsVpnVrfRouteProto value. If this metric is not used, its value should be set to −1.
mplsVpnVrfRouteMetric3	An alternative routing metric for this route. The semantics of this metric are determined by the routing protocol specified in the route's mplsVpnVrfRouteProto value. If this metric is not used, its value should be set to −1.
mplsVpnVrfRouteMetric4	An alternative routing metric for this route. The semantics of this metric are determined by the routing protocol specified in the route's mplsVpnVrfRouteProto value. If this metric is not used, its value should be set to −1.
mplsVpnVrfRouteMetric5	An alternative routing metric for this route. The semantics of this metric are determined by the routing protocol specified in the route's mplsVpnVrfRouteProto value. If this metric is not used, its value should be set to −1.
mplsVpnVrfRouteRowStatus	Row status for this table. It is used according to row installation and removal conventions. This object contains the row status for

Table 11.5 continued

Object name	Object description
	this entry. Agents should not allow entries to become **active (1)** in this table unless a corresponding mplsVrfTable entry exists for the primary index of mplsVrfName. Not doing so could result in inconsistent configuration and/or confused management systems.
mplsVpnVrfRouteStorageType	Storage type value.

11.9 MplsVpnRouteTargetTable

The MPLS VPN Route Target Table (MplsVpnRouteTargetTable) contains route target entries associated with each VRF on a device. As described earlier (see Section 11.1.3), route targets specify the policies governing the import and export of routes to and from a VRF. The route target community identifies one or more routers that may receive a set of routes (that carry this community) carried by BGP. This is transitive across the autonomous system boundary. Route target entries are encoded as BGP extended community route targets (Sangli et al. 2002). This draft specifies that the BGP extended community route target be encoded as an 8-byte value that includes a 1- or 2-byte type field followed by a value field for the remaining bytes. The value of the type field for the route target community can be 0x00, 0x01, or 0x02. The value of the low-order octet of the extended type field for this community is 0x02. When the value of the type field is 0x00 or 0x02, the standard requires that the use of the local administrator subfield in the value field be unique within the autonomous system carried in the global administrator subfield. We will not go into the additional specifics of the format here, and instead will refer you to Sangli et al. (2002). All routes associated with a particular extended communities attribute belong to the communities listed in the attribute.

11.9.1 MplsVpnVrfRouteTargetTable Indexing

This table is indexed by three objects. First, mplsVpnVrfName serves as a primary index into this table and allows its contents to be grouped by VRF. Next, the mplsVpnVrfRouteTargetIndex is an arbitrary index assigned to a route target entry. An implementation may choose this to be some unique identifier such as a memory address or another unique index onto the data structure that holds the route target information. The last index provides a third grouping of the entries into groups of mplsVpnVrfRouteTargetType. This last index seems as if it could have been left out. However, some of the service providers who gave input to the MIB felt that it gave them quicker access to the route target most important to them at the time they queried the table. For example, if an operator discovers a routing problem with a particular VPN site not being able to reach other sites (PEs), the operator will first go to the PE that is directly connected to that site and examine the VRF for that VPN. If that does not check out, chances are that the route target table is not importing routes correctly from other sites. A quick GET-NEXT of only the mplsVrfName will return a list of route target entries, each with its import/export policy specified.

Table 11.6 enumerates and defines each of the objects contained within the MplsVpnRouteTargetTable.

Table 11.6 MplsVpnVrfRouteTargetTable tabular objects.

Object name	Object description
mplsVpnVrfRouteTargetIndex	This object is used as the secondary index for route target entries configured for a particular VRF. This value must be unique. Implementations should take heed to store this value in NVRAM if entries are configured using the SNMP interface because an NMS will expect this index to remain the same across reboots.
mplsVpnVrfRouteTargetType	This object contains the route target distribution type. Three values are possible: **import (1), export (2), both (3).** The third value was added as a convenience for configuration purposes. Many implementations require that two entries be configured: one for import and one for export, and thus may not support **both (3).** We refer you to the agent capability statement for the particular implementation in question, as it will definitively state whether this is supported.
mplsVpnVrfRouteTarget	This object contains the route target value that is distributed to other PEs or imported by other PEs. The specifics of the format can be found in Sangli et al. (2002). It should also be noted that the length of the MplsVpnRouteDistinguisher textual convention will be reduced to 8 bytes in future revisions of the PPVPN-MPLS-VPN MIB. The current length of 256 octets is incorrect.
mplsVpnVrfRouteTargetDescr	This object contains a human-readable description of the route target (hence the DisplayString type). If one does not exist,

Table 11.6 continued

Object name	Object description
	implementations are encouraged to set this to a zero-length string.
mplsVpnVrfRouteTarget-RowStatus	This object contains the row status for this entry. Agents should not allow entries to become **active (1)** in this table unless a corresponding mplsVrfTable entry exists for the primary index of mplsVrfName. Not doing so could result in inconsistent configuration and/or confused management systems.

11.10 MplsVpnVrfBgpNbrAddrTable

The MPLS VPN VRF BGP Neighbor Address Table (mplsVpnVrfBgpNbrAddr-Table) specifies per-interface MPLS/eBGP neighbor information for each VRF. Entries are created in this table for every VRF capable of supporting MPLS/BGP VPN that requires eBGP configuration. This implies that VRFs not requiring eBGP configuration should not have corresponding entries in this table. Agents allowing configuration of the information in this table should coordinate the entries created or modified in this table with those in the eBGP process on the device as well as the BGP-4 MIB, if it is supported. More specifically, agents should not allow entries to become active in this table unless a corresponding entry is also active in the BGP-4 MIB. Similarly, entries in this table should be destroyed if entries in the corresponding BGP-4 MIB tables are deleted. Entries in this table also need to be coordinated with the MplsVpnVrfTable and mplsVPNIfConfTable for similar consistency.

11.10.1 MplsVpnVrfBgpNbrAddrTable Indexing

This table is indexed by three values. As with all tables in this MIB, the primary index for this table is the mplsVpnVrfName. This allows these entries to be created on a per-VRF basis. The second index—mplsVpnIfConfIndex—specifies which VRF interface this configuration applies to. Finally, mplsVpnVrfBgpNbrIndex provides a unique tertiary index allowing more than one eBGP neighbor to be configured per VRF-enabled interface.

Table 11.7 lists and defines each of the objects contained within the MPLS VPN VRF BGP Neighbor Address Table (mplsVpnVrfBgpNbrAddrTable). Each object is shown with a corresponding description. In general, the description shown is composed of two parts. First, it explains the definition of the object as defined in the PPVPN-MPLS-VPN MIB standard. Second, it may also contain suggestions on implementing or interpreting the object.

Table 11.7 MplsVpnVrfBgpNbrAddrTable tabular objects.

Object name	Object description
mplsVpnVrfBgpNbrIndex	This object contains a unique tertiary index for an entry in the mplsVpnVrfBgpNbrAddrEntry Table allowing one or more eBGP neighbors to be configured per VRF-enabled interface.
mplsVpnVrfBgpNbrRole	This object denotes the role played by this eBGP neighbor with respect to this VRF. Valid values are **ce (1)** or **pe (2).** A neighbor cannot play the role of both a PE and a CE.
mplsVpnVrfBgpNbrType	This object specifies the address family of the PE address specified in mplsVpnVrfBgpNbrAddr.
mplsVpnVrfBgpNbrAddr	Denotes the eBGP neighbor address given the type in mplsVpnVrfBgpNbrType.
mplsVpnVrfBgpNbrRowStatus	This object is used to create, modify, and/or delete a row in this table. Note the warnings and guidelines for implementations above.
mplsVpnVrfBgpNbrStorageType	This object contains the row status for this entry. Agents should not allow entries to become **active (1)** in this table unless corresponding mplsVrfTable and mplsVrfIfConf entries exist for the primary mplsVrfName and secondary mplsVpnIf-ConfIndex indexes. Not doing so could result in inconsistent configuration and/or confused management systems.

11.11 MplsVpnVrfBgpNbrPrefixTable

The MPLS VPN VRF BGP Neighbor Prefix Table (mplsVpnVrfBgpNbrPrefix-Table) specifies per-VRF VPN-V4 multiprotocol prefixes supported by BGP for each VRF. Entries are created in this table for every VRF capable of supporting MPLS/BGP VPN that requires BGP configuration. Agents allowing configuration of the information in this table should coordinate the entries created or modified with the BGP process on the device as well as the BGP-4 MIB if it is supported.

11.11.1 MplsVpnVrfBgpNbrPrefixTable Indexing

This table is indexed by four indexes. The primary index—mplsVpnVrfName—specifies the VRF that this configuration entry is associated with. The remaining indexes are taken verbatim from the current BGP-4 MIB and include bgp4PathAttrIpAddrPrefix, bgp4PathAttrIpAddrPrefixLen, and bgp4PathAttrPeer. *A note to implementations that future versions of the PPVPN-MPLS-VPN MIB may remove this table entirely because the BGP-4 MIB may be modified to incorporate a VRF name.* This will enhance existing implementations, as a table that essentially duplicates pieces of the BGP-4 MIB will not be necessary. This table was added originally because such a mechanism did not exist at the time, but operators wanted to be able to manage this BGP information on a per-VRF basis.

Table 11.8 enumerates and defines the objects contained in the mplsVpnVrfBgpNbrPrefixTable.

Table 11.8 MplsVpnVrfBgpNbrPrefixTable tabular objects.

Object name	Object description
mplsVpnVrfBgpPAtrPeer	This object defines the IP address of the BGP peer where the path information was learned.
mplsVpnVrfBgpPAtrIpAddr-PrefixLen	This object contains the number of bits of the IP address prefix in the Network Layer Reachability Information field.
mplsVpnVrfBgpPAtrIpAddr-Prefix	An IP address prefix in the Network Layer Reachability Information field. This object is an IP address containing the prefix with length specified by mplsVpnVrfBgpPAtrIpAddrPrefixLen. Any bits beyond the length specified by mplsVpnVrfBgpPAtrIpAddrPrefixLen are zeroed.
mplsVpnVrfBgpPAtrOrigin	This object contains the ultimate origin of the path information specified by this entry. Valid values are **igp (1),** which specifies networks are interior; **egp (2),** which specifies that networks are learned via EGP; or **incomplete (3),** which specifies that the origin of the route is as yet undetermined.
mplsVpnVrfBgpPAtrASPath-Segment	The sequence of AS path segments. Each AS path segment is represented by a triple <type, length, value>. The type is a 1-octet field that has two possible values: 1 (AS_SET), an unordered set of ASs a route in the update message

continued

Table 11.8 continued

Object name	Object description

	has traversed, and 2 (AS_SEQUENCE), an ordered set of ASs a route in the update message has traversed. The length is a 1-octet field containing the number of ASs in the value field.
	The value field contains one or more AS numbers. Each AS is represented in the octet string as a pair of octets according to the following algorithm:
	first-byte-of-pair = ASNumber / 256;
	second-byte-of-pair = ASNumber & 255;

mplsVpnVrfBgpPAtrNext-Hop	This object contains the address of the border router that should be used for the destination network.
mplsVpnVrfBgpPAtrMulti-ExitDisc	This object contains the metric that is used to discriminate between multiple exit points to an adjacent autonomous system. A value of −1 indicates the absence of this attribute.
mplsVpnVrfBgpPAtrLocal-Pref	The originating BGP-4 speaker's degree of preference for an advertised route. A value of −1 indicates the absence of this attribute.
mplsVpnVrfBgpPAtrAtomic-Aggregate	Whether or not the local system has selected a less specific route without selecting a more specific route. Valid values are **lessSpecificRouteNotSelected (1)** or **lessSpecificRouteSelected (2).**
mplsVpnVrfBgpPAtr-AggregatorAS	The AS number of the last BGP-4 speaker that performed route aggregation. A value of 0 indicates the absence of this attribute.
mplsVpnVrfBgpPAtr-AggregatorAddr	The IP address of the last BGP-4 speaker that performed route aggregation. A value of 0.0.0.0 indicates the absence of this attribute.
mplsVpnVrfBgpPAtrCalc-LocalPref	The degree of preference calculated by the receiving BGP-4 speaker for an advertised route. A value of −1 indicates the absence of this attribute.
mplsVpnVrfBgpPAtrBest	An indication of whether or not this route was chosen as the best BGP-4 route.
mplsVpnVrfBgpPAtr-Unknown	One or more path attributes not understood by this BGP-4 speaker. Size zero indicates the absence of such attribute(s). Octets beyond the maximum size, if any, are not recorded by this object.

11.12 MplsVpnVrfSecTable

The MPLS VPN VRF Security Table (mplsVpnVrfSecTable) specifies per-VRF security-related information. This table AUGMENTS the MplsVpnVrfTable; therefore each entry in this table corresponds to exactly one entry in the MplsVpnVrfTable.

This table contains two objects that are used to track the number of times illegal labels are received on a VRF interface, as well as configure the threshold at which the MPLSNumVrfSecIllegalLabelThreshExceeded notification will be emitted. This notification should be issued infrequently, but when it is issued, it is important for the operator to heed its warning because either a configuration or security problem has been detected, neither of which should be treated lightly.

The MPLS VPN VRF Security Table (mplsVpnVrfSecTable) contains the objects in Table 11.9.

Table 11.9 MplsVpnVrfSecTable tabular objects.

Object name	Object description
mplsVpnVrfSecIllegalLblVltns	This object indicates the number of illegally received labels on this VPN/VRF since the VRF was created.
mplsVpnVrfSecIllegalLblRcvThrsh	This object contains the number of illegally received labels that must be received before the MPLSNumVrfSec-IllegalLabelThreshExceeded notification is issued.

11.13 Notifications

This section concerns itself with the notifications that an LSR supporting the PPVPN-MPLS-VPN MIB may emit. In particular, this section enumerates and defines each of the notifications and explains under which circumstances each notification should be emitted. We wrap up the section with a detailed example that shows under which circumstances the mplsNumVrfRouteMidThreshExceeded and mplsNumVrfRouteMaxThreshExceeded notifications are issued.

11.13.1 mplsVrfIfUp

This notification is generated either when the ifOperStatus of an interface associated with a VRF changes to the **up (1)** state or when an interface with ifOperStatus = **up (1)** is associated with a VRF. This notification is emitted with the corresponding mplsVpnInterfaceConfIndex and mplsVpnVrfName objects.

11.13.2 mplsVrfIfDown

This notification is generated either when the ifOperStatus of an interface associated with a VRF changes to the **down (1)** state or when an interface with ifOperStatus = **up(1)** is disassociated with a VRF. This notification is emitted with the corresponding mplsVpnInterfaceConfIndex and mplsVpnVrfName objects.

11.13.3 mplsNumVrfRouteMidThreshExceeded

This notification is generated when the number of routes contained by the specified VRF exceeds the value indicated by mplsVrfMidRouteThreshold. This notification is emitted with the corresponding mplsVpnVrfName and MPLSVpnVrfPerfCurrNumRoutes objects.

11.13.4 mplsNumVrfRouteMaxThreshExceeded

This notification is generated when the number of routes contained by the specified VRF reaches or attempts to exceed the maximum allowed value as indicated by mplsVrfMaxRouteThreshold. This notification is emitted with the corresponding mplsVpnVrfName and MPLSVpnVrfPerfCurrNumRoutes.

Figure 11.10, repeated here, illustrated how both mplsNumVrfRouteMidThreshExceeded and mplsNumVrfRouteMaxThreshExceeded behave by showing an example of a running count of current routes on a PE. As the routes are added or deleted, the appropriate notifications (when necessary) are emitted.

11.13.5 mplsNumVrfSecIllegalLabelThreshExceeded

This notification is generated when the number of illegal label violations on a VRF as indicated by mplsVpnVrfSecIllegalLabelViolations has exceeded mplsVpnVrfSecIllegalLabelRcvThresh. Although this notification should be emitted rarely, when one is emitted, an operator should take quick notice because it could mean that the PE-CE connection is misconfigured, or a true security violation

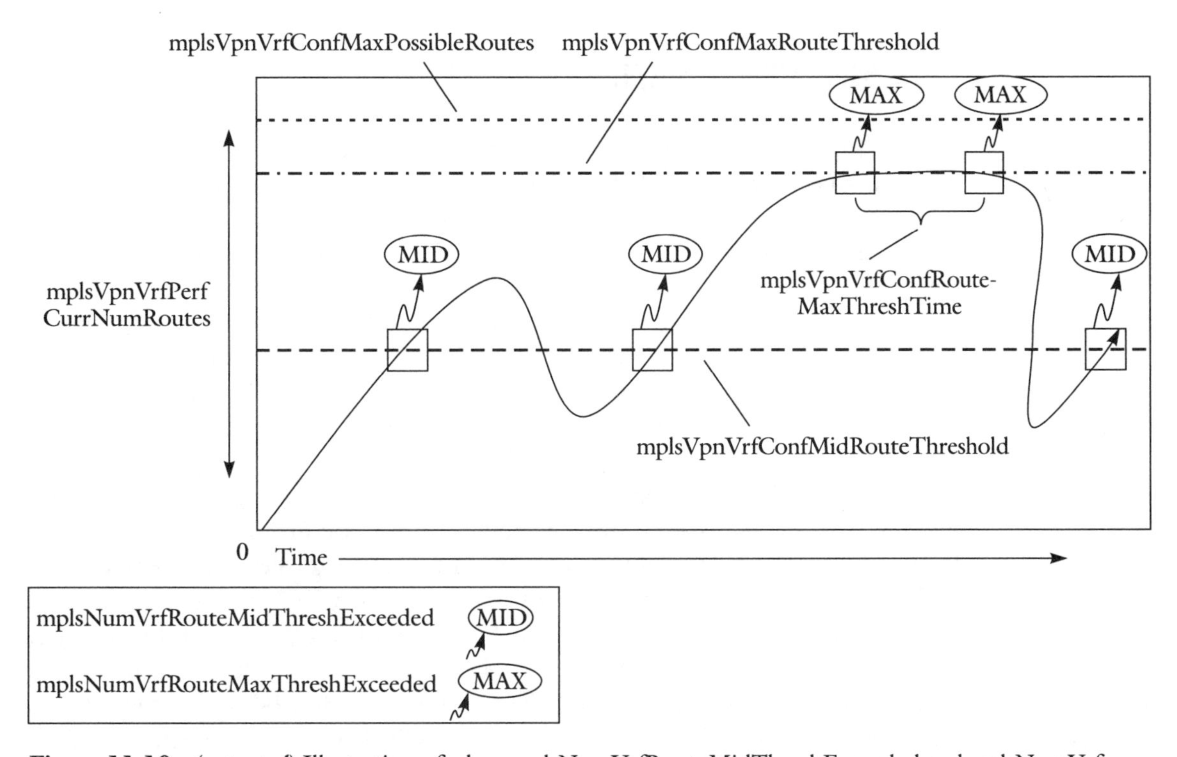

Figure 11.10 (*repeated*) Illustration of when mplsNumVrfRouteMidThreshExceeded and mplsNumVrf-RouteMidThreshExceeded notifications are issued, as well as how they relate to mplsVpnVrfConf-MaxPossibleRoutes, mplsVpnVrfConfMaxPossibleRoutes, and mplsVpnVrfConfMaxRouteThreshold.

has just happened. In the case of the latter, it is likely that someone is trying to spoof the labels in a CsC connection to illegally gain access to the CsC VPN connection.

This notification is emitted with the corresponding mplsVpnVrfName and mpls-VpnVrfSecIllegalLabelViolations objects. These objects are sufficient to allow an operator to take action. A typical question asked about this notification has been why the threshold is not included in the varbind of the notification. The reason this is not included is that the value of mplsVpnVrfSecIllegalLabelViolations should be one greater than the threshold at the time this notification is issued. If the threshold is not changed between the times when the notification is sent, then the operator receiving the notification can simply subtract one from the value in the notification to arrive at this value.

11.14 Enterprise VPN Example

The following example network demonstrates how the PPVPN-MPLS-VPN MIB can be used to manage the PEs associated with a particular enterprise VPN deployment. Figure 11.13 contains a simple service provider MPLS-enabled core network. The core network must have MPLS enabled in order to support MPLS VPN. This network contains four P MPLS LSRs as well as two PE LSRs configured to support two customer enterprise VPNs (VPN1 and VPN2). Each P or PE router is listed with its loopback interface address. Both PEs have routing distinguishers configured, which are shown below their loopback addresses. Both PEs have iBGP sessions established between each other. Three customer sites are installed for VPN1, while two are installed for VPN2. PE routers PE1 and PE2 are connected to three and two CE routers for each VPN, respectively. Each site is noted by a particular CE that is attached to a PE. The route distribution protocol running on the PE-CE links is noted next to each link. For instance, the top-left CEs are both connected to PE1 via OSPF, while the bottom-left CE for VPN2 is connected via RIP.

Each PE router is configured to run MPLS and MPLS VPN. Each PE router is configured with VRFs to attach each VPN site. In the case of PE1, it contains two VRFs named "VPN1" and "VPN2." The former VRF attaches to the two sites in VPN1 via interfaces Eth2 and Eth1, and has an RD of 100:1. PE1 attaches "VRF2"

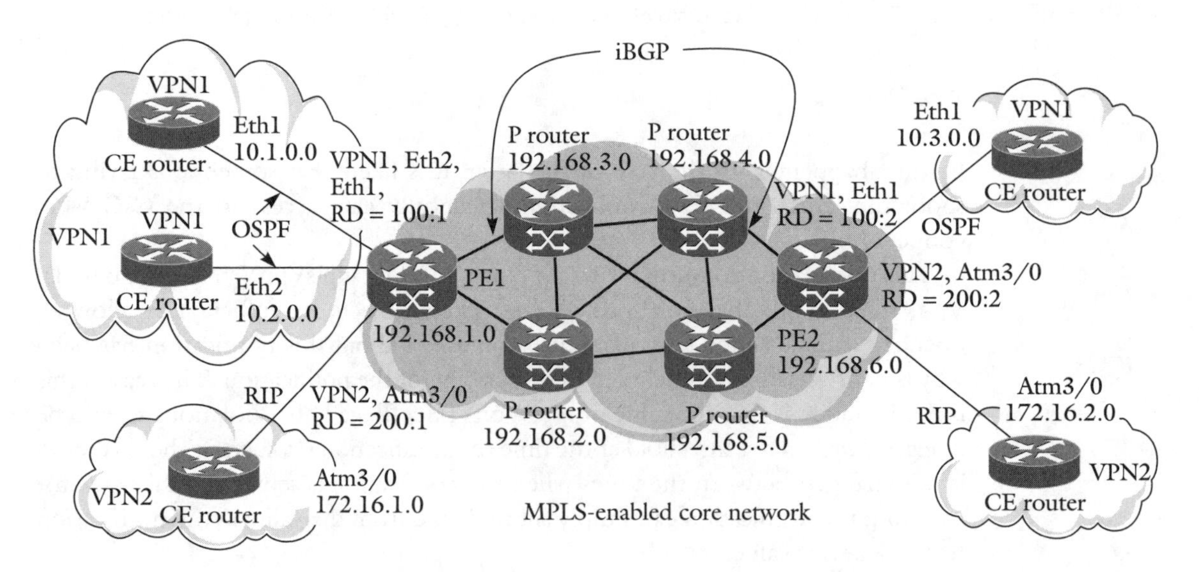

Figure 11.13 Sample enterprise MPLS VPN network configuration.

to the single site in VPN2 via interface Atm3/0. This VRF uses the RD of 200:1. Note that the interface names are equivalent for both sites of the PE-CE link only by coincidence; they can be different depending on the desired configuration. PE2 also is configured with two VRFs named "VPN1" and "VPN2," one for each VPN. VRF "VPN1" attaches to the CE at the VPN1 site via Eth1 and is configured with an RD of 100:2. The second VRF attaches to the CE at the customer site for VPN2 via interface Atm3/0 and uses an RD of 200:2. The routing protocols for these PE-CE connections are arranged in the same manner as those defined between PE1 and its CEs.

If we go through the example and examine the routes associated with VPN1, we will find that 10.1.0.0, 10.2.0.0, and 10.3.0.0 must be exchanged between PE1 and PE2 for VPN1 to function correctly. Similarly, we find that the routes associated with VPN2 are 172.16.1.0 and 172.16.2.0. The routes associated with the provider network are 192.168.1.0, 192.168.2.0, 192.168.3.0, 192.168.4.0, 192.168.5.0, 192.168.6.0 and must be taken into account in this example.

11.14.1 MplsVpnVrfTable for PE1 and PE2

Let us now examine the mplsVpnVrfRouteTargetTable on the PEs configured in the example. Due to the configuration, both PEs will contain two VRFs, one for each VPN. These entries will be reflected in the MPLS VPN VRF Table. In fact, the only differences between the entries found on PE1 and PE2 will be that VRF1 on PE1 will have two interfaces (Eth1 and Eth2) attached to it, while the same VRF on PE2 will only have a single interface attached. The route distinguishers for the VRFs will also be slightly different.

Let us first examine the configuration for PE1. PE1 is configured with two VRFs; thus the MPLS VPN VRF Table will contain two entries, one for mplsVpnVrfName = "VPN1" and one for "VPN2." Table 11.10 shows the columnar entries for index 5.56.80.78.49, which represents the octet string "VPN1," and index mplsVpnVrfName.5.56.80.78.50, which represents the octet string "VPN2."

Table 11.10 MPLS VPN VRF Table configuration for PE1.

Object name	mplsVpnVrfName . . 5.56.80.78.49.49	mplsVpnVrfName . . 5.56.80.78.49.50
mplsVpnVrfName	5.56.80.78.49.49 ("VPN1")	5.56.80.78.49.50 ("VPN2")

continued

Table 11.10 continued

Object name	mplsVpnVrfName . . 5.56.80.78.49.49	mplsVpnVrfName . . 5.56.80.78.49.50
mplsVpnVrfVpnId	5.56.80.78.49.49 ("VPN1")	5.56.80.78.49.50 ("VPN2")
mplsVpnVrfDescription	5.56.80.78.49.49 ("VPN1")	5.56.80.78.49.50 ("VPN2")
mplsVpnVrfRouteDistinguisher	64.00.00.00.00.00.00.01 (RD = 100:1)	0c.00.00.00.00.00.00.01 (RD = 200:1)
mplsVpnVrfCreationTime	0x0051	0x0071
mplsVpnVrfOperStatus	up (1)	up (1)
mplsVpnVrfActiveInterfaces	2	1
mplsVpnVrfAssociatedInterfaces	2	1
mplsVpnVrfConfMidRoute-Threshold	400	400
mplsVpnVrfConfHighRoute-Threshold	4000	4000
mplsVpnVrfConfMaxRoutes	8000	8000
mplsVpnVrfConfLastChanged	0x0001	0x0001
mplsVpnVrfConfRowStatus	active (1)	active (1)
mplsVpnVrfConfStorageType	nonvolatile (3)	nonvolatile (3)

Let us now examine the configuration for PE2. PE2 is configured with two VRFs; thus the MPLS VPN VRF Table will contain two entries, one for mplsVpn-VrfName = "VPN1" and one for "VPN2." VRF "VPN1" is configured with RD = 100:2, and VRF "VPN2" is configured with RD = 200:2. Table 11.11 shows the columnar entries for index 5.56.80.78.49, which represents the octet string "VPN1," and index mplsVpnVrfName.5.56.80.78.50, which represents the octet string "VPN2." Notice that the entry for "VPN1" also differs in the entry from PE1 in the number of associated and active interfaces.

Table 11.11 MPLS VPN VRF Table configuration for PE2.

Object name	mplsVpnVrfName . . 5.56.80.78.49.49	mplsVpnVrfName . . 5.56.80.78.49.50
mplsVpnVrfName	5.56.80.78.49.49 ("VPN1")	5.56.80.78.49.50 ("VPN2")
mplsVpnVrfVpnId	5.56.80.78.49.49 ("VPN1")	5.56.80.78.49.50 ("VPN2")
mplsVpnVrfDescription	5.56.80.78.49.49 ("VPN1")	5.56.80.78.49.50 ("VPN2")
mplsVpnVrfRouteDistinguisher	64.00.00.00.00.00.00.02 (RD = 100:2)	0c.00.00.00.00.00.00.02 (RD = 200:2)
mplsVpnVrfCreationTime	0x0020	0x0023
mplsVpnVrfOperStatus	up (1)	up (1)
mplsVpnVrfActiveInterfaces	1	1
mplsVpnVrfAssociatedInterfaces	1	1
mplsVpnVrfConfMidRoute-Threshold	400	400
mplsVpnVrfConfHighRoute-Threshold	4000	4000
mplsVpnVrfConfMaxRoutes	8000	8000
mplsVpnVrfConfLastChanged	0x0001	0x0001
mplsVpnVrfConfRowStatus	active (1)	active (1)
mplsVpnVrfConfStorageType	nonvolatile (3)	nonvolatile (3)

11.14.2 mplsVpnVrfIfConfTable for PE1 and PE2

Now that we have established the VRF configuration, let us now examine how the interface configuration table for each VRF would be configured on *both* PE1 and PE2 given the topology in the example from Figure 11.13. Let us start with PE2. The conceptual configuration is depicted in Figure 11.14. Notice that multiple interface entries can be associated with more than one VRF due to the multiple indexes in the mplsVpnIfConfTable. Also note that the interfaces associated with the VRFs are also present in the IF-MIB's ifTable. If the interfaces were not found in the ifTable, this relationship would not function correctly.

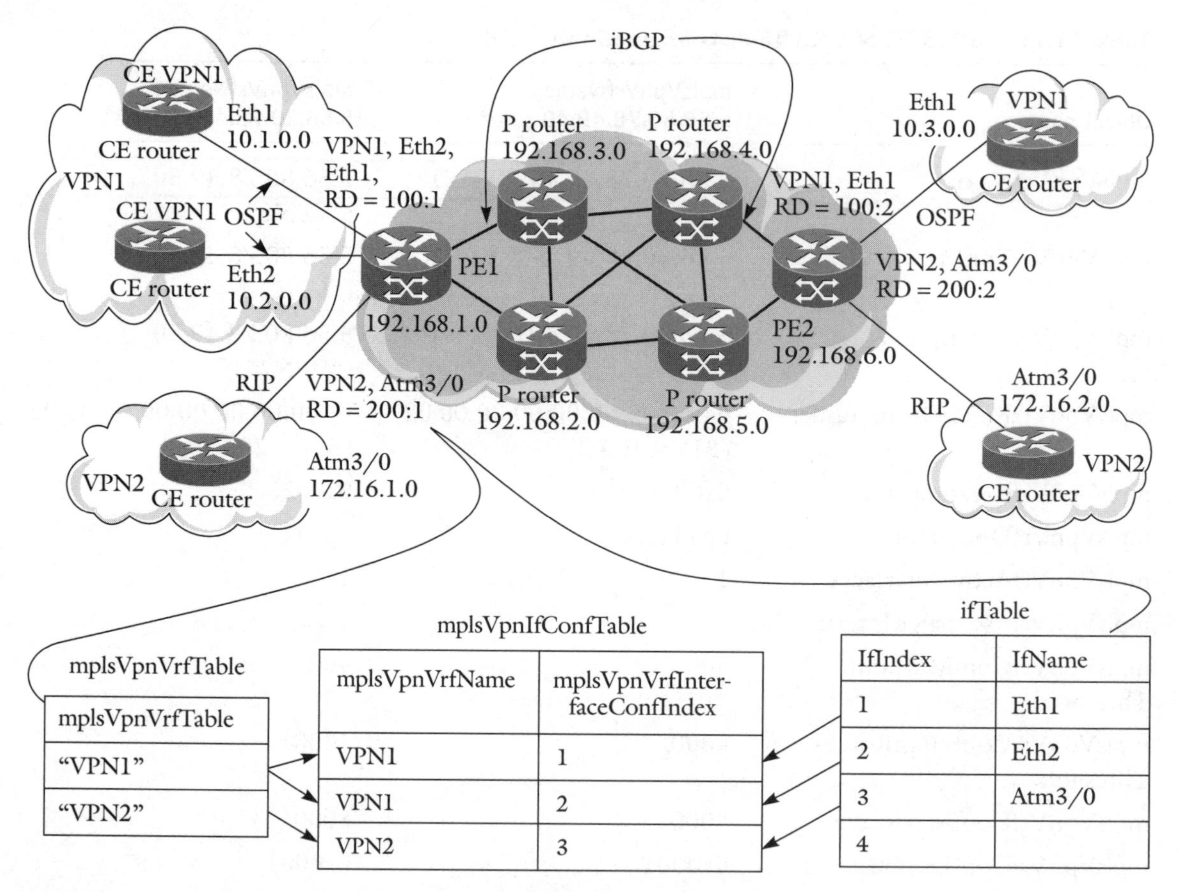

Figure 11.14 Conceptual MPLS VPN VRF Configuration Table for PE1.

We will not list the table for PE2 explicitly simply because it only differs from the one shown for PE1 in that a single interface is associated with each VRF. Specifically, VRF1 is associated with Eth1 (ifIndex = 1), and VRF2 is associated with Atm3/0 (ifIndex = 3). Note that the clean ifIndex assignment in this example is purely coincidental and is only given this way for the purposes of this example. Other combinations can and normally do exist. The MIB tables will work for any valid ifIndex assignment.

Table 11.12 shows the complete mplsVpnVrfConfTable for PE1 given the conceptual picture shown above.

Table 11.12 The mplsVpnVrfIfConfTable for PE1.

Object name	mplsVpnVrfName. 5.56.80.78.49 ("VPN1"), mplsVpnVrfIfConfIndex = 1
mplsVpnIfConfIndex	1
mplsVpnIfLabelEdgeType	customerEdge (2)
mplsVpnIfVpnClassification	enterprise (2)
mplsVpnIfVpnRouteDistProtocol	ospf (2)
mplsVpnIfConfStorageType	nonvolatile (3)
mplsVpnIfConfRowStatus	active (1)
Object name	**mplsVpnVrfName. 5.56.80.78.49 ("VPN1"), mplsVpnVrfIfConfIndex = 2**
mplsVpnIfConfIndex	2
mplsVpnIfLabelEdgeType	customerEdge (2)
mplsVpnIfVpnClassification	enterprise (2)
mplsVpnIfVpnRouteDistProtocol	ospf (2)
mplsVpnIfConfStorageType	nonvolatile (3)
mplsVpnIfConfRowStatus	active (1)
Object name	**mplsVpnVrfName. 5.56.80.78.50 ("VPN2"), mplsVpnVrfIfConfIndex = 3**
mplsVpnIfConfIndex	3
mplsVpnIfLabelEdgeType	customerEdge (2)
mplsVpnIfVpnClassification	enterprise (2)
mplsVpnIfVpnRouteDistProtocol	rip (3)
mplsVpnIfConfStorageType	nonvolatile (3)
mplsVpnIfConfRowStatus	active (1)

11.14.3 MPLSVpnVrfRouteTargetTable for PE1 and PE2

Given the example network shown in Figure 11.13, the route target table for PE2 would conceptually appear as shown in Figure 11.15. Notice how each route target from both PE2 and PE1 must be explicitly imported and exported for both

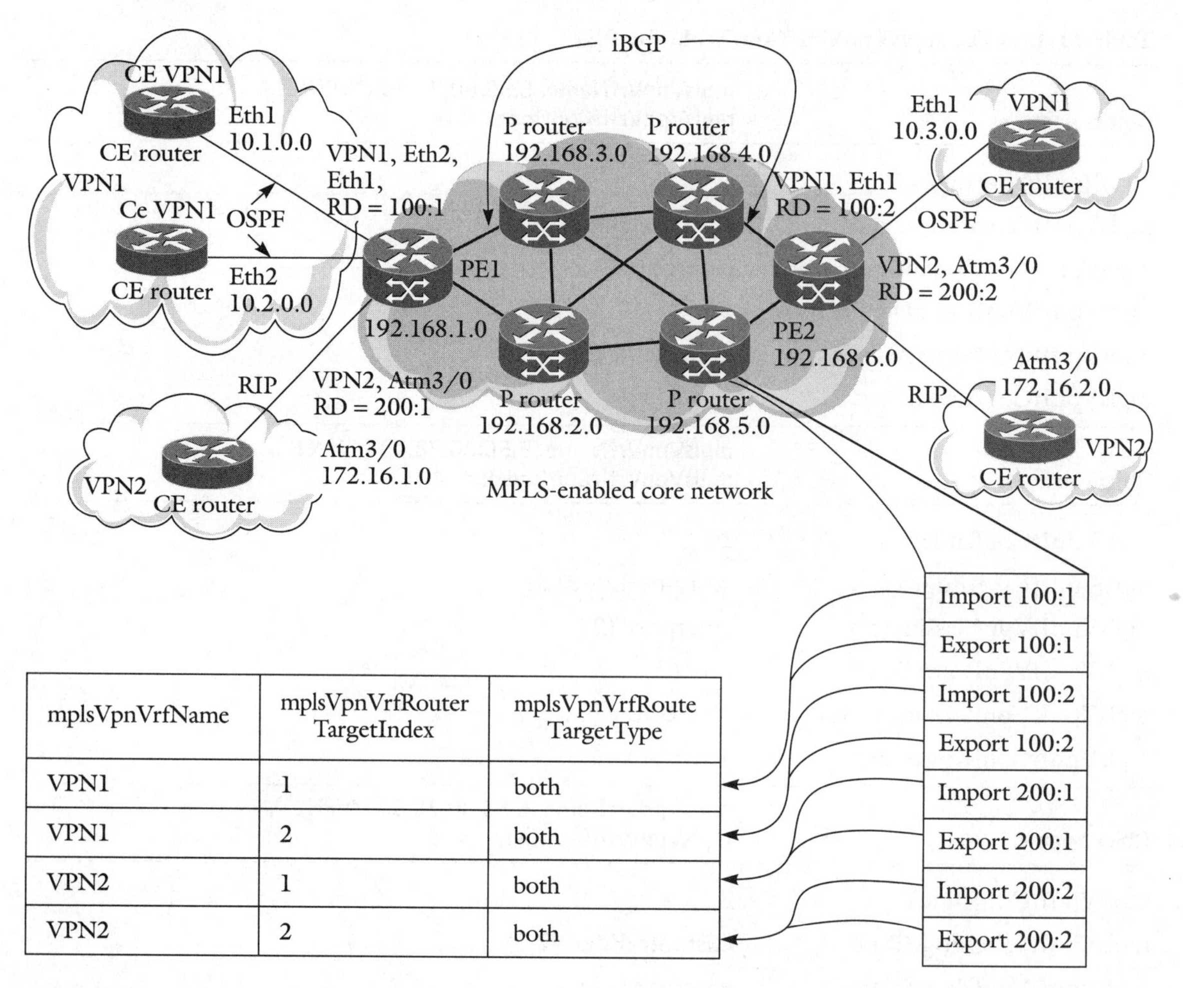

mplsVpnVrfName	mplsVpnVrfRouter TargetIndex	mplsVpnVrfRoute TargetType
VPN1	1	both
VPN1	2	both
VPN2	1	both
VPN2	2	both

Figure 11.15 Conceptual Route Target Table for PE2.

VPN1 and VPN2 to function correctly. Also notice that as an optimization, the mplsVpnVrfRouteTarget table allows for a shorthand route target entry type notation of both to be given to like pairs of the same import and export target. This abbreviated notation reduces redundant entries from appearing in the table.

Let us now examine how the actual mplsVpnVrfRouteTarget table would be configured given this conceptual figure. Each route target entry would be configured on PE1 and PE2 as is shown given the example network from Figure 11.13. Both tables would be identical on PE1 and PE2, so we will only show one copy of the table (Table 11.13).

Table 11.13 MPLS VPN VRF Route Target Table configuration for PE1 and PE2.

Object name	mplsVpnVrfName. 5.56.80.78.49 ("VPN1"), mplsVpnVrfRouteTargetIndex.1, mplsVpnVrfRouteTargetType.3(both)
mplsVpnVrfName	5.56.80.78.49 ("VPN1")
mplsVpnVrfRouteTarget	0x0200000000001001
mplsVpnVrfRouteTargetDescr	"100:1"
mplsVpnVrfRouteTargetRow-Status	active (1)

Object name	mplsVpnVrfName. 5.56.80.78.49("VPN1"), mplsVpnVrfRouteTargetIndex.1, mplsVpnVrfRouteTargetType.3(both)
mplsVpnVrfName	5.56.80.78.49 ("VPN1")
mplsVpnVrfRouteTarget	0x0200000000001002
mplsVpnVrfRouteTargetDescr	"100:2"
mplsVpnVrfRouteTargetRow-Status	active (1)

Object name	mplsVpnVrfName. 5.56.80.78.50 ("VPN2"), mplsVpnVrfRouteTargetIndex.1, mplsVpnVrfRouteTargetType.3(both)
mplsVpnVrfName	5.56.80.78.50 ("VPN2")
mplsVpnVrfRouteTarget	0x0200000000002001
mplsVpnVrfRouteTargetDescr	"200:1"
mplsVpnVrfRouteTargetRow-Status	active (1)

Object name	mplsVpnVrfName. 5.56.80.78.50 ("VPN2"), mplsVpnVrfRouteTargetIndex.2, mplsVpnVrfRouteTargetType.3(both)
mplsVpnVrfName	5.56.80.78.50 ("VPN2")
mplsVpnVrfRouteTarget	0x0200000000002002
mplsVpnVrfRouteTargetDescr	"200:2"
mplsVpnVrfRouteTargetRow-Status	active (1)

11.14.4 mplsVpnVrfBgpNbrAddrTable for PE1

Both PE1 and PE2 will have similar BGP Neighbor Tables configured, as shown in Table 11.14, given the network we have given in Figure 11.13. In particular, PE1 will list PE2 as an iBGP neighbor, and PE2 will list PE1 as its neighbor. Given a more complicated example, all that would differ would be the number of entries in the table; thus we have avoided additional BGP neighbors for the purposes of this example. The single entry is simple, as it includes only a single entry for PE2. The mplsVpnVrfBgpNbrRole indicates that PE2 is another peer PE device. PE2 is indicated by the IPv4 address type and address specified as an octet string.

Table 11.14 The MplsVpnVrfBgpNbrAddrTable for PE1.

Object name	mplsVpnVrfName. 5.56.80.78.49 ("VPN1"), mplsVpnVrfIfConfIndex = 1, mplsVpnVrfBgpNbrIndex
mplsVpnVrfBgpNbrIndex	1
mplsVpnVrfBgpNbrRole	pe (2)
mplsVpnVrfBgpNbrType	ipv4 (1)
mplsVpnVrfBgpNbrAddr	05.C0.A8.06.00 (IPv4 addr of PE2)
mplsVpnVrfBgpNbrRowStatus	nonvolatile (3)
mplsVpnVrfBgpNbrStorageType	active (1)

11.14.5 MplsVpnVrfRouteTable for PE1 and PE2

The example in Figure 11.13 contains a configuration of two VRFs on both PE1 and PE2. Each VRF on the PE requires routes to be imported and exported to and from the VRF to other PEs. This has been shown earlier in the route target configuration. Now we need to focus on exactly which routes will be programmed into the VRF and global routing tables given the sample network. These routes will ultimately be exchanged between PE1 and PE2 using BGP. The routing tables on both PE1 and PE2 are first shown conceptually in Figure 11.16. When the network stabilizes and all routes have been distributed, both PE1's and PE2's tables will be identical, at least conceptually. The primary difference between PE1's and PE2's routes will lie in the next-hop addresses.

One additional configuration option that we must mention at this time is the IP address configuration for the interfaces on PE1. These interface addresses must

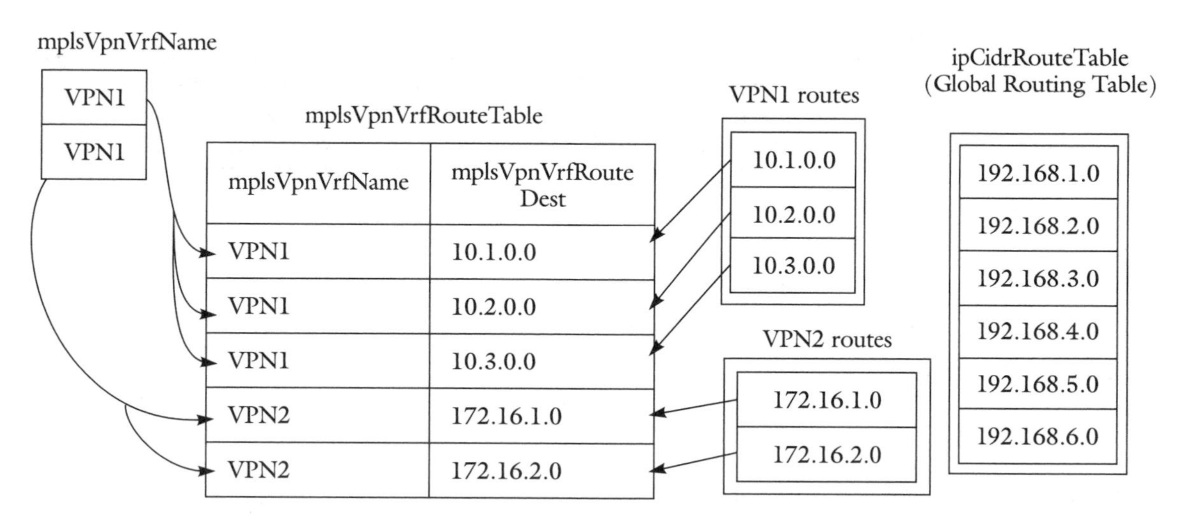

Figure 11.16 Conceptual VRF Route Table for PE1 and PE2 from Figure 11.13.

be used when the next-hop address is a local route (versus remote). For the purposes of this example, assume that Eth1 is assigned IPv4 address 10.1.0.1; Eth2, 10.2.0.1; and Atm3/0, 172.16.1.1. These addresses will appear in the next-hop field for the corresponding routes that must be delivered locally over those links.

Now let us examine how the actual table will appear on PE1 assuming that the network has stabilized and all routes have been distributed as shown in Figure 11.15. Due to the large size of the tables, we have broken the tables into two pieces: one for the routes on PE1 for VRF1 and one for the routes for VRF2. We will begin with the routes for VRF1 in Table 11.15.

Table 11.15 The MplsVpnVrfRouteTable for PE1.

Object name	mplsVpnVrfName. 5.56.80.78.49 ("VPN1"), mplsVpnVrfRoute-Dest. 5.10.1.0.0, mplsVpnVrfRouteMask.5.255.255.0.0, mplsVpnVrfRouteTos.0, mplsVpnVrfRouteNextHop. 5.10.1.0.1
mplsVpnVrfRouteDest	5.10.1.0.0
mplsVpnVrfRouteDestAddr-Type	ipv4 (1)
mplsVpnVrfRouteMask	5.255.255.00.00(255.255.0.0)

continued

Table 11.15 continued

Object name	mplsVpnVrfName. 5.56.80.78.49 ("VPN1"), mplsVpnVrfRoute-Dest. 5.10.1.0.0, mplsVpnVrfRouteMask.5.255.255.0.0, mplsVpnVrfRouteTos.0, mplsVpnVrfRouteNextHop. 5.10.1.0.1
mplsVpnVrfRouteMaskAddr-Type	ipv4 (1)
mplsVpnVrfRouteTos	0
mplsVpnVrfRouteNextHop	5.10.1.0.1 (IP address of Eth1)
mplsVpnVrfRouteNextHop-AddrType	ipv4 (1)
mplsVpnVrfRouteIfIndex	1 (Eth1)
mplsVpnVrfRouteType	local (3)
mplsVpnVrfRouteProto	ospf (13)
mplsVpnVrfRouteAge	43
mplsVpnVrfRouteInfo	0.0
mplsVpnVrfRouteNextHop-AS	0
mplsVpnVrfRouteMetric1	0
mplsVpnVrfRouteMetric2	0
mplsVpnVrfRouteMetric3	0
mplsVpnVrfRouteMetric4	0
mplsVpnVrfRouteMetric5	0
mplsVpnVrfBgpNbrRow-Status	nonvolatile (3)
mplsVpnVrfBgpNbrStorage-Type	active (1)

Object name	mplsVpnVrfName. 5.56.80.78.49 ("VPN1"), mplsVpnVrfRouteDest. 5.10.02.0.0, mplsVpnVrfRouteMask.5.255.255.0.0, mplsVpnVrfRouteTos.0, mplsVpnVrfRouteNextHop. 5.10.02.0.1
mplsVpnVrfRouteDest	5.10.2.0.0
mplsVpnVrfRouteDestAddr-Type	ipv4 (1)

Table 11.15 continued

Object name	mplsVpnVrfName. 5.56.80.78.49 ("VPN1"), mplsVpnVrfRoute-Dest. 5.10.1.0.0, mplsVpnVrfRouteMask.5.255.255.0.0, mplsVpnVrfRouteTos.0, mplsVpnVrfRouteNextHop. 5.10.1.0.1
mplsVpnVrfRouteMask	5.255.255.0.0
mplsVpnVrfRouteMaskAddr-Type	ipv4 (1)
mplsVpnVrfRouteTos	0
mplsVpnVrfRouteNextHop	5.10.2.0.1 (IP address of Eth2)
mplsVpnVrfRouteNextHop-AddrType	ipv4 (1)
mplsVpnVrfRouteIfIndex	2 (Eth2)
mplsVpnVrfRouteType	local (3)
mplsVpnVrfRouteProto	ospf (13)
mplsVpnVrfRouteAge	43
mplsVpnVrfRouteInfo	0.0
mplsVpnVrfRouteNextHop-AS	0
mplsVpnVrfRouteMetric1	0
mplsVpnVrfRouteMetric2	0
mplsVpnVrfRouteMetric3	0
mplsVpnVrfRouteMetric4	0
mplsVpnVrfRouteMetric5	0
mplsVpnVrfBgpNbrRow-Status	nonvolatile (3)
mplsVpnVrfBgpNbrStorage-Type	active (1)

Object name	mplsVpnVrfName. 5.56.80.78.49 ("VPN1"), mplsVpnVrfRouteDest. 5.10.3.0.0, mplsVpnVrfRoute-Mask. 5.255.255.0.0, mplsVpnVrfRouteTos.0, mplsVpnVrfRouteNextHop. 5.192.168.3.0
mplsVpnVrfRouteDest	5.10.3.0.0

continued

Table 11.15 continued

Object name	mplsVpnVrfName. 5.56.80.78.49 ("VPN1"), mplsVpnVrfRoute-Dest. 5.10.1.0.0, mplsVpnVrfRouteMask.5.255.255.0.0, mplsVpnVrfRouteTos.0, mplsVpnVrfRouteNextHop. 5.10.1.0.1
mplsVpnVrfRouteDestAddr-Type	ipv4 (1)
mplsVpnVrfRouteMask	05.255.255.00.00
mplsVpnVrfRouteMaskAddr-Type	ipv4 (1)
mplsVpnVrfRouteTos	0
mplsVpnVrfRouteNextHop	5.192.168.3.1 (IP address of link to 192.168.3.0)
mplsVpnVrfRouteNextHop-AddrType	ipv4 (1)
mplsVpnVrfRouteIfIndex	4 (link between PE1 and 192.168.3.0)
mplsVpnVrfRouteType	remote (4)
mplsVpnVrfRouteProto	bgp (14)
mplsVpnVrfRouteAge	43
mplsVpnVrfRouteInfo	0.0
mplsVpnVrfRouteNextHop-AS	0
mplsVpnVrfRouteMetric1	0
mplsVpnVrfRouteMetric2	0
mplsVpnVrfRouteMetric3	0
mplsVpnVrfRouteMetric4	0
mplsVpnVrfRouteMetric5	0
mplsVpnVrfBgpNbrRow-Status	nonvolatile (3)
mplsVpnVrfBgpNbrStorage-Type	active (1)

Next, let's examine the VRF route table entries related to VRF2 on PE1 in Table 11.16.

Table 11.16 The MplsVpnVrfRouteTable for PE1.

Object name	mplsVpnVrfName. 5.56.80.78.49 ("VPN1"), mplsVpnVrfRoute-Dest. 05.172.10.01.00, mplsVpnVrfRouteMask. 5.255.255.0.0, mplsVpnVrfRouteTos.0, mplsVpnVrfRouteNextHop.05.172.10.01.01
mplsVpnVrfRouteDest	05.172.10.01.00
mplsVpnVrfRouteDestAddr-Type	ipv4 (1)
mplsVpnVrfRouteMask	5.255.255.0.0
mplsVpnVrfRouteMaskAddr-Type	ipv4 (1)
mplsVpnVrfRouteTos	0
mplsVpnVrfRouteNextHop	05.172.10.01.01
mplsVpnVrfRouteNextHop-AddrType	ipv4 (1)
mplsVpnVrfRouteIfIndex	3 (Atm3/0)
mplsVpnVrfRouteType	local (3)
mplsVpnVrfRouteProto	rip (8)
mplsVpnVrfRouteAge	43
mplsVpnVrfRouteInfo	0.0
mplsVpnVrfRouteNextHop-AS	0
mplsVpnVrfRouteMetric1	0
mplsVpnVrfRouteMetric2	0
mplsVpnVrfRouteMetric3	0
mplsVpnVrfRouteMetric4	0
mplsVpnVrfRouteMetric5	0
mplsVpnVrfBgpNbrRow-Status	nonvolatile (3)
mplsVpnVrfBgpNbrStorage-Type	active (1)

continued

Table 11.16 continued

Object name	mplsVpnVrfName. 05.ac.10.02.00 ("VPN1"), mplsVpnVrfRouteDest.04.172.10.2.0, mplsVpnVrfRouteMask. 05.255.255.0.0, mplsVpnVrfRouteTos.0, mplsVpnVrfRouteNextHop. 05.192.168.2.0
mplsVpnVrfRouteDest	04.172.10.02.00
mplsVpnVrfRouteDestAddr-Type	ipv4 (1)
mplsVpnVrfRouteMask	05.255.255.0.0
mplsVpnVrfRouteMaskAddr-Type	ipv4 (1)
mplsVpnVrfRouteTos	0
mplsVpnVrfRouteNextHop	192.168.2.1 (IP address of link to 192.168.2.0)
mplsVpnVrfRouteNextHop-AddrType	ipv4 (1)
mplsVpnVrfRouteIfIndex	5 (link between PE1 and 192.168.2.0)
mplsVpnVrfRouteType	remote (4)
mplsVpnVrfRouteProto	bgp (14)
mplsVpnVrfRouteAge	43
mplsVpnVrfRouteInfo	0.0
mplsVpnVrfRouteNextHop-AS	0
mplsVpnVrfRouteMetric1	0
mplsVpnVrfRouteMetric2	0
mplsVpnVrfRouteMetric3	0
mplsVpnVrfRouteMetric4	0
mplsVpnVrfRouteMetric5	0
mplsVpnVrfBgpNbrRow-Status	nonvolatile (3)
mplsVpnVrfBgpNbrStorage-Type	active (1)

Given the simplicity of the example network, the number of routes contained in any of the VRFs is limited. However, this does not detract from the effectiveness of this example, as an example with a larger number of routes per VRF would only differ in the number of entries in the MplsVpnVrfRouteTable. The manner in which entries are added and managed in this table remains.

11.15 Summary

This chapter introduced you to the PPVPN-MPLS-VPN MIB. The chapter began with an introduction to MPLS layer-3 VPNs. This introduction included an explanation of the basic premises behind MPLS L3 VPN, as well as the basic mechanisms used to deploy the technology. In particular, the specifics of route targets, route distinguishers, and virtual private routing and forwarding tables were discussed. Next, the chapter went on to introduce how this service can be managed using the PPVPN-MPLS-VPN MIB. A brief introduction to the structure of the MIB, its tables, and scalars was given. Next, we covered each of the MIB's tables in detail by going over each table's purpose, its indexing, and the definitions of each of the objects within each table. Along the way, tips and pitfalls for implementations and managers were given.

Finally, we introduced an example of an enterprise VPN deployment and gave the details of the topology and other configuration parameters. We then went back through the MIB's tables and demonstrated how each table would appear if queried on either of the PEs defined in the example.

Further Reading

Rosen, E., et al. "BGP/MPLS VPNs." IETF Internet Draft. January 2002. *www.ietf.org/internet-drafts/draft-ietf-ppvpn-rfc2547bis-01.txt.*

Sangli, E., et al. "BGP Extended Communities Attribute." IETF Internet Draft. March 2002. *www.ietf.org/internet-drafts/draft-ietf-idr-bgp-ext-communities-03.txt.*

RFC 2858. Bates, T., Y. Rekhter, R. Chandra, and D. Katz. "Multiprotocol Extensions for BGP-4." June 2000. *www.ietf.org/rfc/rfc2858.txt.*

Davie, B. S., and Y. Rekhter. *MPLS: Technology and Applications.* First edition. San Francisco: Morgan Kaufmann Publishers. 2000.

Gray, E. W. *MPLS: Implementing the Technology.* Reading, Mass.: Addison-Wesley Professional. 2001.

To find out more about the IETF, visit their Web page at *www.ietf.org/*.
For more information about IANA, check out their Web site at *www.iana.org/*.
To locate information about the ITU: *www.itu.int/*.
To locate information about the MPLS Forum: *www.mplsforum.org/*.
To locate additional information about the OIF: *www.oiforum.com/*.
Cisco's Web site also provides a great deal of information regarding MPLS: *www.cisco.com*.

Cheenu Srinivasan

has over a decade's experience in networking and telecommunications. He received his B.Tech. degree in electrical engineering from the Indian Institute of Technology, Madras, in 1991, and an M.A. in 1993 and a Ph.D. in 1995, both in electrical engineering from Princeton University, New Jersey. He has since worked in various lead architecture and development roles in building ATM, IP, MPLS/GMPLS, and SS7 based products at Lucent Technologies (Bell Labs and Network Systems), Tachion Networks, and Alphion Corp. He has been an active participant and one of the primary contributors to MPLS and GMPLS network management standards at the IETF. He is currently a senior member of the technical staff in the systems and software group at Parama Networks in New Jersey.

Cheenu, as one of the original co-authors of the base MPLS MIBs, you have a keen insight into how these MIBs were formed and have progressed to the state that they are today. What have been some of the challenges that the MIBs have undergone, and do you think that they are not up to the task of being used in real production networks?

We started our efforts to develop network management standards for MPLS over a year after the standardization of MPLS technology itself had been initiated. It is always a little more challenging to define a management layer atop a predefined technology rather than have the two defined hand in hand. However, we have come a fair distance since that time in terms of completeness and maturity. During this first phase of development, the feedback has been mainly from implementers of the technology. We still await widespread deployment of this work and more detailed feedback from service providers who will actually have to build and run MPLS networks and ensure that there is adequate network management support to ensure smooth operation.

Has the IETF been a good place to standardize these documents?

The IETF philosophy is "rough consensus and working code." This implies that the standardization process is iterative rather than one shot, and convergence happens through implementation experience. The emphasis on implementation meant that by the time the drafts went through a few revisions there were several implementations to test the work. The feedback derived from implementers is invaluable in ensuring a useful and robust standard. Of course, as I mentioned already, we still await widespread deployment, and I expect the work to go through some more revisions as it works its way through the

IETF standards process (RFC 2026). Overall, yes, the IETF has been a very good place to standardize our work.

What areas of MPLS do you see as being deficient in terms of manageability? Generally speaking, what challenges do you see today to the complete management of MPLS-enabled networks, and in the future of GMPLS-enabled networks? What solutions do you envision for these problems?

There are five broad areas of management—fault, configuration, accounting, performance, and security—together referred to as "FCAPS." The existing MPLS management standards quite adequately address the fault, configuration, and performance aspects of MPLS. I don't believe there is any new security concern introduced by MPLS per se that needs to be addressed explicitly by the management standards documents. However, based on service provider requirements and feedback, we may need to add better accounting support.

Policy-based management of MPLS networks is an important area that is yet unaddressed. Work is needed to define and integrate this with the rest of the MPLS management standards.

Efforts to extend the MPLS management standards to address GMPLS networks have just begun. The main challenge will be to ensure that this activity does not result in two separate standards but instead to define one coherent set of documents to address both MPLS and GMPLS network management.

12

Future Directions for MPLS Network Management

"If knowledge can create problems, it is not through ignorance that we can solve them."

—Isaac Asimov

Introduction

This chapter discusses future directions of MPLS management. This includes tools, techniques, and technologies that might not have been mature enough to cover in the current text due to the fact that they would invariably be quite different by the time this book was printed. These items may be included in updated revisions of this book, or may be found in other volumes. It is, however, worth mentioning them now so that you might get a head start on these topics with your favorite vendor(s) or perhaps in their evolution within the standards community.

This chapter provides a final wrap-up of the chapters presented in this book. We include a discussion and summation of the interviews at the end of each chapter of this book. Finally, at the end of this chapter, we have included an extensive list of resources for the new and upcoming topics introduced and discussed in this chapter. We hope that these references allow you to investigate the topics introduced in much more detail than we were able to cover them here. These resources may also allow you to follow the standards related to the topics presented in this chapter in their most up-to-date form.

12.1 Generalized MPLS (GMPLS)

Not long ago many people and companies began to work on extending MPLS in such a way that it could be used to control future data and transmission networks (i.e., optical) as well as existing ones. This work first began to be standardized within the IETF MPLS Working Group, but was later moved to a new Working Group called the Common Control and Measurement (CCAMP) Working Group. Work on GMPLS continues today within CCAMP.

The benefit of GMPLS is that a common control plane can be realized that can control many different types of transport networks. The advantage to this approach is that much of the existing code bases currently deployed by vendors to support packet- or cell-based networks with MPLS as the control plane can be easily extended to support TDM and physical optical or other future (gravity waves?) networks as well. This provided an opportunity for vendors to save money in development costs, as well as to reduce the time to market of this solution. Service providers deploying this equipment benefited from this approach: they could leverage much of their existing integrated management infrastructure because generalized MPLS is designed with MPLS as its base; therefore, it can be managed much as "classic" MPLS networks are.

To this end, the existing MPLS MIBs have been extended or *generalized* to support GMPLS as well as MPLS. In an effort to preserve existing MIB implementations, the GMPLS flavors of the MPLS MIBs do not obviate the "classic" MPLS MIBs, but instead just extend them in ways required to support GMPLS. In essence, the existing MIBs remain as they are so that they may function as they exist in current implementations. However, for implementations that support both MPLS and GMPLS, some of the tables in the MIBs are extended or augmented to provide the additional functionality required for GMPLS.

Several drafts were proposed that extend the existing MPLS MIBs as just described. These should be accepted as CCAMP Working Group documents by the time this book is published, so consult the CCAMP Working Group's Web site at *www.ietf.org/html.charters/ccamp-charter.html*. The drafts are currently listed in the Internet Draft repository as shown in Table 12.1.

Table 12.1 Proposed GMPLS MIBs.

Draft name	Description
draft-nadeau-ccamp-gmpls-tc-mib-01.txt	Common textual conventions and object identities for GMPLS MIBs
draft-nadeau-ccamp-gmpls-lsr-mib-01.txt	Extensions to the MPLS-LSR MIB for GMPLS
draft-nadeau-ccamp-gmpls-te-mib-01.txt	Extensions to the MPLS-TE MIB for GMPLS

Another draft appears in the draft repository called draft-nadeau-ccamp-gmpls-label-mib-01.txt. This document was originally proposed to replace the labels used by the MPLS-LDP, MPLS-TE, and MPLS-LSR MIBs by having them point at the new table found in this document. Fortunately, the authors of this document were able to have this new table integrated into the existing MPLS-LSR MIB before its publication as an RFC, resulting in a much cleaner approach. The GMPLS version of this new table allows labels to be represented as octet strings, allowing them to be very long and of private or unknown format. The new table also supports arbitrary concatenation of labels, a feature that is very useful in SONET and SDH networks. The existing MPLS-LSR MIB, MPLS-LDP MIB, and a new GMPLS-TE MIB will point at labels found in this document if the labels represent GMPLS labels. However, since the index into this table is an Unsigned32, it can also be used to continue to represent the MPLS type label in implementations that do not support GMPLS.

12.2 Pseudo-Wire Edge-to-Edge Emulation

The Pseudo-Wire Edge-to-Edge (PWE3) IETF Working Group was started to standardize the emulation of circuits (such as T1, E1, T3, E3, and SONET/SDH) and services (such as ATM and Frame Relay) over packet switched networks (PSNs) using IP, L2TP, or MPLS. Current implementations from various vendors allow the emulation of various networking types over MPLS and L2TP, with some promising the use of pure IP as a transport in the future. The architecture approved by the PWE3 Working Group describes a layered model where a service is emulated by attaching that service to the constructs available in the various PSN transport technologies. For example, a point-to-point Frame Relay virtual circuit can be

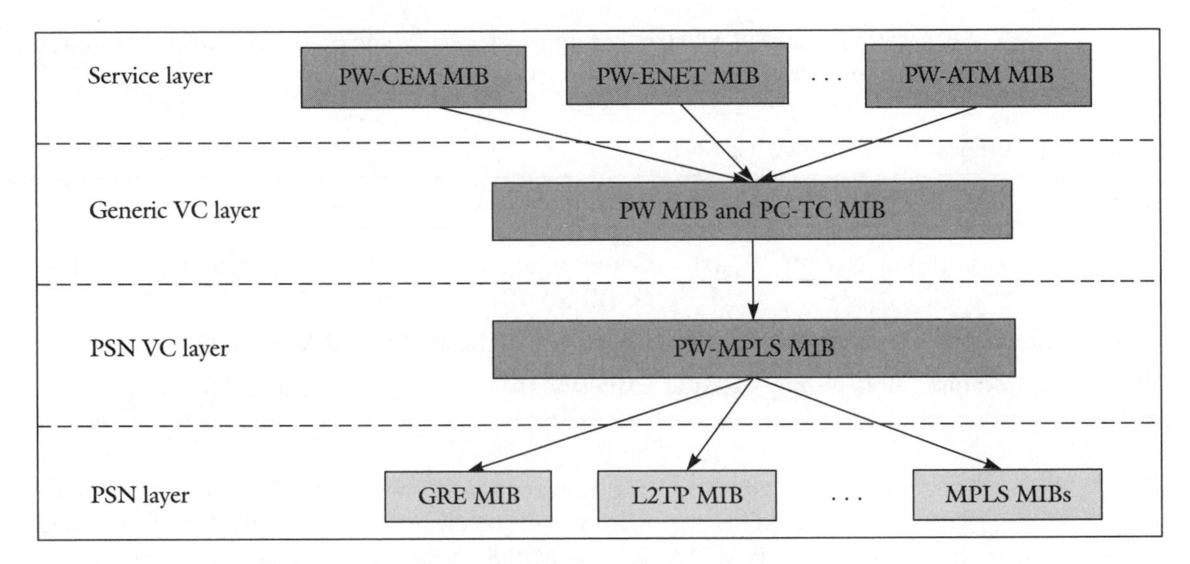

Figure 12.1 IETF PWE3 management model.

established across an MPLS network by carrying the Frame Relay circuit inside of an MPLS LSP that is established using LDP or RSVP-TE. The advantages to this approach are that service providers can now reuse existing L2TP or MPLS-enabled networks to carry a variety of traffic that in the past required separate parallel networks to be deployed.

In accordance with this framework, the PWE3 Working Group has also specified a framework for managing PWE3 services, illustrated in Figure 12.1. This framework defines a four-layered approach whose goal is to "glue" together existing service layer and PSN layer MIBs into a coherent model of management. This allows management stations that currently understand how to manage service layers such as Ethernet or Frame Relay to continue managing these services as they did in the past. This is achieved by having them continue to manage these services through their existing standard MIBs. The same is true for the accepted transports MPLS, IP, and L2TP. These transports can be managed as they are without any PWE3 services running. The PWE3 management framework then associates these two layers together using a general virtual circuit (VC) abstraction. This is in actuality realized by "gluing" the PW VCs to their corresponding transport layer entity via a PSN VC layer. This layer provides the actual mapping between the general VC and the specific transport entity (i.e., LDP-signaled LSP).

The drafts currently available are listed in Table 12.2. At the time of the printing of this book, the drafts listed in Table 12.2 were still individual contributions to the IETF, and thus did not appear as PWE3 Working Group documents. In the future, please look for updated drafts on the PWE3 Working Group Web page at *www.ietf.org/html.charters/pwe3-charter.html.*

Table 12.2 Proposed PWE3 Management Information Bases.

Draft name	Description
draft-zelig-pw-enet-mib-00.txt	Describes managed objects for modeling of Ethernet pseudo-wire services
draft-zelig-pw-mib-02.txt	Describes managed objects for modeling of pseudo-wire services on a general PSN
draft-zelig-pw-mpls-mib-01.txt	Describes MIB module for PW operation over MPLS LSR
draft-nadeau-pw-tc-mib-02.txt	Describes textual conventions and object identities used for managing pseudo-wire services
draft-sathappan-pw-atm-mib-00.txt	Describes managed objects for modeling an adaptation of ATM VCs over a PSN
draft-danenberg-pw-cem-mib-02.txt	Describes managed objects for modeling an adaptation of SONET/SDH circuits over a PSN

12.3 New Developments in MPLS

Since the writing of this book began, several important documents pertaining to MPLS network management have been proposed in the IETF MPLS Working Group. The first document of interest is an overview of the Management Information Bases available for MPLS management. The goal of the document was to provide the reader with an overview or road map of all MPLS-related MIBs and to act as a high-level review. The reason such a document was published was that, simply put, there are many MIBs available to manage MPLS. For the person just starting to try to get their hands around the bigger picture of MPLS management (or for the experienced professionals sometimes), it is quite a task indeed to collect all of these MIBs together. This document can be found on the MPLS WG Web page as draft-ietf-mpls-mgmt-overview-01.txt, but bear in mind that the version number of the document may be updated in the future.

The next document of interest is entitled "Detecting Data Plane Liveliness in MPLS" and is available currently on the MPLS Working Group's Web page as draft-ietf-mpls-lsp-ping-00.txt. This document describes a mechanism that can be used by network managers to detect data plane failures in MPLS LSPs. This mechanism defines an MPLS "echo request" and "echo reply" that can be sent by an LSR for the purposes of fault detection and isolation. The document also describes several mechanisms for transporting the echo reply. The mechanism proposed in this document is quite useful for LDP-based MPLS networks, as well as those networks deploying the various MPLS applications such as L3 VPN or PWE3 (including L2 VPN).

Two MIBs have been proposed recently in the MPLS Working Group. The first, entitled "Link Bundling Management Information Base," is defined to manage traffic engineering link bundles as they are defined in the corresponding TE Link Bundling IETF draft (soon to be an RFC). This MIB defines the "knobs and buttons" necessary to configure and manage links that participate in TE link bundling. This MIB also is designed to work with the CCAMP's LMP MIB (see Section 12.2). The other MIB defined in the MPLS Working Group is entitled "Multiprotocol Label Switching (MPLS) Traffic Engineering Management Information Base for Fast Reroute." This MIB defines extensions to the MPLS-TE MIB (see Chapter 8) that fully expose the MPLS TE Fast Reroute capabilities of an LSR. These functions go beyond those defined in the MPLS-TE MIB, as the Fast Reroute features were not fully defined at the time the MPLS-TE MIB was completed.

It is clear that although the standard management picture for MPLS has improved markedly over the past few years, it is still not complete, especially in terms of configuration. Furthermore, we predict that ongoing refinements of MPLS will continue to be made within the IETF and other standards bodies, and thus corresponding management tools and techniques will be produced to manage these new features. Some of these will be incorporated into GMPLS, while others that are MPLS-specific will remain in the base MPLS. The existing MIBs are likely to be extended, augmented, or revised to encompass these functions. Additional Policy Information Bases (PIBs) or X.500 schemas may also be produced for these new features as well.

12.4 IETF PPVPN Working Group VPN Management Standardization

The IETF Provider Provisioned Virtual Private Networks (PPVPN) Working Group has produced the PPVPN-MPLS-VPN MIB (see Chapter 11), which can be used to manage MPLS VPN deployments. In addition to this work, the PPVPN Working Group has produced a document that contains common textual conventions for PPVPN-related MIBs and that is used as a common denominator for the MIBs produced by this Working Group. This document, entitled "Definition of Textual Conventions for Provider Provisioned Virtual Private Network (PPVPN) Management," can be found on the PPVPN Working Group's Web page. In addition, a management framework containing an object and information model is being produced. This forthcoming document will provide guidance for those implementing MIBs and other management interface documents such as Policy Information Bases (PIBs) for the PPVPN Working Group. This document will also provide a baseline of objects that management stations will need to provide to operators managing networks deploying PPVPN services. This is also a guide of requirements for devices being deployed in those networks.

12.5 DMTF

The Distributed Management Task Force (DMTF) is currently working on information and data models for MPLS. The purpose of this work is to provide operators and network managers alike with a common picture of how an MPLS network is managed by providing a common set of objects that both devices and management stations should support. The object and information models can be found at *www.dmtf.org* for members of the DMTF. The Web site provides information about becoming a member.

12.6 Concluding Remarks

At this point, we have discussed many of the tools and techniques that can be used to fill an operator's conceptual management toolbox. By providing these tools, device vendors and third-party software vendors make an MPLS deployment a complete package that a provider can use to manage their network to its fullest potential. We have discussed many tools and techniques, both standards-based and proprietary, which have and are being implemented by MPLS device vendors and third-party software vendors. The applications discussed in this book have been deployed by many dozens of vendors, thus making their approaches reasonable. These tools can be viewed in several general areas. First, we discussed management tools and techniques that cover basic MPLS-enabled networks where LDP is used as the sole means for binding labels to packets using the routing rules established by the routing protocol(s). Next, we discussed those tools and techniques that could be used to manage (or actively engineer) a traffic-engineered MPLS network. These tools were shown to allow an operator to not only manage fault conditions, but also to allow them to fully (or partially) control and provision the flow and contour of traffic through their networks. Finally, we discussed a tool that could be used to manage an MPLS VPN network. Along the way, we tried to show how each new tool built upon, or could be used in conjunction with, those that were already presented in the text. In all cases, the tools presented allow a service provider to use them as an integral part of their management strategy.

The complexity of managing MPLS networks grows almost daily with the introduction and deployment of new applications such as L2 VPN and pseudo-wire services. When MPLS was originally introduced in 1996, the production of standards-based management interfaces and tools such as NetFlow and offline traffic engineering systems were nonexistent or available only as proprietary tools that only worked (or worked well) with that particular vendor's devices. The reluctance of vendors to produce management tools and solutions can be attributed to two things. First, MPLS was not long ago considered a new or emerging technology by service providers. At that time, providers were more interested in deploying the technology to prove that it would work. Thus, management interfaces were not perceived as being a crucial requirement at that time. As a result, device vendors, typically with limited resources, were more interested in providing these service providers with devices that were "feature rich" enough for them to successfully run in a large service provider environment. To this end, network management took a backseat in the larger scheme of purchasing and development priorities.

However, as the technology has grown in maturity and acceptance by being deployed in many hundreds of provider networks, so has the pressure from these same providers to have device vendors produce management tools and techniques that could be used to help them deploy this service in a scalable and profitable manner. It was only at this time that the development of management tools by device vendors quickly became a high priority. Development of tools that could be used for customer billing, troubleshooting, provisioning, or other aspects of network management soon became a "must have" requirement, simply because operators needed to deploy this technology on a large scale and for profit—it was no longer an experiment or trial in their minds. Following on the heels of this demand has been an explosive growth in tools and applications from device vendors, and there has been a proportional offering from a small but growing collection of third-party network management application vendors. In many cases, these small niche vendors have fortunately filled in the gaps left by device vendors who were either unable or unwilling to push the limits of network management for MPLS.

It is clear that, given the growing demand from service providers for standards-based as well as proprietary management solutions, the trend of improved coverage of network management tools and techniques will continue to increase. It is a relief for the customers that we deal with on a nearly daily basis that this process has also accelerated to the point where new features are deployed with network management either in the same release as the firmware or shortly thereafter. In fact, we know of several vendors who have taken this approach and are implementing some of the tools and techniques described earlier hand in hand with the router/switch features they manage. This would not be possible without a strong demand from service providers who purchase MPLS-enabled devices from those device vendors.

These points were brought up and discussed in many of the interviews given in this book. Interviewers commented that the technology continues to grow to meet the demands of service providers. This means not only new features and applications of MPLS, but also its emergence as a completely manageable protocol.

This strong demand for management has also resulted in a strong growth of interest in operations and management in all technical areas related to MPLS within the standards bodies such as the IETF. In particular, most of the standards Working Groups that we participate in now have management interfaces such as MIBs defined as an explicit goal of the work done within that standards group. Therefore, it has recently been the case that as soon as a protocol standard is agreed upon, its management interface or interfaces are agreed upon shortly thereafter.

The future is bright for MPLS. There are many hundreds of deployments today, which shows its clear utility to service providers. Many new applications of MPLS exist today, and we are sure that several more will be invented in the not-so-distant future. These applications will increase MPLS's utility, which can only drive the demand for the technology further. Management tools and techniques are now developed and deployed hand in hand with these new applications, as they are now a top-priority requirement from those deploying the technology. We feel this is a crucial step in positioning MPLS as a fully mature and "prime time" approach. Making MPLS a manageable technology has not only increased the demand for MPLS, but also increased its utility manyfold. It is our hope that this trend will continue in the future.

Further Reading

Mannie, E., et al. "Generalized Multi-Protocol Label Switching (GMPLS) Architecture." IETF Internet Draft. September 2002. *www.ietf.org/internet-drafts/draft-ietf-ccamp-gmpls-architecture-02.txt.*

Lang, J., et al. "Link Management Protocol (LMP)." IETF Internet Draft. March 2002. *www.ietf.org/internet-drafts/draft-ietf-ccamp-lmp-03.txt.*

Dubuc, M., et al. "Link Management Protocol Management Information Base." IETF Internet Draft. February 2002. *www.ietf.org/internet-drafts/draft-ietf-ccamp-lmp-mib-01.txt.*

Kompella, K., et al. "Link Bundling in MPLS Traffic Engineering." IETF Internet Draft. May 2002. *www.ietf.org/internet-drafts/draft-ietf-mpls-bundle-01.txt.*

Papadimitriou, D., et al. "GMPLS Signalling Extensions for G.709 Optical Transport Networks Control." IETF Internet Draft. March 2002. *www.ietf.org/internet-drafts/draft-ietf-ccamp-gmpls-g709-00.txt.*

Zelig, D., and T. Nadeau. "Ethernet Pseudo Wire (PW) Management Information Base." IETF Internet Draft. February 2002. *search.ietf.org/internet-drafts/draft-zelig-pw-enet-mib-00.txt.*

Zelig, D., S. Mantin, T. Nadeau, and D. Danenberg. "Pseudo Wire (PW) Management Information Base." IETF Internet Draft. February 2002. *search.ietf.org/internet-drafts/draft-zelig-pw-mib-02.txt.*

Zelig, D., S. Mantin, T. Nadeau, D. Danenberg, and A. Malis. "Pseudo Wire (PW) over MPLS PSN Management Information Base." IETF Internet Draft. February 2002. *search.ietf.org/internet-drafts/draft-zelig-pw-mpls-mib-01.txt.*

Nadeau, T., D. Danenberg, D. Zelig, and A. Malis. "Definitions for Textual Conventions and Object-Identities for Pseudo-Wires Management." IETF Internet Draft. February 2002. *search.ietf.org/internet-drafts/draft-nadeau-pw-tc-mib-02.txt.*

Sathappan, S., M. Venkatesan, and T. Nadeau. "PW ATM Pseudo Wire (PW) Emulation Network Management Information Base Using SMIv2." IETF Internet Draft. February 2002. *search.ietf.org/internet-drafts/draft-sathappan-pw-atm-mib-00.txt.*

Danenberg., D., S. Park, T. Nadeau, D. Zelig, S. Mantin, and A. Malis. "SONET/SDH Circuit Emulation Service Over Packet (CEP) Management Information Base Using SMIv2." IETF Internet Draft. May 2002. *search.ietf.org/internet-drafts/draft-danenberg-pw-cem-mib-02.txt.*

Nadeau, T., C. Srinivasan, and A. Farrel. "Multiprotocol Label Switching (MPLS) Management Overview." IETF Internet Draft. December 2001. *www.ietf.org/internet-drafts/draft-ietf-mpls-mgmt-overview-01.txt.*

Cetin, R., S. De Cnodder, D. Gan, and T. Nadeau. "Multiprotocol Label Switching (MPLS) Traffic Engineering Management Information Base for Fast Reroute." IETF Internet Draft. February 2002. *search.ietf.org/internet-drafts/draft-ietf-mpls-fastreroute-mib-00.txt.*

Dubuc, M., S. Dharanikota, T. Nadeau, and J. Lang. "Link Bundling Management Information Base." IETF Internet Draft. February 2002. *search.ietf.org/internet-drafts/draft-ietf-mpls-bundle-mib-01.txt.*

Schliesser, B., and T. Nadeau. "Definition of Textual Conventions for Provider Provisioned Virtual Private Network (PPVPN) Management." IETF Internet Draft. *search.ietf.org/internet-drafts/draft-ietf-ppvpn-tc-mib-01.txt.*

IETF Common Control and Measurement Plane (CCAMP) Working Group: *www.ietf.org/html.charters/ccamp-charter.html.*

IETF Pseudo-Wire Edge-to-Edge Emulation Working Group: *www.ietf.org/html.charters/pwe3-charter.html.*

IETF Multi-Protocol Label Switching: *www.ietf.org/html.charters/mpls-charter.html.*

IETF Provider-Provisioned Virtual Networks Working Group: *www.ietf.org/html.charters/ppvpn-charter.html.*

To find out more about the IETF, visit their Web page at *www.ietf.org/.*

Appendix A: IETF and Other Standards Bodies

Introduction

This appendix will provide information about Internet Engineering Task Force (IETF) Request for Comments (RFC) documents, as well as the drafts that may become RFCs in the future.

The Internet Engineering Task Force can be located at *www.ietf.org*. Please visit their Web site for more information about the standards organization or other related information.

A.1 The IETF

The IETF is a large open international community of network designers, operators, vendors, and researchers concerned with the evolution of the Internet architecture and the smooth operation of the Internet. It is open to any interested individual. The actual technical work of the IETF is done in its Working Groups, which are organized by topic into several areas (e.g., routing, transport, security, etc.). Much of the work is handled via mailing lists. The IETF holds meetings three times per year.

The IETF Working Groups are grouped into areas and managed by area directors (ADs). The ADs are members of the Internet Engineering Steering Group (IESG). Providing architectural oversight is the Internet Architecture Board (IAB). The IAB also adjudicates appeals when someone complains that the IESG has failed. The Internet Society (ISOC) charters the IAB and IESG for these purposes.

The general area director also serves as the chair of the IESG and of the IETF, and is an ex officio member of the IAB.

The Internet Assigned Numbers Authority (IANA) is the central coordinator for the assignment of unique parameter values for Internet protocols. The IANA is chartered by the ISOC to act as the clearinghouse to assign and coordinate the use of numerous Internet protocol parameters.

A.2 How Standard MIBs Are Progressed at the IETF

Management Information Bases (MIBs) are generally treated and progressed in the same manner as are other IETF documents. A MIB is usually created after a particular technology (or piece of technology) is fairly well understood and is actually being developed by vendors. It is at this point that either most vendors begin to think about how to manage these new features using a standard management interface, or they are told by customers that they must provide a standard management interface that will be (mostly) the same regardless of which vendor that customer chooses to utilize in their networks. It is usually at this point that one or several authors get together and create a MIB for this technology. This MIB is hopefully based on several (or at least one) working implementations. The MIB is then proposed to the Working Group that is responsible for hosting the technology in question. If the Working Group decides to adopt the MIB as a Working Group document, the draft progresses with the blessing, guidance, and input (and sometimes political influences) from the Working Group. After some period of time, the Working Group decides that the MIB is sufficiently capable of managing the feature that it was originally chartered to manage, and what is referred to as a "Working Group last call" is raised by the Working Group's chair. It is at this point that IETF members are given a last opportunity to comment on the draft. There may be several "last calls" depending on how complex the last round of modifications have been. Once this period is complete, the MIB is then brought before the IESG. It is this group that gives the document a review within the larger context of all existing IETF standards. The IESG may reject the document, which will extinguish it at that point. If work is to progress on the document, a new document must progress through the original process explained earlier. However, if the document is received (usually after several rounds of corrections), it is raised to Draft Standard status and the draft is then officially published by the IETF as what is called a Request for Comments (RFC) document. The document is also assigned a unique number (RFC 1233, for example). It is at this point that the document must be demonstrated to be interoperable by several full implementations of the technology. This is a long and slow process, but once complete, the document may be raised to full Internet Standard status.

For more details on this process, please visit the IETF's Web site, where this and many other procedures are explained in detail.

A.3 How to Obtain an Internet Draft

An Internet Draft (ID) is a document that a specific IETF Working Group chooses to adopt and work on. When the Working Group feels that the document is finished and ready for advancement, the Working Group will forward the document to the IESG. The document is then reviewed by this group and, if approved, will be assigned an RFC and processed to standards track or informational status. It is important to note that not all Internet Drafts become RFCs. There are various reasons why an Internet Draft may not advance, which are beyond the scope of this text. Please visit the IETF's Web site for more information.

Internet Drafts are readily available electronically from many repositories on the Internet. However, the first place you should look when seeking out an ID is the place where they are first published and are always maintained: *www.ietf.org/ID.html.*

A.4 How to Obtain an RFC

As described earlier, Internet RFCs are documents that have progressed through the process of Working Group *consensus* as well as having been given approval by the IESG. RFCs may progress further and eventually become full-standard documents; sometimes they may not. Typically, this process lasts for several years while interoperability testing is undergone.

Internet RFCs are readily available electronically from many repositories on the Internet. Several books have printed hardcopies or compact disks of a particular snapshot in time of the RFC database. Access to the RFCs in this manner is problematic as the database changes as new documents are added. The search of such a CD or printed copy of the database may not locate a particular RFC simply because it was published after the snapshot of the database was taken. Therefore, the first place you should look when seeking out an RFC is the place where they are first published and maintained: *www.ietf.org/rfc.html.*

A.5 How to Find an Internet Working Group

The IETF is divided into Working Groups (WGs). Each Working Group is chartered with a specific goal or goals that it must achieve within a certain period, or be shut down. Working Groups are usually assigned to standardize a specific technology and will progress all or most of the documents related to that technology throughout the duration of that WG's existence. All Working Groups are eventually shut down or put into a dormant state to await further work, so a search of the IETF's Web site for a particular Working Group may not bear any fruit.

Currently active Working Groups can be found at *www.ietf.org/html.charters/wg-dir.html.*

A.6 Other Standards Bodies

MPLS was originally standardized at the IETF. This work continues today. This has happened because MPLS came as a new routing protocol. However, several other standards bodies have attempted to, or are actively attempting to, provide a means for the progression of the technology, its development, and/or its standardization. These include the International Telecommunication Union (ITU), the MPLS Forum, and the Optical Interworking Forum (OIF). Each standards body has a different mission, which in some cases conflicts with that of the IETF. The reason is that in the past the standards related to networking have been delegated to the various standards bodies along OSI layer boundaries. However, with the advent of MPLS, which operates at layer "2½," it has become clear that it is not necessarily clear which standards body should work on which portion of the technology. This is further compounded by the recent proposal to use MPLS as a general signaling mechanism for optical networking (GMPLS). What is clear is that the majority of the work on MPLS has taken place in the IETF. We hope that, with the cooperation between the various standards bodies, the work on MPLS will continue there, as coordination among standards groups is usually difficult at best. Information about the other standards bodies can be found in the following places:

- For more information about IANA, check out their Web site at *www.iana.org/*.
- To locate information about the ITU: *www.itu.int/*.
- To locate information about the MPLS Forum: *www.mplsforum.org/*.
- To locate additional information about the OIF: *www.oiforum.com/*.

Appendix B: MPLS-TC MIB

This appendix will provide information about the Internet Engineering Task Force (IETF) MPLS Working Group's document entitled "Definitions of Textual Conventions and OBJECT-IDENTITIES for Multi-Protocol Label Switching Management." This document is most colloquially known as the MPLS-TC MIB. This MIB module is a little different from typical MIBs in that it only contains textual conventions, which are imported by all of the other MPLS MIB modules. Although not warranting a complete chapter because it contains no objects that are read or written to manage an MPLS network, we felt that this document should be explained here for completeness and as a reference.

The MPLS-TC MIB was introduced soon after the IESG review of the MPLS-LSR, MPLS-LDP, and MPLS-TE MIBs had begun. It was clear that all of the MIBs shared a common set of SNMP textual conventions, and it was thus viewed that it was important to align them. It was important to unify the MIBs behind a common set of textual conventions for several reasons. First, there was an overlap in some of the specifics of some of the textual conventions among a few of the MIBs. This meant that in one MIB the definition of a particular object was slightly different from another. In other cases, they were noticeably different. Second, it is important for these things to be referenced in a common place so that designers of new MIB modules would be able to find them in a single location. Note that as of the publication of this book, the MPLS-TC MIB has changed several times in an effort to refine it as well as align it with the latest modifications requested by the IESG to the base MPLS MIBs. Therefore, an examination of the current version of this document will reveal it to be a bit different from the one shown in this appendix. However, we show the older version of the document here because it was made to work with the revisions of the MIBs discussed in the chapters above. A subsequent edition of this text will be updated to reflect the RFC versions of all of the MPLS MIBs, including the MPLS-TC MIB.

Let's now examine each of the textual conventions (TCs) and explain their meaning in detail. A note to the reader: although the TCs presented here are used by the standard MIBs, it is both an accepted and strongly recommended practice that proprietary extension MIBs import and utilizes these TCs.

MPLSBitRate

The MPLSBitRate defines a TC that is used when specifying bitrate limits or thresholds for MPLS interfaces. For example, when configuring an MPLS TE tunnel, it is important to specify the mean, minimum, and maximum bitrates.

```
MPLSBitRate ::= TEXTUAL-CONVENTION
 DISPLAY-HINT "d"
 STATUS current
 DESCRIPTION
  "An estimate of bandwidth in units of 1,000 bits per second.
  If this object reports a value of 'n' then the rate of the
  object is somewhere in the range of 'n-500' to 'n+499'. For
  objects which do not vary in bit rate, or for those where no
  accurate estimation can be made, this object should contain
  the nominal bit rate."
 SYNTAX Integer32 (1 .. 2147483647)
```

MPLSBurstSize

The MPLSBurstSize TC is used to specify appropriate burst values for data rates.

```
MPLSBurstSize ::= TEXTUAL-CONVENTION
 DISPLAY-HINT "d"
 STATUS current
 DESCRIPTION
  "The number of octets of MPLS data that the stream may send
  back-to-back without concern for policing."
 SYNTAX Unsigned32 (1 .. 4294967295)
```

MPLSExtendedTunnelId

The MPLSExtendedTunnelId TC is used as one of the four unique identifiers for an MPLS TE tunnel.

```
MPLSExtendedTunnelId ::= TEXTUAL-CONVENTION
 STATUS current
```

```
DESCRIPTION
 "A unique identifier for an MPLS Tunnel. This MAY represent an
 IpV4 address of the ingress or egress LSR for the tunnel.
 This value is derived from the Extended Tunnel Id in RSVP or
 the Ingress
Router ID for CR-LDP."
SYNTAX Unsigned32
REFERENCE
 "1. Awduche, D., et al., RSVP-TE: Extensions to RSVP for LSP
      Tunnels, draft-ietf-mpls-rsvp-lsp-tunnel-08.txt,
      February 2001.
  2. Constraint-Based LSP Setup using LDP, Jamoussi, B., et
     al., draft-ietf-mpls-cr-ldp-05.txt, February 2001."
```

MPLSLabel

```
MPLSLabel ::= TEXTUAL-CONVENTION
 STATUS current
 DESCRIPTION
  "This value represents an MPLS label. The label contents are
  specific to the label being represented.
    The label carried in an MPLS shim header (for LDP, the
  Generic Label) is a 20-bit number represented by 4 octets.
  Bits 0-19 contain a label or a reserved label value. Bits 20-
  31 MUST be zero.
    The frame relay label can be represented by either 10-bits
  or 23-bits depending on the DLCI field size and the upper 22-
  bits or upper 9-bits must be zero, respectively.
    For an ATM label the lower 16-bits represents the VCI, the
  next 12-bits represents the VPI and the remaining bits MUST
  be zero."
 REFERENCE
  "1. MPLS Label Stack Encoding, Rosen et al, RFC 3032, January
       2001.
   2. Use of Label Switching on Frame Relay Networks, Conta et
      al, RFC 3034, January 2001.
   3. MPLS using LDP and ATM VC switching, Davie et al., RFC
      3035, January 2001."
 SYNTAX Unsigned32 (0 .. 4294967295)
```

MPLSLdpGenAddr

```
MPLSLdpGenAddr ::= TEXTUAL-CONVENTION
 STATUS current
 DESCRIPTION
```

```
    "The value of a network layer or data link layer address."
  SYNTAX OCTET STRING (SIZE (0 .. 64))
```

MPLSLdpIdentifier

```
MPLSLdpIdentifier ::= TEXTUAL-CONVENTION
 STATUS current
 DESCRIPTION
   "The LDP identifier is a six octet quantity which is used to
   identify an Label Switching Router (LSR) label space.
      The first four octets encode an IP address assigned to the
   LSR, and the last two octets identify a specific label space
   within the LSR."
 SYNTAX OCTET STRING (SIZE (6))
```

MPLSLdpLabelTypes

```
MPLSLdpLabelTypes ::= TEXTUAL-CONVENTION
 STATUS current
 DESCRIPTION
   "The Layer 2 label types which are defined for MPLS LDP/CRLDP
   are generic(1), atm(2), or frameRelay(3)."
 SYNTAX INTEGER {
          generic(1),
          atm(2),
          frameRelay(3)
          }
```

MPLSAtmVcIdentifier

```
MPLSAtmVcIdentifier ::= TEXTUAL-CONVENTION
 STATUS current
 DESCRIPTION
   "The VCI value for a VCL. The maximum VCI value cannot exceed
   the value allowable by atmInterfaceMaxVciBits defined in ATM-
   MIB. The minimum value is 32, values 0 to 31 are reserved for
   other uses by the ITU and ATM Forum. 32 is typically the
   default value for the Control VC."
 SYNTAX Integer32 (32 .. 65535)
```

MPLSLdpID

```
MPLSLSPID ::= TEXTUAL-CONVENTION
 STATUS current
 DESCRIPTION
  "An identifier that is assigned to each LSP and is used to
  uniquely identify it. This is assigned at the head end of the
  LSP and can be used by all LSRs to identify this LSP. This
  value is piggybacked by the signaling protocol when this LSP
  is signaled within the network. This identifier can then be
  used at each LSR to identify which labels are being swapped
  to other labels for this LSP. For IPv4 addresses this results
  in a 6-octet long cookie."
 SYNTAX OCTET STRING (SIZE (0 .. 31))
```

MPLSLsrIdentifier

```
MPLSLsrIdentifier ::= TEXTUAL-CONVENTION
 STATUS current
 DESCRIPTION
  "The Label Switching Router (LSR) identifier is the first 4
  bytes or the Router Id component of the Label Distribution
  Protocol (LDP) identifier."
 SYNTAX OCTET STRING (SIZE (4))
```

MPLSInitialCreationSource

```
MPLSInitialCreationSource ::= TEXTUAL-CONVENTION
 STATUS current
 DESCRIPTION
  "The entity that originally created the object in question.
  The values of this enumeration are defined as follows:
      other(1)  This is used when an entity which has not been
                enumerated in this textual convention but which
                is known by the agent.
       snmp(2)  The Simple Network Management Protocol was used
                to configure this object initially.
        ldp(3)  The Label Distribution Protocol was used to
                configure this object initially.
       rsvp(4)  The Resource Reservation Protocol was used to
                configure this object initially.
```

```
          crldp(5)  The Constraint-Based Label Distribution
                    Protocol was used to configure this object
                    initially.
    policyAgent(6)  A policy agent (perhaps in combination with one
                    of the above protocols) was used to configure
                    this object initially.
        unknown(7)  The agent cannot discern which component
                    created the object."
     SYNTAX INTEGER {
            other(1),
            snmp(2),
            ldp(3),
            rsvp(4),
            crldp(5),
            policyAgent(6),
            unknown (7)
            }
```

MPLSPathIndex

```
MPLSPathIndex ::= TEXTUAL-CONVENTION
 STATUS current
 DESCRIPTION
  "A unique identifier used to identify a specific path used by a
  tunnel."
 SYNTAX Unsigned32
```

MPLSPathIndexOrZero

```
MPLSPathIndexOrZero ::= TEXTUAL-CONVENTION
 STATUS current
 DESCRIPTION
  "A unique identifier used to identify a specific path used by a
  tunnel. If this value is set to 0, it indicates that no path
  is in use."
 SYNTAX Unsigned32
```

MPLSTunnelAffinity

```
MPLSTunnelAffinity ::= TEXTUAL-CONVENTION
 STATUS current
 DESCRIPTION
```

```
  "Include-any, include-all, or exclude-all constraint for
  link selection."
 SYNTAX Unsigned32
```

MPLSTunnelIndex

```
MPLSTunnelIndex ::= TEXTUAL-CONVENTION
 STATUS current
 DESCRIPTION
  "Index into mplsTunnelTable."
 SYNTAX Integer32 (1 .. 65535)
```

MPLSTunnelInstanceIndex

```
MPLSTunnelInstanceIndex ::= TEXTUAL-CONVENTION
 STATUS current
 DESCRIPTION
  "Instance index into mplsTunnelTable."
 SYNTAX Unsigned32 (0 .. 65535)
```

Glossary

ATM (Asynchronous Transfer Mode) A network technology based on transferring data in cells of a fixed size that are relatively small compared to those utilized by other technologies. The constant cell size results in simplified forwarding that enables ATM equipment to transmit voice, video, and computer data over the same converged network—something not possible before its time.

BGP (Border Gateway Protocol) The most widely used interdomain routing protocol used in IP-based networks.

Command line interface (CLI) An interface that typically provides the user with ASCII screens of information for viewing specific device functions or configuration, as well as a structured (albeit proprietary) syntax for interacting with it. The prevalent method of connecting to a network device's CLI is to use a Telnet network connection, although other means exist.

Common Object Request Broker Architecture (CORBA) Defines a distributed object-computing infrastructure that has several uses. In particular, CORBA provides an architecture that automates many common network-programming tasks such as object registration, location, and activation of network objects.

Constraint-based routing The process by which the most desirable routes in a network can be chosen based on constraints such as the minimum available bandwidth available over a particular route.

Constraint-based Routing Label Distribution Protocol (CR-LDP) An extension to the LDP that enables constraint-based routing and QoS reservation in an MPLS network similar to that provided by RSVP-TE.

CSPF (Constrained shortest path first) An extension of the shortest path first (SPF) algorithm. It considers each link for inclusion in the shortest path tree only if it meets certain specified constraints.

Differentiated Services The QoS (Quality of Service) architecture defined by the IETF. This architecture divides traffic into a small number of classes in order to provide QoS to large aggregates of traffic.

Edge LSR A label switching router (LSR) that first imposes a label.

Explicit route object (ERO) An object exchanged by an LSP setup protocol such as RSVP or LDP that is used to specify the sequence of network nodes (e.g., hops) that an explicitly routed LSP must traverse.

eXtensible Markup Language (XML) A popular management interface that provides an encoding method for individual and batched management commands that allows for the creation of a command-specific syntax that can be automatically parsed and checked.

FEC-to-NHLFE (FTN) The mapping from a FEC (set of equivalent forwarding information) to an NHLFE (MPLS label).

Forwarding The process by which a network node transfers a data unit (e.g., packet or cell) from one or more input interfaces to one or more output interfaces.

Forwarding Equivalency Class (FEC) A set of packets that can be handled equivalently for forwarding. This set of packets may be bound to a single label. For example, a set of packets destined for the same destination address prefix may be considered as the same FEC, and thus bound to the same label.

Forwarding Information Base (FIB) A forwarding table in a router or switch.

IETF (Internet Engineering Task Force) The major standards-setting organization for Internet and IP protocols, consisting of more than 80 working groups.

Inform *See* Notification.

Interface The boundary between adjacent layers of the ISO model. Interfaces may be physical connectors on a network node such as an Ethernet port, or may be virtual sub-interfaces such as a PPP interface.

Interior Gateway Protocol (IGP) A protocol for exchanging routing information between gateways (hosts with routers) within an autonomous network. The routing information can then be used by a network protocol to specify how to route traffic. An example of an IGP is the OSPF protocol.

Internet Protocol (IP) The predominant networking protocol used throughout the world. IP specifies the format of packets (datagrams) as well as the addressing scheme used to identify them. Most networks combine IP with a higher-level protocol called Transport Control Protocol (TCP), which establishes a virtual connection between a destination and a source. IP by itself is something like the postal system. It allows you to address a package and drop it in the system, but there's no direct link between you and the recipient. TCP/IP, on the other hand, establishes a virtual connection between two end points that allows them to send messages back and forth until the connection is broken.

Label A short, fixed-length identifier with local significance used to identify an MPLS label switched path.

Label merging The replacement of multiple incoming labels belonging to a particular FEC with a single outgoing label. This is sometimes used in an effort to share resources.

Label Swapping/Switching A forwarding paradigm allowing streamlined forwarding of data by using labels to identify classes of data packets that are treated indistinguishably when forwarding.

Label switching router (LSR) A device that implements label switching and is capable of forwarding packets encoded with an MPLS header.

Layer 2 (L2) The protocol layer residing directly under layer 3.

Layer 3 (L3) The protocol layer at which IP and its associated routing protocols operate.

LDP (Label Distribution Protocol) The IETF protocol that defines the distribution of label bindings in MPLS networks.

Longest match The forwarding algorithm most often used for IP forwarding. A fixed-length IP address is compared against a set of variable-length entries in a routing table. These entries are matched for the entry with the most leading bits in the route's address.

LSP (label switched path) The path that will be followed by a labeled packet over several network hops.

Management Information Base A database of object definitions used by SNMP-enabled devices (managers/agents). The definition specifies whether an SNMP manager can monitor the object.

Management Information Base (MIB) Database of network management information that is used and maintained by a network management protocol such as SNMP. The value of a MIB object can be changed or retrieved by means of SNMP commands, usually through a network management system. MIB objects are organized in a tree structure that includes public (standard) and private (proprietary) branches.

Management interface An interface that allows network operators to manage devices in their networks by providing remote access to each device's control, configuration, and status information. Many different types of management interfaces exist, but in general, a management interface is composed of two parts: a protocol describing the communication rules between the operator and the device, and the format of the information that will be exchanged using that protocol.

Maximum transmission unit (MTU) The largest packet size that can be transmitted on a data link without fragmentation.

MPLS domain A contiguous set of nodes that operate MPLS routing and forwarding.

MPLS edge node An MPLS node that connects an MPLS domain with a node that is outside of the domain, either because it does not run MPLS and/or because it is in a different domain.

MPLS egress node An MPLS edge node that handles traffic as it leaves an MPLS domain by removing the MPLS header and forwards the packet accordingly.

MPLS ingress node An MPLS edge node that accepts unlabeled traffic as it enters an MPLS domain. This node may classify the traffic into a FEC, and then impose an MPLS header if a label binding is available.

MPLS interface An interface on which MPLS traffic is forwarded by an LSR.

MPLS label A label that is carried in a packet header and that represents the packet's FEC.

MPLS node A node that is running one or more of the MPLS protocols.

Multi-Protocol Label Switching (MPLS) The name given by the IETF to the workgroup responsible for label switching. Also the name of the label switching technology it has standardized.

Network Management Station/System (NMS) An application used to manage one or more network devices via one or more management interfaces such as SNMP, CLI, or CORBA. Applications that communicate with devices via SNMP typically contain several components in the IETF SNMP architecture as defined in RFC2571.

Next hop label forwarding entry (NHLFE) The NHLFE is used when forwarding a labeled packet. It contains the packet's next hop and the operation to perform on the packet's label stack. The latter may be either to replace the label at the top of the label stack with a specified new label, or to pop the label stack, replace the label at the top of the label stack with a specified new label, and then push one or more specified new labels onto the label stack.

Notification SNMP provides a means by which agents are able to issue asynchronous messages to managers (or midlevel managers). These messages are called SNMP notifications. SNMP notifications are defined in a MIB module with the NOTFICATION-TYPE macro. Notifications can be sent from a notification-originator to a notification-receiver using two mechanisms. A TRAP can be sent either as a TRAPv1 or a TRAPv2 depending on the version of the Protocol Operations being used, or by using an INFORM (only available with version 2 of the Protocol Operations). The reception of a notification is not always guaranteed, so in cases where it is required, the INFORM notification message can be used.

Object identifier (OID) Specified as an ordered sequence of nonnegative integers written from left to right and separated by a period (i.e., dot). This is referred to as the *dot-notation*. For example, "1.1" represents an OID. The OID space itself does not have any limitation as to how many branches (subIDs) are possible. For SNMP, however, a limit of 128 subIDs has been defined as a limit. Each consecutive integer is separated from the numbers around it by a period (i.e., a dot). The sequence must contain two integers at a minimum and does not have a maximum number (although all implementations will have a specific limit to this size).

Open shortest path first (OSPF) Link-state routing protocol.

Packet switching Refers to protocols in which messages are divided into packets before they are transmitted onto an outgoing interface. Each packet is transmitted individually and can even follow different routes to its destination. Once all the packets forming a message arrive at the destination, they are recompiled into the original message.

Port A physical interface to a switch/router or an identifier used by transport protocols to distinguish application flows between a pair of hosts.

Request for Comment (RFC) A document in a series that is maintained by the IETF that includes all Internet protocol standards.

Resource Reservation Protocol (RSVP) Protocol used to reserve network resources to signal a traffic flow. This mechanism may be used to signal QoS (Quality of Service) parameters for this flow.

Resource Reservation Protocol with Traffic Engineering Extensions (RSVP-TE) Traffic engineering extensions to the RSVP protocol used for reserving network resources to provide Quality of Service guarantees to application flows, which is also used to establish forwarding state for explicitly routed LSPs.

Route distinguisher An identifier used in BGP/MPLS VPNs to ensure uniqueness of address prefixes among VPNs when multiple VPNs use the same address space (e.g., the address space defined in RFC 1918).

Router A device that interconnects any number of LANs or WANs and is capable of forwarding traffic between them.

Route target An extended community that identifies a group of routers and, in each router of that group, a subset of forwarding tables maintained by the router that may be populated with BGP routes carrying that extended community.

Scalar object An object that can only exist as a single instance.

Shortest path first (SPF) The common routing algorithm used by link-state routing protocols, such as OSPF and IS-IS; also known as Dijkstra's algorithm.

Simple Network Mangement Protocol (SNMP) A network management protocol used to manage TCP/IP networks. This protocol provides functions that enable you to access the data object whose definitions are located in the MIB. SNMP provides a means for monitoring and controlling network devices, and for managing configurations, statistics collection, performance, and security.

Structure of Management Information (SMI) A data modeling language used to model management data. The SMI [RFC2578] is the language that is used to write, define and specify a MIB module.

Switch In networks, a device that filters and forwards packets between adjacent networks. Switches operate at the data link layer (layer 2) of the OSI reference model and therefore support any packet protocol.

TCP (Transmission Control Protocol) A transport layer component of the Internet Protocol suite. This protocol provides reliable data delivery services.

Textual convention When designing a MIB module, it is sometimes beneficial to define new types that are derived from those defined in the SMI. Each of these new types has a different name, a similar syntax, but more precise semantics than the type it is derived from in the SMI. These newly defined types are termed *textual conventions* and are defined with the TEXTUAL-CONVENTION data type.

Trap *See* Notification.

Tunnel A sometimes-secure communication path between two peers, such as routers.

Virtual circuit A circuit used by a connection-oriented layer 2 technology such as ATM or Frame Relay, requiring the maintenance of state information in layer 2 switches.

Virtual circuit identifier (VCI) A field in the ATM header used to identify the virtual circuit to which a cell belongs.

Virtual path identifier (VPI) A field in the ATM header used to identify the virtual path to which a cell belongs.

Virtual Private Network (VPN) A group of sites that, as the result of a set of administrative policies, is able to communicate with each other over a shared backbone network. These systems use encryption and other security mechanisms to ensure that only authorized users can access the network and that the data cannot be intercepted.

Bibliography

RFC 2702. Awduche, D., J. Malcolm, J. Agogbua, M. O'Dell, and J. McManus. 1999. "Requirements for Traffic Engineering over MPLS." September.

RFC 3031. Rosen, E., A. Viswanathan, and R. Callon. 2001. "Multiprotocol Label Switching Architecture." January.

Le Faucheur, F., L. Wu, B. Davie, S. Davari, P. Vaananen, R. Krishnan, P. Cheval, and J. Heinanen. 2000. *MPLS Support for Differentiated Services.* draft-ietf-mpls-diff-ext-07.txt. August.

RFC 1155. Rose, M., and K. McCloghrie. 1990. "Structure and Identification of Management Information for TCP/IP-based Internets." May.

RFC 1157. Case, J., M. Fedor, M. Schoffstall, and J. Davin. 1990. "Simple Network Management Protocol." May.

RFC 1213. McCloghrie, K. 1991. "Management Information Base for Network Management of TCP/IP-based Internets: MIB-II." March.

RFC 1212. Rose, M., and K. McCloghrie. 1991. "Concise MIB Definitions." March.

RFC 1215. Rose, M. 1991. "A Convention for Defining Traps for Use with the SNMP." March.

RFC 1901. Case, J., K. McCloghrie, M. Rose, and S. Waldbusser. 1996. "Introduction to Community-based SNMPv2." January.

RFC 1902. Case, J., et al. 1996. "Structure of Management Information for Version 2 of the Simple Network Management Protocol (SNMPv2)." January.

RFC 1903. Case, J., K. McCloghrie, M. Rose, and S. Waldbusser. 1996. "Textual Conventions for Version 2 of the Simple Network Management Protocol (SNMPv2)." January.

RFC 1904. Case, J., K. McCloghrie, M. Rose, and S. Waldbusser. 1996. "Conformance Statements for Version 2 of the Simple Network Management Protocol (SNMPv2)." January.

RFC 1905. Case, J., K. McCloghrie, M. Rose, and S. Waldbusser. 1996. "Protocol Operations for Version 2 of the Simple Network Management Protocol (SNMPv2)." January.

RFC 1906. Case, J., K. McCloghrie, M. Rose, and S. Waldbusser. 1996. "Transport Mappings for Version 2 of the Simple Network Management Protocol (SNMPv2)." January.

RFC 2271. Harrington, D., R. Presuhn, and B. Wijnen. 1998. "An Architecture for Describing SNMP Management Frameworks." January.

RFC 2272. Case, J., D. Harrington, R. Presuhn, and B. Wijnen. 1998. "Message Processing and Dispatching for the Simple Network Management Protocol (SNMP)." January.

RFC 2434. Narten, T., and H. Alvestrand. 1998. "Guidelines for Writing an IANA Considerations Section in RFCs." IBM, Maxware. October.

RFC 2515. Tesink, K., ed. 1999. "Definitions of Managed Objects for ATM Management." February.

RFC 2547bis. Rosen, E. et al. 2002. "BGP/MPLS VPNs." draft-ietf-ppvpn-rfc2547bis-01.txt. January.

RFC 2570. Case, J., R. Mundy, D. Partain, and B. Stewart. 1999. "Introduction to Version 3 of the Internet-standard Network Management Framework." April.

RFC 2571. Harrington, D., R. Presuhn, and B. Wijnen. 1999. "An Architecture for Describing SNMP Management Frameworks." April.

RFC 2572. Case, J., D. Harrington, R. Presuhn, and B. Wijnen. 1999. "Message Processing and Dispatching for the Simple Network Management Protocol (SNMP)." April.

RFC 2573. Levi, D., P. Meyer, and B. Stewart. 1999. "SNMPv3 Applications." April.

RFC 2574. Blumenthal, U., and B. Wijnen. 1999. "User-based Security Model (USM) for Version 3 of the Simple Network Management Protocol (SNMPv3)." April.

RFC 2575. Wijnen, B., R. Presuhn, and K. McCloghrie. 1999. "View-based Access Control Model (VACM) for the Simple Network Management Protocol (SNMP)." April.

RFC 2578. McCloghrie, K., D. Perkins, J. Schoenwaelder, J. Case, M. Rose, and S. Waldbusser. 1999. "Structure of Management Information Version 2 (SMIv2)." STD 58. April.

RFC 2579. McCloghrie, K., D. Perkins, J. Schoenwaelder, J. Case, M. Rose, and S. Waldbusser. 1999. "Textual Conventions for SMIv2." STD 58. April.

RFC 2580. McCloghrie, K., D. Perkins, J. Schoenwaelder, J. Case, M. Rose, and S. Waldbusser. 1999. "Conformance Statements for SMIv2." STD 58. April.

RFC 2684. Grossman, D., and J. Heinanen. 1999. "Multiprotocol Encapsulation over ATM Adaptation Layer 5." September.

RFC 2685. Fox B., et al. 1999. "Virtual Private Networks Identifier." September.

RFC 3032. Rosen, E., D. Tappan, G. Fedorkow, Y. Rekhter, D. Farinacci, T. Li, and A. Conta. 2001. "MPLS Label Stack Encoding." January.

RFC 3035. Davie, B., J. Lawrence, K. McCloghrie, E. Rosen, G. Swallow, Y. Rekhter, and P. Doolan. 2001. "MPLS Using LDP and ATM VC Switching." January.

RFC 3034. Conta, A., P. Doolan, and A. Malis. 2001. "Use of Label Switching on Frame Relay Networks Specification." January.

Feldmann, A., and J. Rexford. 2000. *IP Network Configuration for Traffic Engineering*. AT&T Research Technical Report 000526-02. May.

Greenberg, A., et al. 2002. *Netscope Traffic Engineering for IP Networks*. AT&T Research Laboratories. *www.nanog.org/mtg-0002/greenberg.html.*

Srinivasan, C., A. Viswanathan, and T. Nadeau. 2000a. *MPLS Traffic Engineering Management Information Base Using SMI v2*. draft-ietf-mpls-te-mib-04.txt. July.

Srinivasan, C., A. Viswanathan, and T. Nadeau. 2000b. *MPLS Label Switching Router Management Information Base Using SMI v2*. draft-ietf-mpls-lsr-mib-06.txt. July.

Nadeau, T., C. Srinivasan, and A. Farrel. 2001. *Multiprotocol Label Switching (MPLS) Management Overview*. draft-ietf-mpls-mgmt-overview-01.txt. December. *www.ietf.org/internet-drafts/draft-ietf-mpls-mgmt-overview-01.txt.*

IANA. 2002. Address Family Numbers. *www.isi.edu/in-notes/iana/assignments/address-family-numbers)*. For MIB, *ftp://ftp.isi.edu/mib/ianaaddressfamilynumbers.mib.*

Boscher, C., P. Cheval, L. Wu, and E. Gray. 2000. "LDP State Machine." draft-ietf-mpls-ldp-state-04.txt. March.

RFC 3037. Thomas, B., and E. Gray. 2001. "LDP Applicability." January.

RFC 2851. Daniele, M., B. Haberman, S. Routhier, and J. Schoenwaelder. 2000. "Textual Coventions for Internet Network Addresses." June.

Nadeau, T., J. Cucchiara, C. Srinivasan, A. Viswanathan, and H. Sjostrand. 2001. "Definitions of Textual Conventions and Object-Identities for Multi-Protocol Label Switching Management." April.

RFC 3036. Andersson, L., P. Doolan, N. Feldman, A. Fredette, and B. Thomas. 2001. "LDP Specification." January.

Sangli, S. et al. 2002. "BGP Extended Communities Attribute." draft-idr-bgp-ext-communities-03.txt. March.

Jamoussi, B., ed. 2001. "Constraint-Based LSP Setup Using LDP." draft-ietf-mpls-cr-ldp-05.txt. February.

RFC 2119. Bradner, S. 1997. "Key Words for Use in RFCs to Indicate Requirement Levels." BCP 14. Harvard University. March.

RFC 2141. Moats, R., ed. 1997. "URN Syntax." *www.ietf.org/rfc/rfc2141.txt.*

RFC 2279. Yergeau, F., ed. 1998. UTF-8: A Transformation Format of ISO 10646. *www.ietf.org/rfc/rfc2279.txt.*

McCloghrie, K., et al. 1991. "Management Information Base for Network Management of TCP/IP-based internets: MIB-II." STD17. March.

RFC 2858. Bates, T., Y. Rekhter, R. Chandra, and D. Katz. 2000. "Multiprotocol Extensions for BGP-4." June. *www.ietf.org/rfc/rfc2858.txt.*

RFC 2376. Whitehead, E., and M. Murata, eds. 1998. "XML Media Types." *www.ietf.org/rfc/rfc2376.txt.*

RFC 2396. Berners-Lee, T., R. Fielding, and L. Masinter. 1998. "Uniform Resource Identifiers (URI): Generic Syntax." *www.ietf.org/rfc/rfc2396.txt.*

RFC 2732. Hinden, R., B. Carpenter, and L. Masinter. 1999. "Format for Literal IPv6 Addresses in URL's." *www.ietf.org/rfc/rfc2732.txt.*

RFC 2781. Hoffman, P., and F. Yergeau, eds. 2000. "UTF-16: An Encoding of ISO 10646." *www.ietf.org/rfc/rfc2781.txt.*

RFC 2096. Bradner, S. 1996. "The Internet Standards Process—Revision 3." BCP 9. Harvard University. October.

Nadeau, T., C. Srinivasan, and A. Viswanathan. 2001. "Multiprotocol Label Switching (MPLS) FEC-To-NHLFE (FTN) Management Information Base." Internet Draft. draft-ietf-mpls-ftn-mib-03.txt. August.

Cucchiara, J., H. Sjostrand, and J. Luciani. 2001. "Definitions of Managed Objects for the Multiprotocol Label Switching, Label Distribution Protocol (LDP)." Internet Draft. draft-ietf-mpls-ldp-mib-08.txt. August.

IANA. Keld Simonsen et al., eds. *Official Names for Character Sets. ftp.isi.edu/in-notes/iana/assignments/character-sets.*

RFC 1766. Alvestrand, H.,ed. 1995. "Tags for the Identification of Languages." *www.ietf.org/rfc/rfc1766.txt.*

ISO/IEC 10646–1993 (E). *Information Technology—Universal Multiple-Octet Coded Character Set (UCS)—Part 1: Architecture and Basic Multilingual Plane.* Plus amendments AM 1 through AM 7.)

ISO/IEC 10646–1:2000. *Information Technology—Universal Multiple-Octet Coded Character Set (UCS)—Part 1: Architecture and Basic Multilingual Plane.*

Unicode Consortium. 1996. The Unicode Standard, Version 2.0. Reading, Mass.: Addison-Wesley Developers Press.

Unicode Consortium. 2000. The Unicode Standard, Version 3.0. Reading, Mass.: Addison-Wesley Developers Press.

Aho, A. V., R. Sethi, and J. D. Ullman. 1986. Compilers: Principles, Techniques, and Tools. Reading, Mass.: Addison-Wesley. Rpt. corr. 1988.

Berners-Lee, T., R. Fielding, and L. Masinter. 1997. Uniform Resource Identifiers (URI): Generic Syntax and Semantics. (Work in progress; see updates to RFC 1738.)

Brüggemann-Klein, A. 1993. "Formal Models in Document Processing." *Habilitationsschrift.* Faculty of Mathematics at the University of Freiburg. (See *ftp.informatik.uni-freiburg.de/documents/papers/brueggem/habil.ps.*)

Brüggemann-Klein, A., and D. Wood. 1991. Deterministic Regular Languages. Universität Freiburg, Institut für Informatik, Bericht 38, Oktober. Extended abstract in Finkel, A., and M. Jantzen, Hrsg., STACS 1992, S. 173–184. Springer-Verlag, Berlin 1992. Lecture Notes in Computer Science 577. Full version titled "One-Unambiguous Regular Languages," in *Information and Computation* 140 (2): 229–253, February 1998.

Clark, J. 1997. "Comparison of SGML and XML." *www.w3.org/TR/NOTE-sgml-xml-971215.*

IANA. Simonsen, K., et al., eds. 1997. Registry of Language Tags. *www.isi.edu/in-notes/iana/assignments/languages/.*

ISO 639:1988 (E). Code for the Representation of Names of Languages.

ISO 3166–1:1997 (E). Codes for the Representation of Names of Countries and Their Subdivisions—Part 1: Country Codes.

ISO 8879:1986(E). Information Processing—Text and Office Systems—Standard Generalized Markup Language (SGML). First edition.

ISO/IEC 10744–1992 (E). Information Technology—Hypermedia/Time-based Structuring Language (HyTime).

ISO 8879:1986 TC2. Information Technology—Document Description and Processing Languages. *www.sgmlsource.com/8879rev/n0029.htm.*

Bray, T., D. Hollander, and A. Layman, eds. 1999. Namespaces in XML. Textuality, Hewlett-Packard, and Microsoft. World Wide Web Consortium. *www.w3.org/TR/REC-xml-names/*.

Bray, T., J. Paoli, C. M. Sperberg-McQueen, and E. Maler. 2000. Extensible Markup Language (XML) 1.0 (second edition). W3C Recommendation. October. *www.w3.org/TR/2000/REC-xml-20001006.pdf*.

OMG. 1995a. *Common Object Request Broker Archictecture*. July.

OMG. 1995b. *Common Object Services Specification*.

Vinoski, S. 1997. CORBA: Integrating Diverse Applications within Distributed Heterogeneous Environments." *IEEE Communications Magazine*. February.

Katz, D., D. Yeung, and K. Kompella. 2001. "Traffic Engineering Extensions to OSPF." draft-katz-yeung-ospf-traffic-06.txt. October. *www.ietf.org/internet-drafts/draft-katz-yeung-ospf-traffic-06.txt*.

Li, T., and H. Smit. 2001. "IS-IS Extensions for Traffic Engineering." August. draft-ietf-isis-traffic-04.txt. *www.ietf.org/internet-drafts/draft-ietf-isis-traffic-04.txt*.

Index

About the Author

Tom Nadeau works at Cisco Systems in Chelmsford, Massachusetts, where he is a Technical Leader responsible for MPLS network management development. He leads a team that designs and develops MIBs and other management tools for Cisco's routing and switching platforms. He also works on other network-management-related and MPLS-related architecture and activities at Cisco, such as the common optical control plane (GMPLS), COPS, Diff-Serv, L2/L3 VPN, traffic engineering, pseudo-wire emulation, and network operations and management in general. He is also one of a small group of engineers at Cisco, designated as "The MIB Police," who are responsible for overseeing and assisting in the design and approval of all internal Cisco-proprietary MIBs. Tom is an active participant in the IETF, ITU, IEEE, and OIF. He is coauthor of all but one of the MPLS, PWE3, PPVPN, and GMPLS-related IETF MIBs, as well as other IETF MIBs, protocol, and architecture documents. Tom has filed several patents in the area of networking. He sits on the advisory boards of numerous technical conferences, where he is also a frequent speaker and panel member. He received his B.S. in computer science from the University of New Hampshire, and a M.S. from the University of Massachusetts in Lowell, where he has been an adjunct professor of computer science since 2000. He teaches courses in data communications for both undergraduate and graduate students. He is the technical editor of *Enabling VPN Aware Networks with MPLS* (Prentice-Hall Publishers, 2001). Tom's hobbies include woodworking, designing and building hi-fi electronics and speakers, and home brewing. When not working, writing, teaching, or working on hobbies, he is at home with his wonderful new son, Henry.